American Institute of Chemical Engineers

39th Annual Loss Prevention Symposium 2005

AA003141

Held at the 2005 AIChE Spring National Meeting

April 10-14, 2005
Atlanta, Georgia, USA

Printed from e-media with permission by:

Curran Associates, Inc.
57 Morehouse Lane
Red Hook, NY 12571
www.proceedings.com

ISBN: 978-1-60423-458-9

Some format issues inherent in the e-media version may also appear in this print version.

American Institute of Chemical Engineers

39th Annual Loss Prevention Symposium
2005

TABLE OF CONTENTS

SESSION LPS-6
CASE HISTORIES AND LESSONS LEARNED

AIChE Spring Meeting 2005

39th Annual Loss Prevention Symposium

FUELS AND PETROCHEMICALS DIVISION
2004-2005 EXECUTIVE COMMITTEE

OFFICERS

CHAIRMAN
Dennis Griffith
Granherne Inc

FIRST VICE CHAIR
Martin C. Poole
Valero Refining Company

SECOND VICE CHAIR
Harry West

PAST CHAIR
Jeffrey H. Yen
Atofina Chemicals

SECRETARY
Richard G. Mallinson
Institute for Gas Utilization Technologies

TREASURER
Mike C. Oballa
NOVA Chemicals Corporation

COUNCIL LIAISON
Jim Porter
DuPont

STAFF LIAISON
John (Jack) Dean
American Institute of Chemical Engineers

WEBMASTER
Galen Suppes
U of Missouri, Chemical Engineering Dept

PROGRAM COMMITTEE CHAIR
Syamal Poddar
Poddar & Associates

NEWSLETTER EDITOR
Robert F. Dye
Dye Engineering and Technology Company

RESEARCH & NEW TECHNOLOGY COMMITTEE LIAISON
Galen J. Suppes
U of Missouri, Chemical Engineering Dept

DIRECTORS

ONE YEAR
Lori (Boock) McDowelL
Grace Davison

Godwin J. Igwe
DuPont Central Research & Development

Colin P. Bowen
Stone & Webster

TWO YEARS
Belma DemireL
Rentech, Inc.

Stephen J. McGovern
PetroTech Consultants

Lynn Rice
UOP

THREE YEARS
Steve Coleman
Equistar Chemicals, LP, Channelview Complex

Ralph W. Pike
Louisiana State University

Gavin P. Towler
UOP Innovation Group

2004-2005 TECHNICAL PROGRAM COMMITTEE

CHAIR

Symal Poddar
Poddar & Associates

PAST CHAIR

Gavin P. Towler
UOP Innovation Group

VICE CHAIR

Lori (Brock) McDowell
Grace Davison

AREA 16a: PETROLEUM

Belma Demirel, *Chair*
Rentech, Inc.

Kirtan K. Trivedi, *Vice Chair*
ExxonMobil Research and Engineering

Dennis O'Brien, *Vice Chair*
UOP LLC

AREA 16b: PETROCHEMICALS

Bipin Vora, *Chair*
UOP LLC

Marvin Cadwell, *Vice Chair*
BP

AREA 16c: GAS

Richard G. Mallinson, *Chair*
Inst. for Gas Utiliz. Tech.
Dept. of Chem. Eng., U. of OK

Chen-Hwa Chiu, *Vice Chair*
Chevron Texaco - Energy Research & Tech.

AREA 16d NEW TECHNOLOGY & ENERGY DEVELOPMENTS

Galen J. Suppes, *Chair*
U of Missouri, Dpt. of Chem Engr.

Thomas Marrero, Co Chair
U of Missouri, Dept. of Chem Engr.

M. Hossein Hariri, *Vice Chair*
Rose-Hulman Institute of Technology

AREA 16e: NON-TECHNICAL PROGRAMMING

Lori (Boock) McDowell, *Chair*
Grace Davison

Gavin P. Towler, *Vice Chair*
UOP Innovation Group

Inerting, A Reliable and Effective Preventive Measure Against Explosions

Tom Hoppe
Expert Services
Ciba Specialty Chemicals, Inc.
Toms River, New Jersey, USA

Norbert Jaeger
Group Service Law & Environment
Ciba Specialty Chemicals, Inc.
CH-4002 Basle, Switzerland

Prepared for Presentation at
American Institute of Chemical Engineers
2005 Spring National Meeting
39th Annual Loss Prevention Symposium
Atlanta, Georgia
April 11-13, 2005

February, 2005

UNPUBLISHED

Inerting, A Reliable and Effective Preventive Measure Against Explosions

Tom Hoppe
Expert Services
Ciba Specialty Chemicals, Inc.
Toms River, New Jersey, USA

Norbert Jaeger
Group Service Law & Environment
Ciba Specialty Chemicals, Inc.
CH-4002 Basle, Switzerland

Abstract

For an explosion to occur, all three elements of the hazard triangle must be present at the same time:

- combustible substance
- ignition source
- oxygen (air)

When dealing with flammable vapors or easily ignitable dusts, it is not always sufficient to depend only on the avoidance of ignition sources, because electrostatic discharges are inherently possible with certain processes, such as the charging of non-conductive bulk materials into vessels or the spraying of water/solvent mixtures.

As an alternative approach, the use inerting as a preventive measure against explosions will reduce the oxygen concentration to a point at which an explosion is no longer possible, even in case of the presence of an ignition source. The required reduced oxygen concentration is characterized through the Limiting Oxygen Concentration, (LOC).

This paper will give guidance on the use of inerting in the processing industry by presenting several practical examples. It will further explain in depth the different type of inerting methods, the technical requirements on the implementation of this preventive measure, and its limitation.

The aim of this paper is to assist people in production dealing with operations protected by inerting. It reflects the present state of the art and current knowledge of the assessment and requirements associated with inerting.

1. Introduction

1.1 Preventive Explosion Protection Measures

In order to prevent an explosion, one of the components necessary to generate the required combustion reaction must be reliably excluded. Therefore at least one side of the hazard triangle shown in Figure 1 must be eliminated.

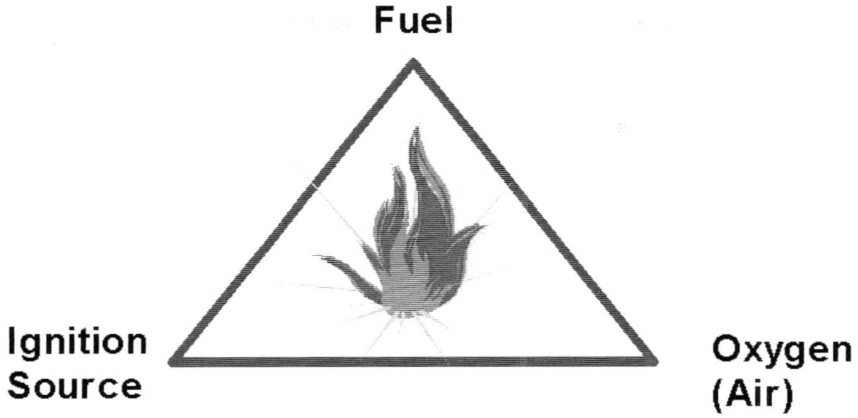

Figure 1: Hazard triangle - Principle of preventive explosion protection [1]

An explosion can thus be excluded with certainty by either:

- avoiding the development of explosible mixtures (combustible dusts, flammable gases); or
- replacing the atmospheric oxygen by inert gas, working in a vacuum, or using inert dust; or
- preventing the occurrence of effective ignition sources.

All three measures are grouped under the heading "preventive explosion protection". By the appropriate application of one of these measures the initiation of an explosion can be prevented.

This paper will give guidance on the use of inerting in the processing industry by presenting several practical examples. It will further explain in depth the different type of inerting methods, the technical requirements on the implementation of these preventive measures and their limitations. It should assist people in production dealing with operations protected by inerting. It reflects the present state of the art and current knowledge of the assessment and requirements associated with inerting.

Inerting reduces the oxygen concentration to the point where an explosion is no longer possible. It is usually employed as a primary measure of preventive explosion protection. The displacement of the oxygen by non-flammable gases or inert solids will avoid the formation of an explosive atmosphere (see Figure 2) [2].

Inert gases dilute the system and reduce the volume percent concentration of the flammable components. As a result, if an ignition occurs, the resulting energy density of the gas phase reaction is insufficient to cause expansion of the reaction zone. As a result, flame propagation and the accompanying pressure wave associated with an explosion is no longer possible.

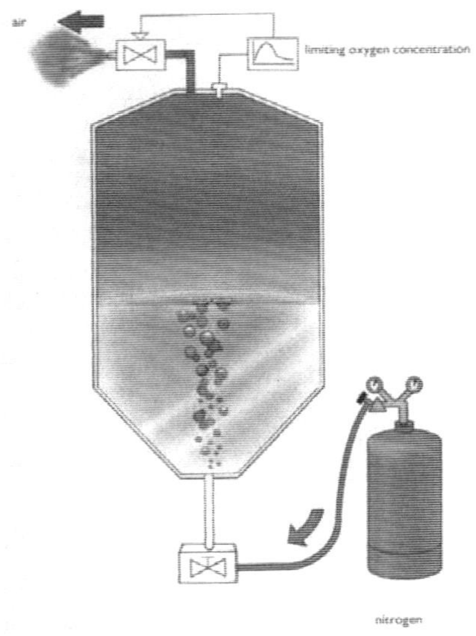

Figure 2: Inerting with a non-flammable gas [3]

1.2 Relevant Safety Characteristics

The use of the preventive measure, inerting, requires the knowledge of the relevant safety (combustion) characteristics that describe the behavior of the fuel in regards to its

- Ignitibility and
- Inertibility

Therefore, knowledge of the following safety characteristics of a material is necessary to evaluate the use of inerting:

- Minimum Autoignition Temperature, MAIT
- Minimum Ignition Energy, MIE
- Limiting Oxygen Concentration, LOC

These properties will be described in more detail in the following sections.

1.2.1 Minimum Autoignition Temperature (MAIT)

The minimum auto ignition temperature, MAIT, is defined as the lowest temperature of a heated surface at which the most readily ignitable mixture of a dust with air just ignites. It provides information on the ignition behavior of a dust suspension when quickly passing over a hot surface. The MAIT can be determined either in the Godbert-Greenwald (GG) furnace [4] or in the BAM (German Federal Institute for the Testing of Materials) furnace [5].

The results can be converted as follows:

$$MAIT_{(BAM)} = 10\ ^0C + 0.9 \times MAIT_{(GG)} \tag{1}$$

$$MAIT_{(GG)} = 1.1 \times MAIT_{(BAM)} - 10\ ^0C \tag{2}$$

1.2.2 Minimum Ignition Energy (MIE)

The minimum ignition energy (MIE) of a dust/air mixture is defined as the lowest value of the electrical energy stored in a capacitor which, at discharge (time-extended by inductivity) over a spark gap, is just sufficient to ignite the most easily ignitable dust/air mixture at atmospheric pressure and ambient temperature. To help assure that equipment is available for a standardized test procedure, a test apparatus known as MIKE 3 has been specially developed and is commercially available. Other test apparatus are also available to determine the MIE and are included in the American Society for Testing and Materials (ASTM) standard on Minimum Ignition Energy of a Dust Cloud in Air [6]. The method for the determination of MIE is also described in the international standard of the International Electrotechnical Commission (IEC) [7].

Table 1: Limiting Oxygen Concentration for Several Gases and Vapors [8]

Combustible Material	**Limiting Oxygen Concentration (Mol Content) Of The Total Mixture Combustible Material/Air/Inert Gas Inerting With:**	
	N_2	CO_2
	O_2 – Limiting Concentration [Vol. %]	**O_2 – Limiting Concentration [Vol. %]**
Benzol	11.2	13.9
Butadiene	10.4	13.0
Butane	12.1	14.5
Cyclopropane	11.7	13.9
Ethane	11.0	13.3
Ethylene	10.0	11.7
Hexane	12.1	14.5
Carbon monoxide	5.0	5.0
Methane	11.6	14.1
Pentane	11.6	14.4
Propane	11.8	14.2
Propylene	11.5	14.1
Gasoline	~ 11.8	~ 14.5
Hydrogen	4.8	4.8

1.2.3 Limiting Oxygen Concentration (LOC)

The limiting oxygen concentration is the concentration of oxygen in a specific fuel/air/inert gas mixture where a fuel explosion will not occur under defined experimental conditions. The LOC is a specific value for the fuel and inert gas used and has to be determined experimentally due to the lack of any other known mathematical correlation. The determination must be done over a large concentration range (see Figure 3) of fuel, by reducing the oxygen concentration in discrete increments in order to ensure accurate measurement. Below the limiting oxygen concentration no explosion is possible irrespective of the fuel concentration. Table 1 summarizes the LOC for several combustible gases. The determination of the LOC is normally carried out in the 20-liter sphere.

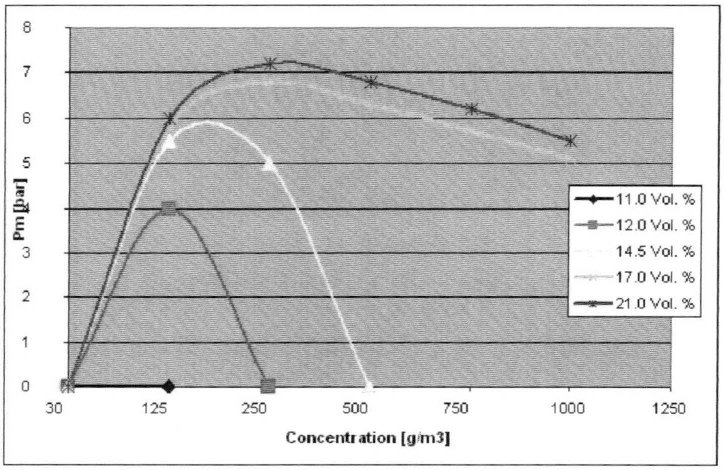

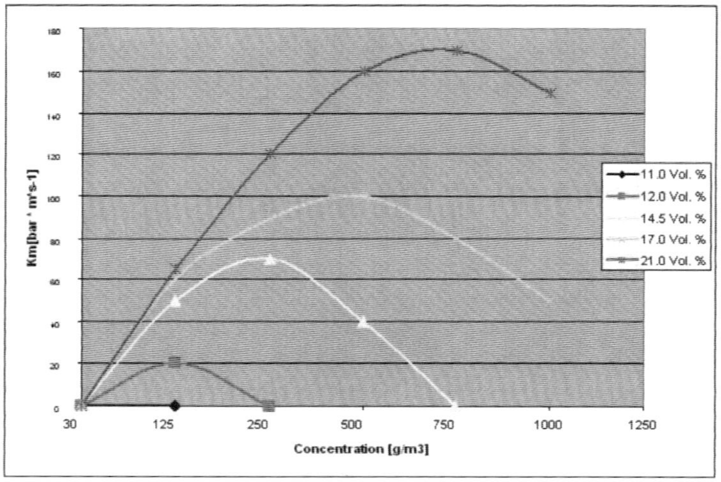

Figure 3: Influence Of The O_2 Concentration In N_2 Upon The Explosion Indices [9]

The LOC is the concentration that must not be exceeded in the installation in order to ensure that an explosion does not occur even if an incendive ignition source is experienced. A safety margin (e.g. LOC – 2%) is normally employed to protect against unanticipated upsets conditions or operating errors.

The knowledge of this safety characteristic is always required when the preventive measure inerting is used.

1.2.3.1 Special Case – Hybrid mixtures

During the processing of solid products containing a flammable solvent or due to the unanticipated generation of a flammable gas from overheating, a combination of a dust and flammable gas or vapor in air can occur. These combinations are commonly referred to as "hybrid mixtures" [5].

Knowledge of the MIE of the combustible dust and the gaseous fuel in addition to the concentration ratios of the mixture allows one to determine the ignition behavior of the hybrid mixtures. It has been observed that the LOC of the hybrid mixture will initially move in the direction of the gas or vapor and then, with higher gaseous fuel concentrations, move back in the direction of the LOC of the dust (see Figure 4).

As can be seen in Figure 4, the lower the LOC of the combustible powder the faster the LOC value for the combustible gas will be reached. This statement applies generally to the presence of arbitrary combustible powders and arbitrary combustible gases [5]. The LOC of the hybrid mixture is directly affected by the fuel component with the lowest limit value.

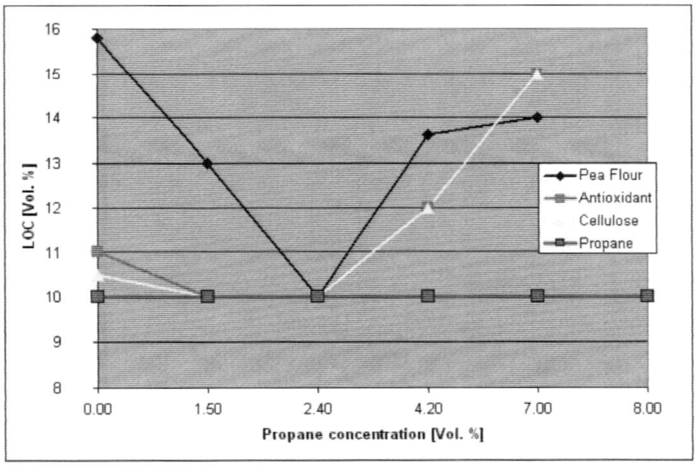

Figure 4: LOC Of Hybrid Mixtures Of Combustible Powders As A Function Propane Concentration [9]

2. Gases Used For Inerting

Typical inerting gases used in industrial practice are nitrogen, carbon dioxide, water vapor, and occasionally a noble gas (e.g., argon). It is always important to take into consideration the dangers of poisoning and/or asphyxiation with the use of inerting gases. The effectiveness of inert gases is directly affected by their molecular structure: Carbon dioxide (O=C=O) is a tri-atomic molecule, nitrogen (N≡N) is a di-atomic, while the noble gases (i.e. Ar) are only mono-atomic. Multi-atomic molecules can take up more energy due to the larger number of free molecular oscillation degrees. Therefore di- and tri-atomic molecules are more effective inert gases (See Figure 5).

$$CO_2 > H_2O\text{-Vapor} > \text{Flue-Gas} > N_2 > \text{Noble Gas}$$

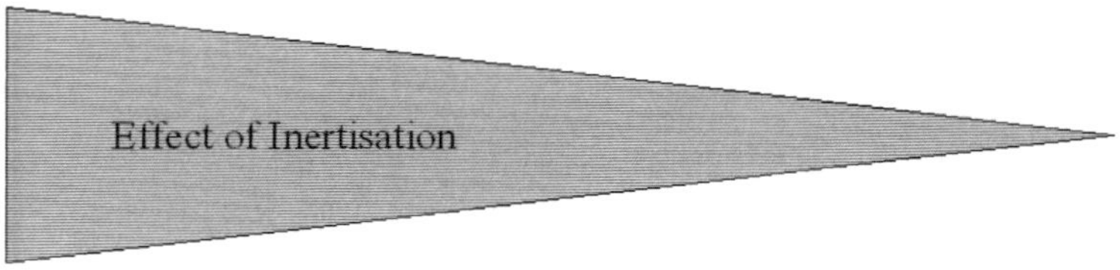

Figure 5: Effectiveness Of Typical Inert Gases

When CO_2 is drawn from high-pressure vessels or from cryogenic tanks, electrostatic discharges can be caused by the formation of dry ice.

Where steam is used for inerting, the possibility of condensation has to be taken into account. During combustion processes, self-inerting can be achieved by controlling the combustion in such a way that the oxygen concentration in the combustion gas is kept below the maximum admissible level.

3. General Requirements For Inerted Equipment

Most combustible gases and vapors are no longer ignitable under atmospheric pressure and ambient temperature if the oxygen content in the inerted mixture is kept below 10 volume % (Vol. %). See Table 1.

The LOC depends directly not only on the material and inert gas-specific characteristic but also on the temperature. A temperature increase will reduce the LOC. As a rule of thumb for gases and vapors the following approximate values can be used in practice:

- An increase of the temperature by 100°C will reduce the LOC by approximately 0.5 to 1.0 Vol. %.

Where no material-specific data are available, a LOC value for gases or vapors of 8 Vol. % should be assumed. For hydrogen sulfide, hydrogen, and carbon monoxide a value of 3 Vol. % should be used.

For organic powders, generally ignition is no longer possible if the LOC in N_2 is kept below 12 Vol. %. Exceptions here are, e.g., paraformaldehyde, 2-naphtanol and biphenyl, where the maximum allowable oxygen concentration should be kept below 4 Vol. %.

As a rule of thumb for combustible powders, the following approximate values can be used in practice:

- An increase of the temperature by 100°C will reduce the LOC by approximately 1.6 Vol. %.

For aerosols and metal dust, the quantification of the LOC should be determined experimentally.

Installations operated under inerted conditions must be adequately protected against the penetration of air and the escape of explosive gases and vapors. For this reason, the integrity of the closed vessel system must be tested regularly for leaks. This is normally done by evacuating the apparatus to a test pressure of $p_o < 0.5$ bar (abs) and then measuring the pressure rise over a defined time period. Reference values are summarized in Table 2. The frequency of the tests depends on operational history and conditions.

Table 2: Reference Values For Pressure – Time Pattern

Δp/Δt [mbar/min]	**Use of application**
< 0.2	Installations operated under atmospheric pressure, such as drying ovens, atmospheric dryers
~ 0.5 – 1.0	Stirred vessels and similar types of closed apparatus
< 0.2	Installations with more stringent tightness specifications, such as vacuum distillations in the 1 – 10 bar (abs) range.
0.002 – 0.2	Installations with highly toxic substances and ultrahigh tightness specifications

When these rule of thumbs figures are applied, it is always necessary to observe the maximum admissible rise of the oxygen concentration per unit time.

The leakage rate $\overset{*}{M}$ can be calculated from the pressure rise per unit time and the vessel volume as follows:

$$\overset{*}{M} = 0.0072 \times \frac{\Delta p}{t} \times V \qquad (3)$$

where:

$\overset{*}{M}$ = Leakage rate [kg/h]
Δp = Measured pressure rise [mbar]
t = Time of measurement [min]
V = Vessel volume [m^3]

4. Limits of Application

Inerting is <u>not</u> a protective measure to avoid exothermic decompositions. For the avoidance of (smoldering) fires, oxygen concentrations lower than the LOC must usually be adhered to and must be determined from case to case.

5. Inerting Methods

5.1 Total Inerting

Total inerting is achieved by diluting the combustible substances with inert gas to the point where an ignitable mixture cannot possibly result from the admixture of air or oxygen, regardless of the quantities added.

In industrial practice, the total inerting approach is applied only to apparatus from where gases and vapors are emitted or released sporadically to the exhaust air system.

5.2 Partial Inerting

Partial inerting uses the fact that the ignitable range of a mixture of a combustible substance and air shrinks as the oxygen concentration declines and ultimately becomes non-explosive at a certain substance-specific oxygen level.

5.2.1 Safety Margins

As a rule of thumb, the fuel/air mixture is diluted with inert gas to obtain an oxygen concentration 2 Vol. % below its LOC.‡ The German VDI Guideline 2263 uses a more conservative approach, which is described in Figure 6 [10].

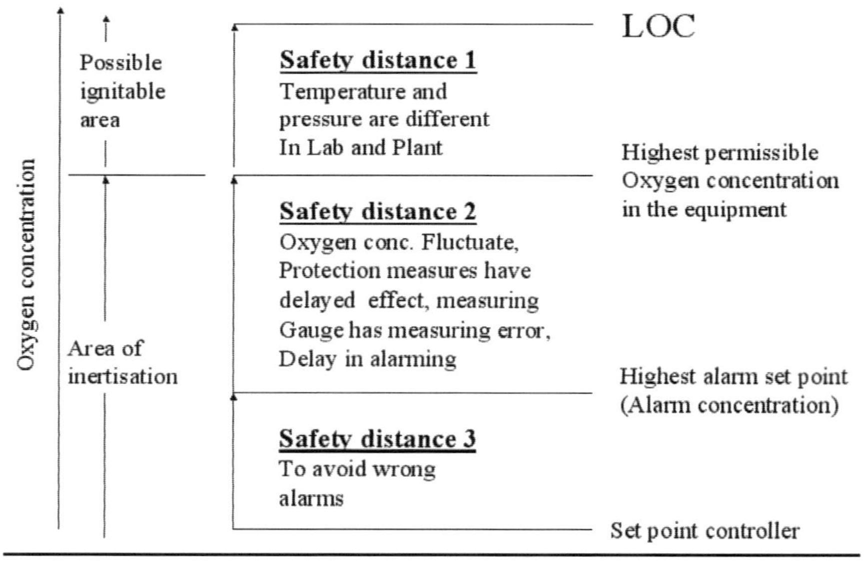

Figure 6: Oxygen Concentration Levels To Be Observed When Inerting Dust-Containing Equipment

‡ Editor's note: NFPA 69, *Standard on Explosion Prevention Systems,* provides safety margin requirements for inerting for those situations where National Fire Protection Association standards are applicable.

6. Methods of Partial Inerting

6.1 Techniques For Producing Partially Inert Atmospheres

6.1.1 Pressure Change Technique

6.1.1.1 Vacuum Inertisation

Vacuum is placed on the vessel lowering its internal pressure to p_1. The vessel is returned to its initial pressure p_2 by relieving the vacuum with inert gas. Figure 7 indicates the pressure ratio $\left(\frac{p_1}{p_2}\right) < 1$ required to obtain a desired residual oxygen concentration.

When establishing a pressure ratio, be sure to take into account both the vessel's pressure and vacuum rating. If necessary, the prescribed residual oxygen concentration can be achieved by a greater number of repetitions using a higher-pressure ratio.

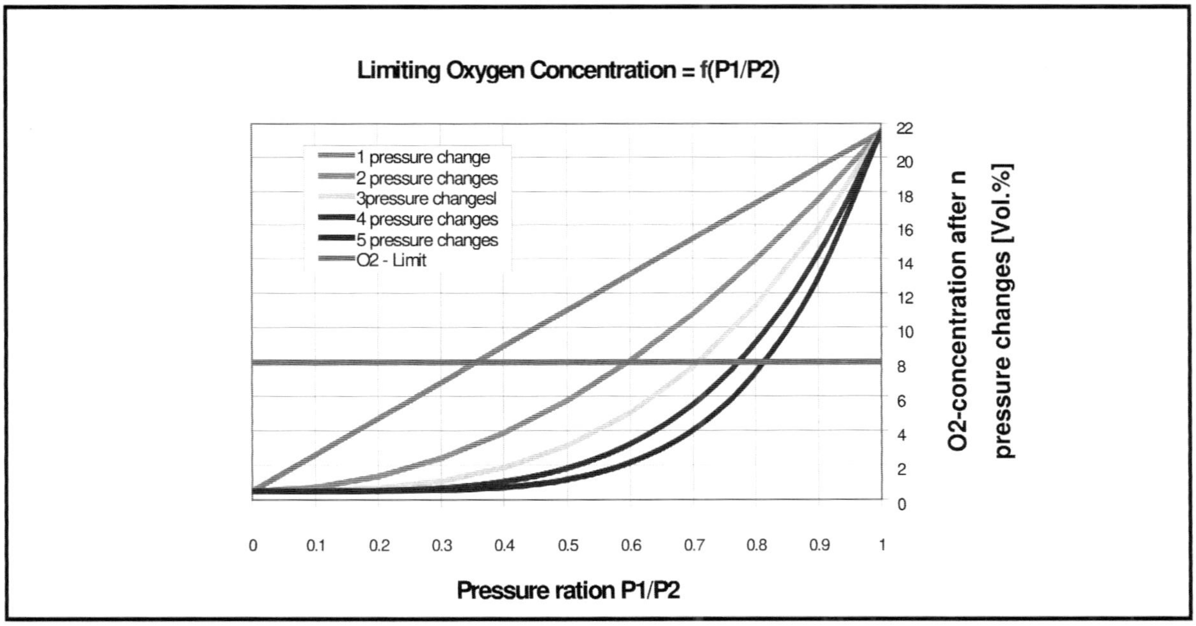

Figure 7: Residual Oxygen Concentration In The Case Of Isothermal Pressure Change Inerting (O_2 Concentration Of The Inert Gas: 0.5%) [8]

Alternatively the determination of the residual oxygen concentration can be calculated as follows:

$$Y_s = \left(\frac{p_1}{p_2}\right)^n \times (y_0 - a) + a \tag{4}$$

where:

Y_s	=	O_2 concentration after n pressure changes (Vol. %)
y_o	=	Initial O_2 concentration of mixture (Vol. %)
a	=	O_2 concentration in the inert gas (Vol. %)
p_1/p_2	=	Pressure ratio, < 1
n	=	Number of pressure changes

Example: Starting from atmospheric pressure (p_2 = 1 bar) the vessel is evacuated to $p_1 = 0.5$ bar. The initial O_2 concentration is $y_o = 21$ Vol. % O_2. The O_2 concentration in the inert gas is a = 0.5 Vol. %. Either one or two pressure changes will be performed. What is the O_2 concentration (Ys) after $n = 1$ and $n = 2$ pressure changes?

$n = 1$

$$Y_s = \left(\frac{0.5}{1.0}\right)^1 \times (21.0 - 0.5) + 0.5$$

Y_s = 10.75 Vol. %

$n = 2$

$$Y_s = \left(\frac{0.5}{1.0}\right)^2 \times (21.0 - 0.5) + 0.5$$

Y_s = 5.62 Vol. %

The effective inert gas consumption V_{N_2} is

$$V_{N_2} = V_{app} \times f \times n \qquad (5)$$

where:

V_{N_2}	=	Effective inert gas consumption [m^3]
V_{App}	=	Vessel Volume [m^3]
f	=	Inert gas consumption (pressure change ratio)
n	=	Number of pressure changes

The inert gas consumption can be calculated as follows:

$$f = \left(1 - \frac{p_1}{p_2}\right), \quad \frac{p_1}{p_2} < 1 \qquad (6)$$

where:

f	=	Inert gas consumption (pressure change ratio)
p_1	=	Pressure before inerting (bar abs.)
p_2	=	Pressure after inerting (bar abs.)

Example: Starting from atmospheric pressure (p_2 = 1 bar) the vessel is evacuated to p_1 = 0.5 bar. The initial O_2 concentration is $y_o = 21$ Vol. % O_2. Either one or two pressure changes will be performed. What is the effective inert gas consumption (V_{N_2}) for a vessel volume of 10 m^3?

n = 1

$$V_{N_2} = \left[\left(1 - \frac{0.5}{1.0}\right) \times 10\right] \times 1$$

$$V_{N_2} = 5m^3$$

n = 2

$$V_{N_2} = \left[\left(1 - \frac{0.5}{1.0}\right) \times 10\right] \times 2$$

$$V_{N_2} = 10m^3$$

The reduced O_2 concentration value should be checked by measurement when a new process is introduced or process modifications have been carried out. A solvent wet vessel should be evacuated gradually to reduce the loading peaks in the vacuum system's exhaust gas flow. The introduction of inert gas should also be closely controlled in order to avoid excessive pressure fluctuations in the inert gas network.

6.1.1.2 Overpressure Inertisation

In this case the closed vessel is raised to an overpressure p_2 and then relieved into the exhaust gas system to the normal pressure level p_1. Figure 7 can also be used for applying this inerting technique. When the pressure ratio is established, the allowable operating pressure (Maximum Allowable Working Pressure (MAWP)) of the vessel system should be taken into account.

The inert gas consumption (pressure change ratio) for pressure inerting can be calculated as follows:

$$f = \left(\frac{p_1}{p_2} - 1\right), \quad \frac{p_1}{p_2} < 1 \qquad (7)$$

Figure 8 illustrates the required inert gas consumption for vacuum and pressure inerting.

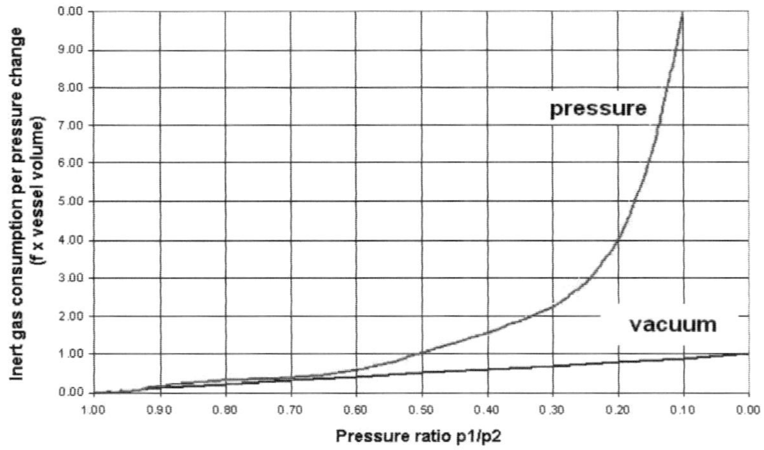

Figure 8: Inert Gas Consumption In The Case Of Isothermal Pressure Change [11].

6.1.2 Displacement Inerting

In certain cases the air in the vessel can be displaced by water and, when the vessel is emptied, inert gas is introduced. If the vessel cannot be completely flooded with water, it may be necessary to repeat this procedure a number of times. The premissble weight loading of the vessel and supporting structure should be taken into account.

6.1.3 Purge Inerting

Purge inerting is applied when the previously described inerting procedures cannot be applied. Inert gas flows continuously through the vessel. The theoretical gas changes required assuming ideal mixing can be calculated as follows:

$$L/V = \ln[(y_0 - a)/(y_s - a)] \tag{8}$$

where:

L	=	Inert gas volume
V	=	Vessel volume
y_o	=	Initial O_2 concentration (Vol. %)
y_s	=	Final O_2 concentration (Vol. %)
a	=	O_2 concentration of the inert gas

Figure 9 illustrates the theoretical gas changes required for purge inerting.

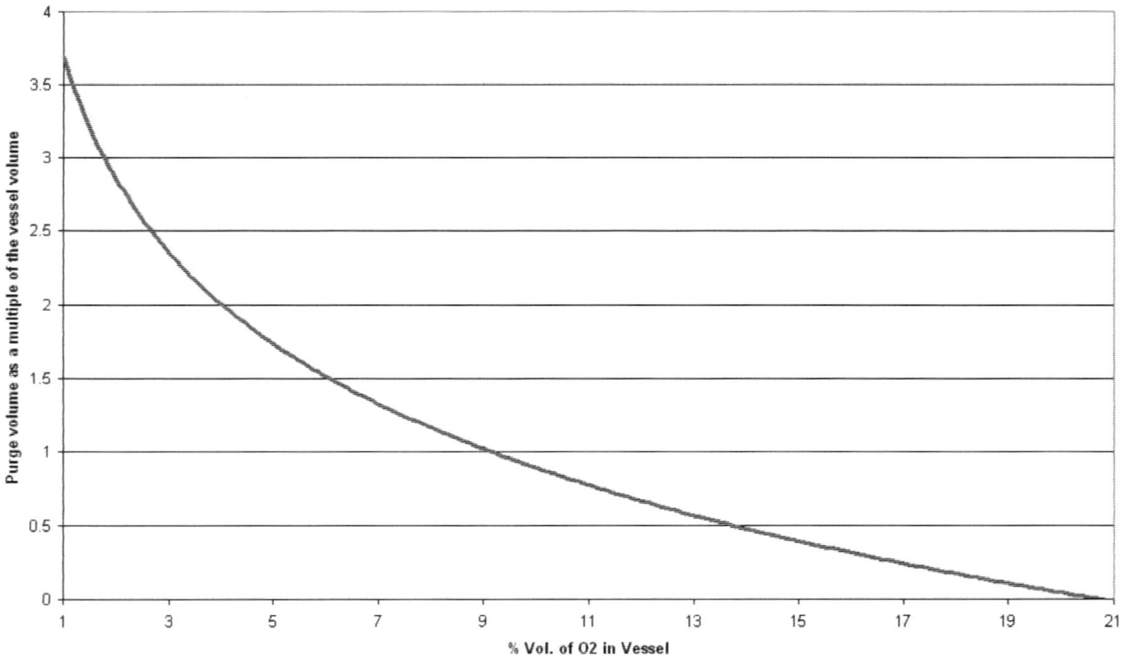

Figure 9: Theoretical Gas Changes Required For Purge Inerting [11]

The gas changes actually required should be determined by conducting a number of O_2 measurements in the vessel. Purging should be effective throughout the entire vessel chamber. This is achieved by injecting the inert gas at a high velocity and either placing the inlet or outlet nozzles far away from each other as possible or by using several inlet and outlet nozzles.

Purge inerting has the highest insert gas consumption. It can be reduced by using:

- Extremely gas tight vessel
- Pressure- or flow-controlled inert gas supply.

6.2 Techniques For Maintaining An Inert Atmosphere

6.2.1 Respiration (Breathing) Of Non-Pressure Rated (Atmospheric) Vessels

Pressure respiration (gas replacement) can be used for vessel inerting if the vessel is being operated at a slight overpressure (see Figure 10). The inerting system should be dimensioned in such a way that the pressure will be maintained inside a defined range.

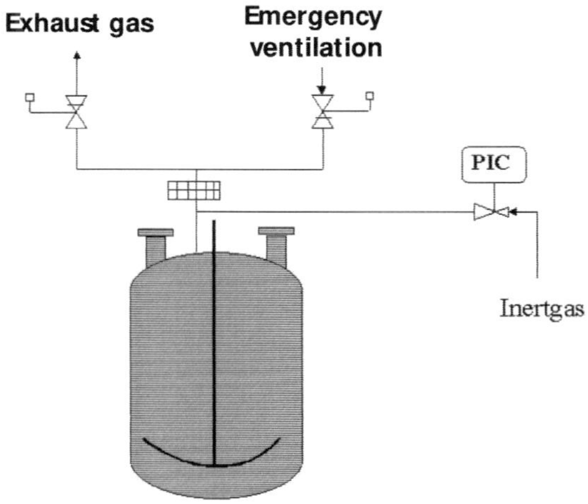

In cases of vessels that are not vacuum rated, the normal response limits for pressure variations are roughly –10 mbar to –35 mbar underpressure and +40 mbar to +65 mbar overpressure. For exceptional operating conditions (e.g., very high filling or discharge rates, a rapid temperature change of the gas blanket, inert gas supply failure), a backup gas make-up system should be installed (e.g., air fed through a combined over-/underpressure valve (conservation vent)). A flame arrestor should be provided in the vent pipe.

For determining the inert gas supply requirements for a vessel, knowledge of the minimum values for the inert gas volume $\dot{V}_1$ flow and the inert gas volume, V_I are necessary. These requirements must be specified to determine the maximum ventilation- and pump flow rates $\dot{V}_A$ and $\dot{V}_P$ [10].

6.2.2 Respiration (Gas Make-Up) On Pressure Vessels

To reduce the exhaust gas flow, most pressure rated vessels are operated at a slight overpressure and vented through a pressure holding valve. The inert gas supply is activated through a pressure holding valve PIC only when the pressure in the vessel drops excessively (see Figure 11). Depending on the vessel utilization, an emergency gas make-up/exhaust system as used for atmospheric vessels may not be needed in this case.

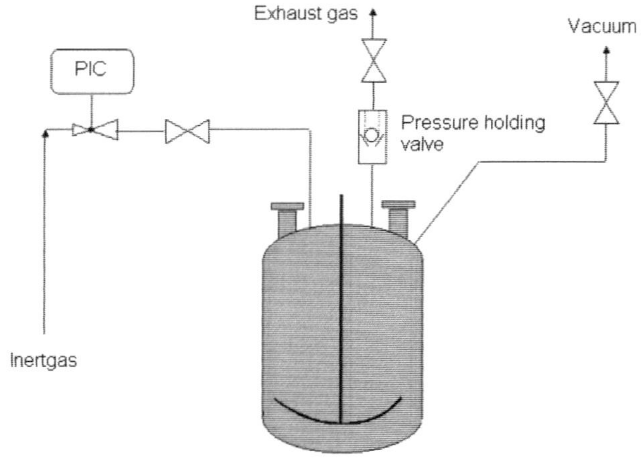

Figure 11: Inert Gas Respiration Of A Pressure Vessel

7. Case Studies

The selection of appropriate preventive measures against explosions requires one to determine the type of explosible atmosphere (i.e. combustible gases or vapors, dusts, or hybrid mixtures) most likely present in the processing system.

Due to fact that the MIE of gases and vapors are extremely low (<< 1mJ) the use of the preventive measure "Avoidance of ignition sources" as the sole protective measure is not safe. In this case "Inerting" as the primary protective measure should be used. For handling of flammable solvents, it is often required that avoidance of ignition only be applied as a secondary protective measure. The reduced oxygen concentration level must be ensured through technical measures. Typical examples of equipment that normally require inerting when processing combustible materials are:

- Centrifuges
- Suction filter apparatus
- Stirred vessels / reactors.

When combustible dusts are handled, avoiding an explosive atmosphere by keeping the dust concentration outside the explosive range is rarely possible due to sedimentation or whirling up of the material being handled. In addition, avoidance of effective ignition sources may not be possible depending on the type of operation and the MIE of the dust. Thus, in some cases, prevention of an explosion can only reasonably be avoided with certainty by reducing the oxygen concentration, e.g., inerting.

Typical examples of dust handling equipments where inerting may be required are:

- Mills
- Dust collectors
- Fluid bed dryers
- Silos

In the following examples, the use of the preventive measure "Inerting" will be described.

7.1 Charging Of Bulk Materials Into Inerted Vessels

In the vast majority of cases, charging a combustible powder into a vessel with a precharged flammable solvent that is at a temperature near or above its flash point requires that the bulk materials be fed into the vessel under closed, inerted conditions. This can be realized by using charging equipment that ensures that the inert atmosphere will not be compromised.

The reasons that the charging of the bulk material through an open manhole into a flammable solvent is not acceptable can be summarized as follows:

- Entrainment of atmospheric oxygen through the open manhole leads to a hazardous increase of the oxygen concentration inside the vessel
- Generation of an explosible hybrid mixture in the near vicinity of the manhole can occur
- There is a higher dust concentration in the vicinity of the manhole where manual operations occur
- Manual operations can generation potential ignition sources (e.g. electrostatic discharges)

It must be kept in mind that the addition of powders through an open manhole into a flammable solvent is one of the most common causes of flash fires in the Chemical Processing Industries (see Figure 12). If possible, it should always be avoided.

Investigations have shown, that an inert atmosphere inside a vessel can be compromised in a short time period once the manhole is opened: For example:

A 2.25 m^3 vessel was inerted and the reduced oxygen concentration was measured showing a concentration of 3.5 Vol. %. During the measurement the manhole was closed and the stirrer was not running. After opening the manhole (diameter of 0.46m) it was noticed, that after 10 minutes the oxygen concentration was increased to a level of 12 Vol. % and after 20 minutes to 16 Vol. % [12].

The reduction of the inert atmosphere inside the vessel is even more critical if a powder is charged trough the open manhole. Figure 13 shows the change of the O_2 concentration inside a vessel during a powder charging operation using 7 drums. As one can see clearly the safe reduced O_2 concentration is no longer present after a short time period. The sole protection of "Inertisation" is no longer valid!

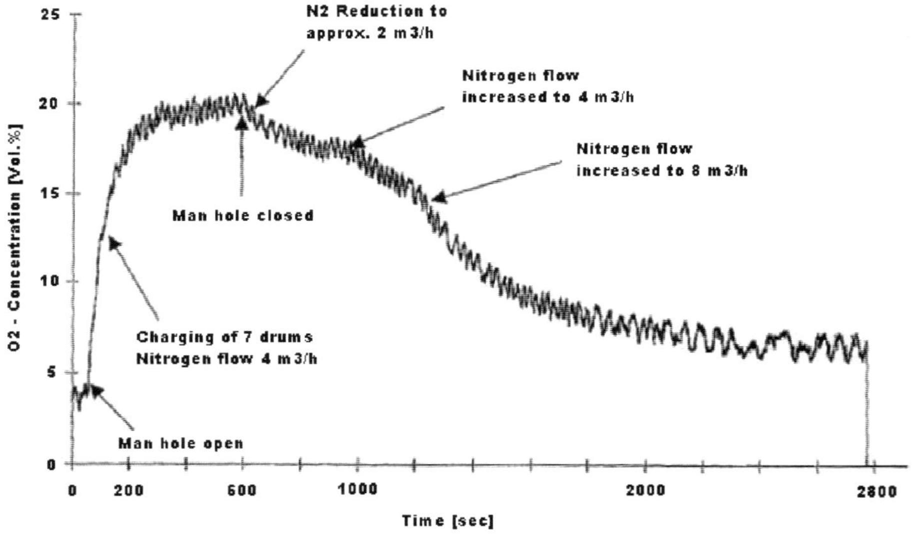

Figure 13: Change Of The O_2 Concentration While Charging Powder Through An Open Manhole

Today, solids are delivered in a large variety of different types of packing forms (e.g. bags, drums, FIBC's). In all cases it is necessary that the equipment used for the powder charge will ensure that the inert atmosphere is maintained inside the vessel. Examples for typical charging stations are shown in Figure 14.

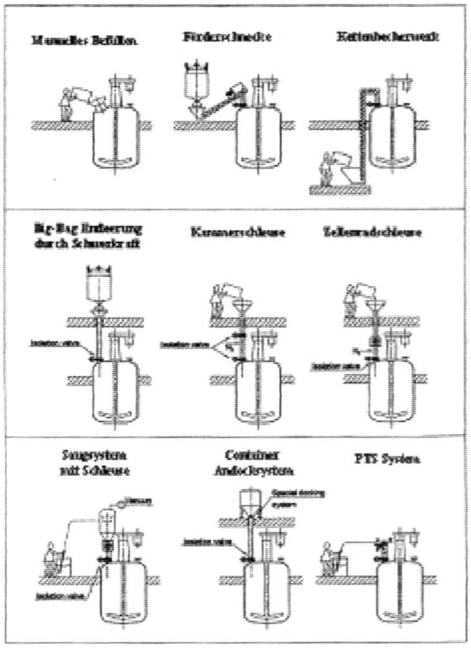

7.1.1 Inert Powder-Charge - FIBC Through A Closed Chute

The following two pictures (Figure 15) illustrate a powder charging station using a chute, which is a fixed installation at the vessel. At the beginning of the charging operation the vessel and the chute will inerted following a defined procedure. Once the defined reduced O_2 concentration has being approached, the flap at the upper end of the chute will be opened. A slight N_2 purge will be maintained during the discharge of the powder out of the FIBC.

Figure 15: Inert Powder Charging Station Through A Closed Chute

The advantages and the disadvantages of the above system are summarized in Table 3.

Table 3: Inert Powder Charging Station Through A Closed Chute

Advantages	Disadvantages
• O_2 controlled charging process • Simple installation and use • Reduction of packaging	• No dosing possible • Powder emissions possible • Empty area around the charging station necessary

7.1.2 Inert Powder Charge Through An Entry Air-Lock

In this case, the controlled discharge out of an FIBC into a vessel is accomplished by using an entry air-lock system installed at the manhole (see Figure 16).

The advantages and the disadvantages of the above-described system are summarized in Table 4.

Table 4: Inert Powder Charge Through The Manhole Using An Entry Air-Lock

Advantages	Disadvantages
• O_2 controlled charging process • Simple installation and use • Reduction of packaging	• No dosing possible • Powder emissions possible • Free space above the vessel

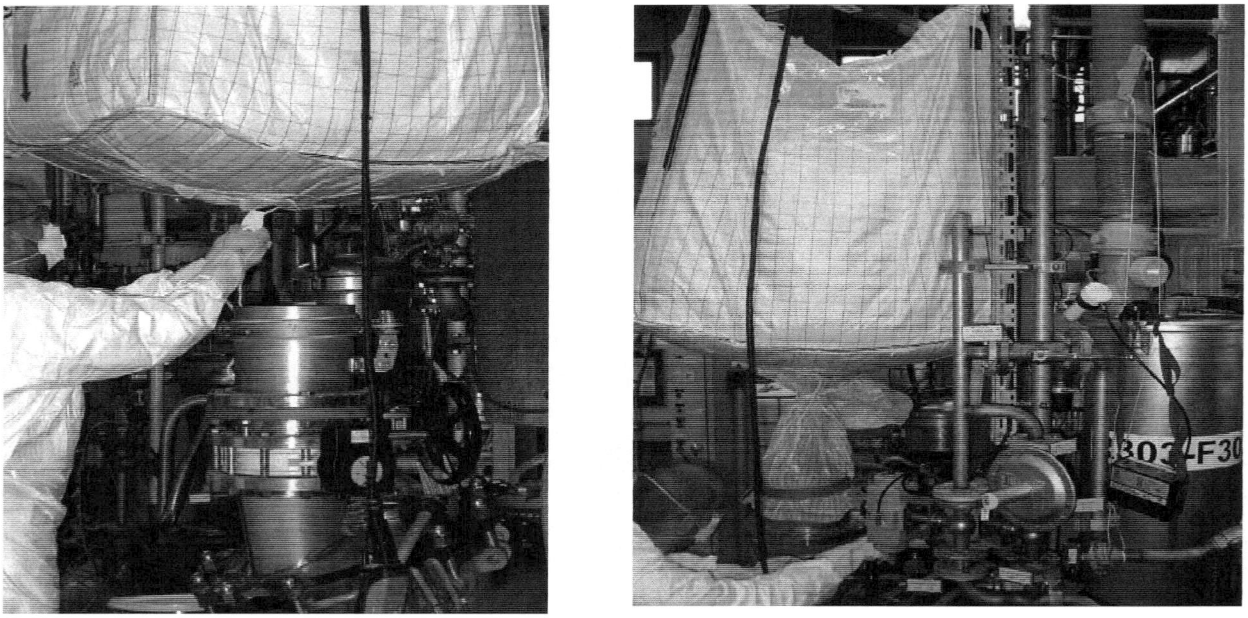

Figure 16: Inert Powder Charge Through The Manhole Using An Entry Air-Lock

7.1.3 Inert Powder Charge Through A Powder Transport System (PTS)

The PTS system does not require a large free space for the installation on a vessel. In addition it can be used either for fine powders and granular material allowing dosing controlled charging into the vessel. Figure 17 shows the logical function mode of the PTS and Figures 18 and 19 two practical practical applications.

The central component of the PTS is the PTS pump. It is composed of the cylindrical PTS body, the PTS filtration membrane and the PTS cover. The PTS is equipped with four automatic valves.

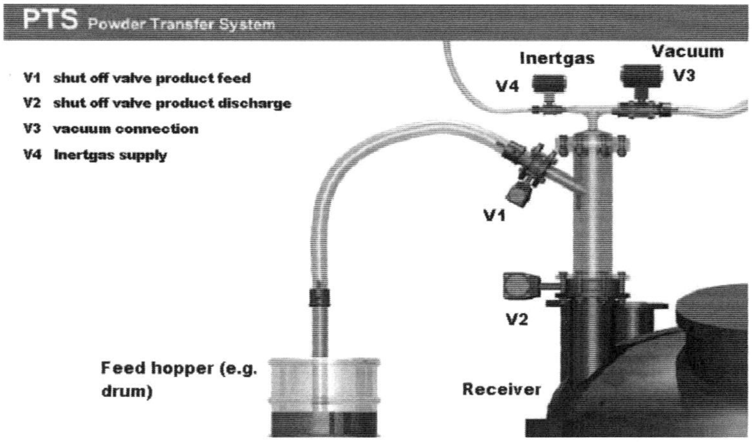

The product inlet valve **(V1)** and the product outlet valve **(V2)** are attached to the PTS body. The vacuum valve **(V3)** and the pressure valve **(V4)** are installed on the PTS cover which is separated from the PTS body by the PTS filtration membrane. The PTS filtration membrane thus separates the product in the PTS body from the conveying air. Any gas, which is suitable for the respective production process, such as compressed air or nitrogen, can be selected as the conveying gas. The valves are connected to an automatic sequencer. They open and close, following a predefined time sequence. This guarantees a gentle transfer with little mechanical stress being exerted on the product. Moreover, product can be discharged into the reactor without oxygen input or dust emissions. Induction, discharging and waiting periods are set in accordance with the production procedures [14].

The advantages of such a system can be summarized as follows:

- Closed system (pressure resistant), dust free
- Separates the vessel from the environment ensuring that the inert atmosphere inside the vessel will be maintained
- Continuous charging rate
- No restriction on the type of packing forms
- Space-saving, compact, durable, mobile design
- Pharmaceutical and chemical industry design, easy to clean

The disadvantages of such a system can be summarized as follows:

- The charging quantity is limited
- Sticking, damp or lumpy material cannot be charged though the PTS system
- High investment costs (design plastic/metal)

Figures 20 and 21 show examples where a PTS system can be used for a variety of different charging situations.

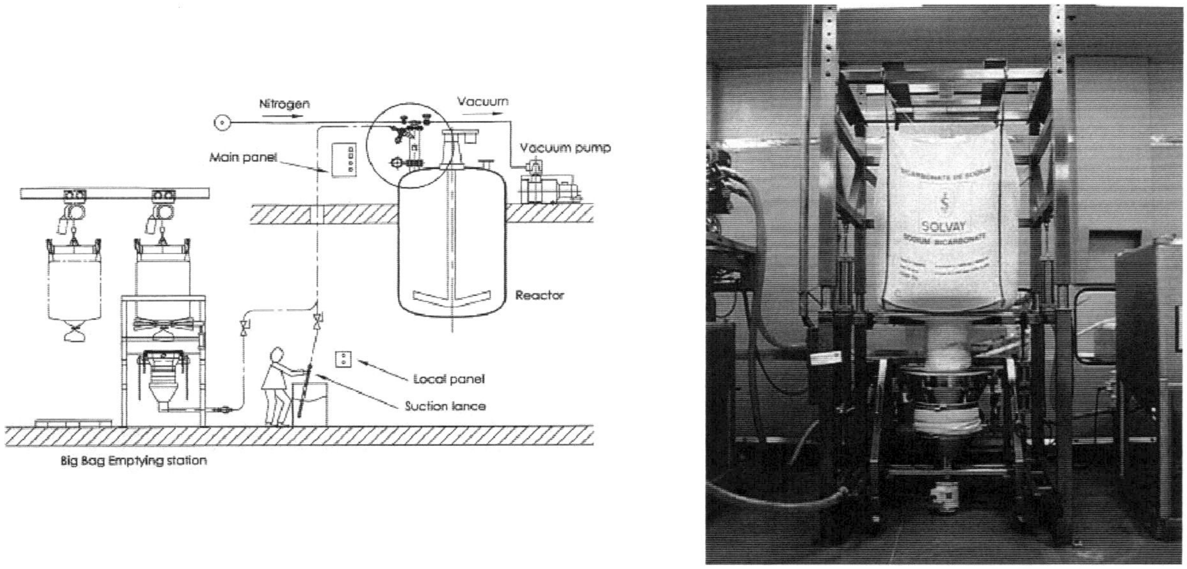

Figure 20: Vessel charging out of an FIBC [13]

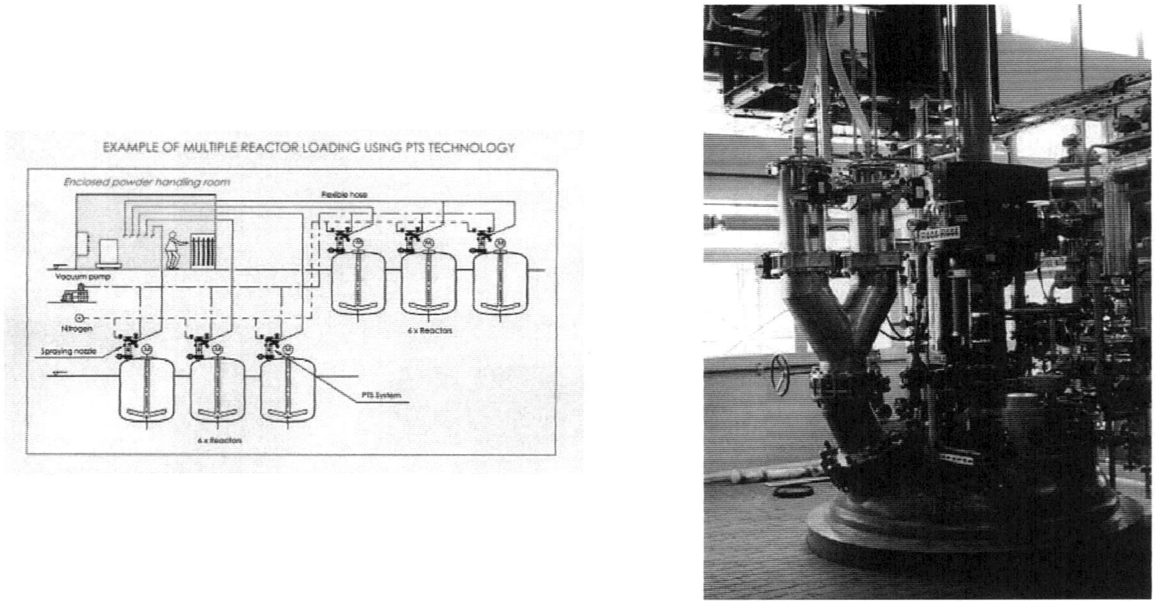

Figure 21: Vessel charging from a commissioning area [13]

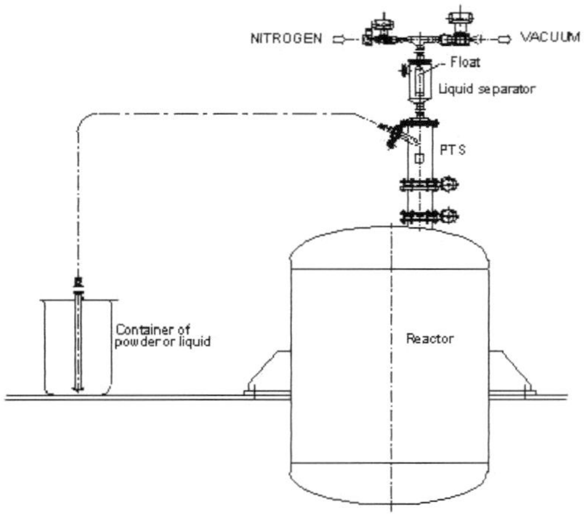

<u>Figure 22:</u> Liquid charging [13]

7.2 Filling And Emptying Of Inerted Vessel With Liquid

It is necessary to ensure that no air enters the inerted vessel when it is filled, emptied, or in the event of vacuum or energy failure. Inerted vessels can be filled or emptied by any of the following methods:

- with a pump (e.g. transfer pump)
- by gravity
- by means of inert gas pressure
- by means of vacuum

Especially when a transfer pump or vacuum is used for filling from a non-inerted vessel, there is a danger that air will flow in after that vessel is completely empty. Any of the following can be applied to prevent this:

- leaving some liquid in the vessel from which it is taken, and/or
- replacing the volume of removed liquid with inert gas
- first inerting the supply vessel and connecting it to the receiver vessel with a gas exchange pipe

8. Oxygen Measurement

Continuous oxygen measurements should be provided wherever sources of ignition are present in the operation and failure of the inerting system would be most likely result in an explosion.

In this case the measurement system fulfills a critical safety function, which means that is has to meet stringent reliability criteria:

- The sensor must be comply with the requirements of the Ex-Zone (i.e., electrical classification requirements)
- The entire measuring system chain must be "fail-safe", i.e. as soon a defect occurs or a piece of equipment fails, the output signal switches securely to the alarm/fault conditions
- The O_2 reading may not be falsified by other gases (response to the inerting gas)
- The sensor must display long-term stability and be easy to calibrate when installed
- The sensor must be usable on or in-line and should measure continuously. Maximum time delay should not be longer than 60 seconds.
- Measuring accuracy of ± 1 Vol. % O_2 is sufficient

9. Disposal Of Inert Gas

When inert gases are passed over a solvent wet surface, they become enriched with the solvent. The quantity of the emitted solvent depends on the degree of saturation of the inert gas and on the quantity of the inert gas.

At 20°C and 100% saturation, the inert gas can absorb the following quantities of solvent, which are summarized in Table 5.

Table 5: Quantity Of Absorbed Flammable Solvents

Flammable Solvent	Absorbed quantity [g/m^3]
Toluene	110
Methanol	171
Acetone	558
Ethyl Acetate	355
Diethyl ether	1909
Methylene chloride	1670
Chloroform	1029

10. Literature

[1] Jaeger, N., and Siwek R., "Prevent Explosions of Combustible Dusts," *Chemical Engineering Progress*, AIChE, 1999.

[2] EN 1127-1, Explosion Prevention and Protection, Basic concept, 1997

[3] IVSS, Section Chemie, Gas Explosions, Compendium for the industrial practice, Heidelberg, 1986

[4] IEC 31 H (CO) 4: *Method For Determining The Minimum Ignition Temperature For Dusts. Part 2: Dust Cloud In A Furnace At Constant Temperature*, 1984

[5] W. Bartknecht, Explosionsschutz, Grundlagen und Anwendungen, Springer Verlag Deutschland, 1993

[6] ASTM 2019-99, *Standard Test Method for Minimum Ignition Energy of a Dust Cloud in Air*, 1999

[7] IEC, *Electrical Apparatus For Use In The Presence Of Ignitable Dust. Part 2 Test Methods, Sheet 2-5, Method For Determining The Minimum Ignition Energy Of Dust/Air Mixtures,* 1995

[8] ESCIS, Booklet, *Inerting, Methods and Measures for the Avoidance of Ignitable Substance-Air Mixtures in Chemical Production Equipment and Plants*, Lucerne, Switzerland, 1994

[9] Jaeger N., and Siwek R., "Safety Characteristics Of Combustible Materials," Lecture documents, EX-I Seminar, German Chemical Trade Organistion, 1992, in German

[10] VDI 2263, Part 2, *Inerting*, 1992

[11] Technical Guidance Note, *Prevention of Explosions by Inerting*, Ciba Specialty Chemicals, 1994, internal document

[12] BGR 117, *Working In Confined Space Areas*, German Chemical Trade Organisation, 2003, in German

[13] Dietrich, Frédéric, Dietrich Engineering Consultants, *Sicherheitsaspekte über Pulverbeschickungen von Reaktoren,* Vortrag beim Ciba Spezialitätenchemie ERFA Pulverhandling 2003, Schweizerhalle, Schweiz

[14] Working Description PTS, Dietrich Engineering Consultants SA, Ecublens, Switzerland

Additional references not cited:

[15] Siwek R., et. al., "Ignition Behavior Of Dusts," Loss Prevention Symposium, Atlanta 1994

[16] Siwek, R. and Cesana, C., "Ignition Behavior of Dusts: Meaning and Interpretation," *Process Safety Progress*, April, 1995

[17] Bartknecht, W., and Siwek, R., "Meaning of Ignition Energy and Minimum Ignition Temperature for the Assessment of the Ignitability of Ignition Sources in Dust/Air-Mixtures, Part 2: Assessment Of The Ignition Efficacy," *Staub-Reinhaltung der Luft* 54 (1994), p. 325-330, Springer-Verlag, 1994, in German

[18] Schampel, K., *Flame Arrestors,* Expert Verlag, 1988, in German

Using Limiting Air Pressure (LAP) As An Alternative To Inertion In Rotary Dryers

Michael Toth *(speaker)*
Staff Engineer
Merck & Co., Inc.
126 E. Lincoln Ave.
Rahway, NJ USA, 07065
732-594-1996 (V)
732-594-1720 (Fax)
michael_toth@merck.com

Dan Muzzio
Project Engineer
Merck & Co., Inc.
126 E. Lincoln Ave.
Rahway, NJ USA, 07065
732-594-2540 (V)
732-594-1720 (Fax)
daniel_muzzio@merck.com

Les Cunningham,
Director
Merck & Co., Inc.
126 E. Lincoln Ave.
Rahway, NJ USA, 07065
732-594-1721 (V)
732-594-1720 (Fax)
les_cunningham@merck.com

Prepared for Presentation at
American Institute of Chemical Engineers
2005 Spring National Meeting
39th Annual Loss Prevention Symposium
Atlanta, Georgia
April 11-13, 2005

UNPUBLISHED

Using Limiting Air Pressure (LAP)
As An Alternative To Inertion In Rotary Dryers

Michael Toth ***(speaker)***
Staff Engineer
Merck & Co., Inc.
126 E. Lincoln Ave.
Rahway, NJ USA, 07065
732-594-1996 (V)
732-594-1720 (Fax)
michael_toth@merck.com

Dan Muzzio
Project Engineer
Merck & Co., Inc.
126 E. Lincoln Ave.
Rahway, NJ USA, 07065
732-594-2540 (V)
732-594-1720 (Fax)
daniel_muzzio@merck.com

Les Cunningham,
Director
Merck & Co., Inc.
126 E. Lincoln Ave.
Rahway, NJ USA, 07065
732-594-1721 (V)
732-594-1720 (Fax)
les_cunningham@merck.com

Abstract

The Limiting air pressure (LAP) of a material is the pressure below which an explosion cannot take place due to the reduced density of oxygen molecules in the vapor phase. This paper addresses:

- The principal of the LAP.
- How the LAP for compounds are determined using a 20 L sphere apparatus.
- How these LAP values are used to design and operate safer process systems.

1. Introduction

A common unit operation in chemical and pharmaceutical manufacturing is the drying of solvent wet solids to obtain a dry, particulate solid. Quite often these are finely divided, combustible solids that present an explosion hazard under the proper conditions, and if a flammable solvent is involved it can increase the risk substantially. Therefore, it is necessary to take practical precautions to either prevent an explosion from occurring or to protect personnel and equipment from its consequences.

An explosion occurs when a fine, combustible dust is dispersed as a cloud at the right concentration in an atmosphere that contains sufficient oxidant to support combustion, and comes in contact with an ignition source of sufficient energy to cause an ignition. By eliminating any one of these conditions an explosion cannot occur. Methods of explosion prevention include inertion, avoiding the formation of a dust cloud, or excluding all possible ignition sources. While prevention is the preferred method, it is not always possible or practical, so some form of explosion protection must be employed. Venting, suppression, and containment are the three main methods of explosion protection.

An explosion can also be prevented, or substantially reduced in severity, by operating under vacuum, or reduced air pressure. The Limiting Air Pressure (LAP) is the absolute air pressure below which an explosion cannot occur due to ignition of a solid powder. In this case an explosion is defined as rapid combustion that generates pressure > 1.2 bara. Since many dryers operate under vacuum, ensuring operation below the LAP is often a logical and cost effective safety precaution. This paper will explain what the LAP is, how it is determined, and how it can be applied in industrial drying applications.

2. The Issue

Many organic powders and some common solvents are highly non-conductive, meaning they are unable to dissipate static electricity that is accumulated during the drying process. In a rotary dryer, tumbling of the wet cake and/or dry powder for extended time periods can build up a static charge, resulting in a high voltage discharge which can ignite the solvent vapor or dust cloud in the dryer. This can result in loss of product, equipment damage, and operator injury.

It is difficult to guarantee elimination of electrostatic discharges when drying non-conductive material in a tumbling dryer. If a flammable vapor or combustible dust cloud is present an explosion can occur unless the process is operated under inertion or vacuum. Inertion involves displacing the oxygen with inert gas, such as nitrogen or carbon dioxide, to a reduced oxygen concentration that can no longer support combustion. The highest oxygen concentration in a dust/air/inert gas mixture at which an explosion fails to exceed a final pressure of 1.2 bara is defined as the Limiting Oxygen Concentration (LOC). Inerting below the LOC prevents an explosion from occurring by two mechanisms. First, it removes the oxygen necessary for combustion to occur. Second, the inert gas acts as a heat sink absorbing energy so the adiabatic flame temperature for propagation of burning is not reached. Operating under vacuum only works by the first mechanism, removing the oxygen necessary for combustion, so it is slightly less effective. Regardless, it is still sufficient as long as low enough pressures are maintained, and often convenient since many dryers operate under vacuum.

Figure 1 illustrates the additional effect nitrogen inertion has on mitigating an explosion versus operating under reduced pressures. The solid tested has a maximum pressure rise (Pmax) of 8.3 bar as determined with an initial pressure (Pi) of 1atm in the 20-L Sphere apparatus. That results in a peak pressure (Ppeak) of 9.3 bara (Pi + Pmax) and a LAP of 0.13 bara. 0.13 bara of air has the equivalent oxygen as 2.68% O_2 at 1 atm (balance N_2). The LOC for this dust was determined in the 20-L sphere apparatus to be 6% O_2 at 1 atm with balance nitrogen. So by inerting (displacing O_2 with N_2) more O_2 needs to be available for an explosion to

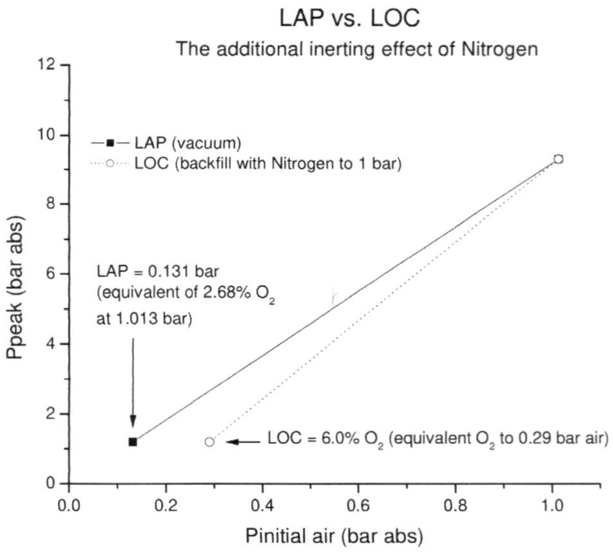

Figure 1. LAP vs. LOC.

occur than just by reducing the air pressure.

3. Theory

The maximum pressure rise generated by an adiabatic dust explosion in a closed vessel is a function of the stoichiometry and heat of combustion. It generally does not exceed 10 bar in equipment operating at atmospheric pressure and is proportionally less for equipment under vacuum following:

$Pmax' = (Pmax/Pi)*Pi'$ (eq. 1) note: all pressure values are absolute

Where:
Pi = initial pressure of 1 atm
Pmax = maximum pressure rise with initial pressure of 1 atm
Pi' = reduced initial pressure (vacuum)
Pmax' = maximum pressure rise with initial pressure Pi'

The peak pressure resulting in the equipment is the sum of the initial pressure and the maximum pressure rise (see Figure 2):

$Ppeak = Pmax' + Pi'$ (eq. 2)

The maximum initial pressure in order not to exceed a given peak pressure can be determined by combining equations 1 and 2 and solving for Pi':

$Pi' = Ppeak/(Pmax/Pi+1)$ (eq. 3)

Since essentially all glass-lined and metal dryers can handle a pressure of 1.2 bar absolute, the LAP is defined based on a Ppeak of 1.2 bara (see Figure 3). It can be misleading to indicate that no combustion takes place because there is a pressure rise from Pi' to 1.2 bara so there may be

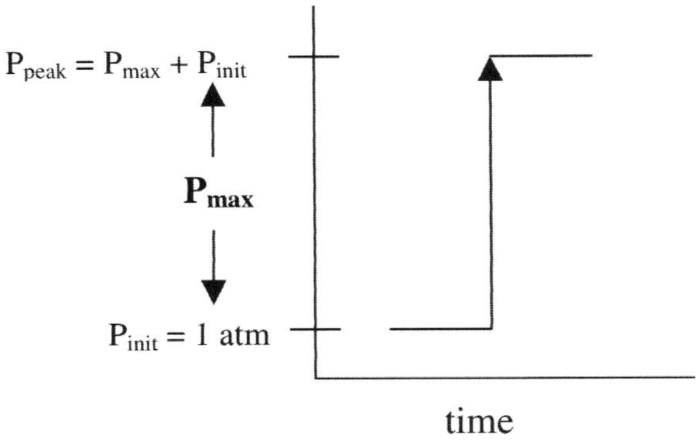

Figure 2. Pmax: Maximum Pressure Rise
Maximum pressure rise due to ignition of a solid at an initial pressure of 1 atm.

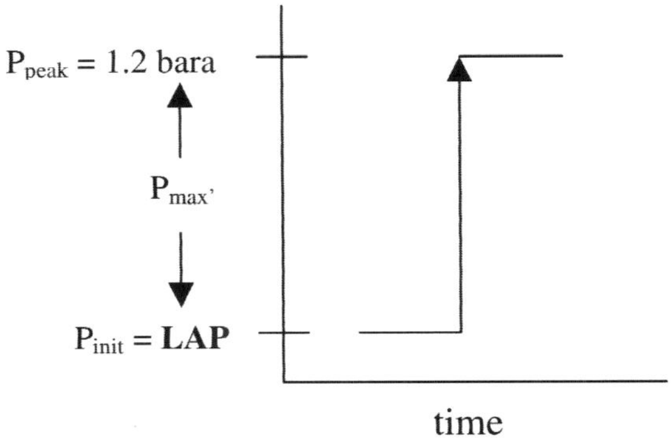

Figure 3. LAP: Limiting Air Pressure
Pressure below which an explosion cannot occur due to ignition of a solid powder.

some product loss/contamination, but there will be no dryer damage should an ignition occur. It should also be noted that this does not account for the possibility of electrostatic damage to the dryer (e.g. pin-holes in glass-lining).

If the dryer cannot maintain vacuum below the LAP to prevent an explosion then the dryer should at least be operated below a maximum initial pressure of Pi,y of the dryer. Pi,y is the initial pressure that a dryer must not exceed to avoid reaching its yield pressure in the event of an explosion (see Figure 4). While this may allow for an explosion to occur and loss of product, as well as potential dryer damage, the explosion will still be contained ensuring the safety of personnel and adjacent equipment.

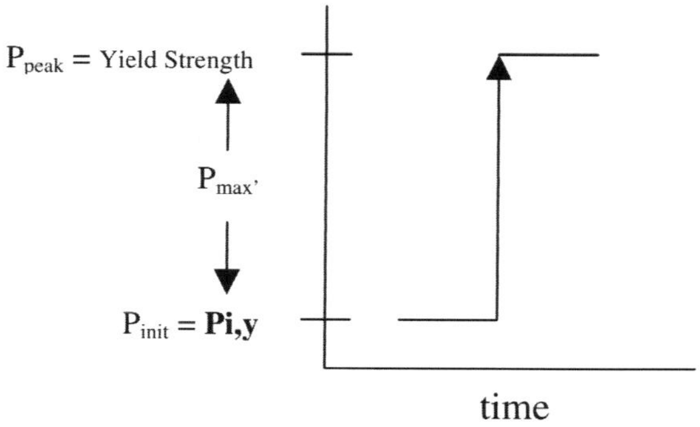

Figure 4. Pi,y: Maximum Initial Pressure
The initial pressure that a dryer must not exceed to avoid reaching its yield pressure in the event of an explosion.

4. Estimation of the LAP Using the 20-L Sphere Apparatus

One use of the 20-L sphere apparatus is to determine the deflagration parameters of a combustible dust-air mixture. These parameters are a function of many factors, such as the turbulence, concentration, and homogeneity of the dust-air mixture, the ignition source, the particle size distribution of the dust, and the initial temperature and pressure of the tested mixture. The standard test to determine the explosion indices Pmax and Kst of the dust are determined over a range of concentrations, at ambient temperature and pressure, with an ignition source of 10,000 Joules, and a feed pressure of 20 bar to disperse the dust cloud. An initial pressure of 1 atm is obtained by pulling vacuum to 0.4 bara within the 20-L sphere and then adding air at 20 bar from the 0.6 L dust storage chamber. By pulling greater vaccum in the 20-L sphere and pressurizing the 0.6 L dust storage chamber to less than 20 bar, reduced initial pressures can be obtained. Initial pressures as low as ~0.4 bara can be obtained without significantly compromising the integrity of the dust-air cloud due to a reduced dispersion pressure.

Figure 5. Limiting Air Pressure (LAP) Determination for a Pharmaceutical Intermediate

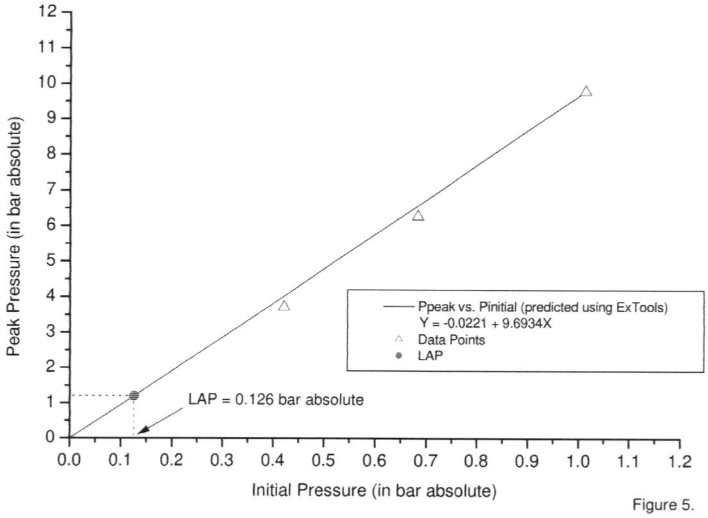

Figure 5.

Figure 5 shows the peak pressures obtained in the 20-L sphere at three different initial pressures for a typical pharmaceutical intermediate. The tests were conducted over a range of concentrations, at room temperature, and using 10,000 J ignitors. At an initial pressure of 1 atm the Pmax is 8.8 bar (Ppeak = 9.813 bara). This extrapolates to an LAP of 0.126 bara using equation 3. The Pmax (and Ppeak) were then determined experimentally at two reduced initial pressures to confirm the theoretical extrapolation. An initial pressure of 0.42 bara resulted in a peak pressure of 3.72 bara, and an initial pressure of 0.68 bara resulted in a peak pressure of 6.28 bara. This data is in good agreement with the theoretical extrapolation.

5. Application

Pressure interlocks should be installed to ensure the dryer is operated below the solid's LAP. The interlocks should stop or prevent dryer rotation at pressures above the LAP or Pi,y. When drying a wet cake, the pressure in the dryer will either be at the solvent vapor pressure, if the removal of solvent is limited by the equilibrium concentration in the vapor phase, or closer to the vacuum pump set point, if the removal of solvent is limited by mass transfer through the cake. If the vapor pressure of the solvent being driven off is greater than the LAP the dryer pressure profiles should be evaluated to ensure the vacuum pump can pull below the LAP (see Figure 6).

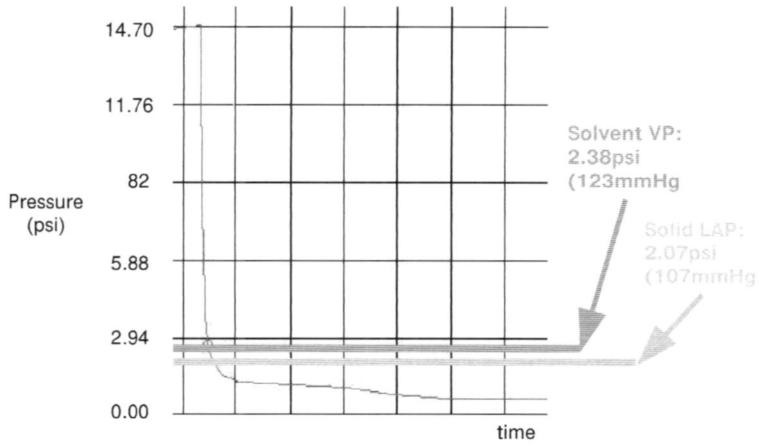

Figure 6. Typical Dryer Pressure Profile for Mass-Transfer Limited Drying

While dusts can have a Pmax as high as 10 bar, solvents generally have a Pmax < 8 bar. For the purpose of determining the LAP, which is based on Pmax, the Pmax of the solid can be used, as the Pmax of a hybrid mixture (dust + flammable vapor) is usually similar to that of the dust. However, it should be noted that a dust cloud in air may become much more susceptible to ignition and the violence of the resulting explosion may increase considerably if the air contains a small amount of flammable vapor, even though the vapor concentration may be well below the LFL. But again, this does not affect the LAP that is based on the powder's Pmax.

6. Assessment Logistics: Information Collection

Each dryer should be evaluated on a case-by-case basis. First, the properties of the solids and solvents should be identified. This includes the conductivity, flammability / explosibility, the solid Pmax, and the solvent vapor pressure. Next, identify the properties of the dryer such as the material of construction (steal, glass-lined, etc.), the yield strength, and MAWP. Finally, gather the process information, such as operating pressures, temperatures, and pressure profiles. Based on this information a logical decision can be made as to what form of explosion prevention/mitigation should be used, or whether it is needed at all.

7. Summary

A review of 26 rotary dryers was conducted. There was a total of 92 different product/dryer combinations analyzed. Of these 92, the LAP was employed as a safeguard in 27 cases. Active inertion was required in five cases, procedural inertion was sufficient for 44 cases, one needed to operate below Pi,y, one needed more data to analyze, and 14 were okay to dry without further safeguards.

8. References

[1] Tarantino, B., "Guideline for Mitigating Static Hazards in Rotary Dryers," Merck & Co., Inc., Process Engineering, August 2003.

[2] Barton, J., *Dust Explosions Prevention and Protection*, Institution of Chemical Engineers, 2002.

[3] Abbot, J., *Prevention of Fires and Explosions in Dryers*, Second Edition, Institution of Chemical Engineers, 1990.

[4] *Standard Test Method for Pressure and Rate of Pressure Rise for Combustible Dusts*, ASTM Designation: E 1226 – 00, 2000.

Loss of Inerting Due to Multiple Exhaust Vents

J. Kelly Thomas, Donald E. Ketchum and Quentin A. Baker
Baker Engineering and Risk Consultants, Inc. (BakerRisk)
3330 Oakwell Ct., Suite 100
San Antonio, TX 78218-3024
(210) 824-5960
kthomas@BakerRisk.com

Prepared for Presentation at
American Institute of Chemical Engineers
2005 Spring National Meeting
39th Annual Loss Prevention Symposium
Atlanta, Georgia
April 11-13, 2005

February, 2005

UNPUBLISHED

Loss of Inerting Due to Multiple Exhaust Vents

J. Kelly Thomas, Donald E. Ketchum and Quentin A. Baker
Baker Engineering and Risk Consultants, Inc. (BakerRisk)
3330 Oakwell Ct., Suite 100
San Antonio, TX 78218-3024
(210) 824-5960
kthomas@BakerRisk.com

Abstract

This paper addresses the potential for a loss of inerting to occur in a vessel with multiple exhaust vents due to air inleakage driven by natural convection. Air inleakage due to natural convection refers to flow induced by the density difference between the gas mixture inside the vessel and that in the exterior environment. Natural convection can result in the inleakage of air into an inerted vessel, despite the positive pressure induced by the purge gas flow.

An air inleakage model was developed to predict the natural convection air inleakage rate and corresponding oxygen concentration as a function of the vessel configuration, operating conditions, exterior environment conditions, and purge gas flow rate. Examples are provided to illustrate the range of oxygen concentrations that can result from air inleakage due to natural convection. Demonstration tests were conducted to illustrate natural convective flow under relevant conditions and provide data for comparison with the model predictions. The results of this work demonstrate that the presence of multiple exhaust vents on an inerted vessel can lead to the development of a significant oxygen concentration. In some cases, the resulting oxygen concentration can exceed the minimum level required to support combustion. If the vessel is inerted to prevent the formation of a flammable gas mixture (i.e., rather than solely for product quality control), then an explosion may result if an ignition source is present.

1. Introduction

This paper addresses the potential for a loss of inerting to occur for a vessel with multiple exhaust vents. Specifically, this paper focuses on the potential for loss of inerting due to air inleakage driven by natural convection. Air inleakage due to natural convection refers to flow induced by the density difference between the gas mixture inside the vessel and air outside the vessel. The term "exhaust vents" as used within the context of this paper denotes any openings of out the vessel through which gas may flow. While engineered vessel exhaust flow paths are obvious candidates in this regard, unintended flow paths (e.g., connections to the atmosphere through ducts or pipes that penetrate into the vessel vapor space, holes in the vessel created by corrosion, broken vessel connections, etc.) will also act as exhaust vents. The impact of these unintended flow paths should be clearly recognized when evaluating the potential for a natural convective flow loop to be established with an inerted vessel.

The National Fire Protection Association Standard *Explosion Prevention Systems*[1] (NFPA 69) provides guidance on the use of inerting systems to prevent explosions. The NFPA 69 guide explicitly cautions against conditions that may lead to a loss of inerting; specific situations discussed by the guide include a rapid decrease in the vessel liquid level or vapor space temperature. However, the guide does not offer an explicit caution with respect to a loss of inerting due to air inleakage driven by natural convection.

It is recognized that a gradual loss of inerting due to air inleakage should be detected if the vessel is equipped with oxygen monitors. However, not all inerted vessels are equipped with oxygen monitors. The process environment in some cases makes direct oxygen monitoring very difficult, if not impossible. Furthermore, the NFPA 69 guide allows periodic oxygen monitoring, and a loss inerting may occur between scheduled oxygen monitoring times. The NPFA 69 guide also does not require oxygen monitoring for padded low-pressure field storage tanks.

The authors are familiar with two cases within the last decade in which a loss of inerting due to air inleakage driven by natural convection has occurred, although neither of these cases is discussed within this paper. The gas mixture inside the vessel was lighter than air in one case and heavier than air in the other. In both cases, the result of the loss of inerting was an explosion that ruptured the vessel. The staff at these plants was unaware that the system configuration and operating parameters could lead to a loss of inerting via this mechanism. The common perception appears to be that the positive pressure induced by purge gas flow will preclude air inleakage. That is, there is a common belief that if inert gas is flowing into the vessel, then the vessel is inerted. This is, unfortunately, not always the case.

This paper is intended to raise awareness for the potential of a loss of inerting due to air inleakage driven by natural convection when multiple exhaust vents are present. While it is acknowledged that this topic may appear to be somewhat esoteric, the fact that multiple vessel explosions have occurred due to this mechanism clearly demonstrates that it is relevant to the design, operation, and maintenance of an effective vessel inerting system. It is the authors' hope that raising awareness in this manner will preclude similar events in the future.

The next section of this paper presents a model that can be employed to determine if air inleakage due to natural convection will occur. The model can be used to predict the air inleakage rate and corresponding equilibrium oxygen concentration as a function of the vessel configuration, operating conditions, exterior environment conditions, and purge gas flow rate. Examples are provided to illustrate the range of oxygen concentrations that can result from inleakage due to natural convection. The results from demonstration tests are presented following the discussion of the air inleakage model. The test results illustrate the natural convection flow mechanism in a relevant geometry and demonstrate that the model provides reasonable air inleakage rate estimates. Conclusions from this work are provided at the end of the paper.

2. Air Inleakage Model

2.1 Model Description

The pressure differential due to the density difference between the vessel vapor space and exterior atmosphere (i.e., the natural convection driving force) can be written as:

$$\Delta P_{nc} = g \cdot \Delta z \cdot \Delta\rho \quad (1)$$

where: ΔP_{nc} = Natural convection (buoyancy) driving force,
g = Gravitational acceleration,
Δz = Effective stack height,
$\Delta\rho$ = Density difference (= $\rho_a - \rho_v$),
ρ_a = Density of the atmospheric air exterior to the vessel, and
ρ_v = Density of the gas in the vessel vapor space.

This expression implicitly assumes that the vessel vapor space is relatively well mixed since a single density value is invoked; although perfect mixing within a vessel vapor space is never achieved, the well-mixed assumption is appropriate under many relevant conditions. The gas mixture density can be calculated based on the ideal gas law.

The applicable value for the effective stack height can be found by applying Bernoulli's equation across all flow paths. To illustrate, consider the vessel vent arrangement shown in Figure 1. It is assumed for the purposes of illustration that the gas mixture in the vessel is lighter than air. The elevation at the top of the vessel is chosen to define the reference atmospheric and vessel vapor space pressures (i.e., P_a and P_v, respectively), although the selection of the reference pressure location is arbitrary. The resulting pressures at the flow path termination points are given on the figure. Bernoulli's equation is applied across each flow path (inlet and outlet), which allows the reference atmospheric and vessel vapor space pressures to be eliminated. The resulting expression gives the total pressure drop in terms of elevation change, which defines the effective stack height (Δz). Unidirectional flow within each flow path is assumed, which is the expected behavior if multiple exhaust vents are present. Bidirectional flow within a flow path is physically possible under a limited range of conditions, but this is of little practical relevance since the corresponding air inleakage rates are generally very low.

For the vessel geometry shown in Figure 1, if the gas mixture in the vessel flows out the upper vent line and air is pulled into the lower vent line, then the natural convection driving force would be:

$$\Delta z = H_{s1} + \Delta H + z_{e2} \quad (2)$$

Conversely, if the gas mixture in the vessel flows out the lower vent line and air is pulled into the upper vent line, then the natural convection driving force would be:

$$\Delta z = H_{s2} - \Delta H + z_{e1} \quad (3)$$

One factor not accounted for by these expressions is that there will be a small increase in the effective stack length due to the extension of the air column into the vessel vapor space. That is, in the absence of the vessel pressurization via the purge flow, natural convective flow would still be established with multiple exhaust paths even if the effective stack height were zero. A reasonable estimate of the increase in the effective stack height is a few pipe diameters. The exact value used is not overly important since, in any practical case, it will not increase the stack height by a significant percentage.

Note that although the natural convection driving force given by eqn.(2) is larger than that given by eqn.(3), this does not ensure that the gas mixture in the vessel will flow out the upper vent line. The flow direction will be that which yields the largest natural convection flow rate, which depends on both the natural convection driving force and the total pressure drop. Since the outlet flow is composed of both the air and purge flows, the total pressure drop will be minimized if the outlet vent line has the smaller hydraulic resistance. A flow reversal could occur (e.g., due to a sudden change in vessel pressure), but the flow direction corresponding to the largest natural convection flow rate will soon be reestablished. Of course, the natural convection driving force must be large enough to overcome the vessel pressure developed by the purge flow, or no natural convective flow will be established.

The pressure drop due to the flow of gas through a vent path can be expressed as[2]:

$$\Delta P = \rho \cdot \frac{V^2}{2} \cdot \left(f \cdot \frac{L}{D_h} + \sum K_j \right) \tag{4}$$

where: ΔP = Pressure differential across flow path,
ρ = Density of fluid,
V = Velocity of fluid,
f = Friction factor (i.e., Moody friction factor),
L = Length of flow path,
D_h = Hydraulic diameter of flow path, and
K_j = Form loss coefficient for j^{th} flow path feature (i.e., entrance, elbow, etc.).

The appropriate friction factor expression for a given flow path depends on the flow regime (i.e., laminar, transition or turbulent). The flow regime can be determined based on the Reynolds number, which can be written as[2]:

$$Re = \frac{\rho \cdot V \cdot D_h}{\mu} \tag{5}$$

where: Re = Reynolds number, and
μ = Absolute viscosity.

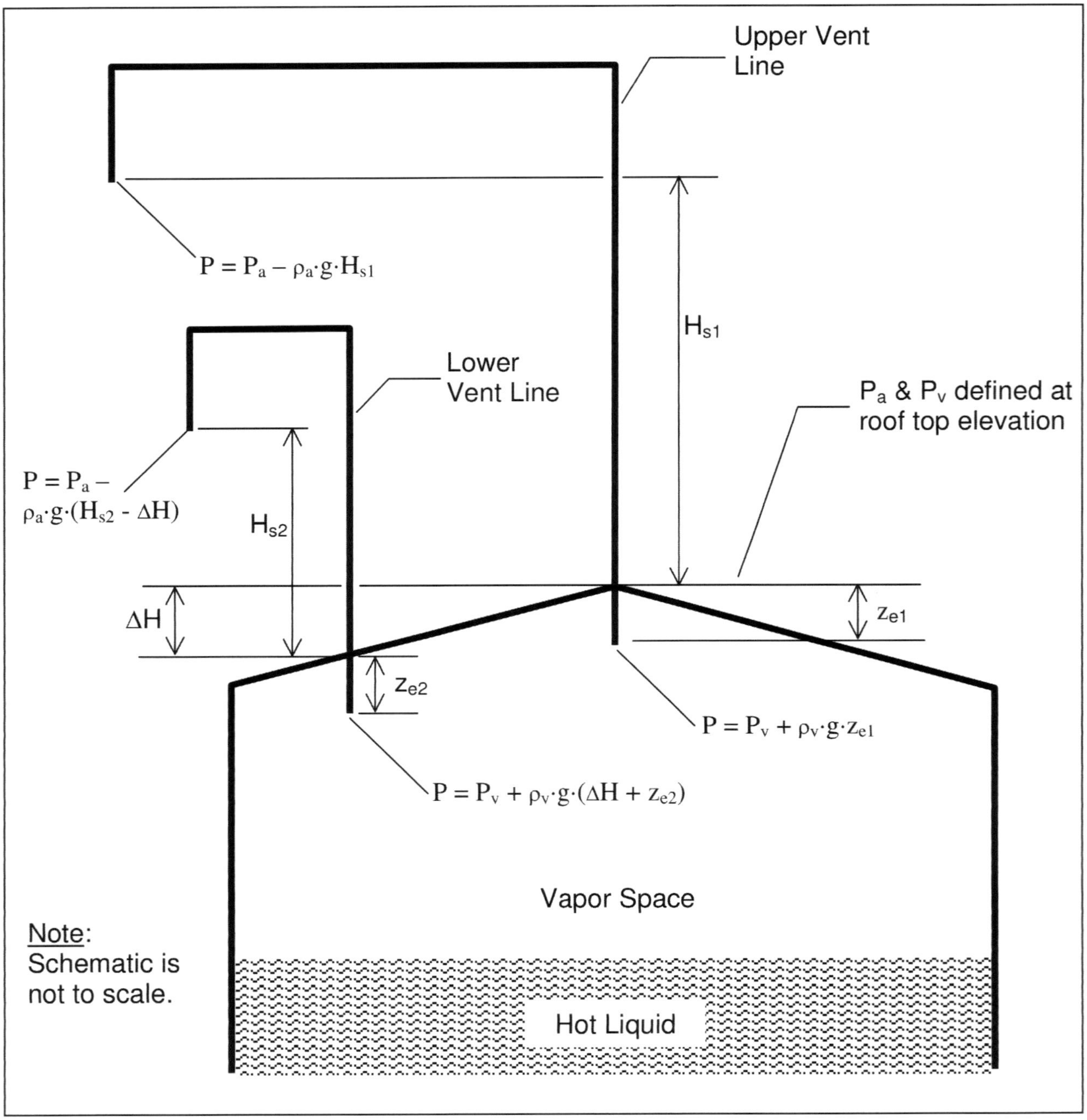

Figure 1. Schematic for Illustration of Model

If laminar flow conditions exist, the friction factor can be expressed as[2]:

$$f = \frac{64}{Re} = \frac{64 \cdot \mu}{\rho \cdot V \cdot D_h} \tag{6}$$

If turbulent flow conditions exist, then the friction factor can be approximated using the expression given below, which is an explicit approximation to the Colebrook-White correlation[3]:

$$f = \left[1.14 - 2 \cdot \log_{10}\left(\frac{e}{D} + \frac{21.25}{Re^{0.9}}\right)\right]^{-2} \tag{7}$$

where: e = Surface roughness.

The laminar friction factor was employed for Reynolds numbers below 2000, the turbulent friction factor was employed for Reynolds numbers above 4000, and linear interpolation between the 2000 and 4000 values was employed for intermediate Reynolds numbers. Form loss coefficients for relevant features (e.g., entrance and exit loss, elbows, tees, etc.) can be found in standard literature references[3,4].

The air flow rate established due to natural convection is that which is required to balance the total pressure drop and the driving force due to natural convection. That is, the air flow rate is found by satisfying the following equality:

$$\Delta P_{nc} = \Delta P_i + \Delta P_o \tag{8}$$

where: ΔP_i = Pressure drop across the inlet paths due to air inflow, and
ΔP_o = Pressure drop across the outlet paths due to gas mixture outflow.

The natural convective driving force is given by eqn.(1) and inlet and outlet pressure drops are given by eqn.(4).

Assuming that the inert gas contains no oxygen and no gas is evolved from the vessel contents, the equilibrium oxygen concentration resulting from a given air inleakage rate can be calculated as:

$$y^e_{O2} = y^a_{O2} \cdot \left(1 + \frac{Q_p}{Q_a}\right)^{-1} \tag{9}$$

where: y^e_{O2} = Equilibrium vessel vapor space oxygen concentration,
y^a_{O2} = Oxygen concentration in air (i.e., approximately 21%),
Q_p = Purge flow rate (referenced to standard conditions), and
Q_a = Air flow rate (referenced to standard conditions).

The transient response of the vessel vapor space can be described as shown below:

$$y_{O2}(t) = y^{i}_{O2} \cdot e^{-\left(t/\tau\right)} + y^{e}_{O2} \cdot \left[1 - e^{-\left(t/\tau\right)}\right] \quad (10)$$

where: y^{i}_{O2} = Initial oxygen concentration (i.e., prior to transient),
t = Time since beginning of transient,
τ = Vessel vapor space time constant ($\tau = V_v / Q_t$),
V_v = Vessel vapor space volume, and
Q_t = Total flow rate into vessel vapor space ($Q_t = Q_p + Q_a$).

The results of calculations using this approach are relatively insensitive to minor variations in the input parameters. A change in a parameter value that increases the natural convective driving force, such as a decrease in the vessel vapor space density for a gas mixture that is lighter than air, increases the air inleakage rate and hence increases the vessel vapor space density. That is, there is a negative feedback loop that limits the system response to changes in parameter values. This is fortunate since there is always some uncertainty and/or variability in the parameter values (e.g., vapor space temperature and composition, ambient temperature, purge flow rate, etc.).

It is noted that this same basic model was successfully applied to evaluate air inleakage into an inerted high level radioactive waste storage tank and to define the minimum inert gas flow rate required to preserve inerting[5]. This same approach has also been employed to design a passive ventilation device to maintain hydrogen concentrations below the flammability limit in radioactive waste storage vessels[6]. BakerRisk also developed a backup passive ventilation system for hydrogen control at a commercial production facility using this approach.

It is recognized that the model presented above is simplified in some respects. Computational fluid dynamics (CFD) codes are available that can potentially provide more accurate results. However, CFD codes typically develop stability problems with low driving forces and low flow velocities, as is often the case with natural convection flow phenomena. Furthermore, it is expected that the model presented above will provide reasonable estimates suitable for engineering level analyses. The comparison between model predictions and test data presented below indicates that this is indeed the case.

2.2 Example Case

Consider again the vessel geometry shown in Figure 1. The parameter values listed in Table 1 are assumed for the sake of illustration. The gas mixture in the vessel vapor space was assumed to be warm (150°C) due to the presence of a hot liquid in the vessel. The purge gas was assumed to be the combustion products of methane, and the composition of the stream was taken to be 88% nitrogen and 12% carbon dioxide. Under the assumed conditions, the densities of the gas mixture inside the vessel (without air inleakage) and in the ambient environment would be 0.86 and 1.16 g/l, respectively.

Note that the upper vent line has a smaller diameter than the lower vent line (3 in. vs. 8 in.). The natural convection driving force for either flow direction (i.e., air flow into the lower or upper vent) would be greater than the vessel pressure established by the purge flow with these system parameter values. However, the natural convection flow rate is maximized if air is pulled into the upper vent and the combined air and purge flow exits the lower vent line. This is somewhat counterintuitive since it might be expected that the "light" gas mixture would flow out the upper vent line, much like a chimney. That natural convective driving force with flow out the upper vent line is roughly double that for flow out the lower vent line. However, since the lower vent line diameter is much larger than that of the upper vent line (i.e., lower hydraulic resistance), the upper vent line pressure drop is roughly two orders of magnitude larger than that for the lower vent line at the same flow rate. The natural convection flow rate is therefore maximized if air is pulled into the upper vent and the combined air and purge flow exits the lower vent line.

Table 1. Example Case Parameter Values

Parameter Description	Assumed Value
Upper vent line:	
Nominal diameter (inches, Schedule 10S)	3
Total length (feet)	15
Number elbows (r/d = 1.2)	2
Discharge elevation above vessel (H_s, feet)	5
Extension into vessel (z_e, inches)	2
Lower vent line:	
Nominal diameter (inches, Schedule 10S)	8
Total length (feet)	8
Number elbows (r/d = 1.2)	2
Discharge elevation above vessel (H_s, feet)	4
Extension into vessel (z_e, inches)	2
Elevation between connection points (ΔH, ft)	1
Pipe surface roughness (mm)	0.01
Atmospheric conditions:	
Temperature (°F)	84
Relative humidity	45
Vessel vapor space temperature (°C)	150
Purge gas flow rate (scfm)	10

The air inleakage rate predicted using the system parameter values given in Table 1 is 13.4 cfm. This air inleakage rate would establish an equilibrium oxygen concentration of approximately 11.6% O_2 (vol.%) at the specified purge flow rate (i.e., 10 scfm). The natural convective driving force is 3.5 Pa (0.014 inches of water column) and 94% of the pressure drop is along the upper vent line (i.e., the air inlet). The flow regime in the upper vent line (air inlet) is turbulent, and that in the lower vent line (gas mixture outlet) is in the transition regime. These air inleakage and purge flow rates would yield a time constant of about one hour for a vessel vapor space volume of 10,000 gallons; hence, once the natural convective flow loop was initiated, the equilibrium oxygen concentration would essentially be established within approximately four hours. In addition, since both the vessel and exterior environmental conditions would be expected to change relatively slowly, a time constant of an hour implies that the vessel vapor space oxygen concentration would be near the equilibrium value under most conditions.

The oxygen concentration required to support combustion is referred to as the Limiting Oxygen Concentration (LOC). The LOC values for most hydrocarbons with nitrogen as the inert species are in the range of 9 to 12% at ambient temperature[1], and slightly lower at the temperature used in the example (i.e., 150°C). The presence of minor concentrations of combustible gases with lower LOC values (e.g., hydrogen, carbon monoxide, etc.) would further reduce the LOC. The oxygen concentration calculated for the example system parameters (i.e., 11.6% O_2) would therefore allow a flammable gas mixture to develop, assuming a fuel concentration in the flammable range.

Figure 2 shows the predicted oxygen concentration as a function of the purge gas flow rate for flow rates up to 25 scfm. Increasing the purge gas flow rate both pressurizes the vessel vapor space, which limits the air inleakage rate, and dilutes the air that does leak into the vessel, which further reduces the resulting oxygen concentration. As can be seen, significant air inleakage due to natural convection would occur up to the upper end of the range shown in the figure (i.e., 25 scfm). However, the resulting bulk vapor space oxygen concentration (i.e., 6.6% O_2) would be well below the LOC for most flammable gas mixtures. A purge gas flow rate of almost 100 scfm would be required under these conditions to completely eliminate air inleakage due to natural convection. Figure 3 shows the predicted oxygen concentration as a function of the vapor space temperature. Vapor space temperature does not have as large of an effect on the predicted oxygen concentration as the purge gas flow rate, as can be seen by comparing Figure 2 and Figure 3. Decreasing the vapor space temperature from 150°C down to 100°C decreases the oxygen concentration by less than one per cent. Higher molecular weight species will be present in many cases due to evaporation of the vessel liquid contents, and this will tend to mitigate the effect of a change in vapor space temperature. That is, a higher vapor space temperature will also increase the average molecular weight of the species in the vessel vapor space, while a lower temperature will also decrease the average molecular weight.

3. Demonstration Test

Demonstration tests were conducted to illustrate natural convective flow in a relevant geometry and to provide data for comparison with the predictions of the air inleakage model described above.

3.1 Test Setup

The rig setup employed for the demonstration test approximates the key thermal hydraulic parameters in the example calculation given in the preceding section. A schematic of the test rig is given as Figure 4. The test vessel was a 55-gallon drum. The larger diameter (lower vent) line was constructed using 2-inch Schedule 20 pipe sections and the smaller diameter (upper vent) line was constructed using 3/4-inch Schedule 40 pipe sections, which approximates the vent line diameter ratio in the example calculation. The vent lines did not penetrate into the vessel vapor space and both vent lines connected to the vessel at the same elevation. The total length of the larger diameter lower vent line was 4.5 feet and its discharge point was 2.25 feet above the vessel. The total length of the smaller diameter upper vent line was 5.5 feet and its discharge point was 3.25 feet above the vessel.

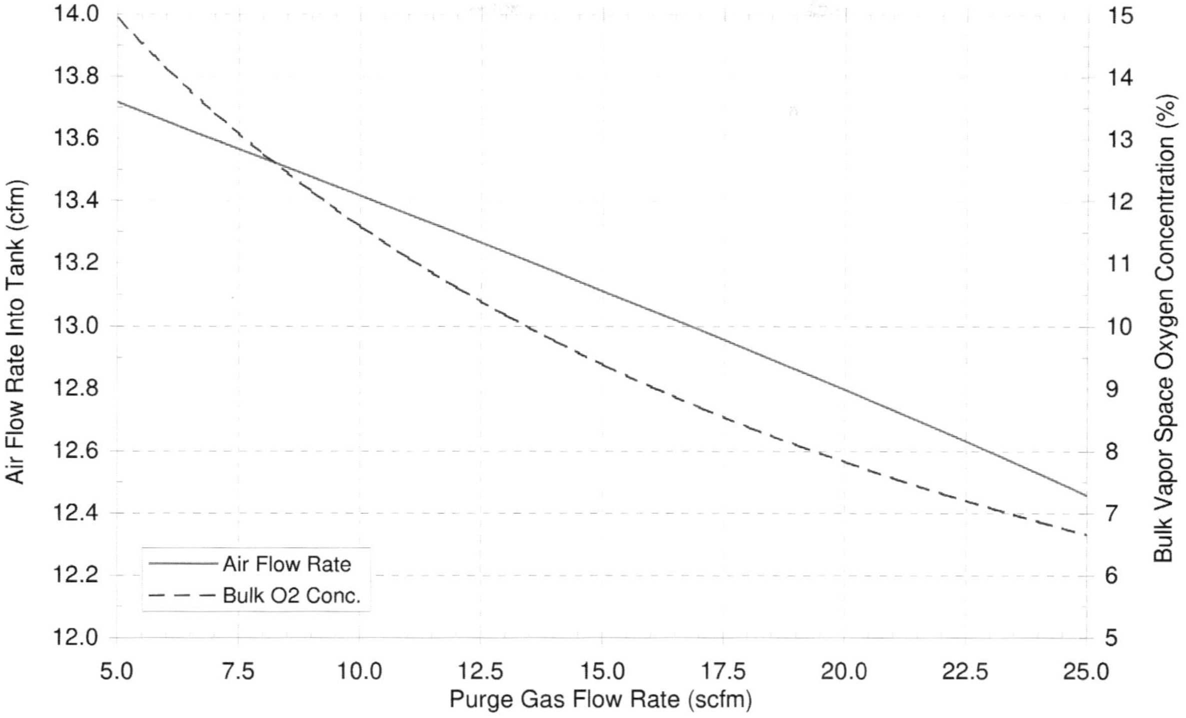

Figure 2. Air Inleakage Rate and Oxygen Concentration vs. Purge Gas Flow Rate

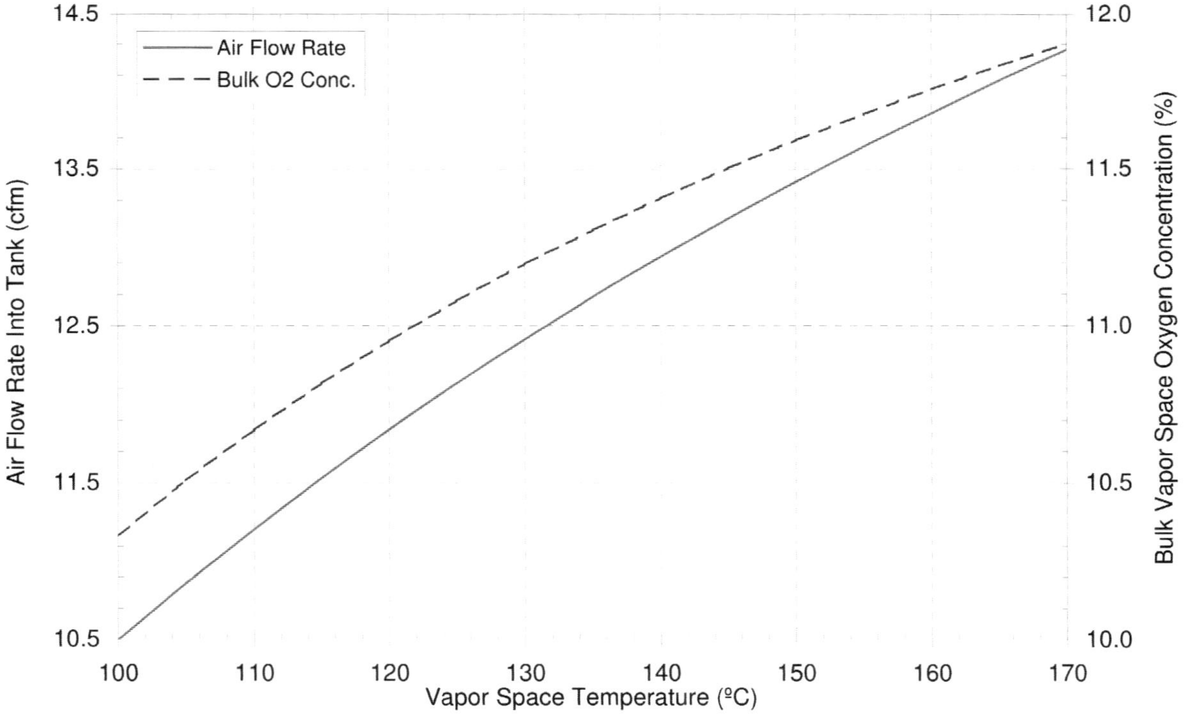

Figure 3. Air Inleakage Rate and Oxygen Concentration vs. Vessel Vapor Temperature

The purge gas was nitrogen introduced to the vessel at a controlled flow rate. An electric heater set located at the vessel bottom was utilized to produce the desired vapor space temperature. A continuous gas flow of 0.03 to 0.05 scfm was drawn from one of three different elevations and pulled through an oxygen monitor (accuracy of ±0.2% O_2 and 90% response in 30 seconds); the maximum sample rate is approximately 5% of minimum total flow rate observed during these tests. The vessel vapor space temperature was also monitored at three elevations. A fan was installed in the vessel to examine the impact of enhanced mixing.

3.2 Test Results

Figure 5 provides the oxygen concentration histories for the demonstration tests run at vapor space temperatures of 150°C and 200°C and purge gas flow rates of 0.5 and 1.0 scfm. The tests were run for a sufficient period of time to approach the equilibrium oxygen concentrations; the system time constant under these conditions was approximately 7 minutes. Air was pulled by natural convection into the test vessel through the smaller diameter upper vent, with the combined purge and air flow out through the large diameter lower vent, as expected based on the results obtained with the air inleakage model. The purge gas flow rate had a much larger effect than the vapor space temperature, which was also expected based on the results obtained with the model. The test vessel was relatively well mixed; both the oxygen concentration and the temperature measurements taken at multiple locations within the vapor space were in good agreement and starting the fan within the vessel vapor space did not change the measured values.

Table 2 provides the measured equilibrium oxygen concentrations along with those calculated using the air inleakage model described in the preceding section. Oxygen concentrations were calculated using the hydraulic resistance values from the literature along with those measured with this test rig. The measured resistances (i.e., pressure drops) were roughly 10% higher than the expected values, so that the corresponding oxygen concentrations are slightly lower. The predicted oxygen concentrations are slightly higher than the measured equilibrium values. This is most likely due to cooling of the gas mixture in the exhaust vent path (i.e., a reduction in the natural convective driving force). It is expected that insulating the exhaust vent line would increase the measured oxygen concentration and hence give better agreement with the predicted values.

Several additional tests were performed to demonstrate the inherent stability of the natural convective flow loop. The purge flow rate was increased sufficiently to pressurize the vessel such that natural convection could not occur (i.e., purge flow was forced out both vent paths). Turning the purge flow rate back down resulted in the natural convective flow loop being quickly reestablished. A second approach was to shut the lower vent path (i.e., the vent through which gas was flowing out of the vessel) so that the exhaust flow was forced out of the upper vent. The natural convective flow loop was quickly reestablished when the lower vent path was reopened.

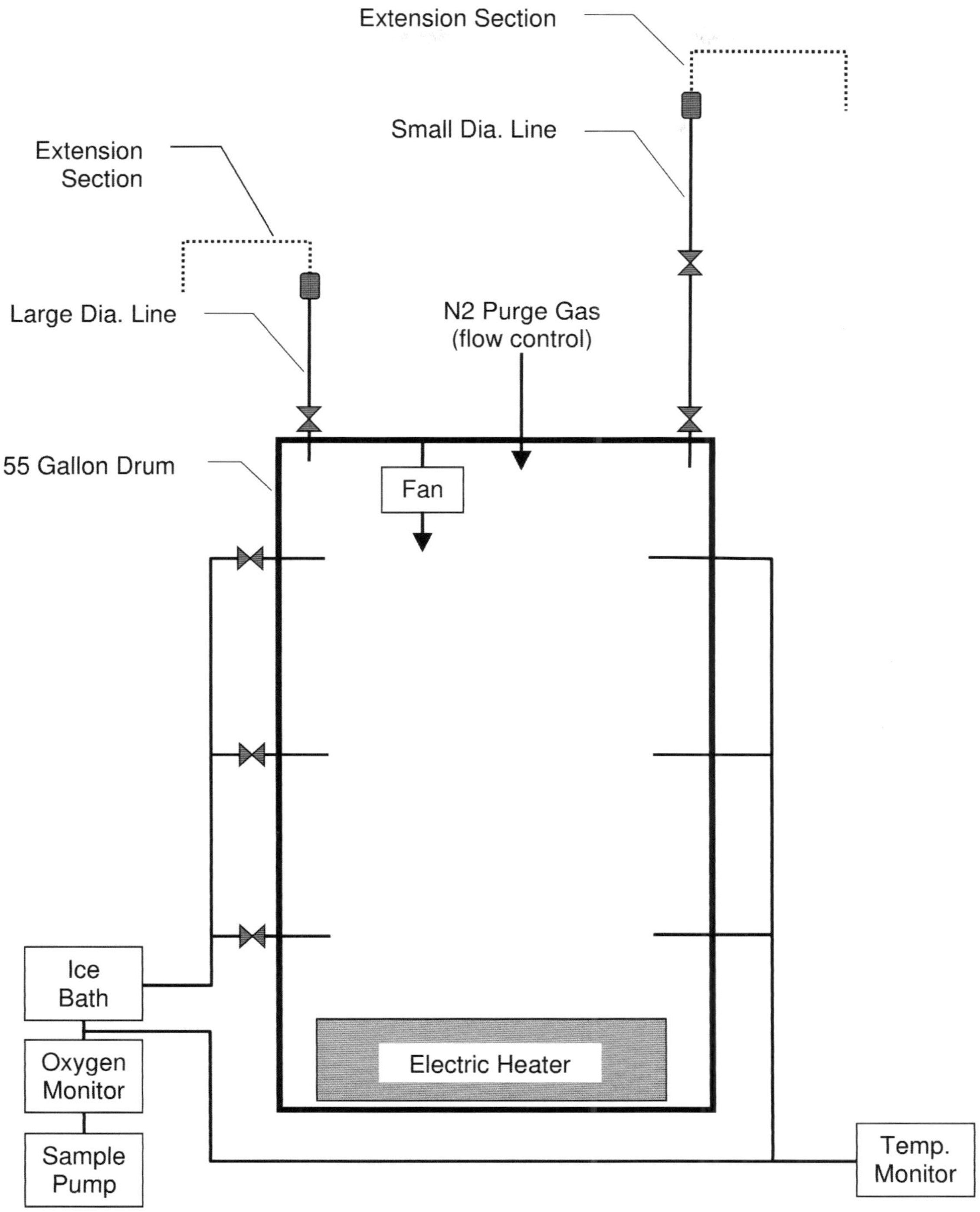

Figure 4. Test Rig Schematic

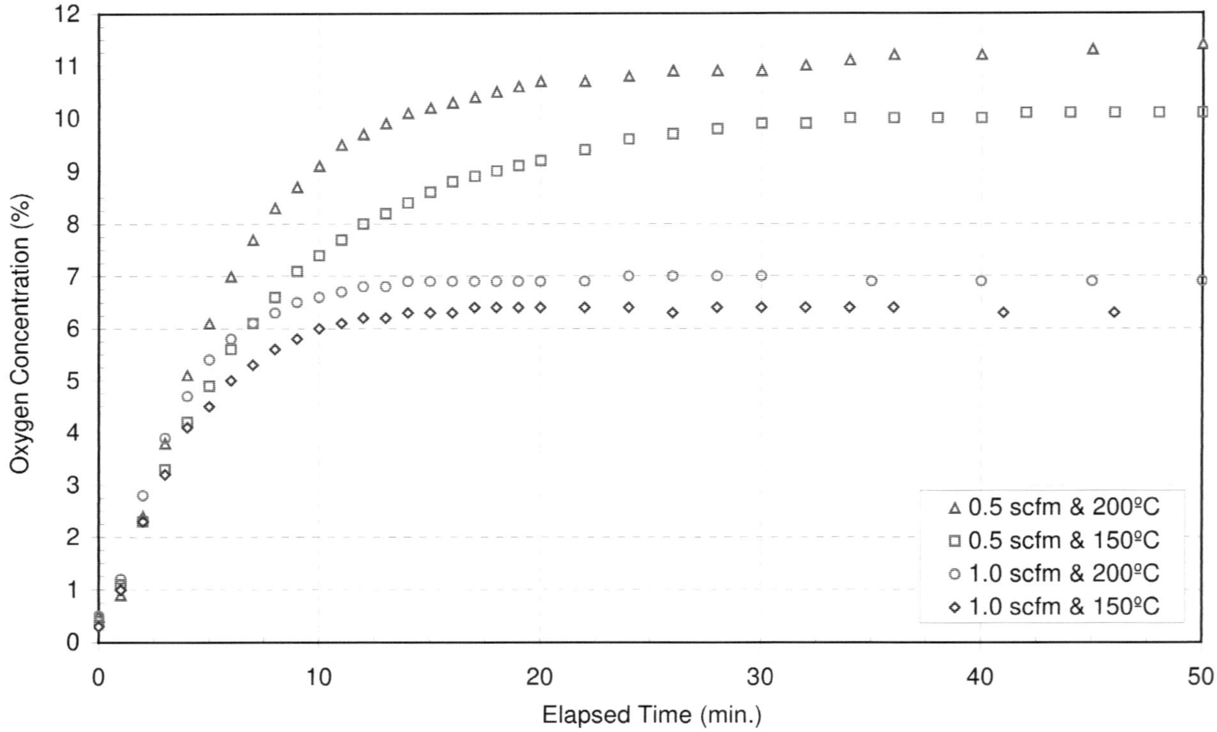

Figure 5. Oxygen Concentration History from Demonstration Tests

Table 2. Results from Demonstration Tests

Vessel Vapor Space Temperature (°C)	Nitrogen Purge Gas Flow Rate (scfm)	Oxygen Concentration (% vol.)		
		Measured Values	Predicted w/ Measured Resistance Values	Predicted w/ Literature Resistance Values
150	0.5	10.1	10.2	10.6
150	1.0	6.4	6.9	7.3
200	0.5	11.5	11.9	12.2
200	1.0	7.0	8.1	8.5

3.3 Flow Visualization

A titanium tetrachloride "smoke stick" was employed to allow the flow field in the vicinity of the vents to be visualized. The smoke was clearly pulled into the small diameter upper vent line (i.e., the inlet) and pushed away from the large diameter lower vent line (i.e., the outlet). Photos illustrating the interaction of the smoke with the gas flow at the ends of the vent lines are given in Figure 6. These photos show that the smoke was pulled into the small diameter upper vent line, demonstrating that air was being pulled into this vent, and that the smoke was pushed away from the large diameter lower vent line, demonstrating that gas was flowing out of this vent.

Figure 7 and Figure 8 show both vent paths in one photo. In Figure 7, it can be seen that smoke is pulled into the small diameter upper vent line, flows into and mixes within the vessel, and then flows out the large diameter lower vent line. Figure 8 shows smoke being pushed away from the large diameter lower vent line due to the gas mixture flowing out this vent line. A vapor space temperature of 150°C and a purge flow rate of 0.5 scfm were being utilized in the test shown in these figures.

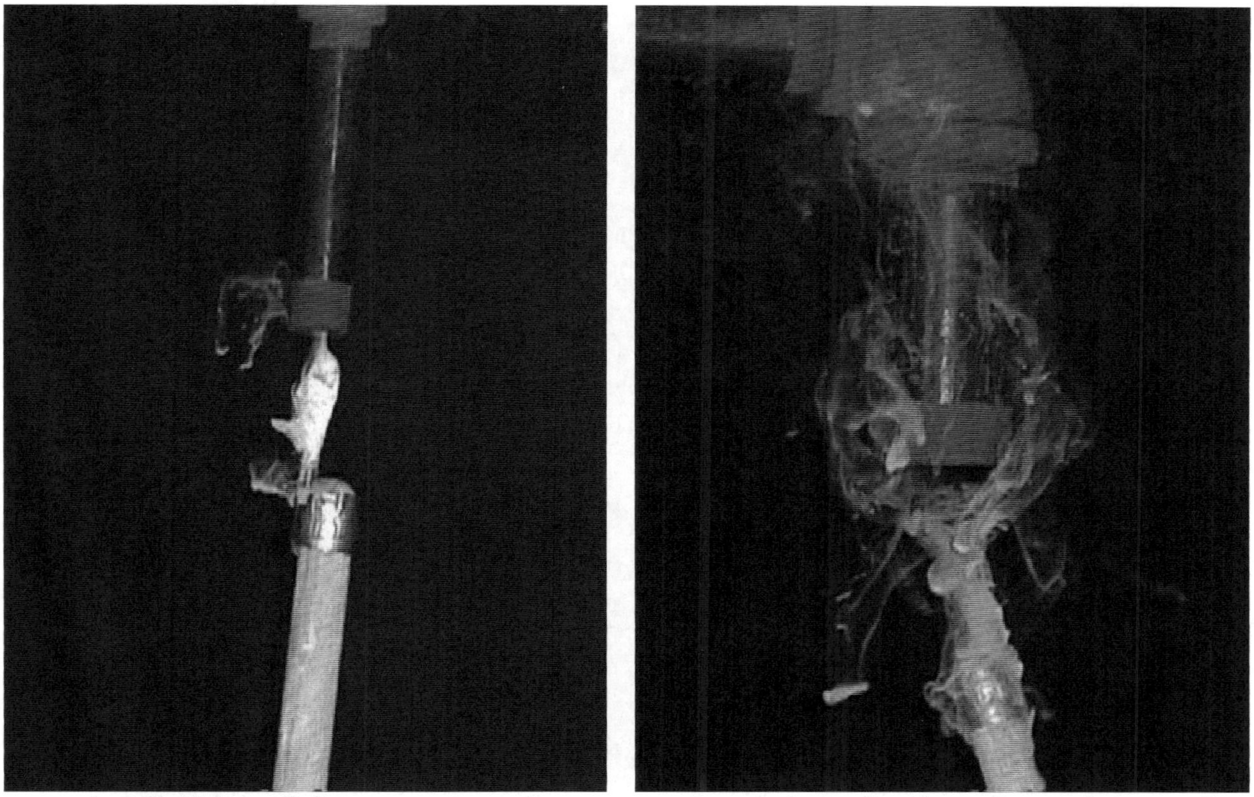

smoke pulled into small diameter upper vent line *smoke pushed away from large diameter lower vent line*

Figure 6. Smoke Interaction with Vent Line Gas Flow

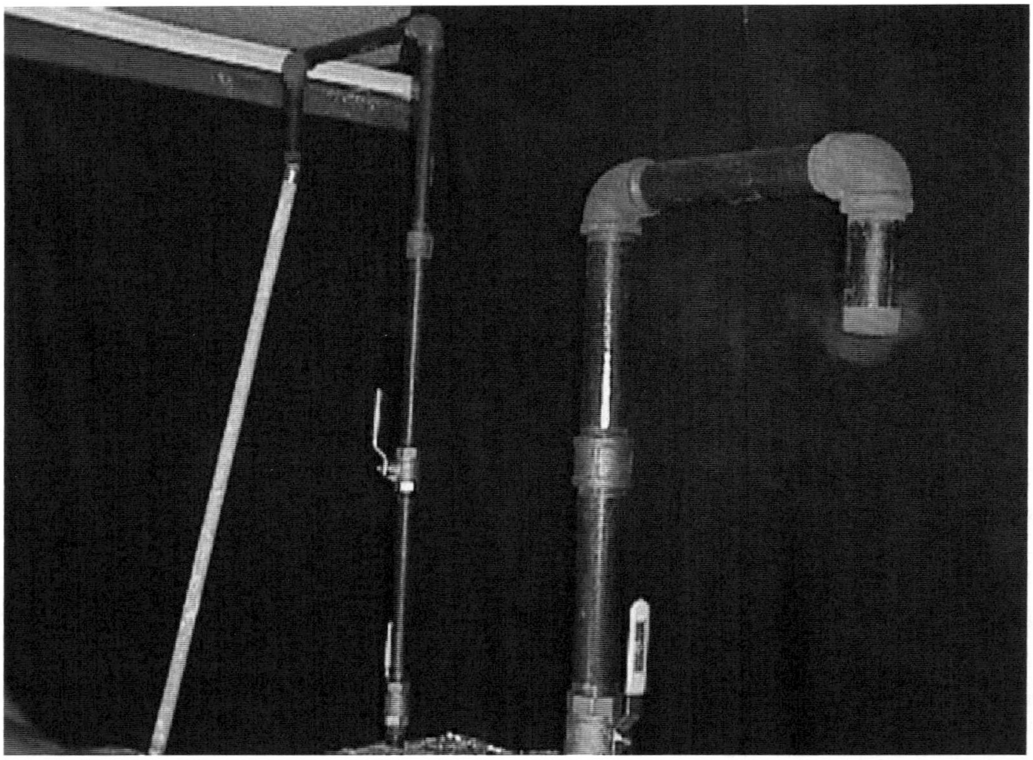

Figure 7. Smoke Flowing Through System and Out Large Diameter Vent Line

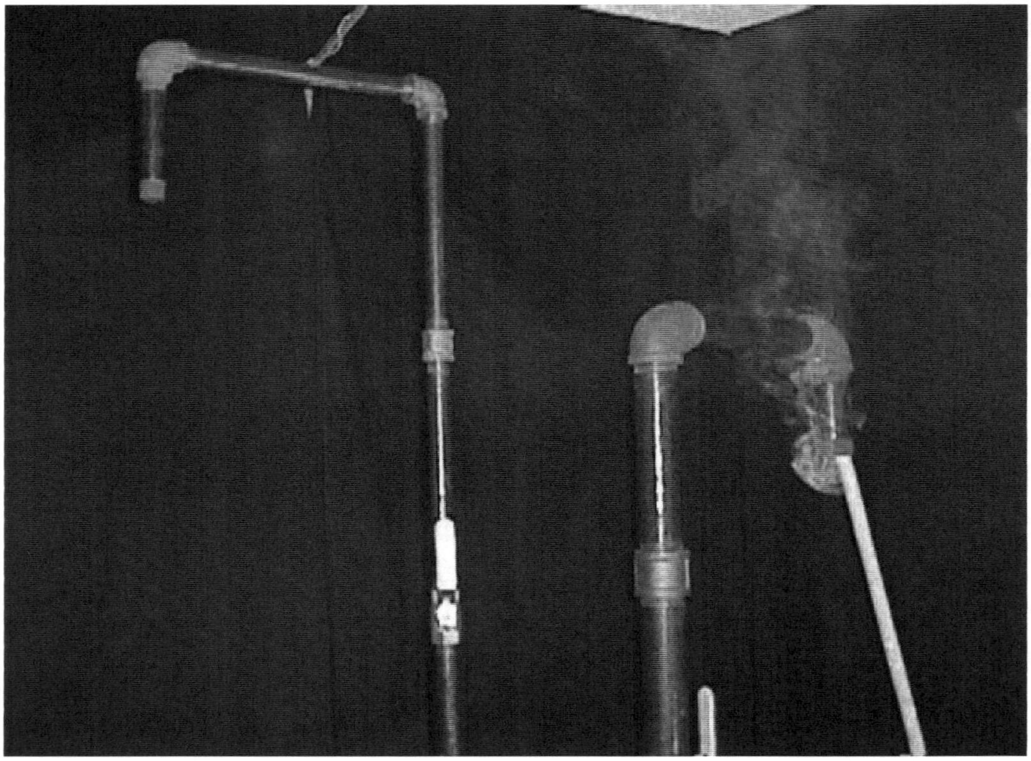

Figure 8. Smoke Being Pushed Out Large Diameter Vent Line

4. Discussion and Conclusions

The results of this work demonstrate that the presence of multiple exhaust vents on an inerted vessel can lead to the development of a significant oxygen concentration due to air inleakage driven by natural convection. In some cases, the resulting oxygen concentration can reach the level required to support combustion. If the vessel is inerted in order to prevent the formation of a flammable gas mixture, then an explosion may result if an ignition source is present. The authors are familiar with two cases within the last decade in which a loss of inerting due to air inleakage driven by natural convection has occurred. In both cases, the staff at the plant was unaware that the system configuration and operating parameters could lead to a loss of inerting via this mechanism. It is the authors' hope that this paper will raise awareness for the potential of a loss of inerting due to air inleakage driven by natural convection when multiple exhaust vents are present, and thereby help to preclude similar events in the future. There are other mechanisms that can result in air leakage into an inerted vessel (e.g., wind effects with multiple vents, rapid decrease in the vessel liquid level or vapor space temperature, etc.) and these other mechanisms should be considered in addition to natural convection when evaluating the effectiveness of a vessel inerting system.

An air inleakage model was presented which can be utilized to determine if air inleakage due to natural convection will occur and, if so, to calculate the resulting air inleakage rate and corresponding equilibrium oxygen concentration. While more sophisticated analysis tools are available to perform such evaluations (e.g., CFD codes), the predictions of the model were in reasonable agreement with the results of demonstration tests. The demonstration tests clearly illustrated the development of a natural convective flow loop under relevant conditions.

5. References

1. Standard on Explosion Prevention Systems, NFPA 69, 2002 Edition, National Fire Protection Association, Quincy, MA.
2. Fox, R.W. and A.T. McDonald (1978) Introduction to Fluid Mechanics, 2nd Ed., John Wiley & Sons, New York, NY.
3. Blevins, R.D. (1992) Applied Fluid Dynamics Handbook, Krieger Publishing Co., Malabar, FL.
4. Crane (1976) Flow of Fluids Through Valves, Fittings, and Pipe, Technical Paper 410, Crane Co. Engineering Division, New York, NY.
5. Thomas, J.K. (1998) "ITP Waste Tank Positive Pressure Inerting System," 1998 Safety Analysis Workshop, 15-19 June 1998, Park City, UT.
6. Thomas, J.K., D.F. Paddleford and J.K. Norkus (1998) "A Passive Ventilation Device for Hydrogen Control," 1998 Safety Analysis Workshop, 15-19 June 1998, Park City, UT.

Inerting Of Centrifuges For Safe Operation

Stanley S. Grossel
Process Safety & Design, Inc.
41 Sussex Road
Clifton, NJ 07012-2017
Telephone/Fax: (973) 779-8579
E-mail: psadi28@aol.com

Prepared for Presentation at
American Institute of Chemical Engineers
2005 Spring National Meeting
39th Annual Loss Prevention Symposium
Atlanta, Georgia
April 11-13, 2005

UNPUBLISHED

Inerting Of Centrifuges For Safe Operation

Stanley S. Grossel
Process Safety & Design, Inc.
41 Sussex Road
Clifton, NJ 07012-2017
Telephone/Fax: (973) 779-8579
E-mail: psadi28@aol.com

Abstract

Many centrifuging operations are hazardous because the feed slurries and wash liquors contain flammable liquids. In addition, ignition sources are frequently present. To avoid fires and explosions in centrifuges, purging is commonly applied before operation begins, and blanketing is done during operation, to lower the oxygen concentration below the limiting oxygen concentration (LOC) of the flammable liquid. This paper will discuss and describe the three commonly used inerting systems: monitoring by flow, monitoring by pressure, and monitoring by oxygen analysis. Details will be given for each method, including schematic drawings and inerting procedures. General requirements for an effective centrifuge inerting system will be discussed. Also, design considerations will be presented for oxygen analyzer systems to avoid problems in their use.

1. Introduction

Many centrifuging operations are hazardous and can result in fires and explosions (deflagrations) unless protective measures are taken. The risk of fires and explosions in centrifuges is much greater than in most process equipment for the following reasons:

1. Liquids in the feed slurry and wash liquor are often flammable and are a source of fuel.

2. Centrifuges operate like a fan drawing ambient air into the centrifuge through any point that is not gas-tight.

3. Many centrifuges have inherent ignition sources (to be discussed below).

To avoid or minimize the potential for fires and explosions, centrifuges are usually purged of air by a continuous purge of an inert gas (usually nitrogen) before introduction of feed, and then blanketed during operation by an inert gas.

2. Sources of Ignition in Centrifuges

Static electricity discharge, mechanical friction causing sparking, and hot surfaces (e.g., bearings) can act as ignition sources in centrifuges. These are discussed below.

2.1 *Static Electricity*

Static electricity discharges can occur from the following causes:

- The feed slurry and wash liquor can become electrostatically charged from flow in the piping, and then discharge when entering the centrifuge.
- High spinning speeds (of up to 1500 rpm) of the centrifuge basket can generate high static charges.

Liquids with low conductivity (K < 100 picoSiemens/meter) such as toluene, hexane, etc. are most hazardous as charges do not drain away (dissipate) quickly even if the centrifuge is grounded. The filter cake can form an insulating layer that prevents good charge drainage.

Static electricity discharges still may be a hazard even when the centrifuge stops. An open centrifuge awaiting digging out of the cake after separation from a flammable slurry liquid or wash liquor may have a flammable vapor/air mixture in or around it. The spark for ignition can come from the operator if he has been isolated from ground by non-conducting footwear or flooring. The bottom discharge chute can also be a source of flammable vapor that could present a hazard during cleaning or maintenance work.

2.2 *Mechanical Friction*

Any mechanical failure can cause friction and sparking. Mechanical failure of a part of a centrifuge can result in ejection of pieces of metal (e.g., feed pipes, wash liquor pipes, ploughs, or cake level devices) at high velocity that can impact with other internal parts and cause frictional sparks to occur. Friction can also be caused by reduction of the clearances in labyrinth seals, for example, due to differential expansion after a change in operating temperature.

Although a centrifuge will withstand a certain degree of vibration, excessive out-of-balance forces can lead to metal/metal contact, resulting in frictional sparks and local hot spots. Prolonged vibration frequently leads to metal fatigue in centrifuge parts, both internal and external (e.g., feed and wash liquor pipes, bolts, and bearings) and can cause major parts of the machine to become loose. This can result in contact between metal parts and generate frictional sparks. Serious problems can occur when using a filtering or peeler centrifuge. This is particularly the case when the cake falls off a horizontal-axis centrifuge due to stopping of the rotor. In this event, any attempt to restart the machine before the solids are cleared is almost certain to cause mechanical damage due to excessive out-of-balance forces, and result in severe damage to the machine.

Bearings can fail suddenly from lack of adequate lubrication, generating enough heat to cause melting in some severe cases. For example, collapse of a main bearing will allow the rotor to drop and result in metal-to-metal contact and frictional sparking.

2.3 Hot Surfaces

Hot surfaces can act as ignition sources and cause fires and explosions in centrifuges. Lack of lubrication to a bearing may allow it to heat up and ignite flammable vapors.

3. Inerting Systems and Procedures

The type of inerting system required for a centrifuge can be based on the level of risk from

Table 1 Guidance on Levels of Risk for Flammable Substances Handled in Centrifuges

Flammable Substances	
High Risk	— Ploughing centrifuges handling liquids at temperatures near to or above their flash points which may be opened frequently.
	— Where pyrophoric substances are present.
	— Centrifuges handling low flash point liquids in confined spaces.
Medium Risk	— Centrifuges handling liquids at temperatures normally below the flash point but which require frequent opening for product removal, cleaning or inspection.
	— Centrifuges handling liquids near to or above their flash point which are infrequently opened.
Low Risk	— Continuous centrifuges processing flammable liquids at temperatures well below their flash points.
	— Centrifuges handling liquids normally below the flash point which are opened only infrequently.
	— Centrifuges handling solids and liquids normally below the flash point where arrangements for product removal have been made which avoid frequent admission of air.

flammable substances involved in the centrifuging operation, as indicated in Table 1 [1].

Before feed slurry is introduced to a centrifuge, an initial inert gas (usually nitrogen) purge needs to be performed to reduce the oxygen concentration in the centrifuge and associated piping and other equipment to a level below the limiting oxygen concentration (LOC) of the flammable liquid or vapor. This procedure is accomplished immediately after the lid of the machine (e.g., a batch centrifuge) has been closed, by admitting nitrogen to the machine at a fairly high flow rate in a reasonable period of time (5-30 minutes), depending of the size of the centrifuge. Typical

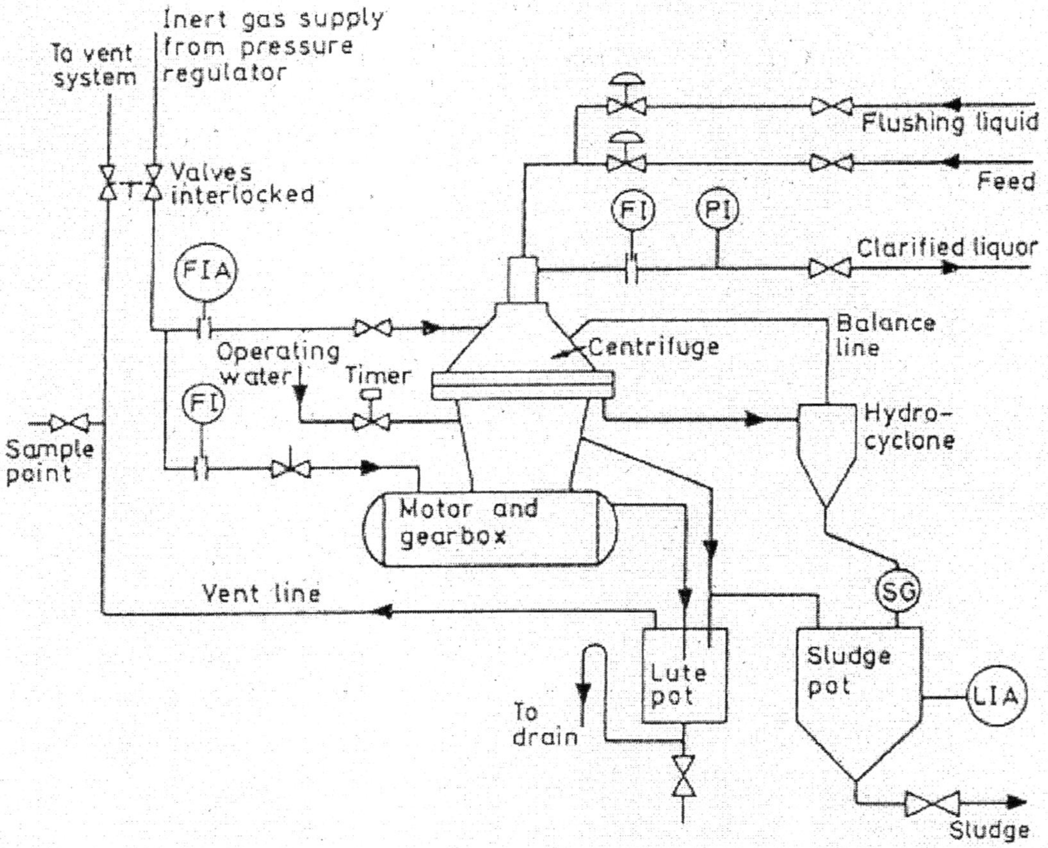

Figure 1 Schematic Drawing of Flow-Monitored Inerting System (for Opening Bowl Clarifier)

Certain centrifuges, such as a continuous pusher and conical basket types generate considerable windage and differential pressures within the machines. Flow-monitored nitrogen blanketing alone is, therefore, not generally suitable for these centrifuges [1].

Schematic drawings of typical flow-monitored systems for continuous peeler and nozzle bowl clarifier centrifuges are shown in the IChemE Guide [1].

3.2 Pressure-Monitored Inerting

A schematic drawing of the pressure-monitored inerting system is depicted in Figure 2. This method is recommended for medium risk installations of batch and continuous machines (see Table 1) [1]. This system is moderately expensive to install, is highly reliable (if properly maintained), and is economical in its use of nitrogen. However, it is more complex than the flow-monitored method because the centrifuge must be completely sealed, and the centrifuge casing and all connecting pipes must be able to withstand a positive pressure of approximately six inches water column (w.c.) (0.22 psig). If leaks develop, nitrogen is wasted trying to maintain

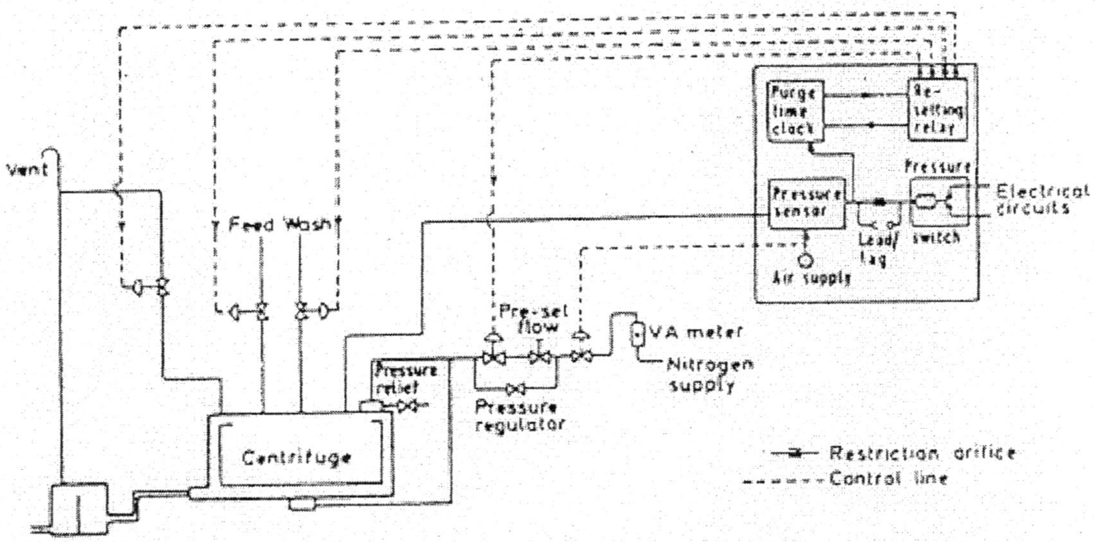

Figure 2 Schematic Drawing of Pressure-Monitored Inerting System (for Batch Centrifuge)

that pressure. The feed and discharge lines to and from the centrifuge must be sealed, and the vent line should be valved so that it opens during the initial purge and is closed at all other times. The vent line should discharge to a safe place outside any building or enclosure.

Initially the system is purged, but prior to the purge, certain interlocks must be satisfied: the lid is closed, the valves in the feed and wash liquor lines must be closed, nitrogen is available at the required pressure, and the vent line valve must be open. After purging for a set time (to reach the desired LOC in the centrifuge), the vent line valve closes automatically, and the continuous nitrogen flow (regulated to a pressure of 3-4 inches w.c. (0.11-0.14 psig) is actuated. This pressure has been found to keep leakage to a minimum while providing adequate pressure to operate the pressure-detecting instruments. The centrifuge can now proceed through normal operation. The self-relieving pressure regulator vents any excess gases when slurry feed enters the centrifuge and maintains the required pressure while discharging the cake and mother liquor.

If problems arise and the pressure rises above 6 inches w.c. or falls below 1 inch w.c., a pressure switch in the vent line will be tripped. This will cause the valves in the feed and wash liquor lines to close, the centrifuge to shut down, and the valves in the vent and nitrogen lines to open, admitting a high flow rate of nitrogen. The initial purge must be repeated before normal operation can resume. If the lid is opened or the nitrogen supply falls below the required pressure, the centrifuge will shut down, requiring the initial purge before starting.

Short-term pressure fluctuations in a centrifuge are quite common. In order to prevent spurious alarms, the pressure switches should have time delay mechanisms. That way, if the proper

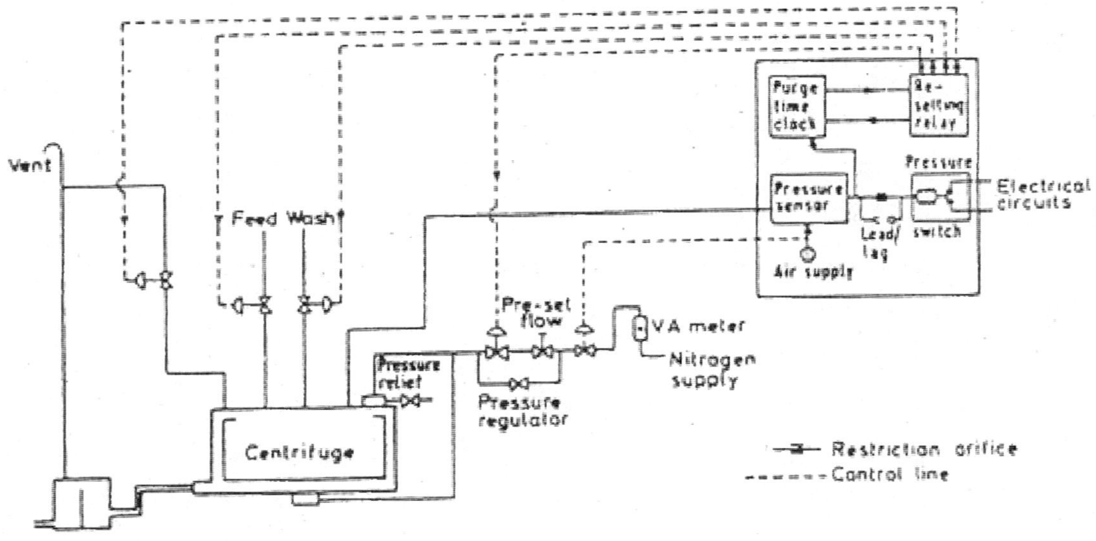

Figure 2 Schematic Drawing of Pressure-Monitored Inerting System (for Batch Centrifuge)

inches water column (w.c.) (0.22 psig). If leaks develop, nitrogen is wasted trying to maintain that pressure. The feed and discharge lines to and from the centrifuge must be sealed, and the vent line should be valved so that it opens during the initial purge and is closed at all other times. The vent line should discharge to a safe place outside any building or enclosure.

Initially the system is purged, but prior to the purge, certain interlocks must be satisfied: the lid is closed, the valves in the feed and wash liquor lines must be closed, nitrogen is available at the required pressure, and the vent line valve must be open. After purging for a set time (to reach the desired LOC in the centrifuge), the vent line valve closes automatically, and the continuous nitrogen flow (regulated to a pressure of 3-4 inches w.c. (0.11-0.14 psig) is actuated. This pressure has been found to keep leakage to a minimum while providing adequate pressure to operate the pressure-detecting instruments. The centrifuge can now proceed through normal operation. The self-relieving pressure regulator vents any excess gases when slurry feed enters the centrifuge and maintains the required pressure while discharging the cake and mother liquor.

If problems arise and the pressure rises above 6 inches w.c. or falls below 1 inch w.c., a pressure switch in the vent line will be tripped. This will cause the valves in the feed and wash liquor lines to close, the centrifuge to shut down, and the valves in the vent and nitrogen lines to open, admitting a high flow rate of nitrogen. The initial purge must be repeated before normal operation can resume. If the lid is opened or the nitrogen supply falls below the required pressure, the centrifuge will shut down, requiring the initial purge before starting.

Short-term pressure fluctuations in a centrifuge are quite common. In order to prevent spurious alarms, the pressure switches should have time delay mechanisms. That way, if the proper pressure is restored after a pre-determined length of time, say 1-3 seconds, operations are not disrupted.

The IChemE Guide [1] has schematic drawings of pressure-monitored inerting systems for batch, decanter, and pusher type centrifuges.

3.3 Oxygen-Monitored Inerting

Continuous measurement of the oxygen concentration in a centrifuge is the only means which can guard against all failures of blanketing (inerting), not only within the centrifuge, but also those which give rise to oxygen contamination of the nitrogen supply system [1]. It is, therefore, the method recommended for use in high risk situations (see Table 1).

Figure 3 is a schematic drawing of an oxygen-monitored method. If the oxygen concentration in the centrifuge system, which is measured continuously, exceeds the safe oxygen level setpoint, nitrogen is automatically fed to the centrifuge. When the oxygen concentration is at a safe level, the flow of nitrogen stops, so that nitrogen consumption is minimized. This system is the most complex and expensive to install of the three methods.

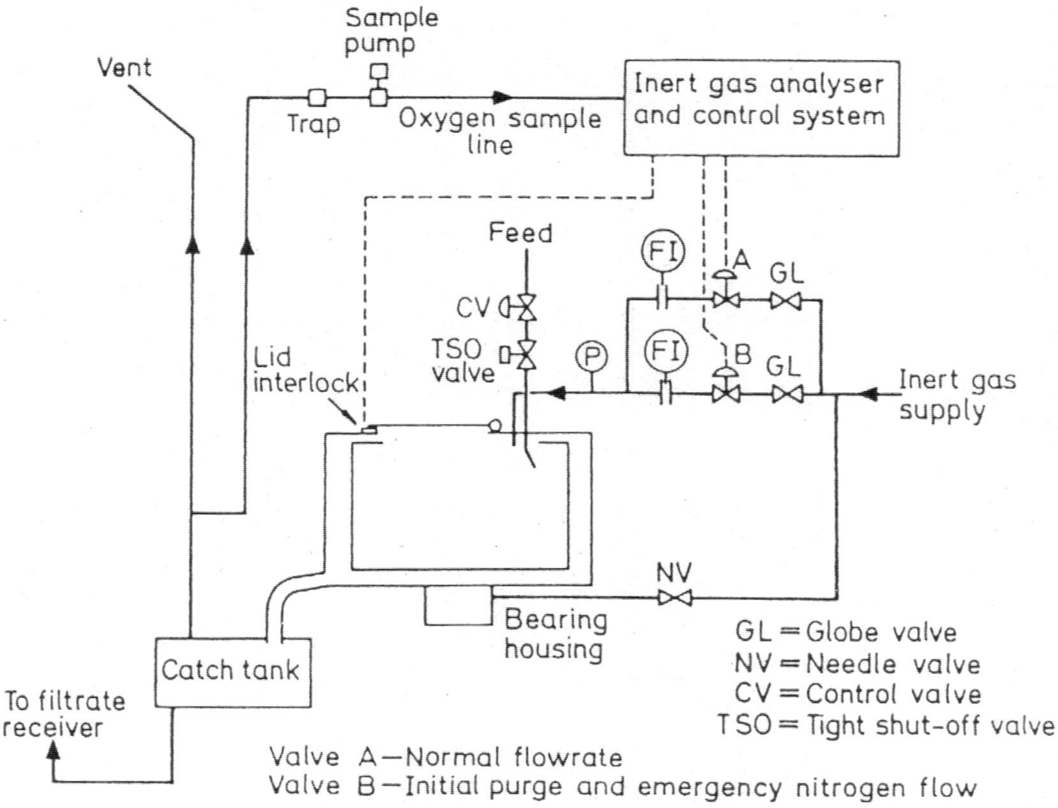

Figure 3 Schematic Drawing of Oxygen-Monitored Inerting System (for Batch Centrifuge)

An oxygen-monitoring system is comprised of the following components: oxygen analyzer cell (sensor), sample conditioning unit (to remove condensable vapors, corrosive vapors, and particulate solids), sample pump or ejector (to draw sample from centrifuge to conditioning unit), shut-off and control valves, sample tubing, and programmable controller. A user can purchase the analyzer cell (sensor) and sample conditioning unit from an oxygen analyzer cell manufacturer and the other components from other vendors and assemble the system. Alternatively, a user can purchase all the components of the system, except for the sample tubing, as a package unit, and assemble the system.

Since continuous flow of nitrogen is a basic feature of this method, the centrifuge is vented to the atmosphere to a safe location or to a VOC control device through a vent line usually provided with a conservation valve and a flame arrester.

Continuous oxygen monitoring is complex and measures have to be taken to avoid spurious alarms and shutdowns. The main component of this system is the oxygen analyzer and the centrifuge's safe operation depends on the analyzer's reliability. Measures to ensure reliability include a high level of preventive maintenance and analyzer system component testing and calibration to ensure that the signals are accurate and the responses are timely (no excessive response lags).

Some potential problems with the analyzer and sampling system are: analyzer power failure, low sample flow rate, analyzer cell failure, and signal drifting. If the nitrogen control system senses a loss of power to the analyzer, it activates an emergency purge and centrifuge shutdown. A low flow switch in the sampling system starts the same actions if the sample line becomes restricted and the flow rate drops.

In order to detect analyzer cell failure and signal drifting, the analyzer should be checked and calibrated by plant personnel daily. The system should incorporate a simple means of doing this, e.g., by admission of ambient air to the cell, which will quickly check the full scale reading. However, alarms and trips should be inhibited (bypassed or locked-out) during calibration, if it is desired that this be done while the centrifuge is in operation. Means must also be provided so that it is impossible to leave alarms and trips in a bypassed or locked-out state.

When the accuracy of the analyzer output is established, centrifuge purging and operation can begin. Again, two lines supply nitrogen to the centrifuge. Gas purging in one branch flows at 15-20 scfm for the initial purge, for the dump sequence, and when an alarm condition occurs. The other line has a low flow rate of nitrogen, just sufficient to maintain a safe condition as indicated by the oxygen analyzer.

Process upsets that can lead to an alarm condition include: high oxygen concentration, low sample flow, low nitrogen flow, and power failure. In the case of a high oxygen concentration in the centrifuge, the nitrogen purge valve opens until the oxygen concentration drops below the desired setpoint (e.g., 5 vol. %), while the centrifuge continues to operate. With the other up-sets, and if the oxygen concentration rises to above the high concentration cutoff setpoint (e.g., 8 vol. %), the centrifuge shuts down and nitrogen flows until the gas is manually shut off.

To minimize sampling lag times and increase speed of response the sample line diameter should be fairly small (1/4" to 3/8") and its length should be fairly short. A good rule of thumb is that 100 feet of ¼" tubing will give a response time of under 30 seconds at a sample flowrate of 5 liters/minute [3].

3.4 Types of Oxygen Analyzers

Two types of oxygen analyzer sensors commonly are used in centrifuge inerting systems: the electrochemical cell sensor and the paramagnetic sensor. Electrochemical cell oxygen sensors are divided into two different technologies/categories, namely: (1) percentage by volume, and (2) partial pressure oxygen sensor. All electrochemical sensors operate on the same principle of having an anode, cathode, and an electrolyte where an output/emf is generated proportional to the percentage of oxygen present. A variety of electrolytes are available to allow the user the option of selecting the best sensor for their particular process conditions. Electrochemical sensors, when exposed to oxygen, are electrolyte-consuming devices, and as such, have a finite operational life. This is typically twelve months, after which either the sensor is replaced or rebuilt in the case of some partial pressure sensors. The raw output signal from these sensors is generally very low and requires amplification/linearization prior to display. They are insensitive to shock/vibration and are not position sensitive. They also have a limited temperature range, and need a preconditioning system when used with centrifuges where the internal atmosphere contains acids, condensable vapors, and particulate solids that could harm the electrolyte. By virtue of electrochemical sensor design they lend themselves to use in hazardous areas by being certified as intrinsically safe.

3.4.1 Percentage by Volume Oxygen Sensor

Percentage by volume oxygen sensors, as their name implies, respond solely to the changes in oxygen concentration. This unique design of these sensors, using a capillary sensing technology, makes them suitable for use on process applications where varying pressures and flows are present. These cells do not require a polarizing voltage and as such have an inherent zero point. They have the following advantages and disadvantages:

- Not sensitive to pressure fluctuations.
- Simplest to calibrate as no zero gas is required.
- Intrinsically safe sensors available.
- Not affected by shock and vibration
- Fails to a danger condition by giving zero output.
- Typical sensor life is one year.

3.4.2 Partial Pressure Oxygen Sensor

Partial pressure oxygen sensors are divided into two categories namely, fixed membrane fuel cell types or polarographic rechargeable type sensors. As the name implies, these sensors measure the partial pressure of oxygen, and as such, are sensitive to pressure variations. When these sensors have expired, the fuel cell type requires complete replacement, while the polarographic sensor requires a new membrane and a recharge of the electrolyte. The polarographic sensor requires a polarization voltage to activate the sensor. These sensors should be calibrated at the same pressure to which they will be exposed. They have the following advantages and disadvantages:

- Intrinsically safe sensors are available.
- Not affected by shock and vibration.
- Fails to a danger condition by giving zero output.
- Typical sensor life is six to 12 months.
- Sensitive to pressure fluctuations.
- Zeroing gas required.

The fuel cell type is the most frequently used. They tend to be relatively unaffected by background gases and are deemed to be robust on chemical applications. They also have a limited temperature range.

One oxygen analyzer manufacturer offers a complete package, called the Inerting Control System, which consists of an electrochemical cell sensor, a sample conditioning unit, block and control valves, and an analyzer/programmable controller [4]. Because an electrochemical cell may start to go bad (the electrolyte is being used up) and there will be no indication that this is happening, two sensors in series are used, and the unit has real-time sensor trending that is designed to alert the user of when it is time to replace the cell. A signal goes from the sensor to the analyzer/controller, which then takes the desired action, depending on the sample analysis (oxygen concentration). When I was working for a pharmaceutical company I had installed several Inerting Control Systems successfully on both batch and continuous centrifuges.

3.4.3 Paramagnetic Oxygen Analyzers

Paramagnetic oxygen analyzers operate on the principle of the paramagnetic tendencies of oxygen. There are two types, namely: (1) the magnetic dumbbell type, and (2) the thermomagnetic type. These sensors have an undetermined life as they are non-depleting (36 to 40 years, depending on the type, according to one manufacturer), However, this can be reduced if it is contaminated by the process gas. When used on applications where solvents are present, this sensor typically requires heated sample lines and a heated measurement cell to prevent the solvent condensing.

Background gases can also affect its measurement, and a strong interaction exists with nitrogen oxides. These sensors have the following advantages and disadvantages:

- Fast response.
- Non-depleting sensor.
- Sensitive to pressure and temperature fluctuations.
- Sensitive to shock and vibration

- Requires zeroing gas.
- Sensitive to background gases.
- Not suitable for in-situ measurement.
- No determined failure mode.

A number of manufacturers make oxygen analyzers using paramagnetic sensors. The author is familiar with one company that offers oxygen analyzer systems for use with centrifuge inerting [5]. The system consists of a sample conditioning unit, and a paramagnetic sensor/analyzer. They do not provide a control system as part of the package, but the analyzer and other system parts are designed to be utilized as an input to an independent control system. I have no experience with such an analyzer system, but a colleague at another pharmaceutical company said that they worked very well.

4. General Requirements for Centrifuge Inerting Systems

Some general guidelines relevant to inerting of centrifuges are as follows:

1. An adequate supply of inert gas (usually nitrogen) must be available at all times when centrifuges are operating, and precautions must be taken to ensure that the gas supply cannot overpressure the centrifuge casing. The desired nitrogen operating pressure is usually controlled by a pressure regulator, which usually has a diaphragm. A pressure relief valve, set at the design pressure of the centrifuge casing, should be installed between the regulator and the centrifuge to relieve if the regulator diaphragm fails open. For high risk systems, it may be desirable to provide a backup supply of nitrogen, e.g., from nitrogen cylinders.

2. The nitrogen inlet line should be provided with either low pressure or low flow sensors and alarms. Low pressure or flow should sound an alarm to alert the operator of a problem. Consideration should be given to interlocking these sensors with the centrifuge motor, and open-shut valves in the slurry feed and wash liquor lines for high risk systems.

3. The nitrogen supply points and vents should be located so as to ensure effective purging. Separate lines often are necessary to feed nitrogen into the casing, the bowl or basket, and the bearing housing.

4. Slurry feed and wash liquor supply vessels, and centrifuge cake receivers should be nitrogen blanketed to prevent air from entering the centrifuge, which can happen either by vortex entrainment or by the vessels running empty.

5. In order to minimize nitrogen consumption, all parts of the centrifuge should be adequately sealed. Particular attention should be paid to the centrifuge lid sealing gasket, the drive shaft seals, and the solids discharge chute.

6. Interlocks should be provided to prevent any outflow of nitrogen when the lid of a batch centrifuge is opened. This is to reduce the risk to operators of asphyxiation during solids removal, and to minimize the use of nitrogen.

5. References

[1] IChemE (1987). *User Guide for the Safe Operation of Centrifuges,* 2nd edition, Institution of Chemical Engineers, Rugby, England, UK.

[2] O'Shea, S. M. (1983). "Nitrogen Blanketing of Centrifuges Reduces Fire Hazards," *Chemical Engineering*, pp. 73-76 (April 4, 1983).

[3] Private communication from Louis Gansky, Servomex Company (July 28, 2004).

[4] Neutronics Inc. (see www.neutronicsinc.com).

[5] Servomex Company Inc. (see www.servomex.com).

Missed Opportunities In Reactive Chemical Hazard Evaluations

R.A. Ogle; D.R. Morrison, III; A.R. Carpenter; Y.S. Su
Exponent, Inc., 2004

Prepared for Presentation at
American Institute of Chemical Engineers
2005 Spring National Meeting
39th Annual Loss Prevention Symposium
Atlanta, Georgia
April 11-13, 2005

UNPUBLISHED

Missed Opportunities In Reactive Chemical Hazard Evaluations

R.A. Ogle; D.R. Morrison, III; A.R. Carpenter; Y.S. Su
Exponent, Inc., 2004

Abstract

Reactive chemicals pose a significant hazard to the process industries due to the inherent nature of these chemicals. This paper examines several reactive chemical accidents caused by self-heating that led to thermal runaway reactions. Although none of these case studies resulted in serious injuries, the runaway reactions caused fires, explosions, or toxic gas releases that resulted in extensive damages and loss of production. This paper will consider the causal factors of these accidents, identify the thermal stability data available prior to the accidents, and present simple hazard calculations indicating the predictability of the accidents. The paper concludes with a discussion of lessons learned from these accidents.

1. Introduction

There continues to be an increased awareness of reactive chemical hazards in industry, government and academia. Many chemicals used in the manufacture of intermediates and final chemical products, both commodity and specialty, can pose a reactive energy hazard given the proper circumstances. The very nature of chemical processing and manufacture relies upon this reactivity of chemical mixtures to economically obtain the final product. The potential consequences posed by uncontrolled chemical reactions – fires, explosions and toxic chemical exposures – are not always readily apparent to those who use, produce, or may be exposed to the chemicals. The diversity of reactive chemicals and potential interactions poses a challenge to process safety engineers seeking to develop hazard management strategies.

The American Institute of Chemical Engineers, along with the Center for Chemical Process Safety, has published several books dealing with the nature of reactive chemical hazards. These useful references include, but are not limited to the following:

- Guidelines for Chemical Reactivity Evaluation and Application to Process Design [1]
- Guidelines for Safe Storage and Handling of Reactive Materials [2]

These references classify reactive chemical hazards into the following categories:

- Polymerizing
- Decomposing (i.e., shock sensitive, thermally decomposing)
- Rearranging (i.e., isomerizing, disproportionating)
- Oxygen reactive (i.e., pyrophoric, flammable, combustible, peroxide former)

- Water reactive
- Specific chemical hazards (i.e., nitrogen reactive, reactive with metals, oxidizing/reducing, acid/base, toxic, etc.).

This paper focuses on the evaluation and control of thermal stability hazards in bulk solids. Several reactive chemical accidents caused by thermal runaway reactions are discussed. Although none of these case studies resulted in serious injuries, the runaway reactions caused fires, explosions, or toxic gas releases resulting in significant property damage or business interruption. Strategies for managing thermal stability hazards from a system safety framework are presented first. Next the case studies are discussed in terms of the causal factors of the accident and the thermal stability data available prior to the accident. A synopsis of the lessons learned concludes the paper.

2. Management Of Thermal Instability Hazards

A substance that undergoes a spontaneous, exothermic decomposition can be considered thermally unstable. Such materials can also be described as self-heating because if the heat released by the exothermic reaction is not dissipated to the environment, it will accelerate the rate of decomposition within the bulk substance. The behavior of a bulk quantity of a thermally unstable material depends largely on the relative magnitudes of the rate of heat generated by decomposition and the rate of heat loss to the environment.[1] All other factors being constant, at a given ambient temperature small quantities of a self-heating material can dissipate heat to the environment more easily than large quantities. If the quantity of material is sufficiently large, the heat generated by decomposition may exceed the heat loss to the environment and the temperature of the material will increase at an exponential rate (Arrhenius law).[1] When the rate of temperature rise becomes nearly vertical, it has reached a state known as thermal runaway.

A key concept in evaluating the propensity for a self-heating material to undergo a thermal runaway is the critical temperature, T_{cr}. The critical temperature is the maximum temperature of the ambient environment at which the material can remain at thermal equilibrium, i.e., the heat generated by decomposition is exactly balanced by the heat dissipated to the environment. The critical temperature for a self-heating material is a function of the quantity of material, its geometry (or the geometry of its container), the ambient temperature, the relative magnitude of external heat transfer rates from the self-heating body to the environment (natural convection, forced convection and radiation) and its time of exposure to the ambient condition. The critical temperature will also depend on whether a package is stacked with others and whether the packages are in intimate contact. Mathematical expressions for the critical temperature were first derived by Frank-Kamenetskii in the 1930s. This work was summarized in Grewer [3] and extensively developed in Bowes.[4] The critical temperature calculated for a given situation determines whether the material will achieve thermal runaway for those specific circumstances. Three possible outcomes can be defined with respect to the critical temperature.[5,6]

- Subcritical State: The ambient temperature is below the critical temperature. The material does not undergo thermal runaway. However, the temperature of the thermally unstable body may be slightly elevated (above the ambient temperature) due to self-heating.

- Critical State: The ambient temperature is equal to the critical temperature. The material temperature is elevated above ambient but does not undergo thermal runaway. This state represents an inherently unstable point and may be empirically determinable only to a certain level of accuracy (e.g., ± 5°C) for a given bulk material. Subcritical and supercritical states are more easily defined.
- Supercritical State: The ambient temperature is above the critical temperature and the material eventually undergoes thermal runaway.

These criteria are best illustrated through the following time-temperature plot (Figure 1).

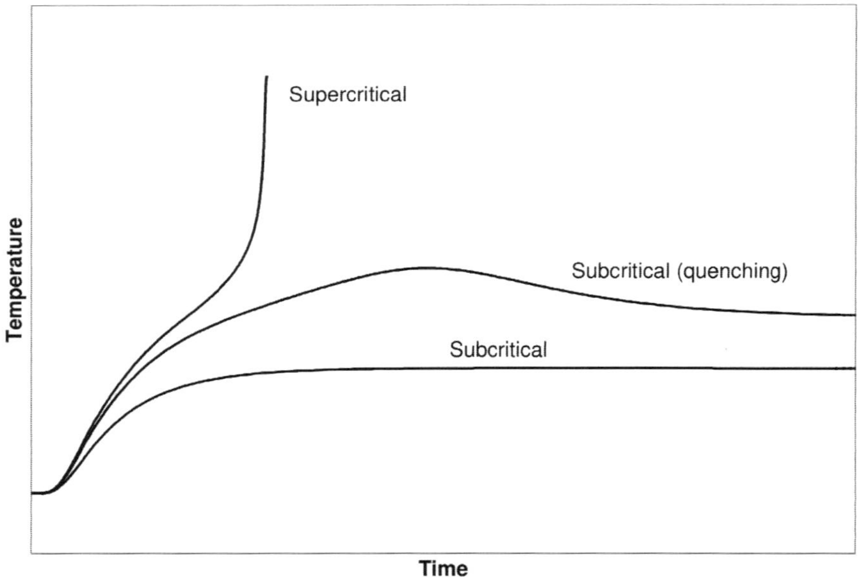

Figure 1. Critical Temperature States.

If the ambient temperature is above the critical temperature, a thermal runaway will eventually occur. Thus, the critical temperature is the threshold criterion for determining the type of self-heating behavior that may be observed for a given set of conditions.

One can choose a variety of strategies for addressing risks from thermally unstable materials. The discipline of system safety suggests the following hierarchy of risk management strategies:

1. Eliminate the hazard.
2. Reduce the risk by implementing engineering controls (passive barriers or active safeguard devices).
3. Reduce the risk by implementing administrative controls (training, procedures, and supervision).

4. Reduce the risk by providing warnings or alarms.
5. Accept the risk.

Drawing from resources such as guidelines published by CCPS, [1,2] a number of specific hazard control actions can be identified for each level in the hierarchy. Examples are shown in Table 1 with an assessment of potential advantages and disadvantages. The choice of which

Table 1. System safety analysis of thermally unstable materials

Hazard Control Strategy	Specific Actions	Advantages	Disadvantages
Eliminate Hazard	Use alternative product	Thermal stability	Feasibility, cost
	Chemical modification (i.e., diluents and inhibitors)	Built in control of reactivity, limit maximum heat release or heat capacity	Feasibility, product effectiveness
Passive Safeguards	Vessel design	Containment/venting	Cost-effectiveness
	Reduce reactor/batch size and storage size	Facilitate heat removal	Cost-effectiveness
	Package design	Facilitate heat removal	Cost-effectiveness, marketing, storage
	Shelf life specifications	Limit duration before use/disposal	Monetary
Active Safeguards	Temperature control	Control bulk temperature	Cost-effectiveness
Procedural Safeguards	Inventory control--First-in, first-out	Limit duration before use/disposal	Human error
	Product inspection	Isolate contaminants, identify "bad" product	Human error
Warnings	Product temperature monitoring	Identify when product undergoes self-heating	Feasibility

strategy to adopt is dictated in part by the critical temperature of the material. Often, the choice is limited due to practical constraints. Furthermore, as seen in the following case studies, implementation of safeguards does not guarantee that thermal runaway will not occur.

There is no single stepwise approach to managing reactive chemical hazards. The individual characteristics of the reactive system must be considered and understood. Only then can logical and appropriate safeguards be designed and implemented to reduce the specific risks. Additionally, the consequences of a reactive chemical accident must be considered in developing safeguards. If the consequences of an accident are minor, a single safeguard may be sufficient, however if the consequences are severe or catastrophic, multiple safeguards may be of greater value.

3. Case Studies

3.1 Manufacturer Recommends a Supercritical Operating Temperature

A user purchased an air pollution control (APC) system for the removal of volatile organic compounds (VOCs) from a circulating process air stream. The APC system consisted of adsorber columns to collect the VOCs, desorber columns to regenerate the adsorption media (polymer beads), and a thermal oxidizer to destroy the VOCs. VOCs generated at the facility were collected and transported to the APC via a process air stream. The process air stream was run through the adsorber columns where the VOCs became adsorbed on polymer beads. The polymer beads would then be carried over to the desorber columns where a hot air stream would strip the VOCs from the polymer beads. The resulting VOC-laden hot air stream was then treated in the thermal oxidizer before the air was vented to the atmosphere.

During the course of normal operation, a fire took place in one of the desorber units. Fuels available for the fire were the VOC vapor-air mixture and the polymer adsorbent. The VOC concentration was determined to be significantly below the lower flammability limit, therefore the polymer adsorbent was the fuel first ignited. Examination of the desorber system did not reveal any competent sources for piloted ignition of the polymer media.

The supplier of the polymer adsorbent had previously warned the APC manufacturer that the adsorbent was capable of undergoing self-heating with an onset temperature of 248°F. Despite these warnings, the APC manufacturer advised the user to increase the desorber inlet air temperature to 275°F to address performance issues. Increasing the air temperature was the direct cause of the accident, because the process temperature was now and thus capable of initiating thermal runaway behavior in the polymer. Using kinetic parameters derived from ARC (accelerated rate calorimetry) data, the time to maximum rate at the operating temperature was predicted to be on the order of days (consistent with the occurrence of the fire). A contributing cause was the mechanical design of the desorber. This allowed polymer beads to spill over and accumulate below the bottom tray in a region of stagnant air at supercritical conditions. Lacking the ability to dissipate sufficient heat through convection, the pile of adsorbent became susceptible to thermal runaway and spontaneously ignited.

3.2 *Scale-Up of a Batch Process Despite Warning Signs*

A facility was involved in the production of a powder mixture consisting of five different components: sodium dichloroisocyanurate (NaDCC), two inert chemicals, and two organic chemicals. Based on bench-scale studies, a ribbon blender was used to mix the powdered ingredients. A ribbon blender consists of two counter-acting blades mounted on a central shaft. Mixing is achieved through the shearing motion induced by the counter-acting blades. Three of the ingredients, including the NaDCC, were introduced to the ribbon blender via a drag conveyor, while the remaining two ingredients were added by hand to the top of the blender. Once all five ingredients were added, the mixture was blended for 15 minutes. Afterwards, a sample of the material was analyzed for quality control. If the sample passed analysis, the batch proceeded to packaging.

The accident occurred during the fourth full-scale run (1,100 lbs) of the mixing process. Prior to this, material had been prepared in much smaller batches of 4 to 110 lbs. No unusual or different observations from the first three runs were noted during the blending operation. After mixing, however, the contents of the blender were allowed to remain in the blender for several hours. When work resumed, the contents of the blender were being transferred to the packaging machine. During this operation, several unusual events were observed:

- Dust emission from the sampling port
- White powder spraying from the gaskets
- White smoke rising from the blender toward the ceiling
- Green smoke traveling toward the floors.

As a result, the building was evacuated.

The primary mixture component, sodium dichloroisocyanurate (60 wt%), is classified by the National Fire Protection Association (NFPA) as a solid oxidizer with an instability rating of 2. It is known to undergo thermal decomposition with an accompanying liberation of a dense white smoke containing chlorine gas. Investigation of artifacts revealed several components filled with a black porous material, similar to char, which possessed a strong chlorine odor. Observed thermal and corrosion damage was consistent with exposure to an exothermic chemical reaction. Differential scanning calorimetry (DSC) and accelerating rate calorimetry (ARC) data confirmed the presence of exothermic decomposition pathways for NaDCC/organic mixtures, albeit at high initiating temperatures. Furthermore, independent tests revealed that the mixture was capable of undergoing self-sustaining thermal decomposition if briefly contacted with a high-temperature, external heat source. Tests resulted in the production of a dense white smoke, and a reaction product similar in appearance to that found at the accident scene.

Several potential initiating sources were considered. The direct cause of this accident was determined to be the formation of hotspots in the batch due to the shearing action of the ribbon blender. Hot spots have been cited as potential ignition source for self-heating. [4] Contributing causes were the two orders of magnitude scale-up of mixing batch size and the additional retention time in the blender due to the work interruption.

During the investigation it was determined that operators, when preparing pilot-scale batches of 88 lbs in a similar mixer, noted significant temperature increases in the mixed product (i.e., the product was uncomfortably warm to touch). By increasing the batch size, the mixture became even more susceptible to thermal runaway due to a decrease in the rate of heat loss.[1] The environment in both the filling and blending rooms was maintained at approximately 14°C. Onset of exothermic decomposition was determined to be 90°C in ARC tests. The self-heating potential of the product was further evaluated by extracting kinetic parameters from the ARC data. An estimate of the time elapsed prior to thermal runaway (~ 3 hr) was used to determine the necessary bulk temperature condition required. Results indicated that a bulk temperature of 75°C was necessary, ruling out bulk temperature of the product as the cause of self-heating. Thus, a more likely scenario is that hotspots generated during the mixing process induced an exothermic decomposition which then propagated throughout the mixture.[1,4] This accident demonstrates that thermal stability tests and analyses may not be sufficient to characterize the ease of initiation of a decomposition reaction.

3.3 Scale-Up of a Shipping Container

A manufacturer of a solid oxidizer used as an active ingredient in consumer cleaning products built a new chemical plant in anticipation of increased demand for its product. The process technology for this plant was transferred from an older overseas facility that had extensive experience with the manufacture and transportation of the product. The commodity produced at the original facility was shipped via truck in super-sacks with typical transit times on the order of a single day. Truck transportation for the new facility, however, was determined to be infeasible, as travel distances were much larger and the anticipated shipping quantities were too large. Instead, rail transportation using pressure differential hopper cars was used.

The manufacturer experienced two successive hazardous material releases during a one-week period, each involving a railcar of commodity. In each case, the certificate of analysis indicated that the shipment had satisfied product specifications. The first accident involved an explosion of a railcar while stationed at a customer site 25 days after production. The customer was using the railcar as temporary onsite storage of the commodity. While located at the unloading station, vapors were observed to vent from the railcar's pressure relief valve, followed by the bursting of the rupture disk. Evacuation of the area was initiated prior to the rupture of the railcar. Based on the observed accident, additional railcars en route to customer locations around the country were found, stopped, isolated and monitored daily. Upon observation of a pressure increase, local emergency response crews vented the pressure. This protocol was changed to hourly monitoring as a second railcar exploded in the period between daily monitoring. Furthermore, it was determined that the best course of action was to open all overhead hatches on each railcar to vent the nonhazardous gas generated by decomposition. The commodity was deemed unsuitable for use at that point and arrangements were made to remove and dispose of all railcar contents.

The material was known to be thermally unstable, water-reactive, and capable of undergoing self-propagating decomposition. Moreover, the decomposition was exothermic, autocatalytic, and generated significant quantities of gas. Per Department of Transportation (DOT) requirements, the primary metric used in assessing the thermal hazard was the self-accelerating decomposition temperature (SADT). The SADT for the material in a railcar was determined to be 75°C. The large value for the SADT makes bulk temperature effects an unlikely cause of the thermal runaway. The direct cause of the accident was the exposure of the commodity to water. CCPS warns that impurities can act as catalysts, reducing the activation energy of a process when present in even small quantities. The material was known to be both water-reactive and susceptible to autocatalytic decomposition. A contributing cause was the use of rail transport instead of truck transport. Rail transit times tend to be longer and rarely permit control over the commodity's environment and exposure times. In one case a customer had used the railcar for temporary onsite storage of the commodity. Moreover, the large increase in the quantity shipped resulted in greater susceptibility to thermal runaway due to lower bulk heat losses. This accident demonstrates that common metrics such as SADT may not be sufficient to account for all circumstances of reaction initiation.

3.4 Scale-Up of Package Size from Subcritical to Supercritical

A distributor of lawn and garden supplies was utilizing a warehouse to store large quantities of pool chemicals for distribution to major retail establishments. Among the pool chemicals was calcium hypochlorite, a solid oxidizer that reacts readily with organic contaminants. Furthermore, calcium hypochlorite is a thermally unstable material that decomposes exothermically. This decomposition is predominantly controlled by the product temperature and occurs slowly at room temperature. At high enough temperatures, however, a thermal runaway reaction can be initiated.

While experiencing abnormally high summer temperatures, employees arriving at the warehouse noticed a strong odor of chlorine. Attempts were made with the assistance of the manufacturers to isolate and remove contaminated products. These efforts, however, focused on a second pool chemical found at the warehouse, trichloroisocyanuric acid. Based on a perceived notion that the smell was dissipating, workers left the warehouse at their normal time. A fire erupted at the warehouse approximately one hour later.

Over time, to accommodate growing retailer needs, increasing quantities of calcium hypochlorite were stored at the warehouse. Pool chemicals were stored in several layers of packaging, which served as thermal barriers to heat release. The intimate stacking of packaged materials resulted in a pseudo-unit of calcium hypochlorite in excess of 1500 lbs. Elevated temperatures due to a recent heat-wave, aggravated by the storage and packaging configurations, contributed to the achievement of a supercritical temperature condition. The resulting fire completely destroyed the warehouse. The direct cause of the accident was the failure to recognize the hazards presented by the storage density of the packaged calcium hypochlorite. From a heat transfer standpoint, this resulted in larger pseudo units of calcium hypochlorite with correspondingly lowered critical temperatures.

The manufacturer of the pool chemical was fully aware of the potential for thermal runaway. This danger was not communicated to the distributor. One aspect of their knowledge was the influence of potential size effects on the critical temperature. Generally speaking, as package size increases, the critical temperature for a package will decrease due to the fact that heat generated by the exothermic decomposition of calcium hypochlorite faces a larger thermal resistance to escape to the package. In Table 2 below, we show the influence of package size. Results were calculated using the method proposed by Frank-Kamenetskii.

Table 2. Frank-Kamenetskii calculations of calcium hypochlorite critical temperature at select sizes (Critical temperatures calculated assuming a cube geometry).

Size (lb)	Tcrit (°F)
100	154.6
250	122.6
500	115.7
1000	109.0
2000	102.4

4. Lessons Learned

The manufacturer must communicate the hazards presented by thermally unstable materials to the user. In several cases, poor recognition of the potential hazards resulted in operating and handling conditions near the critical point of the reactive materials. Self-heating effects, while initially minor, increase over time as the heat released by the exothermic reaction accelerates the rate of decomposition. As a result, noticeable signs of thermal runaway often occur long after initiation. Because of this lag-time, thermal runaway accidents often occur in the distribution and storage and not the synthesis of these compounds.

Evaluation of the thermal reactivity hazards should be done at each stage of scale up. Small-scale laboratory tests are not representative of all conditions encountered at the commercial production scale.[1]

There are two useful metrics for evaluating the risk associated with storing and handling thermally unstable materials: the time-to-maximum-rate (for supercritical packages) and critical temperature (for subcritical packages). The standard formulas for these calculations require estimates for the kinetic parameters of the decomposition reaction. The time-to-maximum-rate calculation, however, is especially sensitive to errors in the estimates of the kinetic parameters. Simply stated, it is easy to be very wrong. The critical temperature calculation is less sensitive and is therefore a more forgiving risk metric.

The critical temperature, however, is not a fixed value and is dependent on a variety of factors, many of which concern the ability of the product to lose heat to the surroundings. Subcritical packages are to be preferred over supercritical packages, but this is not always practical. In modifying product packaging and size, careful scrutiny should be given the possible impacts on the critical temperature. If supercritical packages must be stored or handled, multiple safeguards may be appropriate to manage the hazard of a runaway reaction.

5. Conclusions

In this paper, four case studies of accidents involving thermally unstable materials have been reviewed. The sensitivity of thermal runaway to a variety of factors has been illustrated in the cases and cannot be understated. A broad range of hazard control actions should be utilized to provide an adequate level of safety. Even so, awareness and recognition of signs of thermal runaway is of paramount importance.

6. References

[1] Center for Chemical Process Safety, *Guidelines for Chemical Reactivity Evaluation and Application to Process Design,* New York: American Institute of Chemical Engineers, 1995.

[2] Center for Chemical Process Safety, *Guidelines for Safe Storage and Handling of Reactive Materials*, New York: American Institute of Chemical Engineers, 1995.

[3] Grewer, Theodor, *Thermal Hazards of Chemical Reactions,* New York: Elsevier Science, 1994.

[4] Bowes, P. C., *Self-heating: Evaluating and Controlling the Hazards,* New York: Elsevier Science Publishers, 1984.

[5] Babrauskas, Vytenis, "Chapter 9. Self-heating," *Ignition Handbook,* Issaquah: Fire Science Publishers, pp. 367-443, 2003.

[6] Gray, Brian. "Chapter 2-10. Spontaneous Combustion and Self-Heating," *The SFPE Handbook of Fire Protection Engineering*, DiNenno, P. J., Ed. in Chief. Quincy: National Fire Protection Association, pp. 2-211-2-228, 2002.

Paper TG005 - 2

Rapid Reactivity Screening Using A Nanocalorimeter

Yen-Shan Liu, Victor M. Ugaz, William J. Rogers, and M. Sam Mannan
Mary Kay O'Connor Process Safety Center (MKOPSC)
Texas A&M University
College Station, TX 77843-3122

Sanjeev R. Saraf
ioMosaic Corporation
Salem, NH 03079

Prepared for Presentation at
American Institute of Chemical Engineers
2005 Spring National Meeting
39th Annual Loss Prevention Symposium
Atlanta, Georgia
April 11-13, 2005

February, 2005

UNPUBLISHED

Rapid Reactivity Screening Using A Nanocalorimeter

Yen-Shan Liu, Victor M. Ugaz, William J. Rogers, and M. Sam Mannan
Mary Kay O'Connor Process Safety Center (MKOPSC)
Texas A&M University
College Station, TX 77843-3122

Sanjeev R. Saraf
ioMosaic Corporation
Salem, NH 03079

mannan@tamu.edu

Abstract

Calorimetric techniques, such as Differential Scanning Calorimeter (DSC), are widely used to characterize reactivity hazard potential. The conventional DSC is well established tool but is limited to a macroscopic sample. A key drawback of conventional macro scale DSC technology is the large thermal inertia of the calorimetric cell and its associated hardware for smaller sample size. The conventional technology can impose severe limitations in cases where only minute sample quantities are available for testing (e.g., forensics, detection of trace explosives, process or product development).

A microreactor based calorimeter is being developed at MKOPSC to obtain accurate measurements with smaller sample sizes. Because these systems incorporate a very small thermal mass and use reagent quantities in the nanogram/nanoliter range, rapid and uniform heating and cooing can be achieved while maintaining a high level of temperature homogeneity. These miniaturized nanocalorimeters will offer enhanced sensing capabilities in an inexpensive portable format so that measurements can be made directly in the settings where they are needed.

This paper discusses the fabrication and working of the nanocalorimeter developed at MKOPSC. With the proposed advanced device, a calorimetric analysis can be performed in a few minutes utilizing a minute sample. Therefore, such a nanocalorimeter can be effectively employed for rapid screening of reactivity hazards at relatively lower costs.

1. Introduction

Chemical reactivity can be viewed as a double-edged sword for the chemical industry. On one hand, it allows materials to undergo desired reactions to form products under moderate temperatures and pressures. However, it may also lead to uncontrolled catastrophic incidents if not managed properly. Identifying, testing, and evaluating chemical reactivity hazards are three crucial steps for the implementation of chemical reactivity hazard management [1]. The required costs and expertise increase significantly from hazard identification to evaluation. As part of a systematic approach to hazard identification, we have developed theoretical and experimental screening techniques that incorporate both computational tools and calorimetric data. Although theoretical screening tests are cost effective, they are not robust enough by themselves to analyze numerous combinations of binary interactions in mixtures. Therefore, experimental screening techniques are an indispensable part of chemical hazard identification.

Differential screening calorimetry (DSC) is the primary experimental screening test for gathering reactivity information such as thermal stability, onset temperature, enthalpy, heat capacity, heat of reaction or decomposition, heat generation rates, explosion potential, runaway reaction probability, and consequences of a runaway reaction based on heat rates [2]. In addition, the National Fire Protection Association (NFPA) reactivity rating parameter, an extensively employed hazard indicator, is generated in part based on DSC data. The NFPA value is important for determining storage procedures and is also used for the Material Safety Data Sheets and the Dow Fire and Explosion Index (F&EI) [3].

Conventional DSC systems are tabletop instruments used for analyzing macroscopic scale samples between 0.01 ~10 g and have screening times of approximately 1 hour. Due to the large thermal mass of the calorimetric cell and its associated hardware, however, the required input energy of the conventional DSC is often large compared to the measured energy of thermal transitions or chemical reactions. In cases where only trace amounts of samples are available, conventional DSCs are unable to measure accurately the energy released.

Small-scaled calorimeters are inherently safer for the screening of reactive chemicals, because smaller sample sizes result in lesser effects due to runaway reactions. Furthermore, the scanning rate of a differential scanning calorimeter is inversely proportional to sample size, as shown in Table 1.

Table 1. Estimated Maximum Scanning Rate Vs. Sample Size

Sample size	Maximum scanning rate
1g	1 K/min
1mg	100 K/min
1ug	100 K/s
1ng	10,000 K/s

A small-scaled tabletop calorimeter, 2-drop calorimeter, developed by Dr. Tom Hofelich of Dow Chemical was initially used to determine the reaction energies of hazardous chemicals with a minimum liquid sample size of 50μL and an operating temperature range from 5 to 50 ^{o}C. However, due to its low operating temperature, the 2-drop calorimeter cannot be employed for chemical hazard evaluations that occur at high temperatures [4].

In order to screen small samples of reactive materials for chemical processes, a new generation of miniaturized calorimeter must be developed to provide onsite measurements at extended operating temperatures. We have designed a portable nanocalorimeter that incorporates a very small thermal mass with the ability to achieve uniform heating in a short time while maintaining a high level of temperature homogeneity. Furthermore, this nanocalorimeter provides an inherently safer design, reduced cost of hardware and samples, rapid screening of reactive hazards, and greater ease of use.

2. Nanoreactor Fabrication

The calorimetric cell consists of a thermal control module and a sample encapsulation module. The thermal control module is made of a silicon chip with patterned resistors used as heaters and temperature sensors. The same metal resistors are used to avoid electrical noises. On the top of the thermal control module, an etched sample encapsulation module is used to accommodate either solid or liquid samples, as shown in Figure 1.

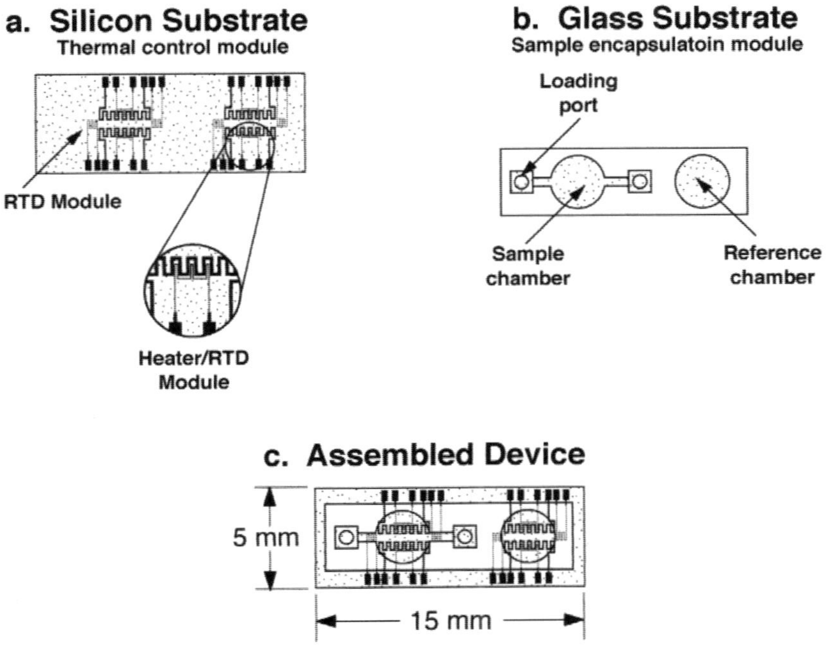

Figure 1. Schematic Illustration Of The Proposed Microfabricated Calorimetric Cell

This modular fabrication approach provides an enhanced level of flexibility by allowing the same thermal control module design to test a variety of samples through the selection of an appropriate encapsulation module. The sample enclosure configuration depicted here is suitable for use with liquid samples.

A prototype of the miniaturized reactor is shown in Figure 2. Following the fabrication of the calorimeter, electronic control algorithms have been developed to implement a variety of thermal analysis protocols.

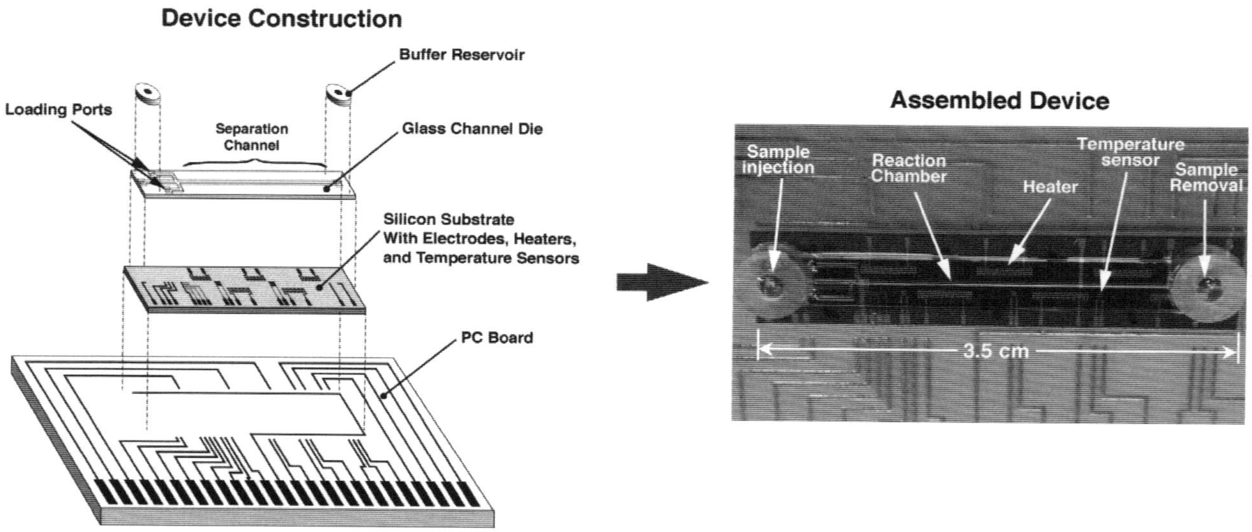

Figure 2. Prototype Of A Miniaturized Reactor

The new prototype of the silicon substrate is shown in Figure 3. At this size scale, over 600 devices can be fabricated on a single 6-inch diameter wafer. The sample size varies with the etched well from 100 nL to 800 nL. For a better heat estimation, we have designed a new smaller chip with dimensions 4.5 x 6 mm and 250 nL volume. A 4-lead arrangement will be used to exclude the resistance of connector leads. The EXCITE and COM leads carry the current through the RTD while the SENSE positive and negative leads measure the voltage developed by the RTD, which is independent of the lead resistance.

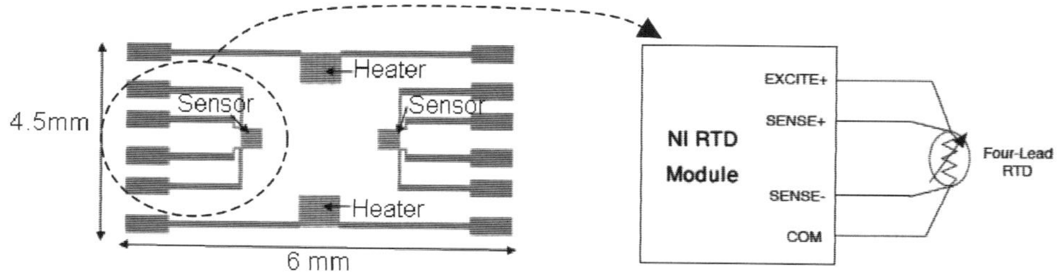

Figure 3. New Design Of The Thermal Control Module With A 4-Lead Arrangement Of Temperature Sensor

Sufficient thermal isolation between the sample and reference chambers is very critical, because residual heat flow between these two locations will reduce the sensitivity of the device. A thermal barrier can be created by etching the substrate or machining trenches on the backside of the wafer using a dicing saw. As shown in Figure 4, an enhanced level of thermal isolation can be achieved using these two strategies. A series of 3-D heat transfer simulations will be performed using FEMLAB (Finite Element Modeling Laboratory) software to decide the optimal design. FEMLAB extends the opportunities of the nanocalorimeter to analyze the thermal behavior under non-adiabatic conditions. This analysis can then be used to determine critical design parameters such as the thickness of the insulation and the input power requirement of the device.

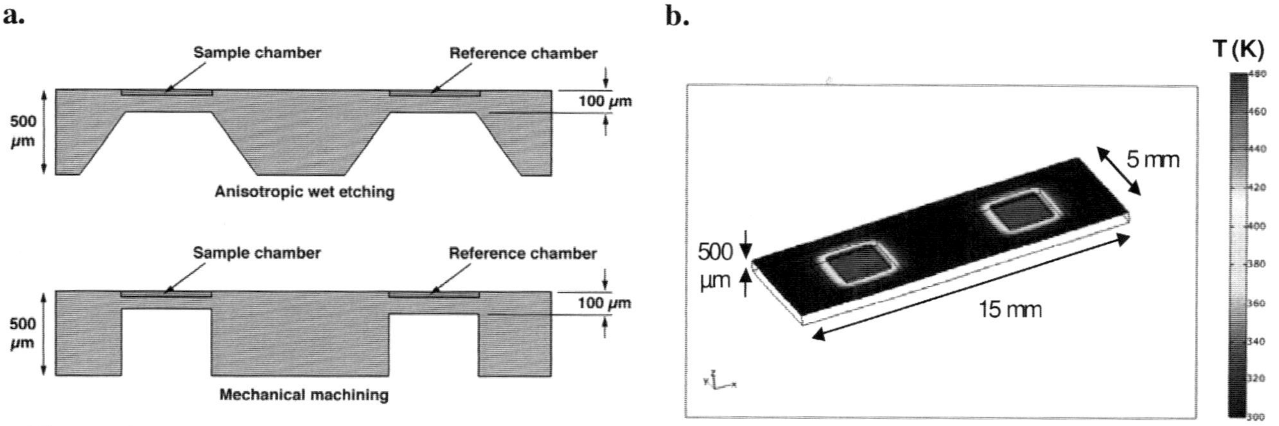

Figure 4. Two strategies to reduce lateral heat flow between the sample and reference chambers. (a) Cross-sectional view depicting removal of material from the backside of the wafer by anisotropic wet etching and mechanical machining processes. (b) 3-D simulation showing thermal isolation between isothermal regions on the surface of a silicon calorimeter with the same dimensions as depicted in Figure 1.

An enhanced level of material fingerprinting can be achieved with simultaneous thermal and optical characterization of samples. Transparent substrates can be employed for the thermal control module so that simultaneous optical, x-ray, spectroscopic, or light scattering measurements can be performed during the course of a calorimetric experiment.

Temperature is controlled using resistive heaters patterned on the thermal control module. Because the resistance of the heaters is known, the electrical output current can be monitored to determine the power supplied to the heater. Similarly, temperatures are measured by monitoring the changes in resistance of similar elements as functions of temperature. Platinum heating elements with resistances ranging from 350 Ω to 2000 Ω provide temperature sensitivity up to 0.2 °C.

The sample enclosure design and loading process mainly depend on the type of samples to be studied. For liquid samples, the enclosure consists of a fluidic channel intersecting a chamber centered over the heaters and temperature control circuitry. Access holes will be machined into the glass substrate to allow individual samples to be loaded with a syringe. An externally pumped fluidic manifold may be constructed to allow automated sample loading and unloading in order to process multiple samples. After loading the sample into the channel, the access ports will be sealed using epoxy, see Figure 5. Solid samples can be inserted directly into an etched chamber, after which the chamber is covered and sealed. Additional mechanical pressure can be applied using a loading fixture to confine the sample.

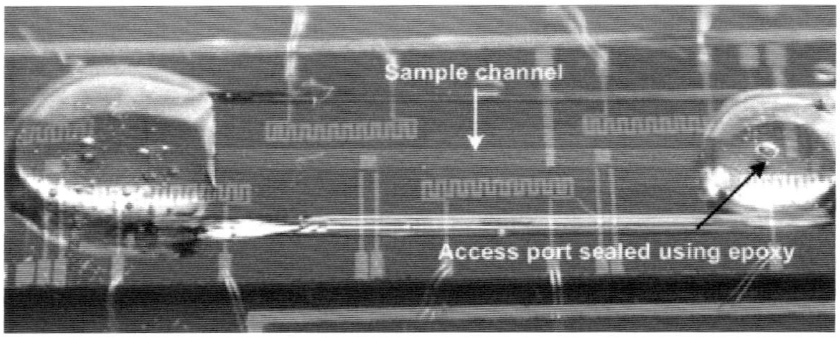

Figure 5. Sample Of Ammonium Iron (III) Sulfate Dodecahydrate Sealed Using Epoxy

3. Operation

The accuracy and sensitivity of the calorimetry system will be measured through the use of well-characterized calibration standards, as shown in Table 2.

Table 2. Typical Calorimetric Calibration Standards

Material (transition)	T_{trans} (°C)	ΔH_{trans} (J/g)
Gallium (s → l)	29.8	79.9
Benzoic acid (s → l)	123	148
Indium (s → l)	156.6	28.6
Tin (s → l)	231.9	60.4

The onset temperatures, reaction enthalpies, and other reaction parameters are measured in the nanocalorimeter with sample sizes of ~ 1 μg and compared with the same parameters measured in an Automatic Pressure Tracking Adiabatic Calorimeter (APTAC) or an Accelerated Rate Calorimeter (ARC®) that use a larger sample size of ~ 2 - 50 g. A library of representative reference chemical systems shown in Table 3 will then be tested to benchmark the performance of the nanocalorimeter relative to other conventional hardware. Further refinements to the device design and control algorithm will compensate for systematic deviations and optimize performance.

Shifts in measured temperature in the nanocalorimeter are expected due to the large sample surface to volume ratios, which leads to more significant surface effects. We have used the APTAC to measure the thermal decomposition behavior of aqueous solutions of hydroxylamine and important derivatives such as hydroxylamine nitrate, hydroxylamine sulfate, hydroxylamine hydrochloride, and have determined the sensitivity of the decomposition onset temperature to ppm quantities of Fe^{2+} and Fe^{3+} [5]. We will test the nanocalorimeter's ability to characterize hydroxylamine samples and estimate a range of ppm iron contaminants based on the depression of the measured onset temperature. Onsite estimation of contaminants in industrial chemical systems (with reactive behaviors sensitive to the presence of contaminants) is an important application of the nanocalorimeter.

Furthermore, we plan to analyze the usefulness of a modulated heating method, which is achieved through the addition of a third time scale. The temperature modulated DSC can be used for the measurement of heat capacity when an irreversible process occurs in the sample.

Table 3. Library Of Energetic Compounds To Be Used For Instrument Testing

Sr. no.	Chemical	Mass gm.	Onset temperature* °C	Max. temperature °C	Heat of reaction cal/g
1	Di-ter-butyl peroxide in toluene[a]	0.71	115	184	49
2	Hydrogen peroxide[a]	8	130	240	700
3	Ammonium nitrate[a]	0.5	201	291	-
4	Styrene[a]	2.2	100	201	172
5	Nitro-styrene[a]	0.95	170	264	297
6	Acrylonitrile[a]	0.79	-	-	389
7	1,3 Butadiene[a]	12	158 ± 10	445 ± 19	260
8	50 wt.% hydroxylamine[b]	2	133	216	420
9	Hydroxylamine nitrate[b]	4.2	171	196	350
10	Hydroxylamine sulfate[b]	4.5	151	185	-

DATA SOURCE: a. ioMosaic, ARC runs b. Texas A&M University, APTAC runs

* The onset temperature is not an intrinsic property of the material, and will vary depending on factors such as sample size and instrument sensitivity.

4. Initial Results

A heating rate of 1 °C/min was applied to measure the boiling point of acetone. The literature value of the boiling point of acetone is 56.2 °C, but our experimental value appears at a lower temperature, as shown in Figure 6, perhaps caused by nanoscale surface effects. Agreement with literature data can be achieved through additional calibrations.

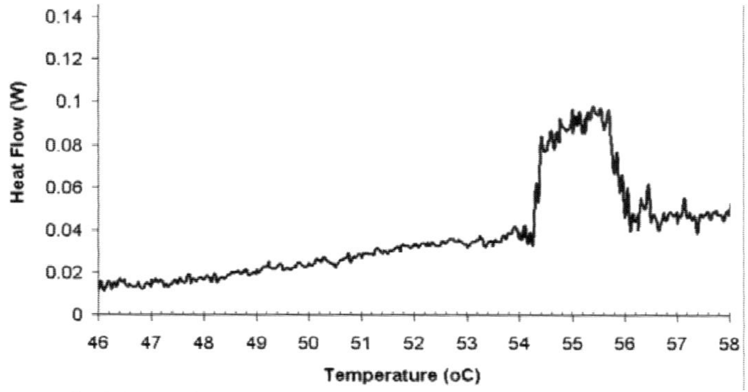

Figure 6. Heat Flow Curve For The Boiling Point Of Acetone

5. Conclusions

Differential scanning calorimetry (DSC) has been extensively used for screening of reactive hazards. The portable nanocalorimeter will serve as a new generation of screening tools for chemical process safety with high efficiency, reduced cost, and enhanced accuracy. Initial work has led to successful fabrication of a nanoreactor capable of sampling nano-sized solid or liquid compounds. Control algorithms incorporating the DSC principle have also been written using LabVIEW. Future work will focus on testing a large variety of samples and modifying the nanoreactor's design and control algorithm to improve its accuracy and sensitivity.

6. References

[1] F. Stoessel, J. Steinbach, and A. Eberz. "Plant and process safety, exothermic and pressure inducing chemical reactions," in *Ullmann's Encyclopedia of Industrial Chemistry*, Wiley-VCH, W.E. Russey, and S. Hawkins, editor; Wiley, John & Sons: 1995; pp. 343-354.

[2] D. J. Leggett. *Chemical Reaction Hazard Identification and Evaluation: Taking the First Steps; Proceedings of the 36th Annual Loss Prevention Symposium,* New York: American Institute of Chemical Engineers.

[3] *NFPA 704: Standard System for the Identification of the Hazards of Materials for Emergency Response, 2001 Edition*, Quincy, Massachusetts: National Fire Protection Association: 2001.

[4] T.C. Hofelich, B.A. Prine, and N.E. Scheffler, "A Quantitative Approach To Determination Of NFPA Reactivity Hazard Rating Parameters," *Process Safety Progress*, 1997, 16 (3), pp. 121 -125.

[5] L.O. Cisneros, W.J. Rogers, and M.S. Mannan, "Comparison of the thermal decomposition behavior for members of the hydroxylamine/water family," *Thermochimica Acta*, 2004, 414 (2), pp. 177-183.

Reactivity Investigation of Mixtures of Propane and Nitrous Oxide

Ronald J. Willey and Shangwei Hu
Northeastern University, Boston, MA 02115

John M. Moses
CF Technologies, Inc., Hyde Park, MA 02136

Abstract

Supercritical and vapor phase reactions of oxidizers and fuels under pressure are of commercial interest. Since these systems have the potential to be explosive, a model mixture was studied with the intent of defining safe operating boundaries. This work investigated the nitrous oxide/propane oxidant/fuel system, diluted with the inert solvent carbon dioxide, and a catalyst as an accelerant, to determine the safe operating regions for the oxidation reaction. Using an ARC® calorimeter for the evaluations, variables included initial system density, catalyst, and mixture compositions. Density of the system ranged from 0.05 to 0.62 g/cc. A number of catalysts were investigated with 10% Pd on carbon being the most active. Strongest reactivity was observed when the mixture was stoichiometric to the complete combustion products of nitrogen, carbon dioxide and water. Three distinct regions of reactivity were observed: mild reactivity (starting at the sensitivity of the ARC of dT/dt = 0.02°C/min to about 0.1°C/min); strong exothermic reaction (dT/dt ~ 1°C/min); and flammable and/or explosive (a distinct and substantial jump in dP/dt in less than 1 second). As a result, a boundary has been proposed for safe mixtures of nitrous oxide, propane, and carbon dioxide, and a basis for determining safe operating compositions for other systems. The computed adiabatic reaction temperature was used to correlate the experimental results to establish the boundary for safe operation.

1. Introduction

The use of the accelerating rate calorimeter, ARC®, for the evaluation of liquid and solid exothermic reaction potential for runaway has been well documented over the past 30 years. The instrument, developed by D. Townsend of Dow Chemical Company, allows evaluation of reactions under conditions close to practice.[1] This includes the ability to run reactions at pressures matching practice, and at low phi factors (the correction factor which is based on the ratio of the total heat capacity of a vessel and the total heat capacity of the vessel contents).[2] The calorimeter uses a heat-wait-search mode advancing in selectable steps of 5 to 25°C per step. The reactants of interest are placed inside a sample container referred to as a bomb. The bomb is then heated to a set point temperature. A period of time, called the wait period, passes until thermal equilibrium is reached. Then, the instrument seeks indication of exothermic reactivity as measured by a thermocouple attached to the bomb outside shell. Reactivity is sensed if dT/dt exceeds 0.02°C/min. When sensed, the instrument allows the reaction to continue, maintaining adiabatic conditions. The underlying thermokinetics expressions for the data acquired by the ARC were published by Townsend and Tou in 1981.[3]

The ARC is regularly used to evaluate exothermic reactions, and determine vent sizing requirements for commercial reactors. Examples of exothermic reactions investigated include: the decomposition of organic peroxides (Tou and Whiting [4], Morisaki [5]), the stability of acrylonitrile (Tartari [6]), the stability of nitrobenzyl halogenides and isoxazole (Cardillo [7,8]), the decomposition of ethylene and propylene oxides (Freeder and Snee [9]), the decomposition of hydrazine (Wedlich [10]), and the reaction of chloronitrobenzene with ammonia to form nitroaniline (Willey et al. [11]).

There is substantial research and commercial interest in vapor phase oxidation with nitrous oxide (N_2O) and in supercritical fluid phase reactions, including oxidations. Prior to building an experimental reactor, we wished to thoroughly investigate the effect of pressure on these exothermic systems. The ARC data collected was the primary source of information for our hazard and operability (HAZOP) study of the reactor system during the design phase.

Recent publications have shown that there is an interest in partial oxidations using N_2O.[12] Solutia (formerly Monsanto) has a pilot program to reduce N_2O emissions, a by-product from adipic acid production, by reacting N_2O with benzene to form phenol using a zeolite catalyst.[13] N_2O safety, due to its positive free energy of formation, must be considered. At elevated temperature it is a metastable compound, self-decomposing exothermically to nitrogen and oxygen.[14] Under certain conditions, explosive decomposition can occur.

Melhem has developed a hazard index based on the heat of reaction and the computed adiabatic reaction temperature (CART).[15] The proposed hazard index breaks down as follows:

- No hazard, $-\Delta H_{rxn} < 0.42$ kJ/g;
- Low hazard, $0.42 \text{ kJ/g} \leq -\Delta H_{rxn} < 1.2$ kJ/g;
- Intermediate hazard, $1.2 \text{ kJ/g} \leq -\Delta H_{rxn} < 3.0$ kJ/g; and
- High hazard, $-\Delta H_{rxn} \geq 3.0$ kJ/g.

Supercritical carbon dioxide (CO_2) has been finding applications in chemical processing as a solvent replacement. Numerous publications and symposia have established its solvent properties, commercial uses, and "green" benefits (Journal of Supercritical Fluids, ACS "Green Processing Symposia").

We have an interest in the partial and complete oxidation of hydrocarbons in supercritical solutions. As mentioned above, these mixtures can be explosive. However, this reactivity can be brought into a controlled region by using an inert diluent such as CO_2. Further, CO_2's critical properties are very similar to N_2O. Finally, little documentation exists on the use of the ARC for gas phase reactions that can be explosive in certain regions. This paper describes the use of the ARC and the results of the investigation of one reaction.

2. Background

The reaction investigated in this work is the oxidation of propane (C_3H_8) by N_2O:

$$C_3H_8 + 10\ N_2O \text{ ---> } 3\ CO_2 + 4\ H_2O + 10\ N_2 \quad -\Delta H_{RXN} = 2{,}890 \text{ kJ per mol } C_3H_8$$

Without diluent, this reaction has a CART of 4,230°K and a $-\Delta H_{RXN}$ of 5.92 kJ/g. Based on Melhem's criteria above, the reaction is in the high hazard regime. For comparison, the combustion of propane in oxygen has a CART of 5,550°K and a $-\Delta H_{RXN}$ of 10.0 kJ/g. This value is well above the flame temperature reported for a propane flame of 3,120°K [16] because it does not take into account heat losses by radiation and dissociation of molecules at higher temperatures.

3. Experimental

3.1 Method of Loading the Sample Bomb

Loading the sample bomb involved first rinsing a ¼" neck, 1" ID diameter, 316 SS, 0.032" wall thickness, sample bomb (TIAX M/N 851-3329 burst pressure of 10,000 psi or 70 MPa) with acetone followed by complete drying and evacuation. Early runs used a Hastelloy-C sample bomb (burst pressure of 15,000 psi or 100 MPa). We encountered difficulty sealing the Hastelloy bomb at the higher temperatures and pressures, and therefore, switched to 0.032" walled, 316 SS. After rinsing and drying, 20 mg of 10% Pd on activated carbon (Aldrich 20,569-9) were placed inside the sample bomb. The sample bomb was then connected to the calorimeter (schematic is shown in Figure 1). The sample thermocouple was attached to the outside of the sample bomb via the clip provided. Next, the system was evacuated to less than 1 mm Hg vacuum via a conventional vacuum pump. The system was leak checked by ensuring that the vacuum held over a 10 minute period. C_3H_8 (commercial grade from Master Mechanic sold under license from BernzOmatic) was slowly metered into the system through a needle valve. The quantity for the initial gas addition was calculated using the Peng-Robinson equation of state (PR-EOS) and the system pressure as measured by the pressure transducer (PT-1 in Figure 1, a Sensotec Transducer Model Number TJE/0743-11 TJA with a range of 0 to 10,000 psi). These estimates were later checked by direct weighing and were found to be within 2% of predicted mass. Next, after evacuation of C_3H_8 from the metering lines connecting the vapor source to the ARC as shown in Figure 1, N_2O (Med-Tech Gases, Medford, MA) was added to the metering line. Next, the system valve was opened briefly to allow equilibrium of pressure and then closed. The amount added was calculated from the pressure change read on P-1 before and after the addition of N_2O and using the PR-EOS. We made a mass check on this approach and found the mass added to be within 3% of predicted.

The addition of CO_2 was the most challenging. Often, we desired system densities above that possible maintaining only gaseous CO_2 in the sample bomb at the loading temperature of 22°C. Thus, we immersed the sample bomb in an ice bath to condense the CO_2 in the bomb. Based on the pressure change in the metering system lines, the amount of CO_2 added was determined. In this case, IUPAC tables [17] were interpolated to determine the amount of CO_2 added. The tables were used because the pressures and temperatures were just below the critical point of

CO_2 (73.8 bar and 31.1°C) where the equation of state does not predict density accurately. This titration was done as many as three times to reach the higher system densities investigated (>0.6 g/cm^3). A weight check on this addition step found the mass to be within 5% of the calculated value. Total system volume including the 1/8" OD transfer tube and pressure transducer was 12.5 cm^3.

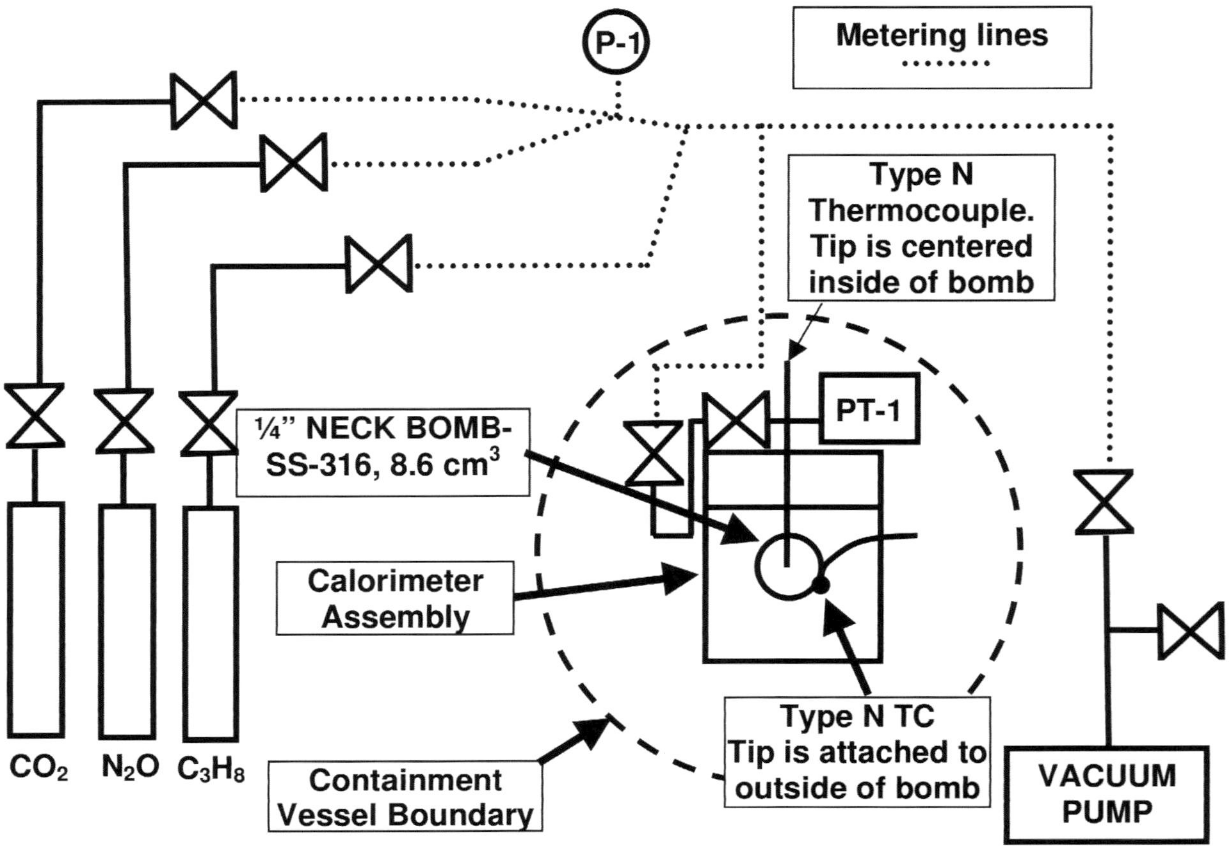

Figure 1. System Schematic for High Pressure Gas Loading Testing

3.2 ARC Runs

Following the vapor loading in the sample bomb, experiments proceeded with heating to the starting temperature, typically 100°C and the heat-wait-search routine. The variables were: a heating step of 25°C, a wait period of 10 minutes, and a search period of 10 minutes. Data were continuously collected via an ARC 2000 program. Maximum pressure limit was set to 9,000 psia or 62 MPa and maximum temperature limit was set as high as 400°C.

3.3 Modifications Made During the Work

As work progressed and experience was gained, we made a number of modifications. One change already mentioned was from using a Hastelloy C to a 316 SS sample bomb. The Hastelloy C bomb was selected for its high pressure rating and corrosion resistance. The difficulty encountered in practice was that the connection of the sample bomb to the ARC frequently leaked at the higher temperatures and pressures. The change to a ¼" neck 1" ID diameter stainless steel sample bomb provided leak-tight seals. Titanium was avoided because of the possible reaction with the N_2O. We did not attempt to verify that reaction.

Another modification was the relocation of the pressure transducer closer to the calorimeter heater. This eliminated some of the dead volume, and maintained the transducer line temperature closer to the calorimeter temperature reducing the likelihood of cooling and condensation in the lines or transducer.

We added an internal thermocouple to monitor temperature at the center of the bomb. This temperature was compared to the wall temperature measured by the thermocouple placed on the clip external to the sample bomb. While differences exist, as shown in Figure 2, the main results and conclusions do not change based on this measurement. The internal temperature tracked very closely the temperature detected by the wall thermocouple, except at the onset of an exotherm when the internal thermocouple responded faster, and indicated higher peak temperature. The temperature difference began to show distinction when the self heating rate exceeded 0.5°C/min and substantial differences appeared when the self heating rate exceeded 1.5°C/min. For the data shown in Figure 2, the heating rate exceeded 40°C/min.

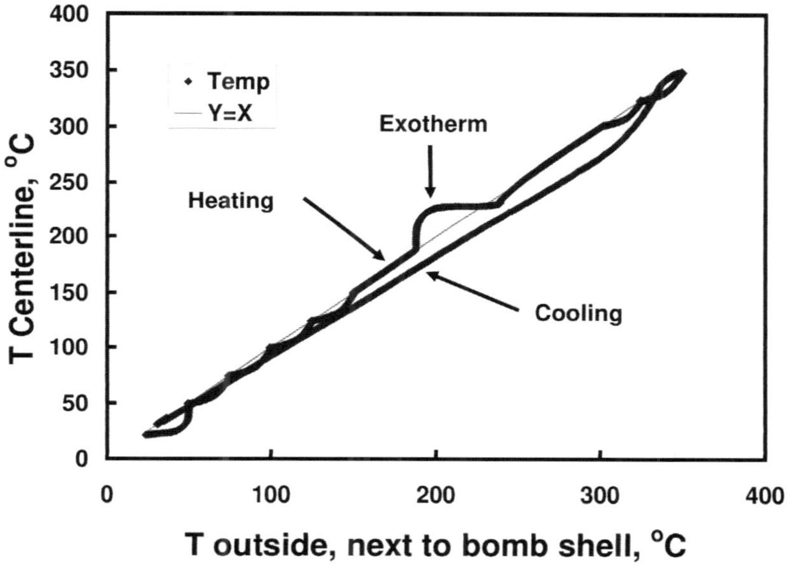

Figure 2. Comparison of temperature measured inside the centerline of the bomb versus the temperature measured on the outside.

4. Results and Discussion

4.1 Initial Verification Runs

Because of the potential hazard associated with oxidant-fuel mixtures, and the need for calibration and baseline data for "high-pressure" vapor phase ARC experiments, initial experiments were with pure CO_2, then mixtures of CO_2 with fuel, and CO_2 with oxidant. System density was gradually built up to the highest value we investigated, 0.62 g/cm^3 (critical density is 0.47 g/cm^3).

Figure 3 shows the temperature and pressure as a function of time for a run done with 0.62 g/cm^3 of CO_2. We see one sustained period at 300°C where the instrument detected a weak exotherm (dT/dt ~ 0.02°C/min). Other than this prolonged search at 300°C, the heat-wait-search routine worked properly, even when the pressure was in the 1,000's of psi.

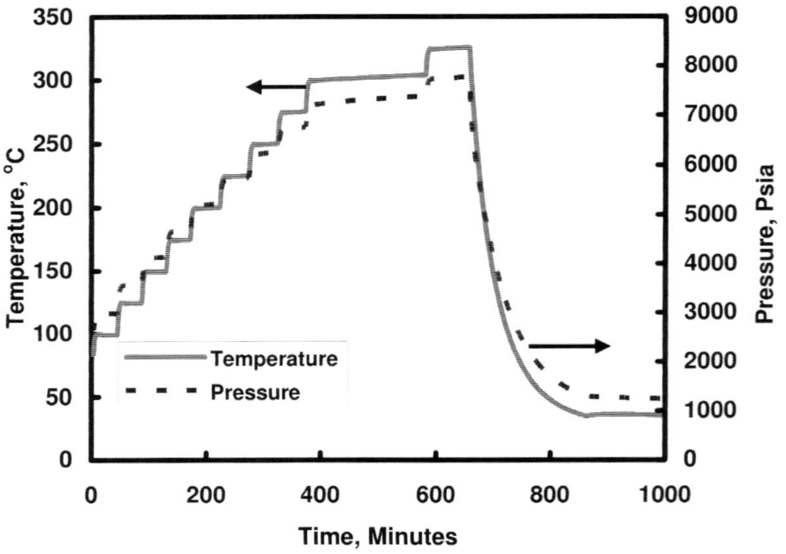

Figure 3. Temperature – Pressure traces for a Hastelloy-C bomb loaded with 0.62 g/cm^3 of CO_2.

Figure 4 shows an excellent pressure versus temperature curve for heating and cooling. The small hysteresis observed is typical for the ARC and is attributed to temperature differences at the pressure transducer during heating versus cooling as compared to the bomb temperature inside the calorimeter. Using a Hastelloy-C bomb in this run (mass of 18.7 g), the phi was 3.0.

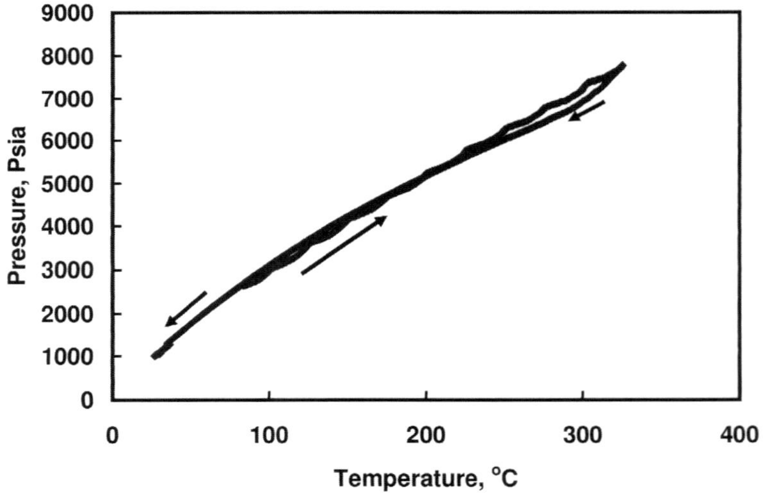

Figure 4. Pressure versus temperature for a pure CO_2 run with system density of 0.62 g/cm^3.

In the next set of experiments, mixtures of C_3H_8 and CO_2 were tested. No exotherms were detected. The initial runs were at a density of 0.05 g/cm^3. In subsequent runs the density was increased to 0.17 g/cm^3. Similarly, with mixtures of CO_2 and N_2O at a density of 0.15 g/cm^3, no exotherms were detected at temperatures up to 400°C.

Three component mixtures, N_2O, C_3H_8, and CO_2, were tested over the density range from 0.05 to 0.15 g/cm^3. Again no exotherms were detected. This completed our initial experiments, and provided the baseline and calibration data required for the reactive systems. When catalyst was added to the three component systems, a wide range of reactions were observed, from controlled, partial oxidations to complete oxidations to ignitions to explosions.

4.2 Data Summary

More than 170 ARC experiments were performed to define the reaction and ignition boundaries for vapor phase fuel-oxidizer-diluent-catalyst system. 55 runs were with the components, C_3H_8, N_2O, CO_2, and palladium on carbon. Of these, 54 runs were valid runs with no leaks or other problems. Of these 54, two resulted in trace reaction or a reaction rate near the detection limit quantifiable with the ARC; 35 resulted in controlled oxidations, these define the "inner" operating boundaries for the reactions of interest, 18 resulted in ignitions and define the area to be avoided, and one resulted in an explosion, confirming the potential hazard. The other 115 runs were for calibration and variations of fuel, oxidizer, diluent, and catalyst.

4.3 Containment of Explosions

Figure 5 is the pressure and temperature history of a run with C_3H_8 and N_2O mixture that exploded. The total initial pressure was 745 psia (5.13 MPa) at 24°C with a system density of 0.15 g/cm^3. 20 mg of a 10% Pd/C catalyst was in place. The ratio of the mixtures was 1 part fuel and 10 parts oxidizer (the stoichiometric ratio). No CO_2 diluent was used in this run. Because the ARC's containment system worked very well, there was no external damage. Inside the ARC, the story is different. Shock waves created by the explosion ruptured the 1/8" stainless

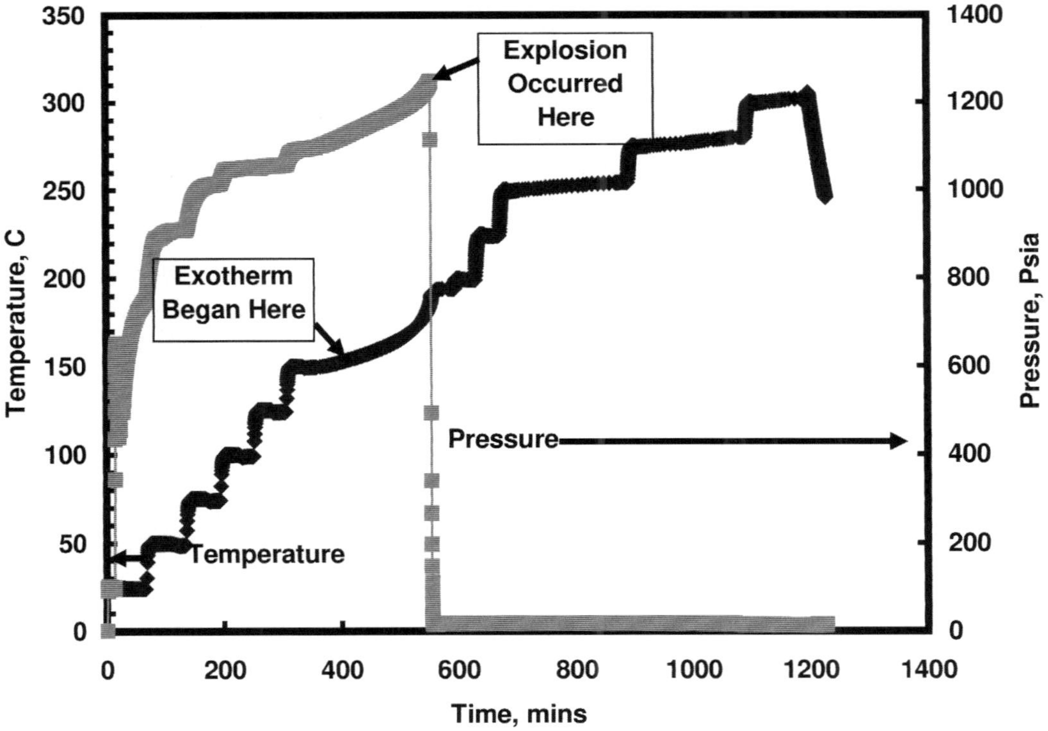

Figure 5. Temperature and pressure profiles for a mixture of N_2O/C_3H_8 that exploded.

steel tubing connector lines (Figure 6). This explosion occurred very rapidly; that is, from the start of the runaway reaction to explosion was less than 4 seconds. As the heat-wait-search figure shows, an exothermic reaction was in progress with the dT/dt rate at 1.5°C/min just before the explosion. The next point acquired showed a sudden jump in dT/dt to 9.4°C/min (approximately 2 second elapsed) then the explosion occurred (at 187.5°C). The run immediately prior to this had the identical mixture but no catalyst. It exhibited no exothermic activity up to a temperature of 350°C. The reader must understand that even though a mixture does not react under certain conditions, it may still have high hazard potential. Trevor Kletz's (author of "What Went Wrong?" and numerous other process safety publications) famous quote is that the "ignition source is always free". In this work, we found a 10% Pd/C catalyst to be a satisfactory ignition source.

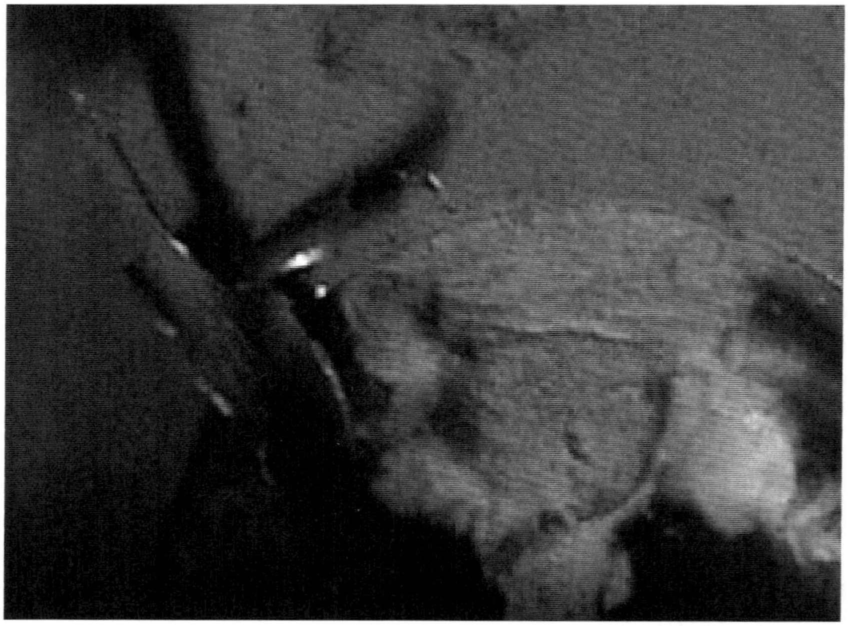

Figure 6. Ruptured 1/8" transfer line caused by explosion of a mixture of N_2O and C_3H_8. A shock wave is indicated by the line rupture.

Shown in Figure 7 is another explosion example, although this was with another catalyst. The mixture was 5% C_3H_8, 47% N_2O, 48% CO_2 with 20 mg of an Fe_2O_3 - SiO_2 aerogel catalyst and a system

Figure 7. Result of explosion with 20 mg of an iron oxide silica aerogel catalyst with a stoichiometric mixture of C_3H_8 and N_2O with about 50% CO_2 present (total vapor mass in bomb approximately 1 gram). Sample bomb was fabricated from 0.032" thick stainless steel with a burst pressure rating of 10,000 psi or 70 MPa.

density estimated at 0.16 g/cm^3. No exothermic activity had been detected before the explosion occurred. Once again, this demonstrates the need for care when using high pressure mixtures of exothermic reactants. The ΔH_{rxn} for this system is approximately 3 kJ/g, placing it at the boundary of intermediate and high hazard as specified by Melhem.[15] However, in view of the explosion, this appears to be a high hazard system, presumably due to the elevated pre-reaction pressure. Therefore, we recommend that initial pressure of the reactants be considered as a factor capable of shifting the boundaries in Melhem's criteria, i.e. higher starting pressure requires less reaction energy to rate a high hazard potential.

4.4 Example of Data Acquired for Ignition Boundary Search

From the ARC runs we observed conventional exothermic reaction behavior, characterized by smooth dP/dt and dT/dt curves, with some mixtures. However other mixtures showed a jump in the curves. This is attributed to ignition of the reactants inside the ARC bomb.

Figure 8 shows a summary of the exotherms recorded as dT/dt for mixtures of $CO_2/N_2O/C_3H_8$ at a density of 0.33 g/cm^3. The N_2O to C_3H_8 ratio was kept constant at 10:1 (stoichiometric for complete combustion). The variable is the concentration of CO_2 added as a diluent. A "jump" appears on three of the four curves shown near a temperature of 189°C. Figure 9 expands upon the observations shown in Figure 8 by showing the resultant dP/dt during this exotherm. We see a distinct discontinuity near 189°C. We have defined this as an ignition event. It is mild because the amount of fuel/oxidizer/inert is near a boundary where no event occurs (see 89% curve in Figure 9). As more CO_2 is added to a mixture maintaining constant density, weaker exotherms appear. The 89% CO_2 mixture is not ignitable, but is reactive. Although this transition for

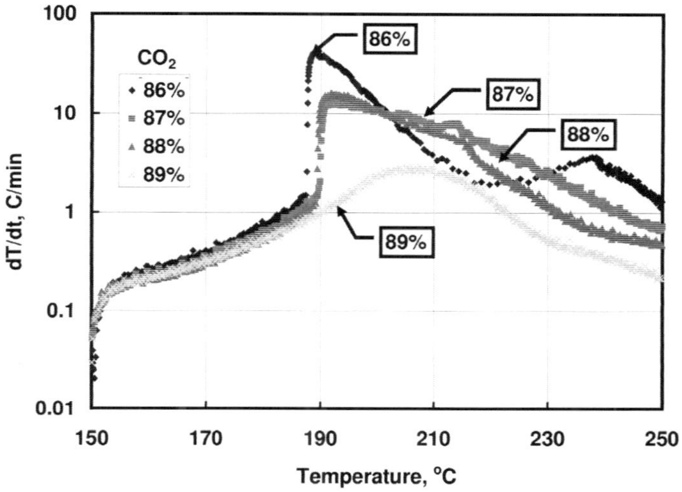

Figure 8. Self heating rate during an exotherm for a series of ARC runs with 20 mg Pd/C catalyst, C_3H_8:N_2O=1:10.

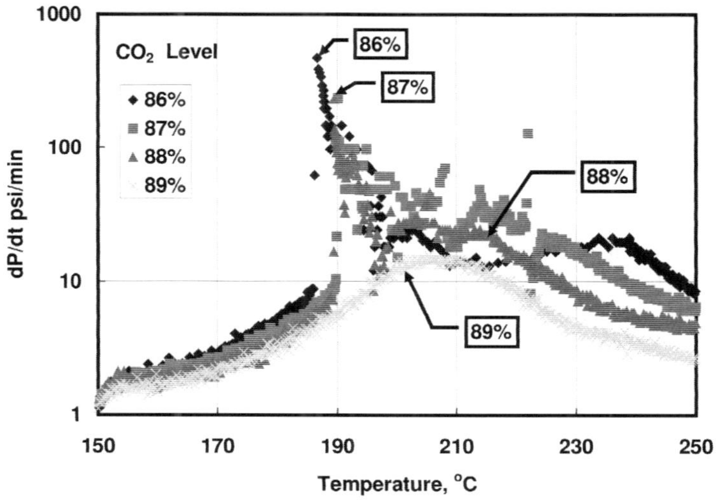

Figure 9. Pressurization rate during an exotherm for a series of ARC runs with 20 mg Pd/C catalyst, C_3H_8:N_2O=1:10.

ignitable to non-ignitable but reactive is the expected behavior, this data locates the boundary for this mixture and density. In the design of our continuous reactor, we make an adjustment to this raw mixture data. The correction accounts for the amount of oxidizer and fuel that have been consumed in the ARC's batch reaction prior to ignition, to estimate the composition at ignition.

4.5 Effect of Catalysts

In the absence of catalyst, stoichiometric mixtures of N_2O and C_3H_8 without diluent were not reactive up to 400°C. The reader should not, under any circumstance, interpret this result to mean that these are safe mixtures, any ignition source with these mixtures would likely result in an explosion. Essentially, the autoignition temperature of the mixture was not reached, and in the absence of catalyst these systems did not have enough energy to initiate reaction. 10 % palladium on carbon was the most active catalyst. Exotherms recorded with other catalytic materials showed that the exotherm onset temperature shifted 15 to 100°C higher.

4.6 Reactivity Diagrams

Figure 10 shows a triangular mixture diagram for N_2O, C_3H_8, and CO_2. The points where "jumps" were observed are solid points. The figure shows that as density increases the points of ignition move closer to the CO_2 apex. The three lines overlaid on the figure are CARTs of 800 K, 1200 K and 1600 K. Up to a density of 0.6 g/cm^3, mixtures with a CART of 800 K or below

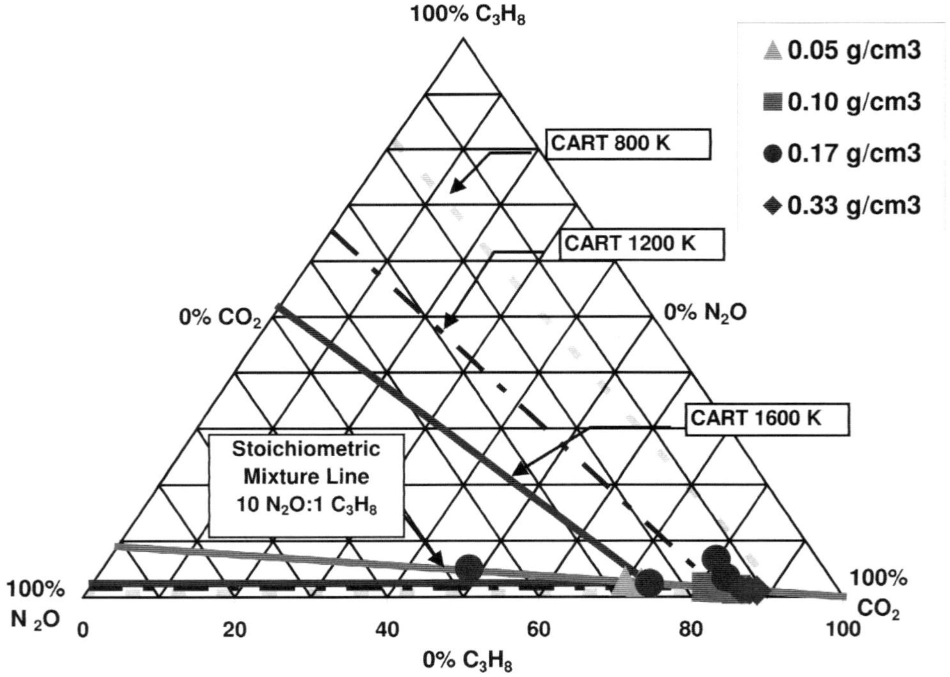

Figure 10. Triangular diagram comparing experimental results to 3 CART's. Solid points represent "jumps" or ignition.

did not ignite. Figure 11 is a magnification of Figure 10 with no ignition points added. We see in this figure overlap of non-ignition with ignition points in portions of the chart. This is due to density; within a particular density there were no overlaps of data. However, it can be seen that the dark diamonds, density of 0.33 g/cm^3, are ignitable at more dilute concentrations than the nonflammable open square symbols, density of 0.1 g/cm^3. Therefore, as expected with these pressurized systems, density must also be considered.

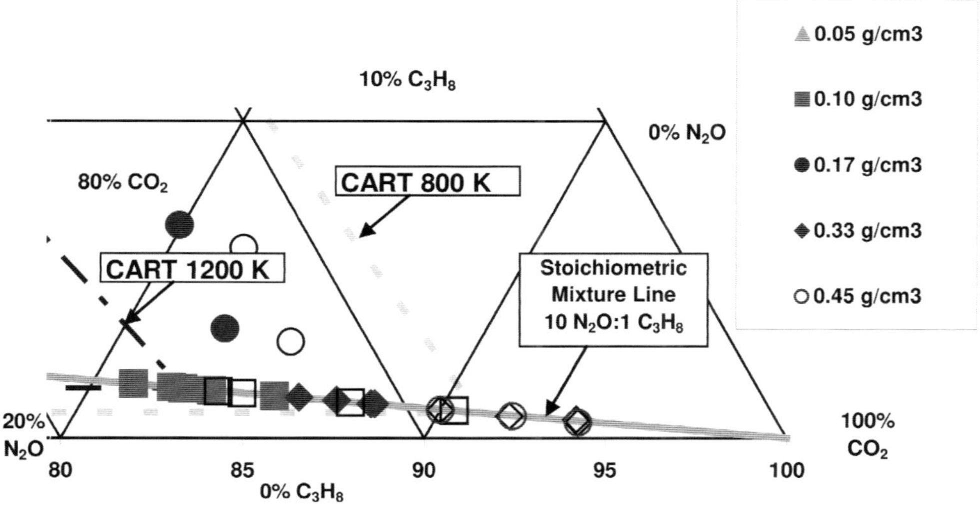

Figure 11. Expanded triangular diagram comparing experimental results to two CART's. Solid points represent "jumps" or ignition, open points represent exothermic reactivity only.

4.7 Phi Factor Discussion

When gases are used the mass of reactant is substantially less than the mass of the bomb. Also, because a diluent is included in the system mass, this must be factored into the calculation. For this work, we defined phi as:

$$\phi = \frac{m_{bomb} c_{p\,bomb} + \sum_{i}^{\text{all inerts \& unreacted species}} m_i c_{pi} + \sum_{i}^{\text{all products}} m_i c_{pi}}{\sum_{i}^{\text{all products}} m_i c_{pi}}$$

This equation accounts for the heat generated by reaction being used to heat up the products, the bomb body, other inerts such as CO_2, and unreacted reactants. Phi equal to one means a stoichiometric mixture of reactants with no diluent, reacting in vessel that has no mass. If a diluent is present, it picks up a portion of the heat and the adiabatic temperature rise rate is dampened, and the final temperature is lower compared to a stoichiometric mixture reacting without a diluent. If excess reactant is present, it too can pick up a portion of the heat. Again, the adiabatic temperature rise rate is lower and the final temperature is lower. The phi quantities are high. The lowest phi factor was 3.0 with a system density of 0.62 g/cm^3. Typically, the phi was 10 to 20 for most runs.

5. Conclusions

1. The thermokinetic safety of pressurized gas phase exothermic reactions can be evaluated using an ARC calorimeter.
2. At densities below 0.5 g/cm^3 explosions can result if the fuel/oxidizer CART is above 800 K, and an ignition source is present.
3. CO_2 can be used as a diluent to render explosive mixtures non flammable and non explosive.
4. The computed adiabatic reaction temperature, CART, assists in defining a boundary between regions of non-flammable exothermic reaction and flammable mixtures.

6. Acknowledgement

The authors acknowledge Mr. Simon Chippett and Mr. S.K. Singh of TIAX, Cambridge, MA for the use of their ARC and discussions of the work; and Dr. Daniel A. Crowl, Department of Chemical Engineering, Michigan Technological University, Houghton, Michigan for the use of his ARC. We acknowledge the assistance of Mr. Mathew Moses, Ms. Mary Parent, and Mr. John Hannon for their assistance in data entry and data analysis. This material is based upon work supported in part by the U. S. Army Research Laboratory and the U. S. Army Research Office under contract number DAAD1900C0093.

7. Literature Cited

[1] Townsend, D.I., *Chemical Engineering Progress. 73*: p. 80, **1977**.

[2] *HarsNet Thematic Network on Hazard Assessment of Highly Reactive Systems Glossary*. 2005, accessed at http://www.harsnet.de/bookmat/Glossary.pdf on 1/27/05.

[3] Townsend, D.I. and Tou, J.C., *Thermochimica Acta. 37*: p. 1-30, **1980**.

[4] Tou, J.C. and Whiting, L.F., *Thermochimica Acta. 48*: p. 21-42, **1981**.

[5] Morisaki, S., *J. Chem. Soc. Japan. 11*: p. 1773-1779, **1982**.

[6] Tartari, V., La Chimica E L'Industria. 65(10), **1983**.

[7] Cardillo, P., *J. Chem Eng. Data. 29*: p. 348-351, **1984**.

[8] Cardillo, P., *J. Loss Prev. Process Ind. 1(1)*: p. 46, **1988**.

[9] Freeder, B.G. and Snee, T.J., *J. Loss Prev. Process Ind. 1*, **1988**.

[10] Wedlich, R.C., *Thermochimica Acta. 171*: p. 1-13, **1990**.

[11] Willey, R.J., Rodrigues, F., Chippett, S., Melhem, G., and Singh, S.K., *Process Safety Progress. 20(2)*: p. 123-129, **2001**.

[12] Panov, G.I., *CATTECH. 4(1)*: p. 18-31, **2000**.

[13] Armor, J. Catalysis, Recent Value, Future Opportunities, 2001, accessed at

http://www.geocities.com/return_science/chemicalthai1.doc on 1/26/2005.

[14] Yaws, C.L., *Matheson Gas Data Book*. Matheson Tri-Gas, Parsippany, NJ 07054: p 624, **2001.**

[15] Melhem, G.A., *Process Safety Progress. 23(2)*: p. 99-107, **2004.**

[16] Lange's Handbook of Chemistry, 12 ed., Table 11-10, **1977**.

[17] "Thermodynamic Tables Project. International Thermodynamic Tables of the Fluid State Carbon Dioxide," Angus, S., B. Armstrong, et al., Eds. Pergamon Press, New York,. **1976**.

Paper TG005 - 4

Chemical Reactivity Hazards and Inherently Safer Technology

B. Venugopal and D.Y. Kohn, Chilworth Technology, Inc.,
250 Plainsboro Road, Building 7, Plainsboro, NJ 08536

Prepared for Presentation at
American Institute of Chemical Engineers
2005 Spring National Meeting
39th Annual Loss Prevention Symposium
Atlanta, Georgia
April 11-13, 2005

UNPUBLISHED

Chemical Reactivity Hazards and Inherently Safer Technology

**B. Venugopal and D.Y. Kohn, Chilworth Technology, Inc.,
250 Plainsboro Road, Building 7, Plainsboro, NJ 08536**

Abstract

The thermal process safety of nitrating an acid, Acid A, to form nitro Acid A, by two different routes is studied. The nitration system #1 uses the classical nitration procedure of Acid A with 70% nitric acid in the presence of concentrated sulfuric acid. The nitration process #2 involves first adding another acid, Acid B, to Acid A, before nitrating the reaction mixture with 70 or 90% nitric acid. The thermal data obtained from the reaction mixture shows that, if the above two nitration processes are run in a batch mode, the nitration process #2 is thermally safer when compared to nitration process #1. For nitration process #1, the assessment of criticality of cooling failure scenario shows that highly hazardous conditions exist, suggesting the process should be redesigned. If the two nitration processes, however, are run in semi-batch mode, there is virtually no difference in the thermal hazards of the processes, and both processes are thermally safe, requiring no special action. The prescribed temperatures, amount of chemicals charged, feed time and agitation rate should be maintained as in this study to reach the above conclusions. Choice between the two processes should be based on economics and environmental considerations and not on safety.

This work clearly demonstrates the advantages of designing an inherently safe process (batch vs. semi-batch mode of operation) and incorporating safety in the early stages of development by comparing the thermal characteristics of two different nitration systems to determine whether one route should be chosen over the other from a thermal process safety point of view.

1. Introduction

The safety aspects of a process are strongly linked to its design. During the course of the development of a new process, at different stages, decisions concerning the design are made. These decisions, even if they are not made for safety reasons, have a strong influence on the process safety. The search for inherently safer processes must begin early in the process life cycle, and never stops. The greatest potential opportunities for impacting process design occur early in the invention and development of processes. In the early stages of a process development, there is a great deal of freedom in the selection of chemistry, solvents, raw materials, process intermediates, unit operations, plant location, and other process parameters. As the process moves through its life cycle, it becomes more difficult and expensive to change the basic process. It is never too late, however, to consider inherently safer alternatives.

The process of designing an inherently safer process, by comparing the thermal characteristics of alternative nitration processes, is illustrated by the following example.

The formation of nitro Acid A involves nitrating Acid A with 70% nitric acid in the presence of concentrated sulfuric acid at 20°C. This route is designated as the "Nitration Process #1". The thermal process characteristics of the above process are compared with an alternative process, designated as "Nitration Process #2. The process involves, adding Acid B to Acid A in solvent, before nitrating the above mixture with 70 or 90% nitric acid.

2. Chemistry

2.1 Nitration Process #1

$$\text{Acid A} + HNO_3 \xrightarrow[\text{Conc } H_2SO_4]{} \text{nitro Acid A} + H_2O$$

2.2 Nitration Process #2

$$\text{Acid A} + HNO_3 \xrightarrow[\text{Acid B Solvent}]{} \text{nitro Acid A} + H_2O$$

The specifics of Acid A and Acid B cannot be given due to client confidentiality.

3. Experimental Procedures

The experiments for both the nitration processes, #l and #2, were conducted using the Mettler Toledo RC1 reaction calorimeter, based on the heat flow principle.[1] A 2-liter glass reactor with an anchor agitator was used. RC1 enables direct measurement of the instantaneous heat output of the reacting system due to chemical or physical processes as a function of the process time. This quantity shows directly whether and how quickly chemical conversions occur in the process phase under consideration. The quantities important for a thermally safe process control, such as the heat dissipation capacities and the expected temperature changes in case of a malfunction, can be derived from the measured heat production. RC1 allows processes to be performed under conditions resembling those in the plant, particularly in regard to temperature profiles, reactant addition, and agitation. RC1 measurement ensures that the determined quantities depend solely on the properties of the reaction mass and that the factors dependent on the scale and on the specific instrument properties have a minimal influence on, or can be eliminated from, the experimental result. A rapid effective control system, a high cooling capacity and a homogeneous jacket temperature allow virtually ideal isothermal measurement conditions. A thermostat with a high heating/cooling capacity and a rapid circulation of the jacket medium ensure that, the heat dissipation is continuously matched to changes in the heat production in the reactor to variation of the jacket temperature.

The measurement system used for the controller cascade of the thermostat simultaneously measures the temperature difference between reactor and jacket. This difference is directly proportional to the instantaneous heat flow between reaction mass and jacket of the instrument. The proportionality factor (heat transfer coefficient × wetted area = k × A) is calibrated during the experiment using a calibration heating system. Due to the rapid control of the jacket temperature, the measured heat flow is identical to the heat production of the reaction mass under investigation.

3.1 Nitration Process #1

Concentrated sulfuric acid (95-98%) and Acid A were charged to the RC1 reactor. The reactor contents were heated to 20°C. 70% nitric acid was then dosed to the reactor contents over 35 minutes. The reactor contents were held for an additional period of about 75 minutes to reach thermal equilibrium conditions. Appropriate calibrations were done (heat transfer coefficient and specific heat determinations) before and after dosing of 70% nitric acid.

Reactant Charges	**Mass** (grams)	**Moles**	**Concentration** (mole/kg of the final reaction mass)
Sulfuric acid (95-98%)	1290	12.76	7.55
Acid A	233	1.62	0.96
Nitric acid (70%)	166	1.84	1.09

3.2 Nitration Process #2 (Using 70% Nitric Acid)

3.2.1 Part A: Acid B Addition

Acid A dissolved in a solvent was charged to the RC1 reactor. The reactor contents were heated to 20°C before dosing Acid B over 30 minutes, maintaining the reactor temperature at 20°C. The reactor contents were held for an additional period of about 55 minutes to reach thermal equilibrium conditions. Appropriate calibrations were done (heat transfer coefficient and specific heat determinations) before and after dosing Acid B.

Reactant Charges	**Mass** (grams)	**Moles**	**Concentration** (mole/kg of the final reaction mass)
Solvent	928	—	—
Acid A	36.8	0.25	0.24
Acid B	102	0.68	0.64

3.2.2 Part B: 70% Nitric Acid Addition

To the above reaction mixture, 70% nitric acid was dosed, maintaining the reactor temperature at 20°C over 30 minutes. The reactor contents were held for an additional period of about 55 minutes to reach thermal equilibrium conditions. Appropriate calibrations were done (heat transfer coefficient and specific heat determinations) before and after nitric acid addition.

Reactant Charges	**Mass** (grams)	**Moles**	**Concentration** (mole/kg of the final reaction mass)
70% Nitric acid	30.7	0.34	0.31

3.3 Nitration Process #2 (Using 90% Nitric Acid)

The experimental procedure and the amounts charged are the same as above, except instead of 70% nitric acid, 90% nitric was used as the nitrating agent. The amount of 90% nitric acid used was 24 grams. 90% nitric acid, though more hazardous than 70% nitric acid, was used to demonstrate the feasibility of the process using a higher concentrated acid.

4. Results And Discussion

The thermal data of the desired reactions from the RC1 runs for the dosing nitration processes are summarized below.

Thermal Data	**Nitration Process #1** (70% Nitric Acid Dosing)	**Nitration Process #2** (Acid B Dosing)	**Nitration Process #2** (70% Nitric Acid Dosing)	**Nitration Process #2** (90% Nitric Acid Dosing)
Total enthalpy evolved, Q_r, (during dosing and hold period), kJ	263	6.3	43.0	36.8
Mass at the and of the reaction, m, kg	1.6869	1.0665	1.0919	1.0928
Specific heat of the final reaction mixture, C_p, kJ/kg K	1.4885	1.1096	1.1724	1.2548
Adiabatic temperature rise, $\Delta T_{ad}(Q_r/mC_p)$ °C,	104.7	5.3	33.6	26.8
Maximum temperature of desired reaction in case of loss of cooling and in case the feed material is charged at once (batch reaction) instead of being dosed, °C (starting from the process temperature of 20°C)	124.7	25.3	53.6	46.8
Accumulation enthalpy, Q_{r1}, (during the hold period), kJ	13.5	0.5	9.8	3.3
Adiabatic temperature rise, ΔT_{ad}, (Q_{r1}/mC_p), (for semi batch reaction), °C	5.4	0.4	7.7	2.4
Maximum temperature of desired reaction in case of loss of cooling, °C (starting from the process temperature of 20°C)	25.4	20.4	27.7	22.4

Note:
For semi-batch reactions, only the accumulation enthalpy needs to be considered for calculating the adiabatic temperature rise. It is assumed that dosing can be stopped immediately in case of runaway situations. For batch reactions, the total enthalpy needs to be considered for calculating the adiabatic temperature rise.

The thermal stabilities of the reaction intermediates were determined by taking samples at various times of dosing and analyzing the samples by differential thermal analysis calorimetry (DTA).

The onset temperatures of the decomposition reactions are summarized below.

	Nitration Process #1 (70% Nitric Acid Dosing)	**Nitration Process #2** (Acid B Dosing)	**Nitration Process #2** (70% Nitric Acid Dosing)	**Nitration Process #2** (90% Nitric Acid Dosing)
	Onset Temperature, °C			
Starting Solution	223.7	230.9	—	314.6
Sample after 1/3 of dosing	166.2	260.5	171.7	168.0
Sample after 2/3 of dosing	113.3	238.7	209.8	255.1
Sample after 100% of dosing	132.4	223.5	217.0	218.0

<u>*Note:*</u>
High-pressure, gold plated crucibles were used for the DTA runs to make sure that during the run no sample was lost, thus, masking the identity of the sample and distorting the results. Also, gold plated crucibles are corrosion resistant to the chemicals tested.

The thermal data of both the desired and undesired reactions as shown above can now be used to assess the criticality of cooling failure scenarios for a semi-batch reaction. By comparing the temperature levels of process, maximum temperature of the desired reaction, boiling point of the solvent and the onsets of decomposition temperatures, the hazardous nature of the cooling failure scenarios can be determined.[2] The results are summarized below.

Temperature Levels, °C	**Nitration Process #1** (70% Nitric Acid Dosing)	**Nitration Process #2** (Acid B Dosing)	**Nitration Process #2** (70% Nitric Acid Dosing)	**Nitration Process #2** (90% Nitric Acid Dosing)
Process Temperature	20	20	20	20
Maximum Temperature of the Desired Reaction	25	20	28	22
Boiling Point of the Solvent	100		40	40
Onset of Decomposition Reaction Temperature	113	224	172	168
Hazard Index	**1**	**1**	**1**	**1**

For comparison purposes, the hazardous nature of the two nitration systems is shown below, if they are conducted in a batch mode.

Temperature Levels, °C	**Nitration Process #1** (70% Nitric Acid Dosing)	**Nitration Process #2** (Acid B Dosing)	**Nitration Process #2** (70% Nitric Acid Dosing)	**Nitration Process #2** (90% Nitric Acid Dosing)
Process Temperature	20	20	20	20
Maximum Temperature of the Desired	125	25	54	47

Reaction				
Boiling Point of the Solvent	100	40	40	40
Onset of Decomposition Reaction Temperature	113	224	172	168
Hazard Index	**4**	**1**	**3**	**3**

Hazard index of 1 means that, after loss of control of synthesis reaction, the boiling point cannot be reached and the decomposition reaction cannot be triggered. Only if the reaction is maintained for a long time under heat accumulation conditions could the boiling point be reached. The evaporative cooling then serves as an additional safety barrier. The process is thermally safe. No special measure is required.

Hazard index of 2 means that, after loss of control of synthesis reaction, the boiling point cannot be reached and the decomposition reaction cannot be triggered. The situation is very similar to Hazard index of 1. Only if the reaction is maintained for a long time under heat accumulation conditions, the decomposition reaction could be triggered and reach the boiling point. In this case, attainment of the boiling point could be a hazard if the boiling rate is too high. If the reaction mass is not kept for a lengthy period of time under heat accumulation conditions, the process is thermally safe. No special measure is required.

Hazard index of 3 means that, after loss of the control of the desired reaction, the boiling point will be reached, but the decomposition reaction cannot be triggered. In this situation, the safety of the process depends on the heat release rate of the desired reaction at the boiling point. Consider using a backup cooling system, dumping of the reaction mass or quenching.

Hazard index of 4 means that, after loss of control of the desired reaction, the boiling point will be reached and the decomposition reaction theoretically can be triggered. In this situation, the safety of the process depends on the heat release rates of both the desired and decomposition reactions at the boiling point. The evaporative cooling may serve as a safety barrier in open systems. Consider using a backup cooling system, dumping of the reaction mass or quenching.

Hazard index of 5 means that, after loss of control of synthesis reaction, the decomposition reaction will be triggered and the boiling point will be reached during the runaway of the decomposition reaction. It is very unlikely that the evaporative cooling can serve as a safety barrier in this case. The heat release rate of the decomposition at the boiling point determines the thermal safety of the process. It is the most critical of all scenarios. It is worthwhile to consider an alternative design of the process in order to reduce the severity or the probability. The following possibilities should be considered; lower concentration, change from batch to semi-batch, optimize semi-batch conditions to minimize the accumulation, etc.

Based on the various hazard classes discussed above, the experimental results show that if the nitration processes are run in a batch mode as opposed to a semi-batch mode, hazardous conditions exist for both the nitration systems. Nitration system #1 is more hazardous than the nitration system #2; however, both of the systems should be redesigned from a safety point of view to a semi-batch mode.

5. Conclusions And Recommendations

Both the nitration routes if run in a semi-batch mode, are thermally safe. There are virtually no differences in the thermal characteristics of the two processes studied. The severity describing the adiabatic profiles of the two routes are both low. The adiabatic profile represents the worst-case scenario. Since risk is defined as (severity × probability), the thermal risks associated with either of the routes are low. The choice between the alternate routes should be based on economy, environmental considerations and not on safety.

In this study, no differences were found from a thermal process safety point of view for the two routes; however, this study clearly shows the importance of studying safety in the early stages of process development. In doing so, the chance of an inherently safe process can be maximized. In an inherently safe process, the severity is reduced, thereby eliminating or reducing the threat of an incident. This approach is preferred in comparison to taking measures to reduce the probability, which requires long-term measures. The approach of including safety studies in the early stages of process development results in an integrated process development. The process will be designed for normal operating conditions, but also for deviations from these optimal conditions. The result is a truly optimized process, since more aspects are considered and a more robust process towards deviations. The development costs will reflect the benefits on a long-term scale during the life of the process.

6. References

[1] Expert Commission for Safety in the Swiss Chemical Industry, "Thermal Process Safety", Booklet #8, Lucerne, Switzerland, 1993. (Copies of the booklet can be obtained from Abteilung Arbeitssicherheit, SUVA, Fluhmattstrasse 1, CH-6002 Lucerne, Switzerland)

[2] Stoessel, F., "What is Your Thermal Risk?", *Chemical Engineering Progress,* Vol. 89, No. 10, pp. 68-75, October 1993.

Paper TG005 - 5

Rapid Identification Of Reactivity Hazards In A Multi-Use Facility

David J. Leggett
Baker Engineering and Risk Consultants, Inc.

Prepared for Presentation at
American Institute of Chemical Engineers
2005 Spring National Meeting
39th Annual Loss Prevention Symposium
Atlanta, Georgia
April 11-13, 2005

February, 2005

UNPUBLISHED

Rapid Identification Of Reactivity Hazards In A Multi-Use Facility

David J. Leggett
Baker Engineering and Risk Consultants, Inc.

Abstract

Multi-use pilot scale plants often run production campaigns in parallel, requiring frequent changes in the chemistry from campaign to campaign and modified procedures to match the new production needs within a short time span. The facility must also match the client's product chemistry and process demands with their equipment capabilities.

These campaigns are likely to occur several times a year with considerable overlap of projects. Several basic but important questions arise for the facility manager. For example, how do the facility personnel:

- Pay adequate attention to all of the safety issues associated with the chemical reactivity?
- Ensure that the facility's equipment and the new process chemistry fit together safely?
- Verify emergency relief and vent sizing, estimate cooling duty needs, and implement the necessary process safety changes?

This presentation introduces a new approach and procedure that assesses each unit's operation of the batch record, identifies potential hazards, and recommends a methodology to evaluate the nature and extent of the hazard. A user-friendly software tool has been developed and is compared with other hazard analysis tools in this presentation.

1. Chemical Reaction Hazard Identification And Assessment

1.1 Introduction

It's Friday afternoon and the phone rings in the Plant Manager's office of Notlob Chemical Inc (NCI), a well-known contract manufacturing facility. The Premier Chemical Company (PCC) wants Notlob to manufacture 100,000 lbs of their new rat poison within the next six months. Dennis Bloodknock, never being one to pass up a challenge, accepts the campaign and the rush is on. A number of predictable outcomes could occur, ranging from a successful completion to tragic failure.

Contract (toll) manufacturers, such as NCI, are required to produce chemical products under less than ideal conditions. NCI must move from project to project: rapidly learning new chemistries; modifying their procedures to fit the next client's needs; matching the technology transfer package with NCI's equipment capabilities; and manufacturing the requested material(s) within a short time span.

Successful navigation through this sequence of events requires reliable hazard information for the target chemical process.

Obtaining reliable estimates of the nature and extent of chemically related hazards in an operation involving the storing, handling, or reactions of chemicals from a variety of readily available sources is possible in a reasonable timeframe. Viewed as a whole, this data gathering exercise provides more than just clues to the risk potential of the new process.

Important data about the process may be found from many sources, such as Material Safety Data (MSD) sheets and the open literature, molecular functional groups, chemical compatibility considerations, thermokinetic information, consequences of planned and unplanned mixing, checklist assessment and calculational tools, and finally hazards testing.

A systematic approach to analyzing the safety of a particular process involves recognizing the various aspects of a manufacturing process that are opportunities for loss of control of that process. They may be summarized as:

- the chemistry and unit operations that together yield the desired result;
- upsets in the chemistry that could lead to an undesired result;
- upsets in the unit operations that could lead to an undesired result;
- upsets in the chemistry, or in a unit operation, that could trigger an upset in the other.

Reviewing each of these general failure scenarios could reveal a serious problem that may occur due to undesired reactivity and/or loss of process control. Typically, information required for this analysis includes knowledge of:

- the desired reactions for the chemicals involved;
- the side reactions that lead to yield loss;
- undesired reactions between reactants, process chemicals and materials;
- reactions that can occur in storage.

This data gathering exercise should be done even if the operation is only blending or mixing. Several serious accidents, including fatalities, have occurred because the consequence of reactions between chemicals involved in blending operations were not known in advance.

1.2 Information Sources for Chemical Reactivity

1.2.1 Material Safety Data Sheets

MSD sheets do not supply the level of information required to perform an information-based hazard analysis and should not be used as the sole source of process safety information. However, MSD sheets do provide guidance on toxicity, physical properties, stability and reactivity, fire fighting and spill management, safety conditions required, and specific chemical interactions to avoid.

1.2.2 Open Literature

There are many excellent sources of information that can provide assistance at all stages of a hazard assessment, such as Bretherick's Handbook[1] for chemical reactivity, Sax[2] for general chemical hazards, and NFPA[3] for fire protection from hazardous materials. The Center for Chemical Process Safety (CCPS) has produced two books[4,5] specifically on managing chemical process safety of reactive materials. These resources are frequently used for the initial assessment.

As the process hazard evaluation proceeds, specialized texts on dust explosions[6], emergency relief design[7], managing runaway reactions[8], and self-heating in solid materials[9] become important. CCPS has at least 40 specialist monographs on a wide range of process safety subjects.

1.2.3 Functional Groups, Chemical Identity, and Reaction Hazard Potential

The structural features of the chemical molecule (bond groupings) are indicative, often strongly, of instability or high reactivity within the molecule. Trends among the constituent atoms of these molecular groupings show, for example, that groups of the non-carbon atoms, such as -N-O- or -N-N≡N-, -NO2, or multiple bonded carbon atoms may be hazardous.

Reactions may be classified as having a high hazard potential for one or more of the following reasons:

- Reactants or products contain functional groups or atom groupings that confer explosive properties;
- The reaction rate becomes explosive if cooling, agitation, addition controllers, etc., fail;
- The reaction has a long and unpredictable initiation period and is strongly exothermic;
- Reactants or reagents are highly toxic or have very low personal exposure limits;
- Reactants react vigorously/explosively with air or moisture, under ambient or near-ambient conditions;
- Reactants or reagents, in particular catalysts, are pyrophoric;
- Side reactions can over-run the main reaction leading to thermal explosion or detonation;
- Product workups involve highly exothermic quenches.

Drawing the line between highly hazardous and moderately hazardous is, to a large degree, linked with the experience that lab personnel have with that particular chemical process. A synthesis that involves the introduction of a functional group known to confer explosibilty, and has only been performed a few times, should be considered as non-routine (Table 1, section 11) and identified as requiring a formal hazards review. [10]

1.2.4 Checklist Tools

Checklists are a straightforward way of reviewing the potential hazards of a chemical manufacturing process. They provide the inexperienced reviewer with a wealth of questions to ask about the intended process, or unit operations, as well as the causes, consequences, and probabilities of an upset condition arising. There are many checklists available.

1.2.5 Chemical Compatibility Charts

Compatibility charts provide a quick reference guide to the consequences of inadvertent mixing and provide an excellent training tool for operators focusing on the areas of the operation that may have a high hazard. The U.S. Coast Guard[11] has a widely used compatibility (inter-reactivity) chart that provides information for about 40 chemical classes or mixtures. CAMEO, developed jointly by the EPA and NOAA, provides on-line chemical compatibility information. Full details are available from the CAMEO website.[12]

1.2.6 Calculational Tools

A hazard analysis based on fundamental parameters requires quantitative information about both the thermodynamics and kinetics of the key reactions. There are some straightforward calculational tools that can estimate these parameters well enough for hazard *assessment* purposes. CHETAH [13] allows estimates of formation heats of molecules in the gas phase to be made. These calculations also lead to reaction heats, estimates of the plosive characteristics of molecules and reactions, and adiabatic temperature rises for reacting mixtures.

If differential scanning calorimetry (DSC) data have been obtained for a pure material or a reaction mixture, several thermal stability indicators[14] may be estimated from the data. These include adiabatic decomposition temperature rise, ΔT_{ADIAB}; critical half thickness, a; critical temperature, T_C; impact sensitivity, SS; explosion potential, EP; time-to-thermal-runaway, t_{NR}; instantaneous power density, IPD; and NFPA instability rating, IR.

1.2.6.1 Initial Screening Calculations

In the absence of real hazards test data, a number of numerical, semi-quantitative methods are available to assist in the assessment of chemical stability. The information provided by these methods should be used to determine the severity of the risk that could be incurred by proceeding without knowledge of reliable values of E_a, A, and ΔH (activation energy and pre-exponential factor in Arrhenius equation, heat of reaction).

Oxygen balance: The oxygen balance (OB) of a compound is the difference between the oxygen content of the compound and that which is required to fully oxidize the carbon, hydrogen and other oxidizable elements present to carbon dioxide, water, etc. It is known empirically that explosives with an oxygen balance close to zero are the most powerful having maximum energy release at this equivalence point. A conservative use of this approach is to be concerned about molecules that show a positive OB, however, concern should also be given to molecules that have a negative OB.

Reaction Heat Estimation: Hess' Law can be used to estimate the energy involved in a reaction, $\Delta H_{reaction}$, if the enthalpies of formation of the reactants and products are known. The theoretical methods include those that use bond contributions, for example average bond summation methods, and ones that use group contributions, (e.g., CHETAH).

1.2.6.2 *Using Adiabatic Information to Derive Specific Engineering Quantities*

Temperature of No Return, T_{NR}: The temperature of no return is the maximum temperature above which the heat generated by the reaction exceeds the cooling capacity available. Therefore, T_{NR} is an extrinsic property depending on the state and properties of the vessel such as the cooling capacity of the vessel, its size and geometry, fill level, rate of agitation, etc. This variable is frequently used to assess the safety of storage tanks and drums where cooling is by natural convection to air. However, if hazards testing has not been performed on the vessel contents, none of these calculations can be performed and "fire-fighting" efforts will be largely a matter of guesswork.

Self-Accelerating Decomposition Temperature, T_{SADT}: This parameter is a measure of the maximum ambient (air) temperature that can be tolerated without thermal runaway, for a selected operating temperature. SADT is influenced by the size of the storage container, similar to T_{NR}. It is a parameter that specifically addresses the storage of potentially thermally unstable materials.

1.2.7 Transforming Information into the Process Hazard Potential

The information gathered from the sources and techniques described in sections 1.2.1 through 1.2.6 are assessed as a whole and documented. Table 1 illustrates one way that the documentation may be accomplished. The table is also used as a decision aid in determining the need to perform a more advanced hazard evaluation. The major part of the hazard review table is outlined in the following sections.

Sections 1 to 5, 10, 13 to 15. Documentation of Hazard Review: These sections record basic information for documentation purposes, including the names of chemicals to be handled.

Sections 6 and 7. Physical Properties of the Chemicals: The physical properties of each chemical represent the basic information that is used for this initial hazard assessment. If all information requested in sections 6 and 7 is available, it will be possible to obtain a reasonable overview of the potential hazards. Realistically, not all the data will be present. The melting and boiling points, flash point and flammable limits of the material constitute a minimum set of physical properties to establish the conditions for safe operations.

Sections 8. Chemical Overview, Mostly from MSD Sheets: This section gathers specific information about each chemical used in the synthesis. Several of the consequences of mishandling common solvents can be assessed from the flash point, the lower and upper flammable limits (LFL and UFL) and minimum ignition energy (MIE).

Section 11: Classification of Process as Routine or Non-Routine: The final step in the process compares the gathered data with the company-defined safety boundaries and risk tolerance, in the form of a Window of Safe Operations (WSO) for the lab. For example, section 11 of the Review Form (Table 1) shows a typical set of flammability, temperature and pressure limits for the reaction conditions that define a typical WSO.

Information provided in sections 11 and 12 help lab personnel decide if any of the proposed operational parameters are outside of the limits that have been pre-defined as constituting a "routine" synthesis. If one or more of these predetermined boundaries are crossed, the synthesis is classified as non-routine and an additional detailed review is required before any laboratory work is started.

1.2.8 A Word of Caution

This approach does not explicitly address the synthesis reaction being performed, except by considering the type of reaction (esterification, nitration, Friedel Craft, etc.) being performed. Evaluation of the thermal hazards of a specific reaction requires a more detailed investigation of the desired reaction(s) and any undesired reactions, together with an analysis of possible equipment upsets or procedural short-comings.[15,16,17,18]

2. Choosing the Appropriate Process Chemistry Hazards Testing

The preceding assessment approach focuses on obtaining a hazard assessment without gathering new experimental hazard test data. However, the assessment may highlight a potentially hazardous situation but without any indication of the severity of the upset. Quantifying the energy and power output resulting from an upset condition or the desired reaction requires specific quantitative data, as detailed in the previous section. If the data is not available, it can be obtained by performing the appropriate hazards testing.

The three types of hazards detailed earlier are thermal stability hazards, chemical reactivity hazards, and operational hazards, as shown in Table 2. Choosing the correct hazard testing to perform is dictated by the information required, the type of process being studied and the chemicals involved. The following sections describe a tool that provides guidance and assistance to answer the questions: "What information is needed to define the nature and extent of the potential hazard?" and "What hazard testing will provide the data necessary to answer this question?"

2.1 Introduction

Consider the chemical reaction that is the basis of a chemical manufacturing process:

$$A_{SOLUTION} + B_{LIQUID} \xrightarrow[85^\circ C]{} \text{Crude Product, AB} \xrightarrow[\text{Vacuum Distill}]{} \xrightarrow[\text{Recrystalize}]{} \text{Pure Product, AB}$$

This chemical description of the process summarizes the chemical reactions and the key condition(s) for reaction. We have already noted that the nature and extent of the reactivity hazard depends upon the specific chemical reaction and the reaction conditions. However, the reaction as written includes a number of engineering steps, each one of which could be hazardous. For example, the use of a solution of reactant A, $A_{SOLUTION}$, requires that the solution is prepared as part of the manufacturing process and will probably raise solvent handling concerns.

Therefore, looking at the chemical reaction from an engineering perspective the manufacture of "Pure Product" comprises the following steps:

1. Dissolve A_{SOLID} in solvent (DMF) using Day Tank DT-101 (Process Steps 1a to 1d)
2. Add B_{LIQUID} to reactor R-101 and heat to 85 °C (Process Step 2)

3. Add $A_{SOLUTION}$ to R-101 over 2 hours (Process Steps 3a – 3c)
4. Stir out reaction mass at 120 °C for 4 hours to form "Crude Product" (Process Step 4)
5. Remove excess solvent (DMF) by vacuum distillation and re-crystallize to "Pure Product"

The first step may be broken down further to the level of the batch record instruction, one action per step:

1. Dissolve A_{SOLID} in solvent (DMF) using Day Tank DT-101
    a. Add solvent (DMF) rapidly to DT-101
    b. Turn on agitator
    c. Add A_{SOLID} slowly from sacks to DT-101
    d. Agitate slurry rapidly until dissolved

A similar detailing of the other main steps results in the complete batch record for the manufacturing process.

2.2 Deriving Hazard Testing Recommendations from the Batch Record Step

This section describes the software package, PHAROS,[‡] developed to support engineers and chemists in deciding if a need exists for process hazard information and in the selection of appropriate hazard testing. The basis of the approach used in PHAROS is the analysis of each step of the batch record and each word within each step. The software recognizes each chemical component, including solvents, the action performed, and the context of the action. Recommendations are made for what hazard assessment information is needed for each step, and the operational, reactivity or thermal stability hazards testing data that may be used to provide the necessary data. The recommendation is in the form of one or more tests; an explanation for the recommendation is also provided.

PHAROS analyzes each word of each batch step systematically because each step is systematically constructed as a sentence of the following form:

Verb + Noun + Adverb +Preposition + Agent + Preposition + Agent

The "sentence " is assembled by selecting "words" from a list of verbs, nouns, adverbs, etc., in a definite order. Some of the verbs, nouns adverbs, prepositions, and agents are:

Verb	Noun	Adverb	Preposition	Agents
Add	Pressure	Immediately	From	Day Tank 1
Agitate	Reactant 1	Quickly	Into	Pump
Pump	Slurry	Slowly	Under	Sack
Transfer	Solvent 1	Sub-Surface	Using	Weigh Scale

[‡] Copyright © Baker Engineering and Risk Consultants, Inc., 2003 – 2005.

For example, the sentences for steps 1 to 4 constructed from the menus of verbs, nouns, etc. are shown below:

Step	Verb	Noun	Adverb	Preposition	Agent	Preposition	Agent
1a	Pump	Solvent1		From	T-1011	To	DT-101
1b	Start	Agitator					
1c	Add	Reactant 1	Slowly	From	Sacks	To	DT-101
1d	Agitate	Slurry	Continuously	Until	Dissolved		
2	Pump	Reactant 2	Continuously	From	Drums	To	R-101
3a	Start	Agitator					
3b	Heat	Reactant 2	Continuously	Using	Heat Exchanger	To	Temperature
3c	Add	Solution	Continuously	From	DT-101	To	R-101
4	Agitate	Reaction Mass	Continuously	At	Constant T		

The software recognizes that Solvent 1 is DMF, that Reactant 1 is a specific chemical, and uses the supplied physical properties data to distinguish between solids, liquids, and gases. The hazards data required for each chemical, and operation involving that chemical, is therefore guided in part by the physical properties data. Information is also entered about the quantity and addition rate of chemicals, agitator speed, and heat-up rates.

As each process step is added the list of recommended tests increases. PHAROS reviews the total set of recommended tests to the point of removing redundancies and adding hazard data and test recommendations. Additional information requirements and test recommendation(s) are added if the software detects that a mixture of chemicals has occurred in a vessel. The system has been designed to produce a conservative set of recommended tests. For example, every use of the noun "reactant" recommends that thermal stability testing (DSC, Adiabatic Calorimetry) is needed. If the material is a solid or liquid, impact and friction sensitivity is also recommended. The recommended test for each of the process steps are shown in Table 3.

Once all process steps have been entered the third part of the software is run, which accesses a large set of rules that address situations where an earlier step may have a consequence to a later step. An example would be the solution of a reactant, dissolved in a solvent in a previous step, is added to a reactor containing another reactant. Reaction calorimetry (Table 4: Entry t) is needed to assess the potential hazard of step 3c but that does not become apparent until the software examines both steps 2 and 3c and detects that step 3c is the actual reaction step. Step 4, initially flagged for reaction calorimetry, is eliminated because this is a stir-out step. Adiabatic calorimetry is prudent at this point to assess the thermal stability of the crude reaction mass. The final report, produced by this phase of the process, describes the type of hazard test information needed based on the complete process and the processing conditions, and recommends the testing that can provide the necessary data.

The final report for steps 1 through 4 is shown in Table 4. The content of the table should be viewed as information requirements that may be satisfied by reference to the open literature, especially in the case of flammability data. The data needs are not equally pressing. For example, it is normal to conduct the operational testing (Table 4:a – f, j - m) in stages. Dust flammability (Table 4:a and b) is usually obtained experimentally; liquid flammability (Table 4:c and m) may often be available in the open literature; MIT, AIT, MOC and charge accumulation are normally performed once the results of a, b, c, and m are known. A detailed examination of the material handling operations for this process did not show a need for friction or sensitivity testing. However, PHAROS suggested that the solids addition to the reactor be reviewed for the static charge accumulation during the pouring-from-sacks operation.

The amount of thermal stability testing was reduced, in this instance, by electing not to perform DSC testing. If both DSC and adiabatic calorimetry are available it is usually reasonable to perform the more comprehensive adiabatic testing only. However, that decision is made on a case-by-case basis; PHAROS will always suggest that both be performed. The reaction mass and crude product are tested using thermal inertia adiabatic calorimetry (Table 4:r and s) to provide data that may be used directly to size, or confirm the sizing of, the emergency relief system of the reactor.

3. Summary

This paper has reviewed a variety of preliminary hazard evaluation procedures that may be used to evaluate rapidly the potential hazard of a chemical manufacturing process. It is often the case that, after this preliminary hazard evaluation, questions remain as to the consequences of a process upset (loss of cooling, agitation, feed pump controllers, etc.). The software system PHAROS has been developed to aid chemists and chemical engineers to determine the hazard potential of each process step by answering the questions: "What information is needed to define the nature and extent of the potential hazard?" and "What hazard testing will provide the data necessary to answer this question?"

4. References

[1] Urben, P.G., *Bretherick's Handbook of Reactive Chemical Hazards*, Vol 1 & 2, 6th Edition, Butterworth Heinemann (1999).

[2] Sax, N. I., *Dangerous Properties of Industrial Materials*, 7th Edition., Van Nostrand Reinhold: New York (1987).

[3] National Fire Protection Association, *Fire Protection Guide to Hazardous Materials*, 10th Edition, NFPA (1991). (Complete text of NPFA 49, 325M, 491M, and 704).

[4] CCPS, *Guidelines for Safe Storage and Handling of Reactive Materials*, American Institute of Chemical Engineers, New York, 1995.

[5] CCPS, *Guidelines for Design Solutions for Process Equipment Failures*, American Institute of Chemical Engineers, New York, 1997.

[6] Bartknecht, W., *Dust Explosions Cause Prevention Protection*, Springer-Verlag, 1989.

[7] Fisher, H.G. et al, *Emergency Relief System Design Using DIERS Technology*, The Design Institute For Emergency Relief Systems (DIERS) Project Manual, American Institute of Chemical Engineers, 1992.

[8] CCPS, *International Symposium on Many Aspects of Runaway Reactions*, American Institute of Chemical Engineers, New York, 1990 – 2002.

[9] Bowes, P. C., *Self-heating: Evaluating and Controlling the Hazards*, 1984.

[10] CCPS, *Guidelines for Hazard Evaluation Procedures*, 2nd Edition, AIChE (1991).

[11] 46CFR150 – Part 150 – Compatibility of Cargoes (COC) with useful background information provided by the Department of Energy at: http://www.eh.doe.gov/chem_safety//chem_comp.html.

[12] CAMEO website http://www.epa.gov/ceppo/cameo/what.htm.

[13] *CHETAH: Computer Program for Chemical Thermodynamics and Energy Release Evaluation,* Version 7.3. Developed by ASTM subcommittee E27.07. See also: http://www.chetah.usouthal.edu/index.html and http://www.astm.org/COMMIT/CUSTOM1/E27.htm.

[14] ASTM Standard "E1231-01 Standard Practice for Calculation of Hazard Potential Figures-of-Merit for Thermally Unstable Materials." Developed by subcommittee E27.02.

[15] Leggett, D.J., "Runaway Reactions: Ignore the Chemistry at Your Peril," *Chemical Engineering*, 107, #8, 78 (August, 2000).

[16] Barton, J. and R. Rogers, *Chemical Reaction Hazards: A Guide to Safety*, 2nd Edition, Gulf Publishing (1997).

[17] CCPS, *Guidelines for Chemical Reactivity Evaluation and Application to Process Design,* AIChE (1995).

[18] Grewer, T., *Thermal Hazards of Chemical Reactions In Industrial Safety Series,* Vol 4, Elsevier (1994).

Table 1. Preliminary R&D Reactive Chemicals Hazard Review Form

1 Run Ref #	________/______	**2. Synthesis Type:**	________________	**3. Date**	____
4. Description:	____________________________		**5. End Use:**	____________________________	

7. Physical Properties	**6. Chemicals to be used – Sample, Reactants, Solvents**		
B Pt / M Pt / VP @ 20 °C	____/____/____	____/____/____	____/____/____
Fl Pt / AI / MIE	____/____/____	____/____/____	____/____/____
LEL / UEL	____/____	____/____	____/____
Dust Explosion Severity			
8. Chemical Overview			
Stability/Reactivity			
Toxicity/Exposure			
NFPA/HMIS Ratings			
Incompatibilities			
Spill Clean-up Material			
PPE Required			
Specific Hazard (Note 1)			
9. Function in synthesis			

10. MDS Sheet Reviewed	Signed: ____________________	Date: ______________________

11. Classified as a Non-Routine Operation (check relevant categories)			
High toxicity / Low exposure limits	☐	Chemical Handled (Air/water sensitive, Corrosive)	☐
Flammability (Fl Pt<10 °C; LFL<10%)	☐	Extreme Reaction Temperature (> 150 °C; < -30 °C)	☐
Flammability (AI<200 °C; MIE < 0.5mJ)	☐	High Temperature Feed (> 50 °C)	☐
High Hazard Reaction or Functional Group	☐	High Pressure Reaction (>10 bar)	☐
High/moderate hazard reaction performed ≤ 2 times	☐	High Pressure Feed (>2 bar)	☐

12. Potential Hazard Level	Routine: Y ☐ N ☐	Non-Routine: Y☐ N☐	
13. Synthesis by	Name: ____________	Sign: ______________	Date: ________
14. Hazard Review by	Name: ____________	Sign: ______________	Date: ________
15. Non-Routine Hazard Review	Schedule Date : ______________	Date Completed : ____________	

Note 1: Enter as many codes as needed to describe the Special Hazard: RDX – Oxidizer/Reducer; PYR – Pyrophoric; POL – Polymerizes; EXP – Reacts explosively; HS – Sensitive to heat; WAT – Water reactive; PER – Peroxide former; INH – Inhibitor required; IMP – Impact/ friction sensitive; TCN – Temperature control needed; GRP – Functional group

Instructions for Use of Form

Use this form as a checklist and review of potential hazards involved in the synthesis. Each section requires responses, as detailed below: **1**. Reference number **2**. Technique **3**. Date filled out **4**. Describe Synthesis **5**. Product end-use **6**. List each chemical used in the run – use a new form as needed; **7**. Physical properties – use MSD sheet, Saxs, Merck Index and "Sources of Ignition" (Bond) **8**. Summarize listed properties as available from reference materials, mostly MSD sheets **9**. Function of each chemical in the synthesis, ie reactant, solvent, catalyst, etc. **10**. MSDS reviewer, when; **11**. What led to a Non-Routine classification? Check all appropriate boxes **12**. What is the assessed potential hazard – Routine or Non-Routine **13**. Who will perform the synthesis; when? **14**.This hazard review; when? **15**. If Non-Routine, when is detailed hazard review scheduled; when completed? Non-Routine hazard review notes must accompany this form with risk reduction recommendations activities as needed.

Table 2. Relating Hazard Testing Technique with Type of Hazard

Thermal Stability Hazards	
Scenarios	*Typical Calorimetric Technique*
Self -reactivity Inappropriate storage conditions Runaway Reactions / Thermal Explosions Loss of utilities leading to loss of cooling, loss of agitation	Differential Screening Calorimetry Adiabatic Calorimetry Dewar Calorimetry
Chemical Reactivity Hazards	
Scenarios	*Typical Calorimetric Technique*
Water and/or air reactivity Inappropriate materials of construction Wrong chemicals added Wrong addition rates, or order of addition Hypergolic and pyrophoric properties of the reactants or products	(Mixing) Isothermal Calorimetry Heat Flow Reaction Calorimetry Iso-Peribolic Calorimetry
Operational Hazards	
Scenarios	*Typical Calorimetric Technique*
Flammability dusts, vapors and gases Mechanically induced hazards Static induced hazards	20-Liter Sphere Dust Explosion Severity MIE, MIT, AIT, COC, LFL, UFL, Layer Ignition Bulk Resistivity / Charge Relaxation Impact and Friction Sensitivity

Table 3. Recommended Tests Generated by PHAROS

Recommended Test	Process Step									Data Generated / Purpose of Test Data
	1a	1b	1c	1d	2	3a	3b	3c	4	
DSC	•		•		•		•	•	•	T_{ONSET}, T_{MAX}, T_{END}, $\Delta H_{ESTIMATE}$
Adiabatic Calorimetry, high ϕ	•		•		•		•	•	•	T_{ONSET}, T_{MAX}, T_{END}, ΔH_{DECOMP} $(T,P)_{TIME}$, $(dP/dt, dT/dt)_{TIME,TEMPERATURE}$,
Adiabatic Calorimetry, low ϕ	•		•		•		•	•	•	T_{ONSET}, T_{MAX}, T_{END}, ΔH_{DECOMP}, ERS Design, $(T,P)_{TIME}$, $(dP/dt, dT/dt)_{TIME,TEMPERATURE}$,
Isothermal Mixing Calorimetry			•					•	•	ΔH_{MIXING}
Reaction Calorimetry								•	•	$\Delta H_{REACTION}$, Power, Energy, Accumulation
MIE	•									Minimum energy required to ignite a dust, vapor or gas cloud
DE, 20 liter			•							Dust Explosion (20 liter) to determine severity of a dust explosion
LFL, UFL	•				•					Lower % of vapor (LFL) and Upper that will ignite in air
MIT	•									Minimum temperature required to ignite a dust, vapor or gas cloud
AIT	•									Temperature for spontaneous combustion of material
MOC	•									Minimum oxygen concentration required to sustain combustion
Charge Accumulation			•		•					Propensity of solids to accumulate and hold static charge during handling
Friction Sensitivity			•		•					Need to determine if any solids or liquids can detonate when processed mechanically
Impact Sensitivity			•		•					Need to determine if any solids or liquids can detonate when processed mechanically

Table 4. Final Set of Data Requirements and Recommended Process Hazard Tests

Recommended Test	Process Step									Data Generated / Purpose of Test Data
	1a	1b	1c	1d	2	3a	3b	3c	4	
DSC			g				p			T_{ONSET}, T_{MAX}, T_{END}, $\Delta H_{ESTIMATE}$
Adiabatic Calorimetry, high ϕ			h				q			T_{ONSET}, T_{MAX}, T_{END}, $(T,P)_{TIME}$, $(dP/dt, dT/dt)_{TIME,TEMPERATURE}$, ΔH_{DECOMP}
Adiabatic Calorimetry, low ϕ								r	s	T_{ONSET}, T_{MAX}, T_{END}, ΔH_{DECOMP}, ERS Design, $(T,P)_{TIME}$, $(dP/dt, dT/dt)_{TIME,TEMPERATURE}$,
Isothermal Mixing Calorimetry			i							ΔH_{MIXING}
Reaction Calorimetry								t		$\Delta H_{REACTION}$, Power, Energy, Accumulation
MIE	a									Minimum energy required to ignite a dust, vapor or gas cloud
DE, 20 liter			b							Dust Explosion (20 liter) to determine severity of a dust explosion
LFL, UFL	c				m					Lower % of vapor (LFL) and Upper that will ignite in air
MIT	d									Minimum temperature required to ignite a dust, vapor or gas cloud
AIT	e									Temperature for spontaneous combustion of material
MOC	f									Minimum oxygen concentration required to sustain combustion
Charge Accumulation			j							Propensity of solids to accumulate and hold static charge during handling
Friction Sensitivity			k		n					Need to determine if any solids or liquids can detonate when processed mechanically
Impact Sensitivity			l		o					Need to determine if any solids or liquids can detonate when processed mechanically

Panel Discussion on Chemical Reactivity Hazards Issues

Robert W. Johnson
Unwin Company
1920 Northwest Blvd., Ste. 201
Columbus, OH 43212-1197

Brian R. Dunbobbin
Air Products & Chemicals, Inc.
7201 Hamilton Boulevard
Allentown, PA 18195-1501

Prepared for Presentation at
American Institute of Chemical Engineers
2005 Spring National Meeting
39th Annual Loss Prevention Symposium
Atlanta, Georgia
April 11-13, 2005

UNPUBLISHED

Panel Discussion on Chemical Reactivity Hazards Issues

Robert W. Johnson
Unwin Company
1920 Northwest Blvd., Ste. 201
Columbus, OH 43212-1197

Brian R. Dunbobbin
Air Products & Chemicals, Inc.
7201 Hamilton Boulevard
Allentown, PA 18195-1501

Abstract

This session will conclude with a panel discussion addressing current issues related to the control of chemical reactivity hazards.

Topics to be addressed may include the on-going efforts of the CCPS Reactivity Management Roundtable.

Audience perspectives on this important issue facing our industry will be solicited.

Explosions In The Manufacturing Industry:
Data from the Hazardous Substances Emergency Events Surveillance (HSEES), 1996–2001

Z. Berkowitz, D.K. Horton, and W.E. Kaye
Agency for Toxic Substances and Disease Registry
2400 Century Center Blvd., 3rd floor
Atlanta, GA USA, 30345
tel. 404-498-0574, fax 404-498-0077
Email: zab3@cdc.gov

Prepared for Presentation at
American Institute of Chemical Engineers
2005 Spring National Meeting
39th Annual Loss Prevention Symposium
Atlanta, Georgia
April 11-13, 2005

UNPUBLISHED

Explosions In The Manufacturing Industry:
Data from the Hazardous Substances Emergency Events Surveillance (HSEES), 1996—2001

Z. Berkowitz, D.K. Horton, and W.E. Kaye
Agency for Toxic Substances and Disease Registry
2400 Century Center Blvd., 3rd floor
Atlanta, GA USA, 30345
tel. 404-498-0574, fax 404-498-0077
Email: zab3@cdc.gov

Abstract

Introduction: Of all manufacturing industry events reported to the Hazardous Substances Emergency Events Surveillance (HSEES) system, acute releases involving explosions accounted for the largest proportion of events with victims. This analysis describes the public health effects of these releases.

Methods: Data reported to the HSEES system were used for the analysis of events involving explosions, with or without fire, during 1996—2001.

Results: Of 39,766 events reported, 19,674 (49.5%) occurred in the manufacturing industry. Of this latter group, 119 (<1%) events involved explosions, and 66 (55.5%) of these events had victims (n=268). (Victims are defined as persons who sustained injuries or symptoms as a consequence of the event). In comparison, 668 (3.4%) of all other manufacturing events had victims. Equipment failure, operator error, and improper mixing were the leading factors contributing to events; process vessel (29.4% of events with explosions/fire events), storage areas (23.5%), ancillary process equipment (10.9%) and material handling (8.4%) were the areas mostly involved. The leading industries that had events with explosions were industrial and miscellaneous chemicals (26.9%), and those industries that made plastics, synthetics, and resins (6.7%). Ammonia (6.2%) and nitrogen (2.8%) were the top substances released. Thirty-two (11.9%) persons died and 51 (19%) victims were hospitalized for treatment. Trauma (25.5%), thermal burns (17.5%), and respiratory symptoms (12.7%) were the most frequent types of injuries.

Conclusions: To reduce the impact of these events, companies should consider addressing the main factors involved in these releases, re-enforcing regulations, and the institutionalizing action plans and safety measures to protect their employees.

1. Introduction

Explosions have the potential to cause severe consequences. These consequences include loss of life, injuries, evacuation, cleanup, medical expenses for treatment, loss of work, and extensive property and equipment damage. The monetary cost associated with a hazardous material transportation event with explosion, based on accidents (i.e., involving a collision) and incidents (i.e. not involving a collision) in the United States was estimated at $2.1 million for the year 1996.[1] The property damage alone for 16 reactive hazards incidents occurring between 1980 and 2001 that were evaluated from available data by the U.S. Chemical Safety and Hazard Investigation Board (CSB) ranged from $10 million to $120 million. This cost did not include other financial losses.[2] The monetary cost of losses from fire and explosions, and safety provisions to prevent them, for the United Kingdom (UK) is £5 billion, or 1% of the UK's Gross Domestic Product.[3]

Explosions involve the release of hazardous materials. Examples include the following: ignition of flammable gas (e.g. ethanol) by heat (e.g. sparks from welding); an interruption in a chemical process operation in a facility (e.g. burst pipe) due to a severe storm causing an electric power outage; excess heat leading to an overpressured vessel as a result of a solid residual plugging the vessel purge lines and reacting with another chemical in the vessel; and methane buildup in water treatment plant, due to poor design of a digester.

The Hazardous Substances Emergency Events Surveillance (HSEES) system captures many of these events and describes the public health consequences associated with them.

The purpose of this analysis is to describe the characteristics, chemicals released, types of industries, factors contributing to the release, and groups of victims (employees, responders, general public) in events involving the release of hazardous substances because of explosions in the manufacturing industry. In addition, we review some measures to reduce chemical incidents and mitigate exposure to humans.

2. Methods

The HSEES is an active state-based system initiated in 1990 to capture data on acute chemical releases and their associated public health consequences, injuries, and evacuations. Since the year 2000, the HSEES system has functioned as a Web-based system.

This analysis includes events reported to HSEES from January 1996 through December 2001. During this time 17 state health departments participated in the system (Alabama, Colorado, Iowa, Minnesota, Mississippi, Missouri, New York, North Carolina, Oregon, Rhode Island, Texas, Washington, Wisconsin, New Hampshire [1996], Louisiana [2001], New Jersey [2001], Utah [1999-2001]). Data sources included state environmental protection agencies, federal databases, fire and police reports, local emergency planning committees, poison control centers, hospitals, and the media.

Hazardous substance emergency events were defined as illegal or uncontrolled releases (or threatened releases) of at least one hazardous substance. A substance was considered hazardous if it might reasonably have been expected to cause an adverse health effect. These releases were required to be removed, cleaned up or neutralized by federal, state, or local law. Data was not collected on events involving petroleum exclusively.

This analysis concentrates on events where the release of at least one hazardous substance occurred because of an explosion (with or without fire), in the manufacturing industry. Events were identified by the industrial coding scheme administered by the US Bureau of Census (1992). Data were included from industries producing nondurable and durable goods. Events involving the manufacturing of methamphetamine were not included because these activities are illegal.

Substances were categorized into 16 groups:

- Acids
- Ammonia
- Bases
- Chlorine
- Other inorganic substances
- Paints & dyes
- Pesticides
- Polychlorinated biphenyls (PCBs)
- Volatile organic compounds (VOC)
- Other
- Mixture
- Formulations
- Hetero-Organics - organic compounds with nitrogen, sulfur, or phosphorous that do not fall into other, more-precise categories. For example: nitrogen (amines, amides, and azides); sulfur (thios and mercaptans) and phosphorus (phosphorous esters)
- Hydrocarbons
- Oxy-Organics - organic compounds with oxygen that do not fall into other, more-precise categories. For example: alcohols, aldehydes, anhydrides, ethers, glycols, ketones, oxides, peroxides, phenolics (salts and esters), and phthalates
- Polymers

The category "mixture" was assigned to substances from different categories that were mixed prior to the release. The category "other inorganic substances" excluded acids, bases, ammonia, and chlorine. If an event involved multiple substances belonging to the same category (including a mixture of substances that were mixed prior to the release), these substances were counted once in that category. If an event involved multiple substances from different categories, substances were counted once in the multiple categories group.

Victims are defined as those persons who sustained injuries or symptoms or died as a consequence of the event. Since a victim could sustain more than one injury or symptom, each one was counted separately; therefore, the number of reported injuries or symptoms exceeded the number of victims.

3. Results

Of all 39,766 HSEES events reported during 1996 through 2001, 19,674 (49.5%) occurred in the manufacturing industry [Figure 1]. Of this latter group, 119 (<1%) events involved explosions with or without fire, and 66 (55.5%) of these events had victims (n=268). In comparison, 668 (3.4%) of all other manufacturing events had 3673 victims. Almost all explosion events occurred in fixed facilities (n=117, 98.3%).

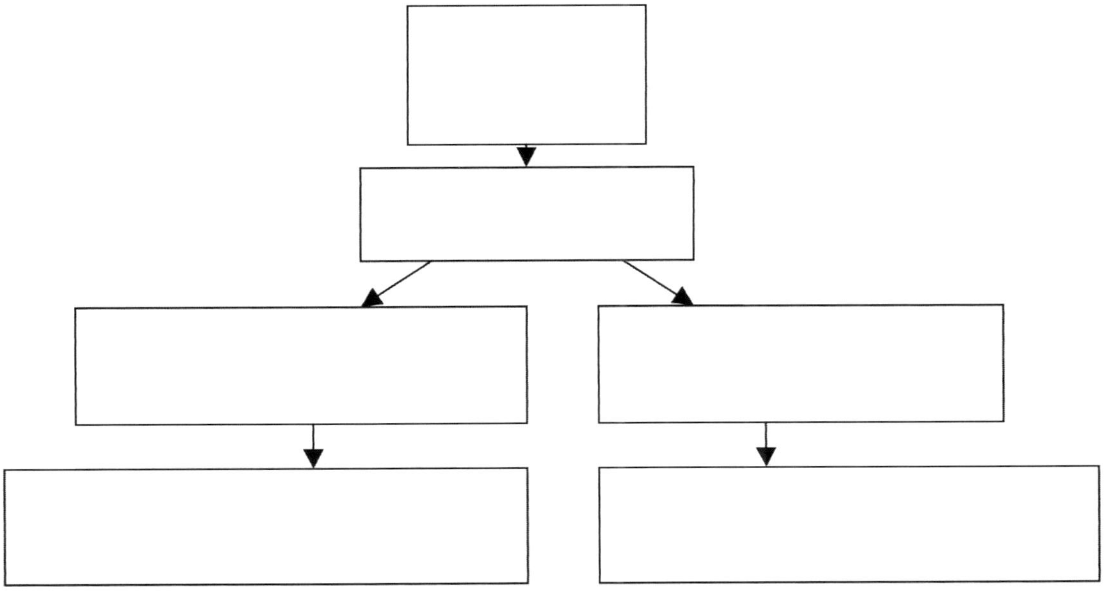

The industries that most frequently had explosion events involved industrial and miscellaneous chemicals (n=32, 26.9%); plastics, synthetics, and resin (n=8, 6.7%); canned, frozen, and preserved fruits and vegetables (n=5, 4.2%); motor vehicles and motor vehicle equipment (n=5, 4.2%); ordnance (n=5, 4.2%); and other primary metal industries (n= 5, 4.2%).

Most events occurred in industrial areas (n=60, 50.8%), next in frequency were commercial areas (n=11, 9.3%) and industrial and commercial areas (n=9, 7.6%). However, some areas were industrial and residential (n=11, 9.3%) or commercial and residential (n=8, 6.8%). More than half of the events (n=70, 58.8%) resulted in releases inside the facility, 42 (35.3%) resulted in releases outside the facility, 5 (4.2%) resulted in releases both inside and outside the facility, and 2 had no information.

The most common primary locations of the events within the facility area were process vessel (n=35, 29.4%), storage above ground (n=28, 23.5%), ancillary process equipment (n=13, 10.9%) and the occurrence of material handling (e.g. loading or unloading) (n=10, 8.4%). Primary factors associated with the events were known for 88 (73.9%) events. Of those events, equipment failure (n=38, 43.2%), operator error (n=20, 22.7%) and improper mixing (n=14, 15.9%) were most frequently reported.

The 119 explosion events involved 188 chemicals of which 177 were released into the environment. The most frequently released substances are shown in Table 1. Of these chemicals, ammonia (n=11, 6.2%) was released most often. In addition to the above most frequently released chemicals, 30 events involved mixtures of chemicals (mostly different from each other).

The category involving multiple substances from different categories had the largest number of explosion events (n=25, 21.0%) [Table 2]. This category also had the largest proportion of events with victims (n=19, 28.8%). Events without explosion were less likely to involve multiple chemicals (2.3%).

Table 1.— The 14 most frequently released substances in explosion events, HSEES 1996—2001.

#	Substance name	Number	%
1	Ammonia	11	6.2
2	Nitrogen NOS*	5	2.8
3	Asbestos	4	2.3
4	Paint or coating NOS*	4	2.3
5	Solvent NOS*	4	2.3
6	Acetylene	3	1.7
7	Black powder	3	1.7
8	Chlorine	3	1.7
9	Explosives	3	1.7
10	Magnesium	3	1.7
11	Nitric acid	3	1.7
12	Styrene	3	1.7
13	Sulfur dioxide	3	1.7
14	Sulfuric acid	3	1.7
		55	**31.1**

* Not otherwise specified

The majority of events (n=91, 76.5%) involved the release of one substance. Two substances were released in 13 (10.9%) events and the remaining events had from 3 to 11 substances released. Of all 177 substances released, the quantity was known for 102 substances. The majority of quantities (n=62, 60.8%) among the 102 substances were reported in pounds and the quantity ranged from 1 lb. to 133,000 lbs., with a median of 289 lbs. The quantity of 35 (34.3%) substances were reported in gallons and ranged from 1 to 18,000 gallons with a median of 100 gallons.

The majority of explosion events with victims involved the release of one substance (n=46, 69.7%). Events with victims involving two substances comprised 8 (12.1%) of the events. More than half of the victims (n=152, 56.7%) sustained injuries in events involving multiple substances from different categories.

Fifteen (12.6%) of the explosion events had fatalities and six events had more than one fatality. In comparison, all other manufacturing events had 18 (<0.1%) events with fatalities and only one had more than one fatality. Of the 15 explosion events involving fatalities, three events had black powder released and two had pyrotechnic chemicals released.

The number of victims in explosion events involving victims ranged from 1 to 90 with a median of 2. Figure 2 displays the severity and disposition of the victims. Most victims were transported to a hospital (n=182, 67.9%), and 51 were admitted. In addition, 32 (11.9%) victims died, either at the scene or after arrival at the hospital. The number of fatalities ranged from 1 to

8 in the 15 events with fatalities. The proportions of victims transported to a hospital and of fatalities in explosion events were much larger than the corresponding proportions in all other manufacturing events (n=1949, 53.1%) of which 251 victims were admitted, and 19 (0.09%) died.

The majority of victims in explosion events (n=227, 84.7%) were employees. The other victims

Table 2.— Frequencies of substance categories in all explosion events and in such events with victims, and the associated numbers of victims and fatalities, HSEES 1996–2001.

	All events		Events with victims		Total victims		
Substance category	**N**	**%**	**N**	**%**	**N**	**Range**	**Fatalities**
Acids	8	6.7	6	9.1	12	1–7	2
Other	16	13.4	9	13.6	27	1–6	9
Mixture	13	10.9	7	10.6	9	1–2	1
Ammonia	8	6.7	4	6.1	9	1–3	0
Bases	1	0.8	0	0.0	0		0
Chlorine	3	2.5	2	3.0	4	2–2	0
Other inorganic substances	14	11.8	4	6.1	9	1–3	2
Paints & dyes	3	2.5	2	3.0	4	2–2	0
Pesticides	4	3.4	2	3.0	13	3–10	10
VOC	13	10.9	6	9.1	18	1–5	4
Multiple categories	25	21.0	19	28.8	152	1–90	3
Hetero-Organics	1	0.8	0	0.0	0		0
Hydrocarbons	1	0.8	0	0.0	0		0
Oxy-Organics	4	3.4	2	3.0	5	1–4	1
Polymers	5	4.2	3	4.5	6	1–3	0
	119	**100.0**	**66**	**100.0**	**268**		**32**

included 7 (2.6%) who were employee members of a company response team, 18 (6.7%) were responders (including career firefighters, police officers, emergency medical technologists and other type of responders), and 16 (6.0%) were members of the general public. Of all 268 victims, the majority were males (n=228, 85.1%). The age of 213 victims was reported and ranged from 2 years to 86 years, with a median of 35 years. Among victims whose age was not reported, 48 were employees, 5 were responders and 2 were members of the general public.

The numbers of the different injuries or symptoms sustained by the victims are shown in Table 3. Trauma was by far the most common injury or symptom reported (n=105, 25.5%) followed in frequency by thermal burns and respiratory irritation. In comparison, the most common injuries or symptoms reported for victims in all other manufacturing events were respiratory symptoms (36%) followed by headache (12%), eye irritation (12%), gastrointestinal symptoms (11%), and dizziness or other central nervous symptoms (10.3%).

Among all victims who died in explosion events (n=32), trauma, thermal burns and smoke inhalation were most common causes of death. Among victims hospitalized for treatment (n=51), the majority sustained thermal burns, trauma, chemical burns, and other symptoms. A large proportion of responders (including members of company response team) (n=9, 36%) and of employee victims (n=41, 18.4%) were not using any personal protective equipment (PPE). Among those persons who wore PPE, their equipment included mostly eye protection, hard hats, steel-toed shoes, level 'D' protection, or some combination of the above.

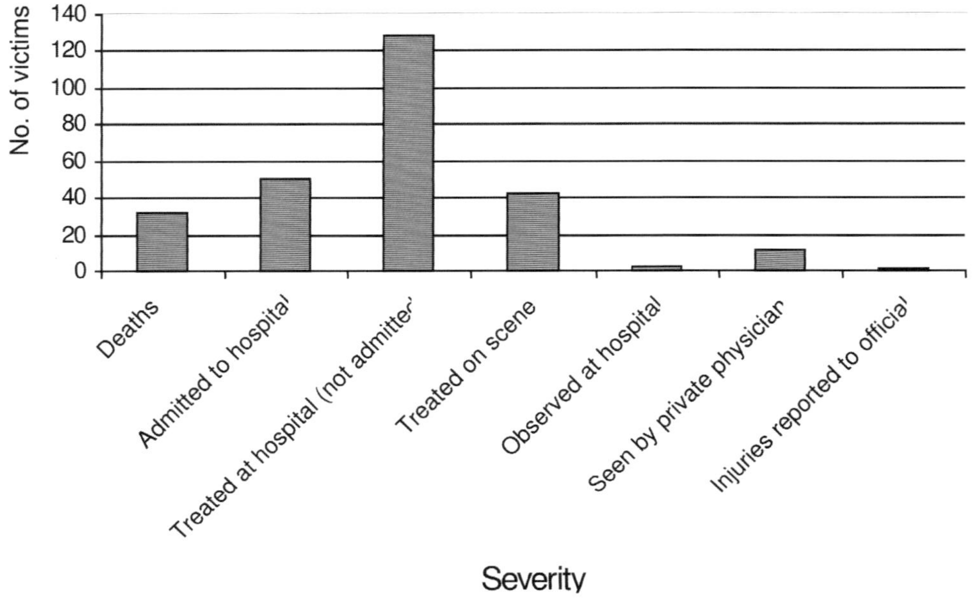

Explosion events were proportionally 9 times as likely to have evacuation ordered as all other manufacturing events (n=62, [52.1%] vs. n=1115, [5.7%] respectively). The number of evacuees was reported for 47 events where an evacuation was ordered and ranged from 1 to 2500 people with a median of 40 people. The length of evacuation in these events ranged from 1 hour to 168 hours with a median of 2 hours. Ninety-one responders and employees were decontaminated at the scene, and an additional 29 employees were decontaminated at a medical facility.

4. Case vignettes

1.— A fire and explosion erupted after a release of coke oven gas at a petroleum refinery as a result of a pipeline burst. Although the number of employees who were working at the facility at the time of the event is unknown, it is likely that most of the 480 facility employees were present, because the event occurred during the day in midweek. Six employees, all males between the ages of 23 and 51, were killed. They sustained thermal burns and were overcome by smoke inhalation. The company followed its standard operating procedures to respond to the event, and ordered an evacuation of the facility.

Table 3.— Number of injuries sustained by victims in explosion events, HSEES 1996—2001.

Injury	Number*	%
Trauma	105	25.5
Thermal Burns	72	17.5
Other	61	14.8
Respiratory Irritation	52	12.7
Chemical Burns	30	7.3
Headache	21	5.1
Eye Irritation	16	3.9
Skin Irritation	15	3.6
Shortness of Breath	13	3.2
Dizziness or other nervous system	8	1.9
Gastrointestinal problems	8	1.9
Heart Problems	8	1.9
Heat Stress	2	0.5
	411	

* The number of injuries exceeds the number of victims because each victim could have sustained more than one injury.

2.— Inappropriate mixing of chemicals in a polyethylene manufacturing facility resulted in an explosion and fire. A letter from the chemical manufacturing plant involved in the explosion to the HSEES program coordinator indicated that a corporate investigative team believed that the most plausible scenario was that a reaction had occurred between residual quantities of butadiene (BD) and popcorn polymer in a butadiene tank. This tank was assumed to be empty, and therefore off line, out of service and under nitrogen purge. However, the purge lines might have been plugged by the popcorn polymer and provided enough heat to overpressure the vessel, which subsequently failed. Ninety persons were injured including one who died. Fourteen victims were admitted to area hospitals, 55 were treated at a hospital and released, and 20 were treated on the scene. Most victims (n=86) were employees, 6 of whom had had hazmat training. Four were members of the general public. The ages of all victims who had their age reported ranged from 17 to 63 years. The majority (n=57) sustained trauma injuries. Thermal burns; headache, respiratory symptoms, shortness of breath and other post traumatic symptoms were most frequently observed. Some of the victims who were hospitalized with burn required skin grafts.

5. Discussion

Although explosion events in the manufacturing industry do not occur frequently, their consequences can be devastating. The above analysis reveals that explosion events were proportionally 16 times as likely to result in victims as all other events in the manufacturing industry. Furthermore, the proportion of events with fatalities among explosion events is more than 100 times larger than the corresponding proportion in all other manufacturing events. The main causes of death in the current analysis were found to be trauma, chemical or thermal burns and smoke inhalation. These findings are supported by the findings of a previous analysis in the manufacturing industry, where the injuries or symptoms sustained by victims involved in explosion events were likely to be more severe.[4]

The current analysis revealed that the majority of explosion events involved equipment failure (e.g. process vessel, storage areas such as tank or drum, and ancillary process equipment such as piping), operator error, and improper mixing of chemicals that resulted in a chemical reaction. This reaction produced gas or vapor that exploded or that over pressurized a chemical vessel releasing volatile material that ignited. Mixing of incompatible chemicals might result from a lack of information or understanding about the properties of the chemicals and their hazards. Some of the explosions may have occurred because of regulation violations.[5]

Explosion events are more likely than other types of releases (e.g. spills only) to affect the public through evacuations, and cause severe damage to property and to the environment.[2] Of 167 reactive hazard incidents evaluated by the US Chemical Safety and Hazard Investigation Board (CSB) during 1980 through 2001, 66% occurred in the chemical manufacturing industry and 16% were associated with fire and/or explosion with toxic releases.[2] An analysis of fire and flame-related occupational fatalities of civilian workers in the United States found that the highest frequency of fatalities in events with fatalities occurred in the manufacturing industry. Many of these events were associated with explosions, as the industry is more likely to have volatile or flammable materials.[6]

The findings in the current analysis suggest that the following activities may reduce explosion events and associated adverse health consequences:

- Provide training of workers on hazardous substances.
- Provide basic safe chemical handling procedures and safety data sheets.
- Label all containers.
- Avoid or minimize heat and ignition sources around flammable chemicals.
- Check drums and pipelines for leakage.
- Check storage and vessel lines for obstruction.
- Encourage the use of proper PPE for workers and employee responders who are at risk for exposure.
- Collaborate with local emergency planning committees (LEPCs) to develop plans for evacuations and sheltering-in-place.

Additional strategies to minimize impact from these types of events suggested by the U.S. Chemical Safety and Hazard Investigation Board[2, 7, 9] through case evaluations and by the US Occupational Safety & Health Administration (OSHA)[8] include:

- Follow OSHA requirements[8].
- Have the facility inspected by OSHA regularly.
- Institute a hazardous materials management plan before the facility embarks on operations.
- Use safe testing methods when modifying equipment.
- Establish hazmat procedures for mitigation and cleanup including the type of PPE necessary to prevent or minimize exposure. In the process of identifying the contaminants employees and responders should not come in contact or smell them, unless the substances are known.

- Use protective measures when administrating decontamination and cleanup. For example, in appropriate circumstances avoid the use of water for cleanup to prevent a reaction. Rather, use talcum powder.
- Establish policies that will accommodate changes to normal operations and procedures at the time of abnormal circumstances in chemical processing. This should require the involvement of skilled personnel who are able to identify potential problems, develop safety measures to protect employees, and establish an action plan. For example, refraining from opening a processing drum above a certain temperature would prevent a release of very hot liquid material.

This analysis has several limitations. First, reporting to the HSEES system is not mandatory. Therefore not all events in the participating states may have been identified. Second, the states might have different minimum reporting quantities for release of hazardous substances. Third, events involving releases of petroleum products only are not included. An inclusion of such releases would have increased the number of events and injuries substantially. Forth, although the surveillance collects information about victims' injuries and symptoms, this formation is mostly limited to the first 24 hours after an event. Also, no information exists about the grades of severity of the injury or symptom. However, victims who were hospitalized with trauma or burns are likely to have sustained severe injuries.

6. Conclusions

The public health consequences associated with events involving the release of hazardous substances that result in explosion with or without fire have the potential to be catastrophic. The primary risk factors associated with explosion events in this analysis were equipment failure, operator error and improper mixing. To reduce equipment failure, companies should consider inspecting maintenance records, equipment, and the production processes. To reduce operator error and improper mixing, companies should consider training employees about hazardous materials and the compatibility of chemicals. Additionally, companies should consider re-enforcing regulations and institution of action plans and safety measures to protect their employees.

Although most events occurred in industrial or commercial areas, approximately 16% occurred in areas where the general public could have been affected. Therefore, local emergency planning committees should work with the community and industry to develop plans for the response to such emergencies including planning for evacuation and sheltering–in–place.

7. References

[1] "Hazardous material risk assessment: final report"
http:/www.fmcsa.dot.gov/safetyprogs/research/briefs/Hazmat_TechBrief.pdf

[2] http://www.csb.gov/completed_investigations/docs/ReactiveHazardInvestigationReport.pdf

[3] http://www.leeds.ac.uk/fuel/msc/mtp/fe01_background.htm

[4] Berkowitz, Z., Barnhart, H. X., Kaye W. E., "Factors Associated With Severity Of Injury Resulting From Acute Releases Of Hazardous Substances In The Manufacturing Industry," *Journal of Occupational and Environmental Medicine*, 45 (7) 734-42, 2003.

[5] http://abclocal.go.com/ktrk/business/100204_APbusiness_chevron.html

[6] Biddle, E.A. and Hartley, D. "Fire – And Flame-Related Occupational Fatalities In The United States, 1980–1994," *Journal of Occupational and Environmental Medicine*, 42(4), 430-37, 2000.

[7] http://www.csb.gov/index.cfm?folder=safety_publications&page=index

[8] http://www.osha.gov/pls/oshaweb/oshadisp.show_document?p_id=1043&p_table=PREAMBLES

[9] http://www.csb.gov/safety_publications/docs/moc082801.pdf

Prediction Of Minimum Flash Point Behavior For Binary Mixtures

M. Vidal, M. S. Mannan, and W. J. Rogers
Mary Kay O'Connor Process Safety Center
Department of Chemical Engineering
Texas A&M University

Prepared for Presentation at
American Institute of Chemical Engineers
2005 Spring National Meeting
39th Annual Loss Prevention Symposium
Atlanta, Georgia
April 11-13, 2005

UNPUBLISHED

Prediction Of Minimum Flash Point Behavior For Binary Mixtures

M. Vidal, M. S. Mannan, and W. J. Rogers
Mary Kay O'Connor Process Safety Center
Department of Chemical Engineering
Texas A&M University

Abstract

Flash point is a primary property used to determine the fire and explosion hazards of a liquid. Minimum flash point behavior (MFPB) is exhibited when the flash point of a mixture is below the flash points of the individual components. The identification of this behavior is critical, because a hazardous situation results from taking the lowest component flash point value as the mixture flash point.

Experimental tests for the complete composition range of a mixture are time consuming, whereas a mixture flash point can be estimated using a computational method and available information. From the infinite dilution activity coefficients of the mixture components, the potential of the liquid mixture to exhibit MFPB can be identified. These coefficients represent the molecular interactions, which affect the vapor pressure of the mixture and consequently its flash point value.

COSMO–RS is used to estimate infinite dilution activity coefficients, and then these parameters are used in a procedure that identify if the mixture will exhibit MFPB or if the flash point value for the mixture will be between the values of the individuals components of the mixture. The purpose of this work is to develop a procedure to predict binary mixture flash points from available information and to study estimation techniques that can be applied when the needed parameters and/or experimental data are not available. The ultimate goal is to predict MFPB even when the experimental data available are limited. The UNIFAC model is used in the prediction of mixtures exhibiting MFPB. The incorporation of quantum chemical calculations and its usefulness in the prediction of mixture flash points is also discussed.

1. Introduction

The flash point of a liquid is one of the most important flammability properties that are used in hazard and risk assessments. It is also the parameter used for regulatory agencies such as National Fire Protection Association (NFPA) to categorize flammability. Flash point is the lowest temperature at which there is sufficient vapor to form an ignitable mixture with air [1].

Good experimental and estimation methods are needed to comply with the current regulations. Most of the methods available are focused primarily on pure compounds; however, most of the chemicals handled in industry are mixtures. Now, the question is, "Can we treat mixtures as pure compounds and use the available methodologies to evaluate the risk associated with it?" The answer to this question is most of time NO. The behavior of mixtures can be extremely different compared to their individual component, and making such assumptions can be very dangerous.

The flash point of a mixture will vary depending on the composition. The total characterization of a mixture requires significant experimental work and time. Probably, that is why the available experimental flash point data of mixtures is scarce. Sometimes mixtures are assumed to have a flash point value equivalent to the minimum flash point value of the individual mixture components and that is considered as a "conservative" measure. This practice can lead to a very hazardous situation if the mixture exhibits a minimum flash point behavior (MFPB) where the flash point of the mixture can be lower than the corresponding flash point value of the individual components. Liaw et al. [2, 3] developed a mathematical model for the prediction of the flash point of binary mixtures that seems to be very promising; however it depends on the availability of some experimental data.

The purpose of this work is to develop a procedure for the estimation of flash point of binary mixtures, especially those exhibiting MFPB. The procedure is intended to guide the user step by step and to highlight the basic information that is needed to evaluate the flash point of binary mixtures. Quantum chemical calculations have been included in the procedure to allow for estimation of parameters and/or properties when no experimental data is available. In addition to that, the UNIFAC model has been used in the prediction of two MFPB mixtures due to its popularity as an estimation and prediction tool in the chemical industry.

2. Background

2.1 Experimental

Flash points are determined experimentally by heating the liquid in a container and then introducing a small flame just above the liquid surface. The temperature at which there is a flash/ignition is recorded as the flash point [4]. Two general experimental methods are the closed cup and open cup. In the closed cup method the fuel is enclosed, while in the open cup the fuel is exposed to open air. The flash points determined with the closed cup method are usually lower than the open cup method values because the vapors are prevented from escaping.

Several standardized test method exist for the experimental determination of flash point. Most of them can be obtained through the American Society for Testing and Materials (ASTM) and are designated as E 502 [5], D 56 [6], D 92 [7], D 93 [8], D 1310 [9], D 3278 [10], D 3828 [11], D 3934 [12] and D 3941 [13]. According to a survey performed by the ASTM Committee E27 [14] the flash point test method E 502, which is a compilation of closed cup methods, is the most frequently used method in the US. The problem is that most of these methods cover primarily the determination of the flash point of pure chemicals and only contain comments on the flammability of mixtures. For example, the ASTM E 1232 [15] specifies that:

> "Care must be exercised in the use of estimated flashpoints for temperature limit testing of mixtures. There are many mixtures (non-ideal solutions) having flashpoints

below the flashpoint of any component. Unless detailed data are available it is difficult to estimate temperature limits of flammability[a] of mixtures."

The availability of flash point experimental values for liquid mixtures is extremely limited. Most of these data are presented in graphs [2, 3, 16-18] or charts [17, 19-27], which can lead to different interpretations. The flash point of a mixture is a function of the mixture composition. Therefore, a complete characterization of the liquid mixture is required.

2.2 Theoretical

The properties of mixtures are hard to predict since a simple mixing rule will not work when the interaction between the components of the mixture are strong. This is the case with liquids that are composed of strongly interacting particles. The flash point of a liquid mixture will depend on the composition of the mixture and on the type of chemicals involved.

Mathematically, the flash point is the temperature at which the vapor pressure is equivalent to the lower flammable limit composition.

$$LFL_i = \frac{P_{i,fp}^{sat}(T)}{P} \quad \textbf{(1)}$$

where $P_{i,fp}^{sat}$ (T) is the vapor pressure at the flash point temperature and P is the ambient pressure. The vapor pressure of the mixtures will be affected depending on the predominant type of interaction (intermolecular forces) among the molecules and consequently the flash point value will change.

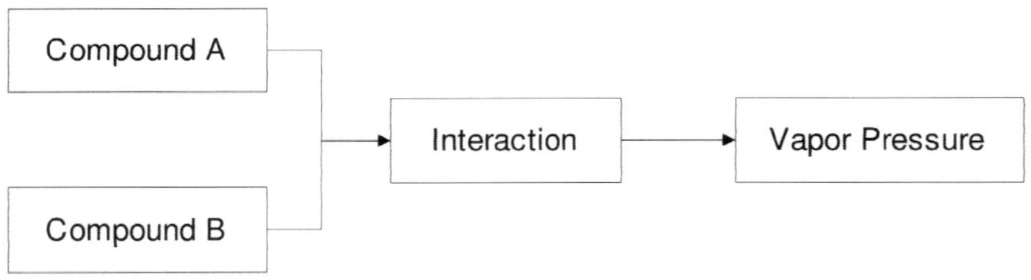

Figure 1: Relation of molecular interaction to macroscopic properties, such as vapor pressure.

For example, a mixture of heptane/octane is expected to have properties that can be predicted with a simple mixing rule that ignores interaction from the values of the individual components because these chemicals are very similar. Both heptane and octane are chain hydrocarbons and the only difference is a $-CH_2$ group. Because both of them are composed of C and H atoms, the interaction among the molecules will be weak. However, when a hydrocarbon is mixed with an alcohol the interaction between them is very strong due to the differences in polarities. The degree of interactions will also depend on the size of the respective molecules, the larger their

[a] The temperature limit of flammability is the minimum temperature at which liquid chemicals evolve sufficient vapors to form a flammable mixture with air under equilibrium conditions.

difference in size, the stronger their interactions. In a liquid mixture of a hydrocarbon and an alcohol the molecules prefer to be alone rather than together. Therefore due to the strong interactions, some of the molecules escape to the vapor phase which raises the vapor pressure of the mixture. Since, in this case both chemicals are flammable, a higher vapor pressure results in a lower flash point value to reach the lower flammable limit (LFL).

In other words, when two flammable compounds are mixed together, the intermolecular forces between them are more important that their respective individual flash point values. On the other hand, when the mixture is composed of a flammable compound and a non-flammable one, the flash point value of the flammable chemical can be estimated as the flash point of the mixture. If the interactions between these two chemicals are so strong that the vapor pressure is higher for the mixture than for the pure components, the non-combustible compound will serve as a diluent in the gas mixture. This is the case of water, which is a non-flammable chemical. When water is mixed with alcohol, for example, the resultant mixture is extremely non-ideal, but because of the non-flammable characteristic of water, the resultant mixture cannot have a flash point value lower than the value of the flammable component.

The flash point of a binary mixture can be estimated by the model developed by Liaw, et al. [2]:

$$\frac{x_1\gamma_1 P_1^{sat}}{P_{1,fp}^{sat}} + \frac{x_2\gamma_2 P_2^{sat}}{P_{2,fp}^{sat}} = 1 \qquad \textbf{(2)}$$

where x_i, γ_i, P_i^{sat}, and $P_{i,fp}^{sat}$ are the mole fraction, activity coefficient, vapor pressure at temperature T and vapor pressure at the flash point T of the mixture components, respectively. This model is a result of the combination of equation (1) with the Le Chatelier's rule for the flammable vapor-air mixture of two components as shown below:

$$\frac{y_1}{LFL_1} + \frac{y_2}{LFL_2} = 1 \qquad \textbf{(3)}$$

where y_1 and y_2 refer to the vapor mole fraction of components 1 and 2 and LFL_1 and LFL_2 refer to the lower flammable limit of component 1 and 2. Another assumption of the model is that the liquid and vapor phase of the mixture are in equilibrium with the vapor phase behaving as an ideal gas. From this equilibrium condition, the compositions of the flammable components in the vapor phase, y_i, are obtained as shown in (4):

$$y_i P = x_i\gamma_i P_i^{sat} \qquad or \qquad y_i = \frac{x_i\gamma_i P_i^{sat}}{P} \qquad \textbf{(4)}$$

The success in the prediction of mixture flash points depends on the accuracy of the parameters used in the model.

The activity coefficient γ_i is a dimensionless parameter that measures the deviation from ideality in a mixture. Non-ideal solutions will have an activity coefficient value smaller or greater than 1. A mixture with a positive deviation from the ideal solution behavior will have an $\gamma_i > 1$, which also means that the partial pressure of the vapor is greater than that predicted by Raoult's law:

$$P = \gamma_i x_i P_i^0 \tag{5}$$

In these mixtures the attractive forces among the unlike molecules are stronger than the attractive forces among like molecules. These are the mixtures with a potential of exhibiting MFPB because of the high vapor pressure generated when the 2 components are mixed. Quantum mechanics, defined as the physical science used to calculate and analyze the energies and spatial distributions of small particles confined to very small regions of space, can be used to estimate the interaction energies and then used as input in activity coefficient models. An excess Gibbs function model is a function of temperature and composition, which allows for the calculation of flash point temperature at different mixture compositions. The relation of the excess Gibbs energy function model and the activity coefficient is presented below (6):

$$RT \ln \gamma_i = \left(\frac{\partial G^E}{\partial n_i} \right)_{T,P,n_j} \tag{6}$$

A review of the available excess Gibbs energy function models can be found in [28].

Liaw et al. [18] developed necessary conditions to identify a binary mixture exhibiting MFPB. The conditions are presented in Figure 2. The mixture component with the lowest flash point value is selected as the component #1 to apply these conditions.

The rules are based on estimation of the vapor pressures of each of the mixture components at its own flash point temperature and at the flash point temperature of the other component. The activity coefficients at infinite dilution are also needed $\left(\gamma_i^\infty = \gamma_i \big|_{x_i \to 0} \right)$, which can be estimated by using any of the G^{ex} models or quantum mechanics methods illustrated in Figure 3.

3. Flash Point Estimation Procedure For Binary Mixtures

The basic information needed for the estimation of the flash point of a binary mixture is:

- flash points of the individual components of the mixture,
- vapor pressure, and
- activity coefficient models.

When experimental data are available for the mixture of interest, the flash point calculation is straightforward by applying the model developed by Liaw et al. [2, 3] and presented in Equation (2). However, this is not always the case, and other ways of obtaining the basic information are needed.

Set $T_{fp1} < T_{fp2}$

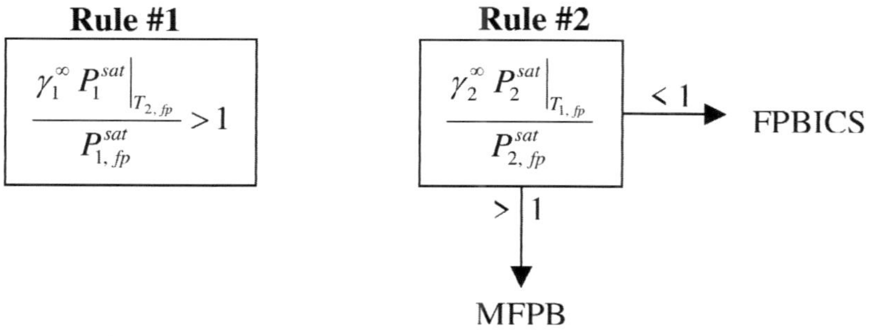

Figure 2: Conditions to identify a mixture that exhibit MFPB as developed by Liaw et al. [18]. (MFPB: Minimum flash point behavior, FPBICS: Flash point between individual components)

A procedure to estimate the flash point of binary mixtures was developed and is presented in Figure 3. The procedure is based on obtaining the data needed for the model of Liaw et al. [2]. It includes the option of determining some of the basic parameters from quantum chemical calculations and/or from correlations.

The flash point values of the individual mixture components can be obtained from experimental measurements or from correlations such as the boiling point and the molecular descriptors. The correlation of the flash point as a function of the boiling point was developed by Satyanarayana and Rao [29] and is presented in Equation (7).

$$T_f = a + \frac{b\left(\frac{c}{T_b}\right)^2 e^{\frac{-c}{T_b}}}{\left(1 - e^{\frac{-c}{T_b}}\right)^2} \qquad \textbf{(7)}$$

where

T_f is the flash point temperature (K),
a, b, and c are constants provided in [29], and
T_b is the boiling point temperature of the material (K).

The equation was fit for over 1,200 compounds with an error of less than 1%.

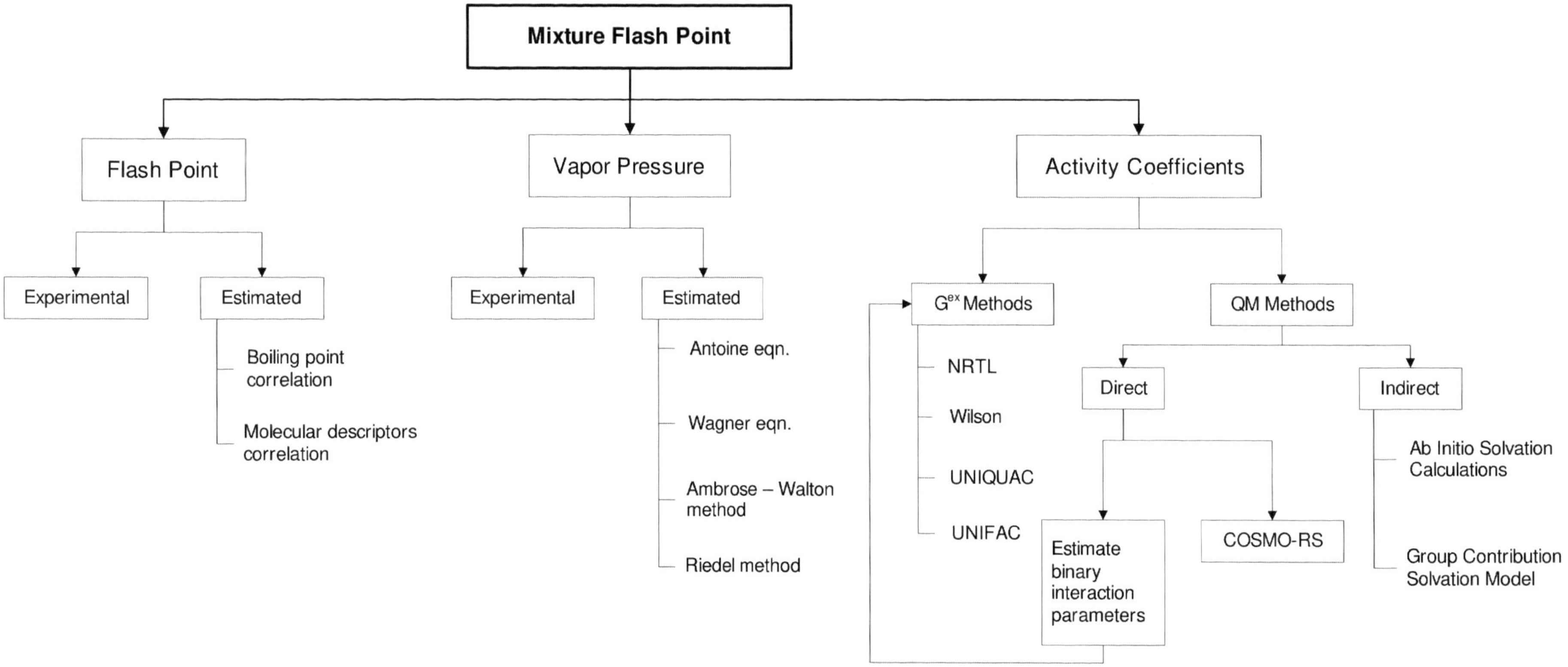

Figure 3: Procedure to estimate the flash point of binary mixtures.

The flash point of pure components can also be obtained from QSPR or quantitative structure property relationships, correlations that relate the flash point value with molecular descriptors. The QSPR approach finds quantitative mathematical relationships between the intrinsic molecular structure and observable properties of chemical compounds. Katritzky, et al. [30] used QSPR to develop the following three-parameter correlation for the flash point:

$$T_F = (44.50 \pm 0.99)^3 \sqrt[3]{G_b} + (16731 \pm 574.45) HDCA + (4.95 \pm 0.73) M_R - (117.70 \pm 9.14) \quad \textbf{(8)}$$

where

$\sqrt[3]{G_b}$ is the cubic root of the gravitational index,
HDCA is a hydrogen donor charged solvent accessible surface area, and
M_R is the molecular weight divided by the number of atoms in the molecule.

The molecular descriptors can be obtained by using a quantum chemical package including drawing and optimizing the molecule. A review of molecular descriptors can be found in [31].

An equation for the saturated vapor pressure, P_i^{sat}, of each mixture component as a function of temperature is needed to estimate the mixture flash point. One of the most common correlations is the Antoine equation:

$$\log P_i^{sat} = A_i - \frac{B_i}{T + C_i} \quad \textbf{(9)}$$

where A_i, B_i and C_i are the parameters of each individual compound. This correlation should not be used outside the temperature range at which the parameters where obtained. Usually in the range of 0.01 to 2 bar, the Antoine equation provides excellent results. The parameters for the Antoine equation can be obtained from collections such as [32, 33], or from online databases such as the NIST Chemistry WebBook [34]. An excellent compilation of data and property calculation websites can be found in [35].

Another correlation that can be used for the estimation of vapor pressure is the Wagner equation presented in (10):

$$\ln P_i^{sat} = \frac{a\tau + b\tau^{1.5} + c\tau^3 + d\tau^6}{T_r} \quad \textbf{(10)}$$

where $\tau = 1 - T_r$, and T_r corresponds to the reduced temperature, which is mathematically defined as $T_r \equiv T/T_c$, where T_c is the critical temperature. This correlation is one of the most accurate, because it is constrained to generate a "reasonable shape" for the vapor pressure curve from a reduced temperature of 0.5 up to the critical point [33]. The parameters for this correlation can be found in [33].

When no experimental data are available, the vapor pressure can be estimated by the Ambrose-Walton and/or Riedel methods. Properties needed for these two methods are the critical temperature (T_c), critical pressure (P_c), and boiling point temperature (T_b) for the Vetere's modification of the Riedel method, and the acentric factor (ω),T_c and P_c for the Ambrose-Walton method. Equations for both methods can be found in [33].

The activity coefficients are the most important parameters of the model because it determines whether a mixture will exhibit a MFPB or not. Values of these coefficients are also the most difficult to obtain due to the vast combination of mixtures that can be created. Some of the excess Gibbs energy function models that can be used to obtain the activity coefficients are: NRTL, Wilson, UNIQUAC and UNIFAC among others. A comparison of these four models is presented in Table 1.

Model	Binary Parameters	Best fit data	Comments
NRTL	$g_{12}-g_{22}, g_{21}-g_{11}$	Aqueous organic mixtures	
Wilson	$\Lambda_{12}, \Lambda_{21}$	Alcohols, phenols, aliphatic hydrocarbons	
UNIQUAC	$u_{12}-u_{22}, u_{21}-u_{11}$	Hydrocarbons, ketones, esters, water, amines, alcohols, nitriles, etc.	Requires pure component molar volumes as well as surface area and volume parameters
UNIFAC	Group interaction parameters, a_{mn} and a_{mn}	Hydrocarbons, ketones, esters, water, amines, alcohols, nitriles, etc.	It is based on the solution of groups concept, and use the group volumes (R_k) and group surface areas (Q_k) as input parameters.

Table 1: Four G^{ex} models

The first three G^{ex} models presented in Table 1 to calculate the activity coefficients depend on experimental interaction parameters. Sometimes the values of these parameters cannot be found for the mixture of interest and alternatives are needed to supply the needed information. The UNIFAC model is the only one from the models above that does not need experimental binary interaction parameters. The contributions due to molecular interactions, parameters a_{mn} and a_{nm}, are obtained from a database using a wide range of experimental results. Some of these parameters can be found in [33]. The advantage of the UNIFAC method is that mixtures composed of the same functional groups can be studied using the same binary parameters.

Quantum chemical calculations can be used to estimate values of the binary interaction parameters to input into any of the activity coefficient models that require experimental data. This approach was recently proposed by Sum and Sandler [36] and is based on a combination of ab initio quantum mechanical methods and the activity coefficient model. Sum and Sandler implemented this approach for the prediction of vapor liquid equilibrium (VLE) for a number of hydrogen bonded binary mixtures. The molecular interactions in the liquids were determined through construction of a molecular cluster model, normally made of eight molecules. If this approach is applied, the cluster size depends on the size of the molecules that compose the mixture. In other words, the critical cluster size must be addressed for each mixture. An assumption that is made when applying this methodology to estimate the activity coefficients is that the binary interactions parameters are not temperature dependant.

Neiman, et al. [37] employed the same methodology, but instead of using quantum mechanical methods, they employed molecular dynamics (MD) simulations to evaluate molecular interactions in the liquid. They claim that MD simulations can describe very well the behavior of liquids if a sufficiently large unit cell is used. In this case, the calculations are dependent upon the size of the unit cell, the simulation time, and the method used to evaluate the interaction energies.

Another alternative is a general theoretical method based on conductor-like screening model for real solvents (COSMO-RS), which was proposed by Klamt and co-workers [38-40]. It is a theory that describes the interactions in a fluid as local contact interaction of molecular surfaces, and the interaction energies are quantified by the values of the two screening charge densities σ and σ', which form a molecular contact. These screening charge densities can be described as molecular descriptors and provide information about the polarity of the molecules. The σ-profile of various compounds is presented in Figure 4.

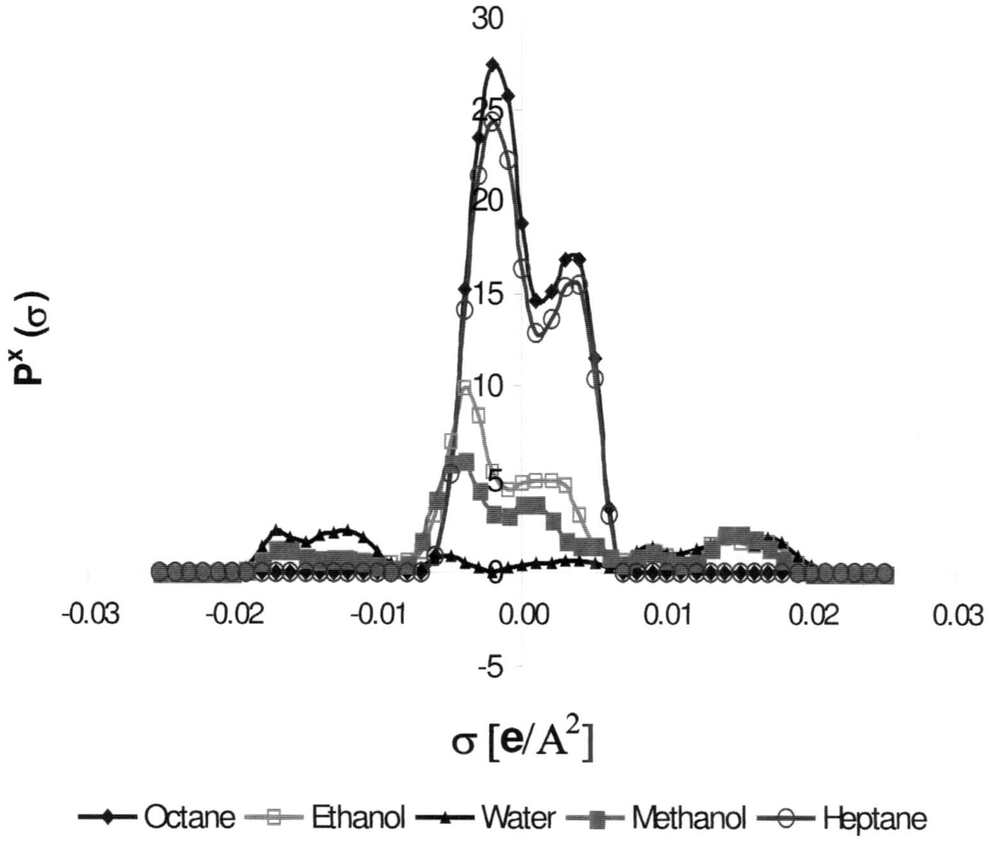

Figure 4: Screening charge profiles

These molecular descriptors in conjunction with statistical thermodynamics are used to calculate the chemical potential and therefore activity coefficient of a component in the mixture. This methodology is very useful in estimating γ_i^∞, infinite dilution activity coefficient. Then the rules presented in Figure 2 are applied to evaluate if the mixture under study has a potential of exhibiting MFPB.

Another approach is an indirect method to obtain infinite dilution activity coefficients from solvation energies. The electrostatic part of the free energy is obtained from various continuum solvation models, and the two energy parameters in the UNIQUAC model are related to the attractive part of the solvation free energy. The fundamental equation that allows the calculation of γ_i^∞ from solvation energies is [41]:

$$RT \ln \gamma_i^\infty = \Delta G_{i/j}^{*sol} - \Delta G_{i/i}^{*sol} + RT \ln \frac{\rho_2^0}{\rho_1^0} \quad \textbf{(9)}$$

where

$\Delta G_{i/j}^{*sol}$ is the free energy change of the solvation of solute i in solvent j,

$\Delta G_{i/i}^{*sol}$ is the free energy change of the solvation of solute i in solvent i, and

ρ_i^0 is the density of liquid i.

More information on this approach can be found from the work of Lin and Sandler [41] and from the work of Nanu and Loos [42].

4. Results And Discussion

Our main purpose in this paper is to show how to estimate the flash point of binary mixtures even in the absence of experimental data and to identify those mixtures that exhibit MFPB. Prediction tools such as UNIFAC as well as quantum chemical calculations have been used and applied to accomplish this goal.

First of all, the general theoretical method COSMO-RS was used to estimate the γ_i^∞, which were then used in the rules presented in Figure 2 to analyze its usefulness in the identification of mixtures exhibiting MFPB. The COSMO-RS calculations were performed using $C^2 \cdot DMol^3$ with input data and are presented in Table 2.

Chemical	Antoine coefficients			Gas phase energy
	A	B	C	[Ha]
Octane	6.90940	1349.820	209.385	-315.718761
Heptane	6.89385	1264.37	216.636	-276.4041106
Ethanol	8.21330	1652.050	231.480	-155.0456301
1-Butanol	7.47680	1362.390	178.730	-233.6744593

Table 2: Input data used For COSMO-RS calculations

The Antoine coefficients were obtained from [43] and the gas phase energies were calculated using Gaussian 03 [44]. The gas phase energies were obtained using the b3lyp level of theory and the 6-31+g(d) basis set. The results obtained for 3 samples mixtures are presented in Table 3.

Mixture	Rule #1	Rule #2	Classification
	$\frac{\gamma_1^\infty P_1^{sat}\big|_{T_{2,fp}}}{P_{1,fp}^{sat}} > 1$	$\frac{\gamma_2^\infty P_2^{sat}\big|_{T_{1,fp}}}{P_{2,fp}^{sat}}$	
Heptane (1) – Octane (2)	2.7555	0.3145	FPBICS
Octane (1) – 1-Butanol (2)	15.5909	16.3828	MFPB
Octane (1) – Ethanol (2)	9.5680	294.4660	MFPB

Table 3: Results using COSMO-RS to obtain γ_i^∞ to identify a mixture that exhibits MFPB based on the conditions developed by Liaw et al. [18].

When compared with the original source [18], where the activity coefficients were calculated using a G^{ex} model, the results obtained with COSMO-RS are lower for Rule #1 and higher for Rule #2. However, the results comply with the inequality in Rule #1 (>1) and they predicted correctly the mixtures that exhibit MFPB. The flash points of the individual components were obtained from [45]. It is important to point out that uncertainties in the parameters used for the vapor pressure cancel out in both rules. Then, the successful identification of a mixture with MFPB depends on the γ_i^∞.

In general COSMO-RS can be used reliably to estimate the γ_i^∞ and then use the rules in Figure 2 to identify mixtures with the potential of exhibiting MFPB. If the results obtained for the γ_i^∞ with COSMO-RS are not accurate, the results in the classification when applying the rules may be the prediction of a mixture with MFPB when in fact it is a mixture with FPBICS. This prediction will not present any hazard, since it will assume that the flash point for the mixture could be lower than the flash point of the individual components. On the contrary, if the prediction were the opposite, the results obtained with COSMO-RS would result in a hazardous misclassification.

The advantage of using COSMO-RS is that the screening charge that is used as an input in the calculations must be calculated just once per chemical, so that the user can create a database of different chemicals and then apply the rules for different mixtures to screen the ones that have the potential of exhibiting MFPB.

The flash points of two aqueous mixtures were estimated using interaction binary parameters that were obtained by quantum chemical calculations and by using experimental interaction binary parameters. The parameters obtained with quantum chemical calculations were used in the UNIQUAC model to estimate the activity coefficients. The results for the water – methanol mixture flash point are presented in Figure 5.

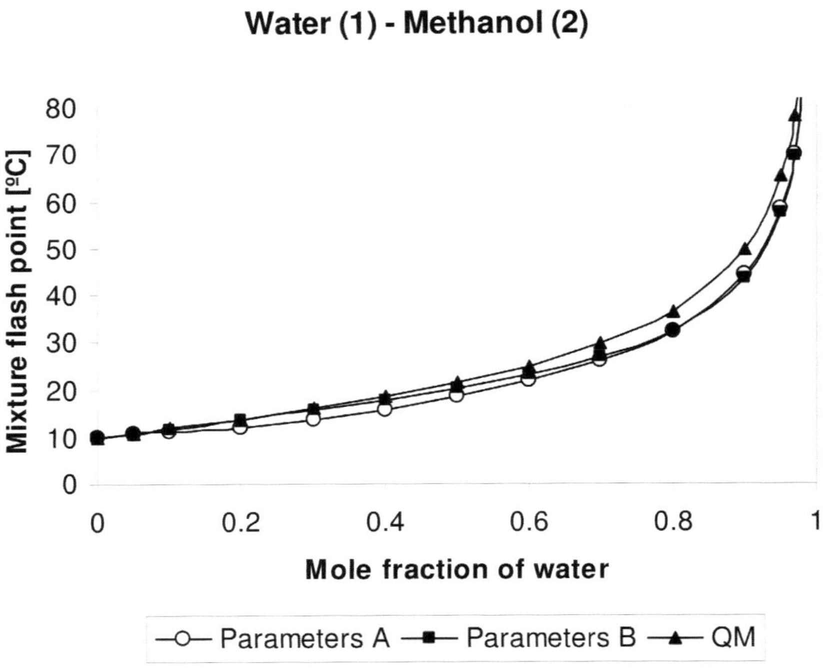

Figure 5: Prediction of the water – methanol mixture flash point

These predictions were made by using different parameters from the literature and by applying Equation 2 with modifications for aqueous mixtures presented in [3]. Each of the parameters, presented in Table 4 was used in the UNIQUAC model for the calculation of the activity coefficients.

Parameters	$A_{12} = (u_{12} - u_{22})/R$	$A_{21} = (u_{21} - u_{11})/R$	Reference
A	-271.26	736.01	[46]
B	180.22	-117.34	[47]
QM	6.37	-47.51	[36]

Table 4: Parameters used in the UNIQUAC model for the prediction of the flash point of the water – methanol mixture

All the predictions agree with the experimental data presented in [3]. The experimental data were not included in Figure 5 because it is presented in graphical form in the original source, but the trend of the experimental mixture flash point is predicted accurately with all of the parameters analyzed.

The same analysis was performed for the mixture of water – ethanol, and the results are presented in Figure 6.

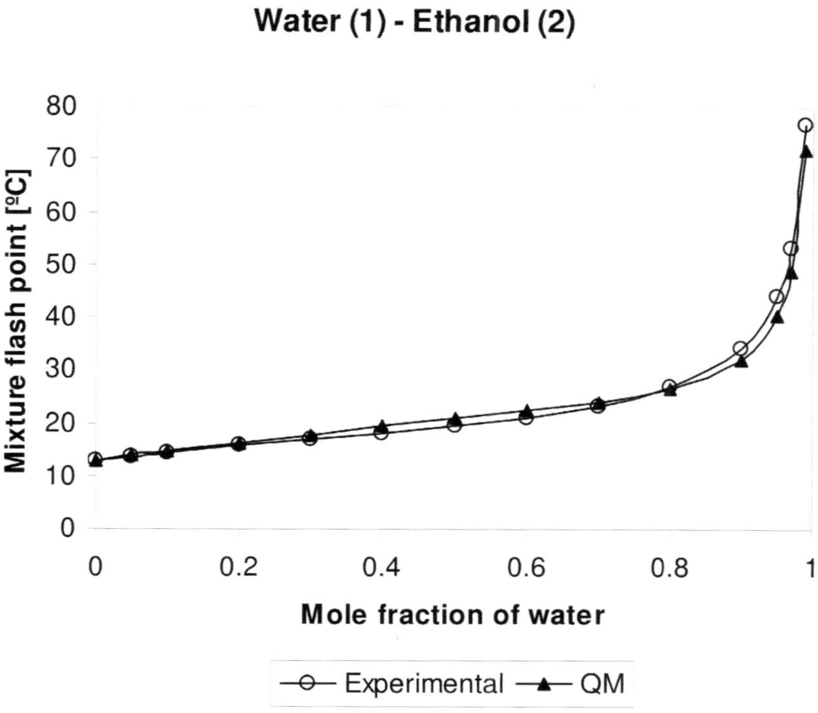

Figure 6: Prediction of the water – ethanol mixture flash point

Using the parameters in Table 5, the predicted results agree with the experimental data presented in [3].

Parameters	$A_{12} = (u_{12} - u_{22})/R$	$A_{21} = (u_{21} - u_{11})/R$	Reference
Experimental	-109.37	299.46	[48]
QM	131.57	-4.49	[36]

Table 5: Parameters used in the UNIQUAC model for the prediction of the flash point of the water – ethanol mixture.

The flash point of two mixtures exhibiting MFPB was predicted using the UNIFAC model for the activity coefficients. The UNIFAC model is composed of two parts: the combinatorial and the residual parts. The combinatorial part represents the differences in size and shape of the molecules in the mixture, and only pure component properties are needed to calculate this part. The residual part represents the energy interactions and is a function of temperature. The temperature that satisfies equation (2), as well as the activity coefficient model and the vapor pressure as a function of temperature, is the one that is selected as the flash point of the mixture. These iterative calculations were performed in an Excel spreadsheet. The mixtures studied were: octane – ethanol, and octane – 1-butanol. These two mixtures were studied previously by Liaw [2, 18] experimentally and theoretically. The experimental data are presented in a graph form in his paper. To avoid any misinterpretation, the theoretical model that best represents the experimental data will be used as a comparison for our calculations. In this work, the UNIFAC model was used for the calculation of the activity coefficients to demonstrate its usefulness in the prediction of mixtures exhibiting MFPB.

The vapor pressures of the components of the mixtures were calculated using the Antoine equation. The parameters used are presented in Table 6.

Compound	*A*	*B*	*C*	Reference
n-octane	6.04394	1351.938	-64.030	[49]
ethanol	7.24222	1595.811	-46.702	[50]
1-butanol	6.54068	1335.028	-96.496	[50]

Table 6: Antoine coefficients used in the prediction of mixtures exhibiting MFPB
$\log\left(P/kPa\right) = A - B/\left[\left(T/K\right) + C\right]$

The parameters used in the UNIFAC model for the studied mixtures are presented in Table 7.

Compound	R_k	Q_k
n-octane	5.8486	4.936
ethanol	2.5755	2.588
1-butanol	3.9243	3.668

Table 7: Group volumes and group surface areas used in the UNIFAC model

The flash points for the octane-ethanol and octane-1-butanol mixtures are presented in Figures 7 and 8.

The UNIFAC model predicts satisfactorily the MFPB of both mixtures. For the octane-ethanol mixture, the predicted flash point is a little lower than the experimental values represented by the NRTL model. However, the MFPB for almost the entire range of compositions ($0.1 < x_1 > 0.9$) is predicted by UNIFAC. In the octane-1-butanol mixture, a minimum flash point value is observed around $x_1 = 0.9$, which was also predicted by UNIFAC.

5. Conclusions

The procedure developed for the estimation of binary mixture flash point is very useful in the assessment of flammability hazards. It saves time and effort by providing the user with techniques that can be used in estimating a parameter that is needed for the evaluation of a mixture flash point. It also allows for the estimation of the mixture flash point when no experimental data are available.

COSMO-RS can be used to estimate the activity coefficients at infinite dilution when the binary interaction parameters for the mixture of interest are not available. Due to its simplicity it allows evaluation and screening several mixtures at the same time. Then, more time and effort can be spent on those mixtures identified as MFPB mixtures.

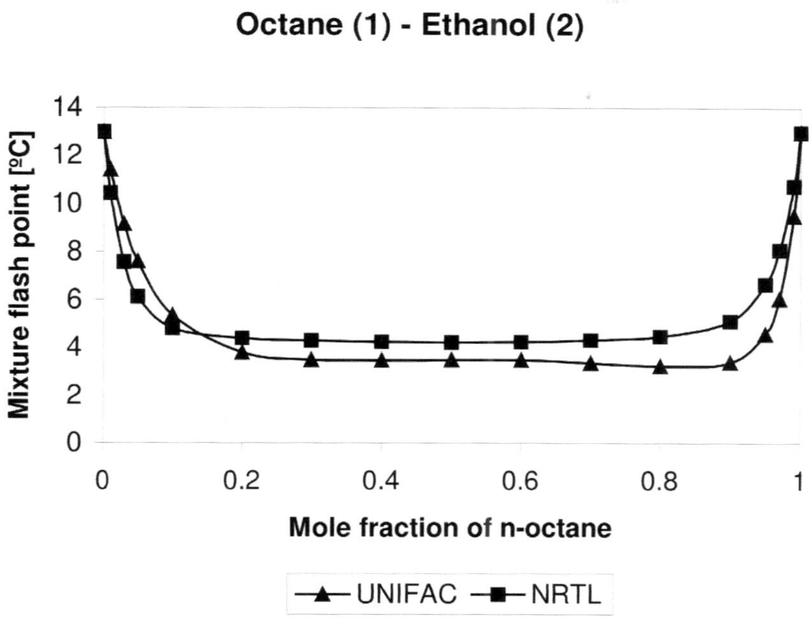

Figure 7: Prediction of the octane – ethanol mixture flash point

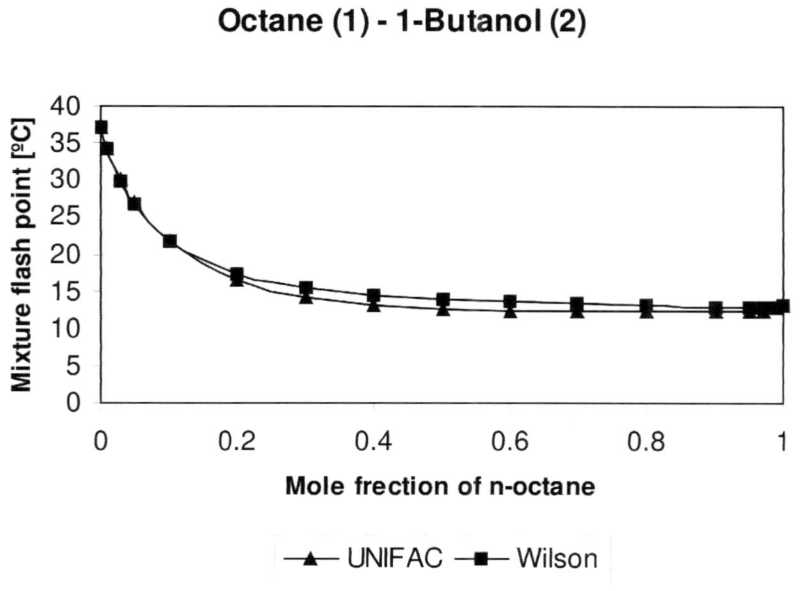

Figure 8: Prediction of the octane –1-butanol mixture flash point

The calculation of binary interaction parameters from quantum chemical calculations appears to be very promising, even when this approach is still in its early stages. Quantum chemical methods provide valuable tools for performing flammability assessments of mixtures that are hard to test experimentally or for mixtures where the basic input data for the evaluation of a mixture flash point are not available.

The UNIFAC group contribution method described very well the MFPB. Therefore, it can be used for the prediction of binary mixtures flash points when the liquid mixture is non-ideal and when the interaction parameters among the mixture components are not available experimentally.

6. References

[1] D. A. Crowl and J. F. Louvar, *Chemical process safety: fundamentals with applications,* 2nd ed. (Prentice Hall PTR, 2002).

[2] H.-J. Liaw, Y.-H. Lee, C.-L. Tang, H.-H. Hsu, and J.-H. Liu, A mathematical model for predicting the flash point of binary solutions, *Journal of Loss Prevention in the Process Industries* 15(6), (2002), 429-438.

[3] H.-J. Liaw and Y.-Y. Chiu, The prediction of the flash point for binary aqueous-organic solutions, *Journal of Hazardous Materials* A101, (2003), 83-106.

[4] ILPI, *The MSDS HyperGlossary: Flash point* (November 15, 2003), retrieved December 2, 2004, from http://www.ilpi.com/msds/ref/flashpoint.html.

[5] ASTM E 502 - 84, "Selection and use of ASTM standards for the determination of flash point of chemicals by closed cup methods," ASTM International.

[6] ASTM D 56, "Standard test method for flash point by tag closed cup tester," ASTM International.

[7] ASTM D 92, "Standard test method for flash and fire points by Cleveland open cup tester," ASTM International.

[8] ASTM D 93, "Standard test methods for flash point by Pensky-Martens closed cup tester," ASTM International.

[9] ASTM D 1310, "Standard test method for flash point and fire point of liquids by tag open-cup apparatus," ASTM International.

[10] ASTM D 3278, "Standard test methods for flash point of liquids by small scale closed-cup apparatus," ASTM International.

[11] ASTM D 3828, "Standard test methods for flash point by small scale closed cup tester," ASTM International.

[12] ASTM D 3934, "Standard test method for flash / no flash test-equilibrium method by a closed-cup apparatus," ASTM International.

[13] ASTM 3941, "Standard test method for flash point by the equilibrium method with a closed-cup apparatus," ASTM International.

[14] E. A. Ural, D. Frurip, L. Britton, W. Fenlon, J. Going, K. Harrison, and J. Niemeier, "The role of ASTM E27 methods in hazard assessment," presented at the 38th Loss Prevention Symposium, AIChE Spring National Meeting, 2004. This paper will appear in two parts in the Dec. 2004 and March 2005 issues of the *Process Safety Progress*.

[15] ASTM E 1232 - 02, "Standard test method for temperature limit of flammability of chemicals," ASTM International.

[16] J. Gmehling and P. Rasmussen, Flash Points of Flammable Liquid-Mixtures Using Unifac, *Industrial & Engineering Chemistry Fundamentals* 21(2), (1982), 186-188.

[17] W. A. Affens and G. W. McLaren, Flammability Properties of Hydrocarbon Solutions in Air, *Journal of Chemical Engineering Data* 17(4), (1972), 482-488.

[18] H.-J. Liaw, T.-P. Lee, J.-S. Tsai, W.-H. Hsiao, M.-H. Chen, and T.-T. Hsu, Binary liquid solutions exhibiting minimum flash-point behavior, *Journal of Loss Prevention in the Process Industries* 16(3), (2003), 173-186.

[19] B. Hanley, A model for the calculation and the verification of closed cup flash points for multicomponent mixtures, *Process Safety Progress* 17(2), (1998), 86-97.

[20] J. M. Lenoir, Predict Flash Points Accurately, *Hydrocarbon Processing* 54(1), (1975), 95-99.

[21] W. H. Ellis, Solvent Flash Points - Expected and Unexpected, *Journal of Coatings Technology* 48(614), (1976), 44-57.

[22] J. Rybicky and J. R. Stevens, Deceptive Closed Cup Flash Points of Solvent Mixtures Containing Chlorinated Solvents, *Journal of Coatings Technology* 53(676), (1981), 40-42.

[23] J. L. McGovern, A Method for Estimating the Flash Points of Coatings Containing Mixtures of Oxygenated and Hydrocarbon Solvents and Petroleum Distillates 2, *Journal of Coatings Technology* 64(810), (1992), 39-44.

[24] J. L. McGovern, 4 Methods for Predicting the Flash-Point of Alkyd Paints Containing Vm-and-P Naphtha and Mineral Spirits 1, *Journal of Coatings Technology* 64(810), (1992), 33-38.

[25] D. T. Wu and R. Finkelman, A Mathematical Model for the Prediction of Closed Cup Flash Points, *Organic Coatings and Plastics Chemistry* 38, (1978), 61-67.

[26] J. Achari and Bhattach.Mm, Flash Points of Mixtures of Acetone with Water, Kerosene and Carbon Tetrachloride, *Indian Journal of Technology* 9(3), (1971), 117-&.

[27] M. H. Choe and H. G. Schecker, Flash Points of Multicomponent Liquid-Mixtures - Comparison of Experimental and Calculated Values, *Chemie Ingenieur Technik* 60(2), (1988), 124-125.

[28] H. Orbey and S. I. Sandler, *Modeling vapor-liquid equilibria: cubic equations of state and their mixing rules* (Cambridge University Press, 1998).

[29] K. Satyanarayana and P. G. Rao, Improved equation to estimate flash points of organic-compounds, *Journal of Hazardous Materials* 32(1), (1992), 81-85.

[30] A. R. Katritzky, R. Petrukhin, R. Jain, and M. Karelson, QSPR analysis of flash points, *Journal of Chemical Information and Computer Sciences* 41(6), (2001), 1521-1530.

[31] M. Karelson, *Molecular Descriptors QSAR/QSPR* (John Wiley & Sons, Inc., New York, USA, 2000).

[32] T. Boublik, V. Fried, and E. Hala, *Vapour pressure of pure substances : selected values of the temperature dependence of the vapour pressures of some pure substances in the normal and low pressure region,* 2nd ed. (Elsevier Science Ltd, 1984).

[33] B. E. Poling, J. M. Prausnitz, and J. P. O'Connell, *The properties of gases and liquids,* 5th ed. (McGraw-Hill, New York, 2001).

[34] *NIST Chemistry WebBook*, retrieved from http://webbook.nist.gov/chemistry/.

[35] Thermodynamics Research Laboratory, *Nano, Quantum & Statistical Mechancis & Thermodynamics Data & Property Calculation Websites,* retrieved from http://tigger.uic.edu/~mansoori/Thermodynamic.Data.and.Property_html.

[36] A. K. Sum and S. I. Sandler, A novel approach to phase equilibria predictions using ab initio methods, *Ind. Eng. Chem. Res.* 38, (1999), 2849-2855.

[37] M. Neiman, H. Cheng, V. Parekh, B. Peterson, and K. Klier, A critical assessment on two predictive models of binary vapor-liquid equilibrium, *Physical Chemistry Chemical Physics* 6(13), (2004), 3474-3483.

[38] A. Klamt, Conductor-like screening model for real solvents - a new approach to the quantitative calculation of solvation phenomena, *Journal of Physical Chemistry* 99(7), (1995), 2224-2235.

[39] A. Klamt, COSMO-RS: A novel bridge from quantum chemistry to fluid phase thermodynamics., *Abstracts of Papers of the American Chemical Society* 226, (2003), U432-U433.

[40] A. Klamt and F. Eckert, COSMO-RS: a novel and efficient method for the a priori prediction of thermophysical data of liquids, *Fluid Phase Equilibria* 172(1), (2000), 43-72.

[41] S.-T. Lin and S. I. Sandler, Infinite dilution activity coefficients from ab initio solvation calculations, *AIChE Journal* 45(12), (1999), 2606-2618.

[42] D. E. Nanu and T. W. D. Loos, Limiting activity coefficients of aqueous flavour systems at 298K by the group contribution (GCS) model, *Molecular Physics* 10(3), (2004), 235-241.

[43] *Antoine.dat* (17-Apr-2002), retrieved 2004, from www.glue.umd.edu/~nsw/ench250/antoine.dat.

[44] Gaussian 03, Revision C.02, M. J. Frisch, G. W. Trucks, H. B. Schlegel, G. E. Scuseria, M. A. Robb, J. R. Cheeseman, J. J. A. Montgomery, T. Vreven, K. N. Kudin, J. C. Burant, J. M. Millam, S. S. Iyengar, J. Tomasi, V. Barone, B. Mennucci, M. Cossi, G. Scalmani, N. Rega, G. A. Petersson, H. Nakatsuji, M. Hada, M. Ehara, K. Toyota, R. Fukuda, J. Hasegawa, M. Ishida, T. Nakajima, Y. Honda, O. Kitao, H. Nakai, M. Klene, X. Li, J. E. Knox, H. P. Hratchian, J. B. Cross, V. Bakken, C. Adamo, J. Jaramillo, R. Gomperts, R. E. Stratmann, O. Yazyev, A. J. Austin, R. Cammi, C. Pomelli, J. W. Ochterski, P. Y. Ayala, K. Morokuma, G. A. Voth, P. Salvador, J. J. Dannenberg, V. G. Zakrzewski, S. Dapprich, A. D. Daniels, M. C. Strain, O. Farkas, D. K. Malick, A. D. Rabuck, K. Raghavachari, J. B. Foresman, J. V. Ortiz, Q. Cui, A. G. Baboul, S. Clifford, J. Cioslowski, B. B. Stefanov, A. L. G. Liu, P. Piskorz, I. Komaromi, R. L. Martin, D. J. Fox, T. Keith, M. A. Al-Laham, C. Y. Peng, A. Nanayakkara, M. Challacombe, P. M. W. Gill, B. Johnson, W. Chen, M. W. Wong, C. Gonzalez, and J. A. Pople, Gaussian, Inc., Wallingford CT, 2004.

[45] NFPA 325, "Fire hazard properties of flammable liquids, gases, and volatile solids," (1994).

[46] B. Khalfaoui, A. H. Meniai, and R. Borja, Thermodynamic properties of water + normal alcohols and vapor-liquid equilibria for binary systems of methanol or 2-propanol with water, *Fluid Phase Equilibria* 127(1-2), (1997), 181-190.

[47] A. Arce, A. Blanco, A. Soto, and I. Vidal, Phase equilibria of water + methanol + hexyl acetate mixtures, *Fluid Phase Equilibria* 128(1-2), (1997), 261-270.

[48] A. Arce, J. Martinez-Ageitos, J. Mendoza, and A. Soto, Water + ethanol + 2-methoxy-2-methylbutane: Properties of mixing at 298.15 K and isobaric vapour-liquid equilibria at 101.32 kPa, *Fluid Phase Equilibria* 141(1-2), (1997), 207-220.

[49] T. Hiaki, A. Taniguchi, T. Tsuji, M. Hongo, and K. Kojima, Isobaric vapor-liquid equilibria of octane+1-butanol, +2-butanol, and +2-methyl-2-propanol at 101.3 kPa, *Journal of Chemical & Engineering Data* 41(5), (1996), 1087-1090.

[50] T. Boublík, V. e. Fried, and E. Hála, *The vapour pressure of pure substances : selected values of the temperature dependence of the vapour pressures of some pure substances in the normal and low pressure region,* 2nd rev. ed. (Elsevier : Distributors for the United States and Canada : Elsevier Science Pub. Co., 1984).

Buoyant Gas Layer Deflagration Effects

Robert Zalosh
Jamie Stern-Gottfried*
Fire Protection Engineering Department
Worcester Polytechnic Institute
Worcester, MA 01609-2280

* currently working for Arup Fire in London, England.

Prepared for Presentation at
American Institute of Chemical Engineers
2005 Spring National Meeting
39th Annual Loss Prevention Symposium
Atlanta, Georgia
April 11-13, 2005

UNPUBLISHED

Buoyant Gas Layer Deflagration Effects

Robert Zalosh
Jamie Stern-Gottfried*
Fire Protection Engineering Department
Worcester Polytechnic Institute
Worcester, MA 01609-2280

* currently working for Arup Fire in London, England.

Abstract

Lighter-than-air flammable gases often form buoyant ceiling layers upon release in an enclosure. A brief review of the dispersion literature for this type of release shows that there is a near-uniform gas concentration in the ceiling layer for many release and enclosure configurations. Delayed ignition of the flammable gas layer produces a deflagration with a pressure hazard if the enclosure is not sufficiently large or open, and a flame radiation hazard to any exposed personnel under the ceiling layer. A simple approximate equation with an empirical burning rate enhancement factor, χ, has been used to compare natural gas ceiling layer peak pressures to previously published data for a 27 m^3 vented enclosure. These comparisons have been used to develop correlations for χ as a function of ceiling layer depth, vent depth, and ceiling height. Equations have also been developed to calculate the propagating flame radiant heat fluxes at potential target locations under the gas layer. These equations account for both the flame propagation speed, layer view factor and the burned gas layer emissivity. Calculated results are shown for a wide variety of methane and hydrogen gas concentrations, layer depths, and ceiling heights to indicate the peak heat flux developed and the time delay between ignition and the development of the minimum heat flux for second-degree burns at that exposure time.

1. Introduction

Ammonia, hydrogen, and methane are examples of lighter-than-air flammable gases that are often stored, handled, and processed indoors. Accidental releases of these gases can occur under a wide variety of release rates and enclosure sizes, geometries, and ventilation conditions. Many combinations of these parameters can produce a buoyant flammable gas layer under the ceiling. Delayed ignition would then produce a deflagration that poses both pressure hazard and flame radiation threat to personnel and equipment in or below the gas layer.

Numerous studies have been devoted to the buoyant dispersion phenomena governing the formation of a flammable gas layer. Highlights of the results and conclusions of some of the most pertinent studies are presented below in order to characterize the types of ceiling layers to be expected with different release conditions. Far less has been published on the pressures and flame radiation resulting from ignition of the flammable gas layer. The main purpose of this paper is to present some relatively simple analytical tools that can be used to estimate deflagration pressures and flame radiant heat fluxes for ceiling layer deflagrations.

2. Dispersion and Ceiling Layer Formation

Baines and Turner [1] first analyzed and observed the buoyant ceiling layer that develops for low momentum releases, i.e. plumes, in the absence of any significant room ventilation. They hypothesized that vertically stratified ceiling layers develop when the stabilizing plume buoyancy force is much greater than the buoyancy-induced plume momentum at the ceiling. The depth of this ceiling layer increases with time as the release continues. Conversely, a well-mixed, relatively deep ceiling layer develops very soon after the release begins when the plume momentum is sufficient to cause overturning of the light gas layer under the ceiling. This well-mixed (both horizontally and vertically) deep layer occurs when the distance between the release elevation and the ceiling, z_r, is large in comparison to the characteristic dimension of the ceiling. Later work by Cleaver et al [2] showed that the simple equation

$$z_r > 0.38\sqrt{A_c} \qquad [1]$$

where A_c is the ceiling area, could be used to delineate the condition under which a uniform ceiling layer would form. The depth of this well-mixed layer remains constant with increasing gas concentration as the release continues. Cleaver et al [2] developed a simple mathematical model and performed full-scale natural gas release experiments to determine the depth and concentration history of the ceiling layer for various release conditions in an enclosure with negligible ventilation, i.e. with no ventilation at the elevation of the ceiling layer. Their model also includes a second layer under the relatively deep (often half z_r or greater) uniform layer. The second layer has a steep concentration gradient.

Cleary et al [2] also conducted tests with downward and horizontally directed releases of natural gas. Their data and analysis indicated that the initial momentum of these releases produced slightly deeper well-mixed ceiling layers than those with upward directed jets and plumes. The momentum of the release dissipated and the jets turned upward after an axial distance L_j given by

$$L_j = \frac{3\, u_0 \sqrt{r_0}}{\sqrt{g(\rho_a - \rho)/\rho}} \qquad [2]$$

where u_0 is the release velocity, r_0 is the release orifice radius, g is the acceleration of gravity, and ρ and ρ_a are the density of the flammable gas and air, respectively.

In the case where the ceiling is either so large that it is effectively unbounded, or where there is a large vent at the top of the enclosure walls, the ceiling layer is relatively thin and the concentration decreases with distance from the plume centerline. Alpert's review [3] of ceiling layers produced by fire products suggests that the layer depth is about $0.075z_r$ at r/z_r of 0.6, and increases to about $0.11z_r$ at $r/z_r > 1$.

Effects of forced ventilation and various free vent configurations on buoyant gas dispersion are often determined via Computational Fluid Dynamics (CFD) calculations. For example, Swain and Swain [4] used CFD calculations to investigate effects of free vent location and forced ventilation rates on the dispersion of hydrogen leaking into a 768 ft^3 (22 m^3) room with a 8 ft (2.4 m) high ceiling. They found that a ventilation rate of 1.5 scfm/ft^2 (7.5 liters/s-m^2) could prevent flammable concentrations from developing through most of the room as long as the steady-state hydrogen leakage rate was less than 100 – 500 liters/min, with the precise value depending on the effective Lower Flammable Limit and extent of room above the LFL. Swain et al [5] extended their CFD calculations to small garage and hallway configurations with various free vent sizes and locations. Some vent configurations produced near-uniform hydrogen concentrations at all elevations (presumably because of internal recirculation), while others produced distinct ceiling layers. In the configurations that produced layering, the hydrogen concentration contours appeared to be flat planes except immediately above and below floor and ceiling vents, respectively, and in the vicinity of wall vents.

One important gas release scenario is the blowdown of a pressure vessel, perhaps through a pressure relief valve. This scenario produces a highly transient release rate with gas flow rates decreasing rapidly as the vessel blows down. However, tests conducted by Zalosh et al [6] with simulated Compressed Natural Gas vehicle fuel tank blowdowns in large parking garages have shown that the gas ceiling layer formed from the initially high release rates persists beyond the duration of the blowdown, even with garage ventilation rates of 1.15 cfm/ft^2 (equivalent to 6.3 volume changes per hour).

This brief review of buoyant flammable gas dispersion analyses and test programs shows some of the many situations in which persistent, near-uniform, flammable gas layers are produced. Some of the gas layers are quite shallow, while others extend down to below the mid-height between the floor and ceiling. We now consider the consequences of igniting these flammable gas layers.

3. Ceiling Layer Deflagration Pressures

The configuration for this analysis is a uniform flammable gas layer of depth h_f in a rectangular enclosure equipped with an upper level explosion vent as shown in Figure 1. We neglect the possible presence of any stratified second layer, and assume that any gas below the ceiling layer is at a concentration below the LFL. Once the gas layer has formed, ignition will produce flame propagation through it, and the enclosure confinement will produce a pressure rise that should be roughly uniform throughout the enclosure as long as the flame speed is small in comparison to sound speed in the gas mixture.

The geometrical shape of the propagating flame depends primarily on the ignition source location and the depth of the layer. As long as the gas concentration is not near the LFL and UFL limit concentrations, the propagation should at least begin as radial propagation away from the ignition site. If the layer is shallow and ignition occurs near the center of the ceiling (perhaps from an electrical fixture on the ceiling), we would expect cylindrical flame propagation with a near-constant cylinder height (that is somewhat larger than the ceiling layer depth) once the flame reaches the bottom of the layer. If the layer is shallow and ignition occurs near the wall furthest from the vent, we would expect the flame to be like a half-cylinder propagating radially, at least until is strikes another wall or is deformed by flow toward the vent that would open at some time after the pressure begins to rise significantly. If the layer is deep, the initial flame propagation should be almost spherical or hemi-spherical depending on the ignition location relative to the nearest wall.

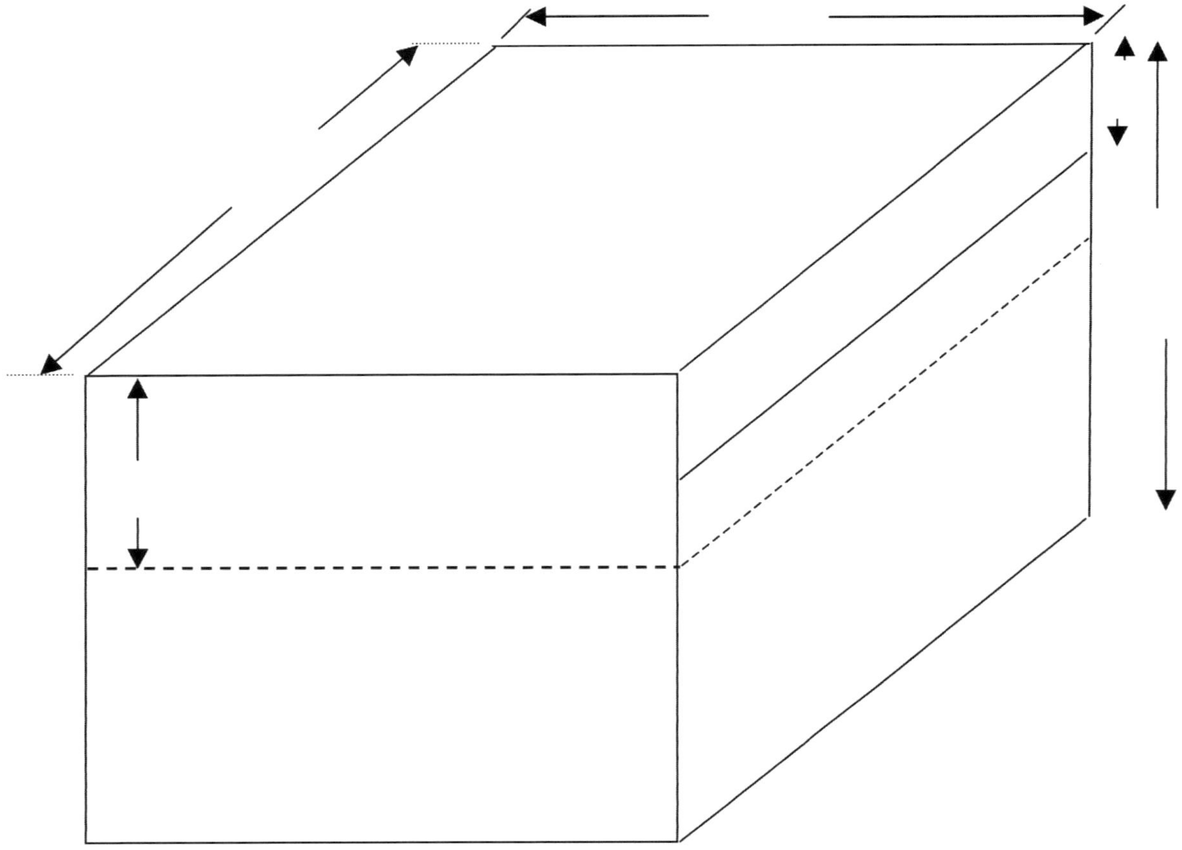

The shape and size of the propagating flame is important because the mass rate of burning and associated rate of pressure increase is proportional to the flame surface area. The various two-zone models of vented gas deflagrations in gas-filled enclosures have been derived to relate the maximum pressure, P_{red}, to the vent area and other parameters use some assumption of the maximum flame surface area, A_{fmax}. In the case of low P_{red} values, the vented flow can be characterized by a Bernoulli equation, and the resulting equation for P_{red} can be written in a form similar to the derivation of Swift and Epstein [7] and others as:

$$P_{red} - P_0 = \frac{\rho_{u0}}{2}\left[(\frac{\rho_{u0}}{\rho_{b0}} - 1)\frac{\chi S_u A_{f\,\max}}{c_d A_v}\right]^2 \qquad [3]$$

where S_u is the laminar burning velocity of the gas-air mixture, χ is the empirical ratio of the turbulent burning velocity to the laminar burning velocity, ρ_{u0} and ρ_{b0} are the unburned and isobaric burned gas densities, respectively, A_v is the vent area, c_d is the vent discharge coefficient, and P_0 is the initial (usually atmospheric) pressure in the enclosure .

Some vented deflagration models, e.g. [8], are derived based on the assumption of spherical flame propagation, and then attempt generalization to non-spherical enclosures by relating the maximum spherical flame surface area to $V^{2/3}$, where V is the enclosure volume. The Swift-Epstein equation, which is the basis for the low pressure gas deflagration guidelines in NFPA 68 [9], equates A_{fmax} to the enclosure inside surface area, A_s, where $A_s = 2(wL + wH + LH)$ using the notation in Figure 1. If we use this approach for flammable gas layers, the value of χ in Equation 8 will have to be quite small for thin layers. Another possible representation for A_{fmax} would be to use the layer surface area, A_{layer}, where $A_{layer} = 2(wL + wh_l + Lh_l)$. This should be a better representation for thin layers. However, large-scale test data are needed for both approaches in order to determine realistic values of χ to use.

The Fire Research Station natural gas layer deflagration venting tests reported by Butlin and Tonkin [10] and by Buckland [11] provide one good database to explore the use of Equation 3. Their tests were conducted in a 3.65 m by 3.05 m by 2.44 m high enclosure with a provision to have polyethylene film vent areas equal to ⅛, ¼, and ½ of the 3.05 m x 2.44 m wall area, i.e. h_v/H ratios of 0.125, 0.25, and 0.50. The uniform gas layers were produced by slowly injecting 10% natural gas in air mixtures under the ceiling while monitoring the layer concentration [12]. Once the desired depth of the gas layer was established, the layer was ignited by a high voltage spark located on the rear wall of the enclosure at the bottom of the gas layer.

Peak deflagration pressures measured by Butline and Tonkin [10] were in the range 0.5 to 12 kPa (0.007 to 1.7 psig) depending on the values of h_l and h_v used in each test. Using these measured values of $P_{red} - P_0$, a value of $S_u = 0.35$ m/s, and calculated values of ρ_{u0} and ρ_{b0} for stoichiometric methane-air mixtures at a pressure of one atmosphere, $c_d = 0.80$, and the corresponding reported values of h_l and A_v, Eqn 3 was used to calculate values of χ for different representations of A_{fmax}. Figure 2 shows the calculated values of χ based on $A_{fmax} = A_{layer}$, while Figure 3 shows the corresponding plot for $A_{fmax} = A_s$.

Results for $h_v/H = 0.125$ in Figure 2 show that $\chi \approx 1$ for all gas layers tested. In other words, there does not same to be any significant flame acceleration or flame surface area increase for these small vent areas. Results for $h_v/H = 0.25$ and 0.50 show larger values of χ that increase linearly with increasing gas layer depth or fill fraction. The average value of χ at $h_v/H = 0.25$ is 1.3, while the average value of χ at $h_v/H = 0.50$ is 1.53. The implied significant flame acceleration or flame surface area enhancement with these larger vent areas may be due to either a flame instability or to external combustion of the vented unburned gas.

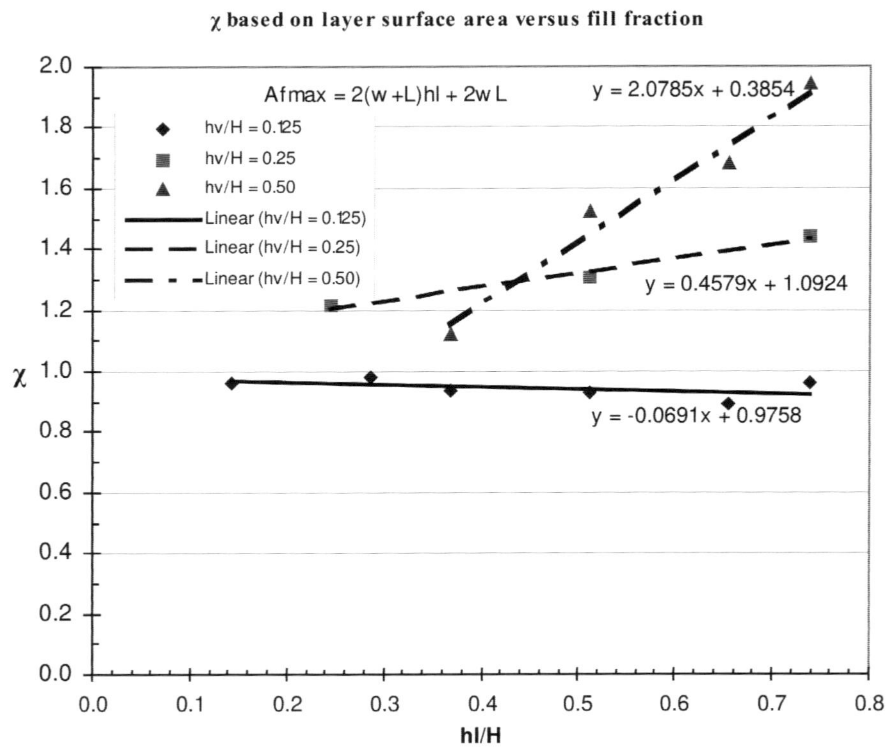
χ based on layer surface area versus fill fraction
Afmax = 2(w +L)hl + 2w L
y = 2.0785x + 0.3854
y = 0.4579x + 1.0924
y = -0.0691x + 0.9758
hv/H = 0.125
hv/H = 0.25
hv/H = 0.50
Linear (hv/H = 0.125)
Linear (hv/H = 0.25)
Linear (hv/H = 0.50)
χ
hl/H
2.0
1.8
1.6
1.4
1.2
1.0
0.8
0.6
0.4
0.2
0.0
0.1
0.2
0.3
0.4
0.5
0.6
0.7
0.8

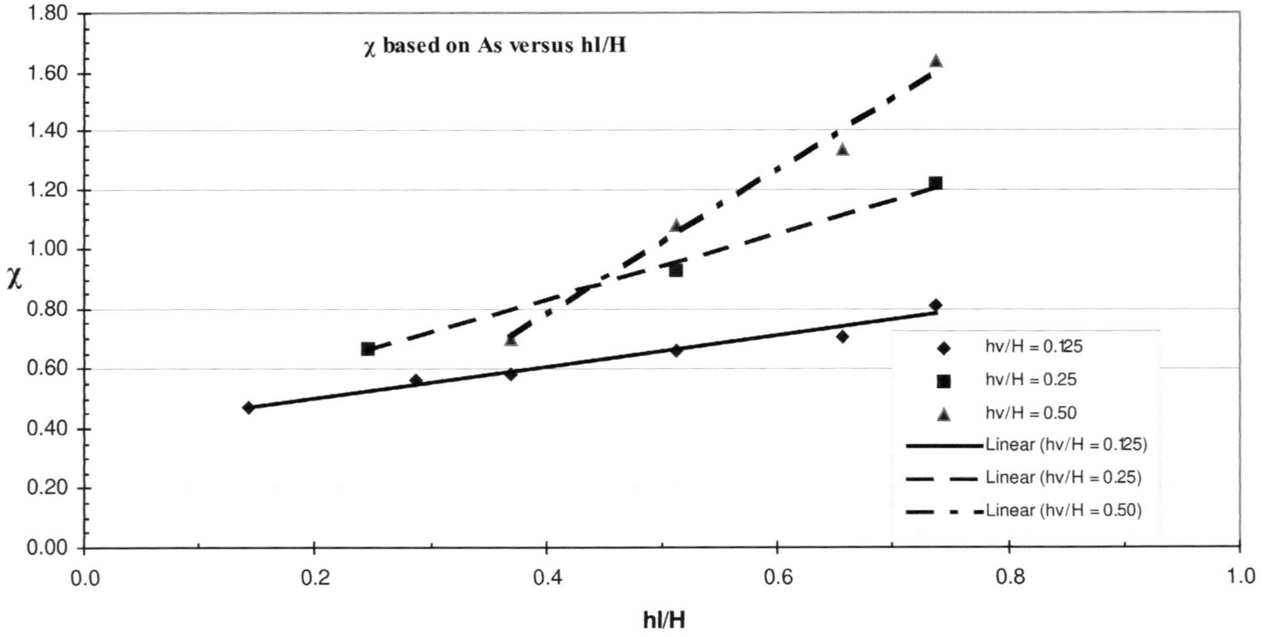
χ based on As versus hl/H
hv/H = 0.125
hv/H = 0.25
hv/H = 0.50
Linear (hv/H = 0.125)
Linear (hv/H = 0.25)
Linear (hv/H = 0.50)
χ
hl/H
1.80
1.60
1.40
1.20
1.00
0.80
0.60
0.40
0.20
0.00
0.0
0.2
0.4
0.6
0.8
1.0

Results in Figure 3 confirm that use of the enclosure internal surface area to compute P_{red} requires χ values significantly lower than 1 when the layer depth is less than half of the enclosure height. The appropriate value of χ for this representation of A_{fmax} increases linearly with layer depth (fill fraction) for all deflagration vent ratios. Since values of $\chi < 1$ imply unrealistic representations of the flame surface area, it is clear that A_{fmax} is significantly less than A_s for fill fractions less than 0.5, and even for larger fill fractions when h_v/H is 0.125. Hence, the association of A_{fmax} with A_{layer} and the use of χ values from Figure 2 is much more realistic.

When the Butlin and Tonkin data [10] for P_{red} are used to calculate an NFPA 68 Swift-Epstein C factor, where $C = \sqrt{P_{red}}\left(A_s / A_v \right)$, the values for all tests are an order-of-magnitude less than the NFPA 68 value for methane (C = 0.14 psi$^{1/2}$ = 0.37 kPa$^{1/2}$). This suggests that gas layer deflagrations can be successfully vented using much less vent area than is needed for full enclosure gas mixture deflagrations. However, it is premature to modify the guidelines without additional testing or data analysis that would include the conditions that are known to produce large flame accelerations. These conditions include highly obstructed enclosures and initially turbulent gas mixtures.

4. Ceiling Layer Flame Radiation

Burned gas produced in a ceiling layer deflagration can pose a significant thermal radiation threat to personnel below. The radiant heat flux, q", from the burned gases to a target below is given by

$$q'' = F_{1-2}\varepsilon\sigma T_f^4 \qquad [4]$$

where F_{1-2} is the geometric configuration factor between the gas layer and target, σ is the Stefan-Boltzmann constant (5.67×10^{-8} W/m^{2}K^4), T_f is the layer flame temperature, and the emissivity, ε, for gaseous combustion products CO_2 and H_2O is given by

$$\varepsilon = \varepsilon_{H2O}(2h_b p_{H2O}, T_f) + \varepsilon_{CO2}(2h_b p_{CO2}, T_f) \qquad [5]$$

where h_b is the depth of the burned gas layer, and p_{H2O} and p_{CO2} are the partial pressures of water vapor and carbon dioxide, respectively, in the burned gas.

The emissivities for H_2O and CO_2 are given by Tien et al [13] in the form of charts based on the gas partial pressures, ceiling layer path length ($2h_b$), and burned gas temperatures. Use of Equation 5 implies that we are ignoring any wavelength band overlap correction, which would cause a slightly lower total emissivity. We are also assuming there is no other radiating gaseous combustion product for the lean mixture gaseous fuels of interest in this paper.

The radiation view factor to a target below depends on the burned gas layer area and ceiling elevation, as well as the target elevation and position with respect to the center of the burned gas layer. In order to evaluate F_{1-2} and ε, we have to derive a model to estimate the area and depth of the burned gas layer as a function of time as the flame propagates through the gas-air mixture. Our approach here is to develop some working equations for two simple configurations so that we can explore the sensitivities to the various parameters without conducting tedious and lengthy calculations for each set of conditions. The configurations selected for this purpose are an unbound ceiling channel formed by adjacent ceiling beams, and a bounded ceiling channel or pocket formed by intersecting beams and joists.

The geometry and nomenclature for the unbounded ceiling channel is shown in Figure 4. Ignition is assumed to result in three stages of flame propagation. Stage 1 is equilateral propagation in all three directions, which ends quickly (order of 0.1 second for most layer and

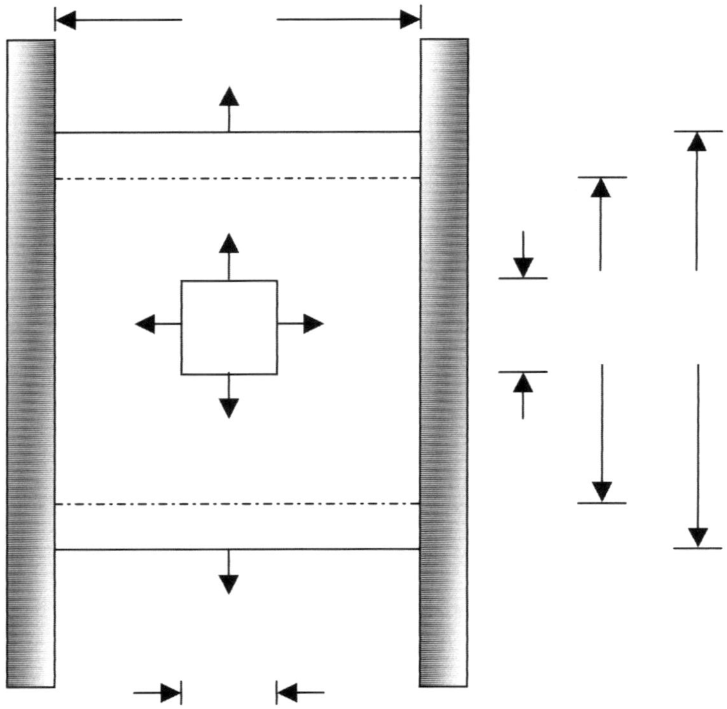

gas mixture conditions) as the flame reaches the bottom of the gas layer. Stage 2 is horizontal propagation both across and along the channel with the unburned gas allowed to flow outward along the length of the channel as illustrated in Figure 4. After the flame propagates across the channel width, W, Stage 3 begins in which the flame only propagates along the channel length. Finally, there is a fourth stage of burned gas spread in which the hot gas spreads along the channel because of a (buoyant) gravity front rather than as a propagating flame front.

Since there is no pressure development in this simple model, the burned gas is allowed to expand without constraint so that the burned gas density, ρ_b, is equal to $\rho_o(T_f/T_\infty)$, assuming the mixture is sufficiently lean that the burned gas molecular weight is roughly equal to that of the unburned gas. The flame speed, v_f, and unburned gas velocity, v_u, with these assumptions are:

$$v_f = \left(\frac{T_f}{T_\infty}\right)\chi S_u \qquad [6]$$

$$v_u = \left(\frac{T_f}{T_\infty} - 1\right)\chi S_u \qquad [7]$$

During Stage 2 of flame propagation, the mass of unburned gas is given by:

$$m_u = \rho_o V_u = \rho_o \left(WL_u - W_f L_f\right)h_u \qquad [8]$$

where h_u is the unburned gas layer depth. The flame dimensions, L_f and W_f, and the unburned gas length, L_u, are given by:

$$L_f = W_f = 2\left(\frac{T_f}{T_\infty}\right)\chi S_u t \qquad [9]$$

$$L_u = 2\left(\frac{T_f}{T_\infty} - 1\right)\chi S_u t + L_o \qquad [10]$$

Using mass conservation, the mass of burned gas, m_b, is:

$$m_b = \rho_b W_f L_f h_b \quad = \quad \rho_0 L_0 W h_0 - \rho_0 V_u \qquad [11]$$

Now substituting Eqn 8 and the burned gas density relation into Eqn 11 and solving for h_b gives:

$$h_b = \left(\frac{T_f}{T_\infty}\right)\frac{WL_o h_o - \left(WL_u - W_f L_f\right)h_u}{W_f L_f} \quad \text{in Stage 2} \qquad [12]$$

Based on the flame area, the burning rate is given by,

$$\frac{dm_b}{dt} = -\frac{dm_u}{dt} = \rho_o \chi S_u 2\left(W_f h_u + L_f h_u\right) \qquad [13]$$

Upon differentiating Eqn 8 and setting it equal to the right hand side of Eqn 13, we obtain the following time derivative for h_u:

$$\frac{dh_u}{dt} = \frac{h_u \chi S_u \left[-8\left(\frac{T_f}{T_\infty}\right)\chi S_u t - 2W\left(\frac{T_f}{T_\infty} - 1\right) + 8\left(\frac{T_f}{T_\infty}\right)^2 \chi S_u t\right]}{W\left[2\left(\frac{T_f}{T_\infty} - 1\right)\chi S_u t + L_o\right] - \left[2\left(\frac{T_f}{T_\infty}\right)\chi S_u t\right]^2} \qquad [14]$$

Numerical integration of Eqn 18 until $t = WT_\infty/(2T_f\chi S_u)$ allows h_u to be calculated during Stage 2 of flame propagation. Results for typical gas layer conditions showed a variation up to about 15% from the initial layer depth, h_0.

During flame propagation Stage 3, the unburned gas mass is given by

$$m_u = \rho_o V_u = \rho_o (L_u - L_f) W h_u \qquad [15]$$

and the equations corresponding to Equations 12 and 13 for Stage 12 now are:

$$h_b = \left(\frac{T_f}{T_\infty}\right) \frac{L_o h_o - (L_u - L_f) h_u}{L_f} \quad \text{in Stage 3} \qquad [16]$$

$$\frac{dm_b}{dt} = -\frac{dm_u}{dt} = \rho_o \chi S_u 2(W h_u) \qquad [17]$$

Using the same procedure as in Stage 2 to solve for h_u now results in $\frac{dh_u}{dt} = 0$ for Stage 3.

Stage 3 ends when $L_f = L_u$, i.e. when all the flammable gas has burned. At that time, the hot gases begin to spread along the channel ceiling as a gravity front. The propagation velocity of the front edge of a buoyant layer in a corridor or tunnel is given by

$$\frac{dL_f}{dt} = C_{gf} \sqrt{g \left(\frac{\rho_\infty - \rho_b}{\rho_\infty}\right) h_b} \qquad [18]$$

where different values of the coefficient C_{gf} have been obtained by various researchers [14, 15]. We have used the value $C_{gf} = 2$, which is close to the original value observed by Thomas [14] in one of the first experiments involving smoke layer spread.

Conserving mass after all unburned gas is consumed and solving for h_b yields:

$$h_b = \frac{\rho_\infty L_o W h_o}{\rho_b L_f W} \quad \text{in Stage 4} \qquad [19]$$

Noting that the ratio of the densities of burned gas to unburned gas is the ratio of the initial temperature to final temperature of the gas (in Kelvin), and substituting Eqn 19 into Eqn 18 gives

$$\frac{dL_f}{dt} = K \sqrt{\frac{1}{L_f}} \qquad [20]$$

where

$$K = 2\sqrt{g\left(\frac{T_f}{T_\infty - 1}\right)L_o h_o} \qquad [21]$$

Noting that at time t_{i4} (beginning of Stage 4), $L_f = L_{fi}$, and integrating [20] gives,

$$L_f = \left[\tfrac{3}{2}K(t - t_{i4}) + L_{fi}^{3/2}\right]^{2/3} \qquad [22]$$

Computations using the preceding equations have been conducted for a variety of conditions using the GasEq computer code to calculate flame temperatures and combustion product compositions, Strehlow's burning velocities [16], and the radiation view factor equations in Seigel and Howell [17]. The baseline case in these calculations has been 6% Methane-Air gas mixture, ho = 0.1 m, $\chi = 2$, ceiling height H = 7 m, and W = L_0 = 5 m, and target elevation = 1.8 m (head height). Calculated values of q"(t) at various target locations for this baseline case are shown in Figure 5.

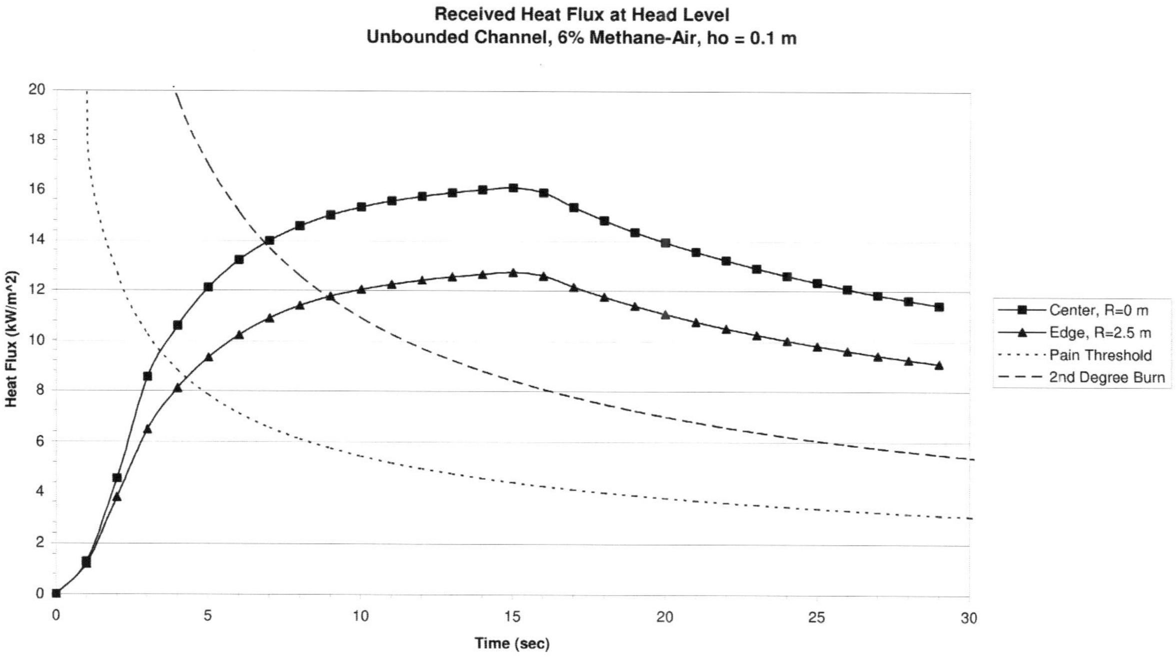

Figure 5. Radiant heat flux versus time for 0.1 m deep, 6% methane-air mixture in unbounded channel.

The heat flux in Figure 5 increases with time up to about 15 seconds as the burned gas area increases. After this time, there is no more unburned gas, and the calculations are conducted for Stage 4, i.e. the buoyant layer spreading stage. The area still increases in Stage 4 but the burned gas depth decreases sufficiently to reduce the gas emissivity. Thus the radiant heat flux gradually diminishes during Stage 4. The heat flux for a target at R = 0 under the center of the gas layer reaches a peak of 16 kW/m^2, while the peak value for a target at R = 2.5 m (at the edge of the gas layer, i.e. under the channel boundary) is about 13 kW/m^2.

The other two curves in Figure 5 show the exposure times associated with second-degree burns and with the pain threshold using correlations developed by Weiczorek and Dembsey [18]. The exposure times in these correlations are for constant radiant heat fluxes. Since the radiant heat fluxes calculated for ceiling layer deflagrations increase with time up to Stage 4, the intersection points with the correlations represent conservative estimates of times required to produce pain or burn injury on bare skin. The intersection points in Figure 5 for a target under the center of the layer are about 10 kw/m^2 at 4 seconds after ignition for pain threshold, and about 14 kW/m^2 about 7 seconds after ignition for second-degree burns. These times are for a stationary observer, whereas mobile personnel could possibly run out from under the 5 m wide channel in this time period to escape serious burn injury for this case.

Results for other calculations with methane-air mixtures of different concentrations and different layer depths are shown in Table 1. The heat fluxes reported in Table 1 are the peak values, q''_{peak}, and the values corresponding to second-degree burns, q''_{2d}. The times, t_{2d}, at which the calculated heat flux curve intersects the second-degree burn correlation, and the times, t_{peak}, corresponding to q''_{peak}, are also reported in Table 1. All other parameters in these calculations corresponded to the baseline case.

Table 1. Key Results for Unbounded Channel Radiant Heat Flux Calculations

CH_4 Concentration	Layer Depth h_0 (m)	t_{2d} (s)	q''_{2d} (kW/m^2)	t_{peak} (s)	q''_{peak} (kW/m^2)
5%	0.1	17	8	30	9
6%	0.1	7	14	15	16
7%	0.1	3	23	7	27
8%	0.1	1.8	35	4	42
6%	0.2	4.5	18	15	26
6%	0.5	2.7	25	15	52

The values of t_{2d} and t_{peak} in Table 1 decrease with increasing methane concentration because S_u increases. The corresponding heat fluxes increase with increasing methane concentration because of the increasing flame temperature and CO_2 and H_2O emissivities. The heat fluxes also increase with increasing layer depth because of the corresponding increasing gas emissivities.

The bounded channel configuration shown in Figure 6 produces a different flame spread rate because the channel walls inhibit flow of the unburned gas ahead of the flame front when L_u becomes equal to L. Beyond this time, the unburned gas velocity is set equal to zero (neglecting any downward flow under the channel wall), and the flame spread rate is:

$$\frac{dL_f}{dt} = 2v_f = 2\chi S_u \qquad [23]$$

which is significantly slower than in the unbounded channel. However, since the burning velocity for a given gas concentration is the same in both cases, the unburned gas is burned in the same time as in the unbounded channel configuration. After the gas is burned, the channel walls prevent any gravity front propagation. Thus, the flame length at a given time is smaller in the bounded channel configuration, and the corresponding burned gas layer depth is larger. This produces a larger emissivity and radiant heat flux to a target directly under the gas layer than the corresponding unbounded channel case.

Bounded Channel

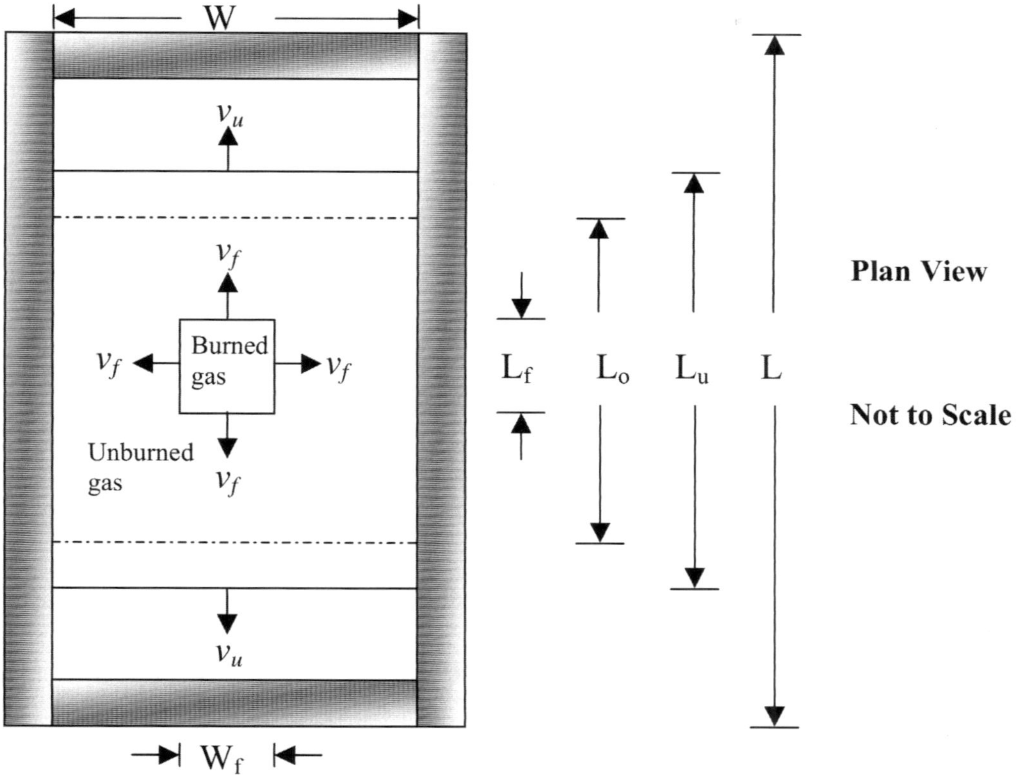

Figure 6. Flame propagation through gas layer in bounded ceiling channel.

Some examples of the calculated heat fluxes for bounded channels are shown in Figure 7 for a 6% methane-air mixture, and in Figures 8 and 9 for hydrogen-air mixtures with hydrogen concentrations of 10% and 20%, respectively. A comparison of these three figures shows that a 10% hydrogen concentration does not produce a large enough radiant heat flux to produce pain or injury, but that a 20% hydrogen concentration produces substantially larger radiant heat fluxes and burn injury threats, and at earlier time, than does a 6% methane concentration.

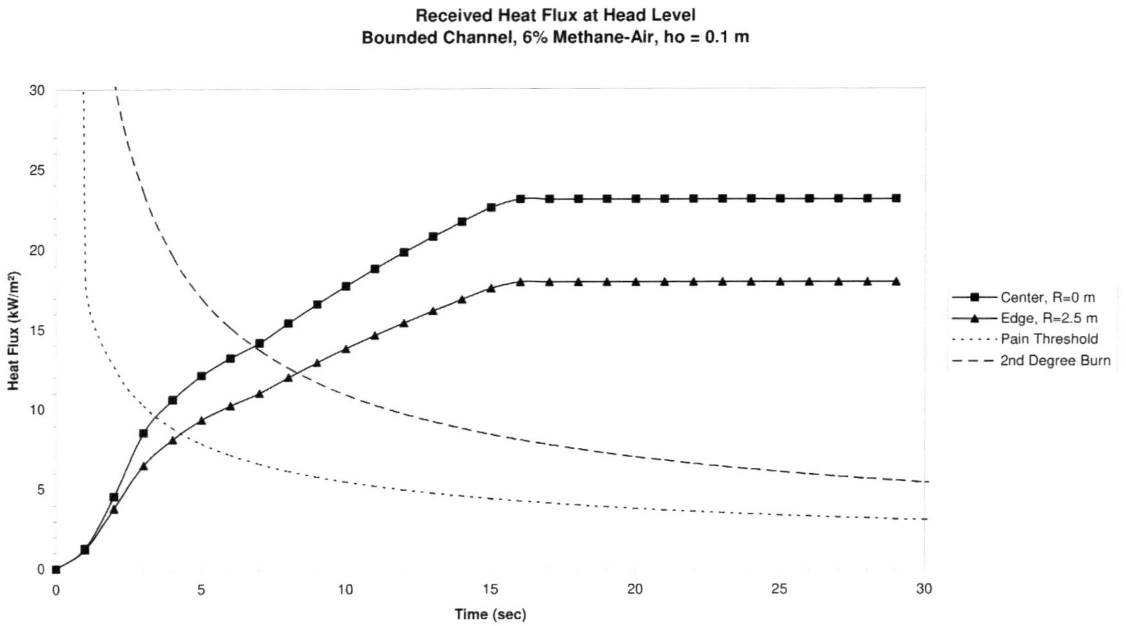

Figure 7. Radiant heat flux versus time for 6% methane-air mixture in a bounded channel.

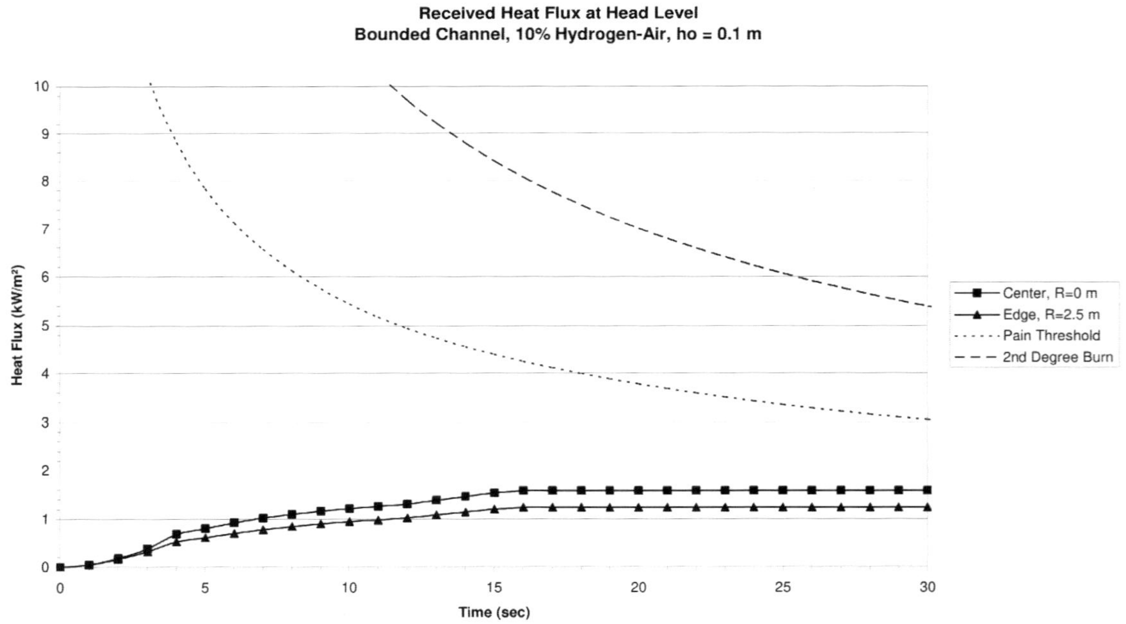

Figure 8. Radiant heat flux versus time for 10% hydrogen-air mixture in bounded channel.

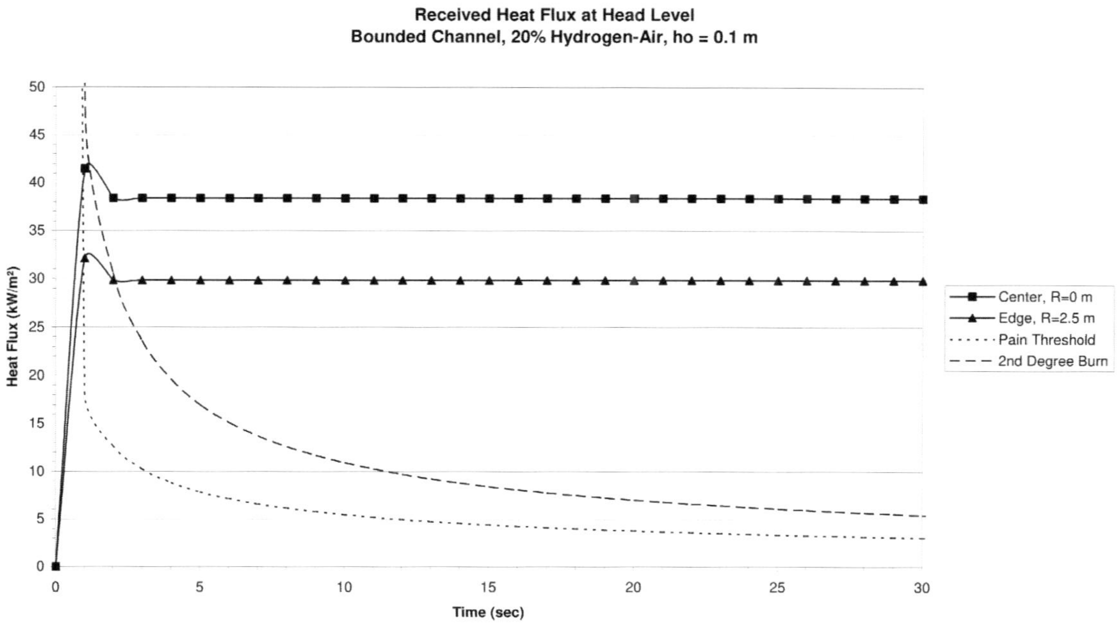

Figure 9. Radiant heat flux versus time for 20% hydrogen-air mixture in bounded channel.

Results for a layer depth, length, and width of 0.1 m, 5 m, and 5m, respectively, and various methane concentrations, ceiling heights, and bounded channel lengths are shown in Table 2. The trend for increasing methane concentration is as it was for the unbounded channel, but the peak heat fluxes are larger in the bounded channel. The peak heat flux decreases with increasing ceiling height, and the escape time before onset of burn injury increases as the ceiling is raised. The peak heat flux also decreases with increasing channel length (for a fixed layer size) because the increasing length allows for a thinner layer of burned gas with a correspondingly lower emissivity.

Table 2. Key Results for Bounded Channel Radiant Heat Flux Calculations

CH_4 Concentration	Ceiling Height (m)	Channel Length (m)	t_{2d} (s)	q''_{2d} (kW/m^2)	t_{peak} (s)	q''_{peak} (kW/m^2)
5%	7	15	17	8	> 30	> 11, < 20
6%	7	15	6	15	15	23
7%	7	15	3	23	7	43
8%	7	15	1.6	37	5	70
6%	5	15	4	19	15	35
6%	10	15	10	11	15	13.5
6%	7	5	6	15	15	33
6%	7	10	6	15	15	28

5. Conclusions

A brief review of buoyant gas dispersion phenomena in an enclosure shows that there are many release and enclosure conditions in which near-uniform concentrations of gas develop in a ceiling layer. Mathematical models described in the literature are available for determining the layer depth and concentration variation with time for simple geometries and release conditions.

The maximum pressure developed when the flammable ceiling gas layer is ignited in a vented enclosure with sufficient area and low vent release pressure to limit pressures to about 12 kPa (1.7 psig) can be estimated using a simple theoretical equation with an empirically determined parameter, χ, to characterize enhanced burning rates. Data from a previously reported [10, 11] explosion test series with natural gas ceiling layers in an otherwise empty 27 m^3 enclosure has been used to develop correlations for χ as a function of the ratios of ceiling layer depth to ceiling height and of vent depth to ceiling height.

A simple set of equations has also been developed to calculate the transient radiant heat flux resulting from ignition of a gas layer under a ceiling in an enclosure that is either sufficiently large or sufficiently open so that there is no significant pressure development as the flame propagates through the gas layer. Calculated heat fluxes for a wide variety of lean mixture methane concentrations and ceiling layer depths show how the peak heat flux heat increases with increasing concentration and depth. Comparisons with correlations for the heat fluxes that can cause second-degree burns as a function of bare skin exposure time show that head burns would be anticipated at minimum exposure times in the range from 3 seconds to 17 seconds for a 1.8 m tall person standing under the center of the gas layer. Similar calculations for hydrogen ceiling layers indicate that the radiant heat flux at this elevation would be below the burn injury threshold for hydrogen concentrations less than 10%, but would be sufficiently high to cause burn injury at exposure times as short as 2 seconds when the hydrogen concentration is 20%.

6. References

[1] Baines, W.D., and Turner, J.S., "Turbulent Buoyant Convection from a Source in Confined Region," *J. Fluid Mechanics,* vol 37, 1969, pp. 51-80.

[2] Cleaver, R.P., Marshall, M.R., and Linden, P.F., "The Build-up of Concentration within a Single Enclosed Volume Following a Release of Natural Gas," *J. of Hazardous Materials,* vol 36, 1994, pp. 209-226.

[3] Alpert, R., "Ceiling Jet Flows," *SFPE Handbook of Fire Protection Engineering*, 3rd Edition, pp 2-18 to 2-31, 2002.

[4] Swain, M.R. and Swain, M.N., "Passive Ventilation Systems For The Safe Use Of Hydrogen," *Int J Hydrogen Energy*, v 21, 1996, pp 823-835.

[5] Swain, M. R.; Filoso, P.; Grilliot, E. S.; Swain, M. N., "Hydrogen Leakage Into Simple Geometric Enclosures," *Int J Hydrogen Energy*, v 28, 2003, pp 229-248.

[6] Zalosh, R., Ho, S-P., Eber, R., "Compressed Natural Gas Dispersion Testing in Parking Garages," WPI Center for Firesafety Studies Report for the Massachusetts Division of Energy Resources, August 2000.

[7] Swift, I. and Epstein, M., "The Performance of Low-Pressure Vents," Proceedings 20th AIChE Loss Prevention Symposium, 1986.

[8] Bradley, D. and Mitcheson, A., "The Venting of Gaseous Explosions in Spherical Vessels," *Combustion and Flame*, v 32, 1978, pp 221-236.

[9] NFPA 68-2002, "Guide for Venting of Deflagrations," National Fire Protection Association, 2002.

[10] Butlin, R.N. and Tonkin, P.S., "Pressures Produced by Gas Explosions in a Vented Compartment," Fire Research Station Note No. 1019, 1974.

[11] Buckland, I.G, "Explosions Of Gas Layers In A Room Size Chamber," Institution of Chemical Engineers Symposium Series N. 58, Rugby, Engl, 1980 p 289-304.

[12] Butlin, R.N., "Production of Gas Layers for Large-Scale Gas Explosion Studies, Part 1. Preliminary Investigations" Fire Research Station Note No. 1004, 1976.

[13] Tien, C.L., Lee, K.Y., Stretton, A.J., "Radiation Heat Transfer," *SFPE Handbook of Fire Protection Engineering*, 3rd Edition, Society of Fire Protection Engineers and National Fire Protection Agency, 2002.

[14] Thomas, P., "The Movement of Smoke in Horizontal Passages Against an Air Flow," Fire Research Note 723, Fire Research Station, 1968.

[15] Kunsch, J., "Simple Model for Control of Fire Gases in a Ventilated Tunnel," *Fire Safety Journal,* v 37, pp 67-81, 2002.

[16] Strehlow, R., *Combustion Fundamentals*, McGraw Hill, 1984.

[17] Siegel, R and Howell, J.R., *Thermal Radiation Heat Transfer*, Hemisphere Pub. Corp., Washington, 1980.

[18] Wieczorek, C.J. and Dembsey, N.A., "Human Variability Correction Factors for Use with Simplified Engineering Tools for Predicting Pain and Second Degree Skin Burns," *Journal of Fire Protection Engineering*, Vol 2, May 2001.

BLEVE Blast
By Expansion-Controlled Evaporation

A.C. van den Berg, M.M. van der Voort, J. Weerheijm and N.H.A. Versloot
TNO Prins Maurits Laboratory
P.O.Box 45
2280 AA Rijswijk
The Netherlands
Email: berg@pml.tno.nl

Prepared for Presentation at
American Institute of Chemical Engineers
2005 Spring National Meeting
39th Annual Loss Prevention Symposium
Atlanta, Georgia
April 11-13, 2005

UNPUBLISHED

BLEVE Blast
By Expansion-Controlled Evaporation

A.C. van den Berg, M.M. van der Voort, J. Weerheijm and N.H.A. Versloot
TNO Prins Maurits Laboratory
P.O.Box 45
2280 AA Rijswijk
The Netherlands
Email: berg@pml.tno.nl

Abstract

This paper presents a new method to calculate the blast effects originating from an exploding vessel of liquefied gas. Adequate blast calculation requires full knowledge of the blast source characteristics, i.e.: the release and subsequent evaporation rate of the flashing liquid. As the conditions that allow explosive evaporation are not entirely clear and the evaporation rate of a flashing liquid is unknown, safe assumptions have been adopted as the starting point in the modeling. The blast effects from a BLEVE are numerically computed by imposing the vapour pressure of a flashing liquid as boundary condition for the gas dynamics of expansion. The numerical modeling is quantitatively explored just for liquefied propane. In addition, this paper demonstrates that often an estimate of BLEVE blast effects is possible with very simple acoustic volume source expressions.

The modeling shows that the rupture of a pressure vessel containing a liquefied gas in free space only develops a blast of significant strength if the vessel near-instantaneously disintegrates. Even if a rupture and the consequent release and evaporation of a liquefied gas extend over just a short period of time, the blast effects are minor.

1. Introduction

The word BLEVE is an acronym for "Boiling Liquid Expanding Vapour Explosion". A BLEVE is the consequence of the rupture of a pressure vessel containing a liquefied gas. The temperature of the liquid exceeds the boiling temperature under ambient conditions. Under atmospheric pressure propane, for instance, has a boiling temperature of 231 K. To keep propane liquid at an ambient temperature of 288 K, it must be stored in a vessel under a pressure of 730 kPa, at 300 K under 1000 kPa and at 326 K under 1800 kPa, the vapour pressures of propane at the respective temperatures.

If due to a sudden rupture of the vessel the liquid pressure suddenly drops, a fraction of the liquid will quickly evaporate. The evaporation process extracts heat from the liquid by which the temperature and the vapour pressure of the liquid fall. The evaporation continues until the liquid temperature has fallen to the boiling temperature at atmospheric pressure and consequently the vapour pressure has fallen to atmospheric pressure.

The quick phase change from liquid to vapour goes hand in hand with the rise of a large volume. One cubic metre of liquid propane, for instance, gives rise to about 260 cubic metres vapour under atmospheric conditions. The volume of vapour pushes the surrounding air aside, which produces a blast wave. The blast wave may do damage up to a substantial distance in the environment. Therefore, the burst of pressure vessels containing liquefied gases is a major scenario considered in hazard analyses in the process industries.

So far however, appropriate methods for the calculation of the blast effects from an exploding pressure vessel of liquefied gas are not available. For lack of better, BLEVE blast effects are often approximated with methods that do not fully apply. The near-field effects are, for instance, calculated with methods for expansion of a pressurised perfect gas while the far-field effects are calculated with TNT-equivalency methods (CCPS, 1994 and CPR 14E, 1997). Such an approach is not very satisfactory.

A literature review in Chapter 2 reveals that present knowledge of the conditions that allow explosive evaporation upon a vessel rupture is insufficient. In addition, the exact evaporation rate of flashing superheated liquids is unknown. Therefore, a safe approach in the modeling of blast effects is the proper way for the time being. The modeling is described in Chapter 3 while the modeling is quantitatively explored in Chapter 4.

2. Some Issues Raised In The Literature

Various authors define a BLEVE in different ways. Birk and Cunningham (1994), for instance, defined a BLEVE as any catastrophic failure of a vessel containing a liquefied gas. Reid (1976 and 1979), on the other hand, defined a BLEVE as a catastrophic failure of a vessel with a liquefied gas followed by explosive evaporation.

The incident record shows that a catastrophic failure of a vessel of liquefied gas is often the consequence of a fire. A fire heats the vessel contents by which the internal pressure rises while at the same time the vessel structure reduces in strength. The consequence of such a scenario is that the flammable vessel contents catch fire directly upon the rupture, resulting in a big fire ball. In the literature this entire sequence of events is often wrongly designated as a BLEVE. In this report the term BLEVE is reserved to just the explosive rupture of a pressure vessel and the flash evaporation of liquefied gas. The subsequent fire ball will not be considered here.

Generally speaking, evaporation of superheated liquid takes place only at the tank wall and is dependent on the temperature in the liquid at boiling nuclei such as impurities or ions. According to Reid (1976 and 1979) explosive evaporation can only occur when the liquid temperature exceeds the super heat limit temperature i.e. the homogeneous nucleation temperature. Only then simultaneous evaporation can occur throughout the liquid. Only such an evaporation process is fast enough to produce significant blast effects. Then throughout the liquid, vapour will develop and powerfully expand entraining the liquid. The liquid will atomize into small droplets, which allows an efficient phase change.

The exact value of the super heat limit temperature or the homogeneous nucleation temperature is not quite clear. Various authors calculated different values on the basis of different equations of state. Reid (1976) predicts the super heat limit temperature to be approximately equal to 89% of the critical temperature.

Other authors (Venart e.a., 1992 and 1993) postulate that explosive evaporation would be possible far below the homogeneous nucleation temperature by a curious interaction between the liquefied gas and a release flow through, for instance, a safety valve or a leak. A release flow leads to a pressure drop in the vessel and a consequent evaporation. The vapour bubbles in the liquid make the liquid swell and increase the liquid level in the vessel by which the vent may start to release liquid. The liquid chokes the release flow after which the internal vessel pressure rises again. The pressure rise would make the vapour bubbles implode. Implosion of vapour bubbles is accompanied by strong local pressure pulses. This would result in a coherent power amplified shock wave through the liquid that may shatter the vessel and explosively evaporate a portion of its contents. The exact conditions however, that would allow such phenomena are not very clear.

This concise literature review indicates that it is not certain when and under what conditions homogeneous nucleation and explosive evaporation are to be expected. The assumption of explosive evaporation seems therefore a very reasonable safe starting point in BLEVE blast modeling. Explosive evaporation is very fast indeed, but an exact rate of evaporation is unknown. Because blast effects are fully determined by the exact vapour source strength, i.e. the flash evaporation rate, a safe approach in the quantification of BLEVE blast is appropriate for the time being.

3. Modeling

3.1 Evaporation

A BLEVE is the explosive evaporation process as a consequence of the rupture of a pressure vessel containing a liquefied gas. Initially, the temperature of the liquid exceeds the boiling point under ambient conditions. The pressure in the vessel is equal to the vapour pressure at the liquid temperature. When the vessel ruptures, the pressure in the liquid falls and a superheated liquid remains. The subsequent flash evaporation process requires heat, which is extracted from the liquid. The decrease of the liquid mass and the fall of the liquid temperature are related by:

$$M_l.C_l.dT = L.dM_l$$

where M_l = liquid mass (kg)
C_l = specific heat liquid (kJ/kg/K)
L = heat of evaporation liquid (kJ/kg)
dT = temperature fall liquid (K)

dM_l = mass decrease liquid (kg)

The evaporation process continues until the liquid temperature has fallen to the boiling temperature under ambient pressure. The liquid mass fraction that evaporates in this way (the flash fraction F) follows from the integration of the above relation over the temperature drop, ΔT, between the initial liquid temperature and the boiling point under ambient pressure:

$$F = 1 - EXP\left[\frac{-C_l.\Delta T}{L}\right]$$

Implementation of the appropriate values for the parameters for propane, for instance: C_l = 2.58 kJ/kg/K and L = 348 kJ/kg (CPR 14E, 1997) shows that if liquid propane flashes from 288 K (ambient) to 231 K (boiling point), about 35 % of the initial liquid mass evaporates. For temperatures of 300 K and 326 K (super heat limit), the flash fractions are 40% and 50% respectively.

3.2 *Blast*

3.2.1 Basic assumptions

The blast effects from a BLEVE are the consequence of the fast evaporation of a fraction of the superheated liquid. The transition of liquid to vapour is accompanied by the rise of a large volume, which has to push the surrounding air away. Obviously, the computation of the gas dynamics requires full knowledge of the volume source strength, i.e. the exact liquid release rate from the vessel and the exact evaporation rate of the flashing liquid as a function of time.

The literature review however, showed that at present the conditions that allow explosive evaporation are not very clear and are subject of speculation. In addition, any knowledge with regard to the exact evaporation rate during explosive evaporation is lacking. For the time being therefore, a safe and conservative approach in the modelling of BLEVE blast is appropriate.

Such an approach consists in the simple assumption that intrinsically the evaporation of superheated liquid can occur infinitely fast. If this is the case, extrinsic circumstances will determine the evaporation rate; extrinsic circumstances such as the liquid release rate and the gas dynamics (inertia) of the vapour and the surrounding air that is being pushed away. This safe assumption of expansion-controlled evaporation constitutes the starting point for the computation of the gas dynamics (blast) produced by a flash-evaporating liquid (Van den Berg et al., 2004).

If the failure of a vessel and the consequent liquid release extends over some length of time, the evaporation rate will be primarily determined by the release rate. The release rate together with the gas dynamics of expanding vapour and air determine the overpressure that develops. The faster the release, the higher the overpressure is. The limit is of course a near-instantaneous release, the consequence of a near-instantaneous disintegration of a pressure vessel. Then, the evaporation rate of the superheated liquid is fully determined by the rate at which the developing vapour can expand by moving the mass of vapour and the surrounding air. The gas dynamics of this expansion process are driven by the vapour pressure of the superheated liquid. During the evaporation the liquid temperature and thereby the vapour pressure drops until the boiling point under ambient conditions has been attained.

3.2.2 Acoustic blast modelling

Sometimes very simple concepts may be helpful in developing understanding and a feel for complicated phenomena. A simple concept, which already has been shown to be a very illuminating analogue for gas explosions (Strehlow, 1981), is the acoustic volume source (Lighthill, 1978). The acoustic volume source analogue consists of a simple expression that relates the acoustic wave overpressure at some distance to the strength of the volume source. The acoustic volume source analogue is a solution of the wave equation and takes a shape, which is dependent on the dimensionality of the space for which it has been formulated (Lighthill, 1978), i.e.:

- a volume source in a half space: $\dfrac{\Delta P}{P_o} = \dfrac{\gamma}{2\pi R.c_o^2}.\dfrac{d\Phi}{dt}$
- a volume source in a tube: $\dfrac{\Delta P}{P_o} = \dfrac{\gamma.\Phi}{2A.c_o}$

where: ΔP = wave overpressure (kPa)
P_o = ambient pressure (kPa)
γ = ratio specific heats
c_o = ambient speed of sound (m/s)
A = cross-sectional area tube (m^2)
Φ = volume source strength (m^3/s)
R = distance from source (m)
t = time (s)

Unfortunately, the acoustic volume source formulation for a space between parallel planes does not lead to an equally simple and practical expression for the wave overpressure.

Although, strictly speaking, the expressions only hold in the low (acoustic) overpressure regime, they demonstrate some interesting fundamental properties. A striking feature is that the blast overpressure from the rupture of a vessel of liquefied gas in a three-dimensional space is proportional to the time-derivative of the source strength, i.e. a steady release of superheated liquid in three-dimensional space does not give rise to any overpressure. Only the start up phenomenon, the very initial phase of growing release flow does this. In a tube, on the other hand, a steady release does give rise to blast.

3.2.3 Gas dynamic blast modelling

Quantitatively, the acoustic approximation fails for higher overpressures which develop for short release times. This is particular the case in a near-instantaneous disintegration of the vessel when the evaporation time is fully determined by the gas dynamics of the expansion.

The gas dynamics of the expanding vapour and the surrounding atmosphere can be computed by numerical solution of the Euler equations, the conservation equations for a compressible, inviscid perfect gas:

$$\frac{\partial \rho}{\partial t}+\frac{\partial}{\partial x_j}(\rho u_j)=0$$

$$\frac{\partial}{\partial t}(\rho u_i)+\frac{\partial}{\partial x_j}(\rho u_j u_i)+\frac{\partial p}{\partial x_i}=0$$

$$\frac{\partial}{\partial t}(\rho E)+\frac{\partial}{\partial x_j}(\rho u_j E)+\frac{\partial}{\partial x_j}(p u_j)=0$$

Where:

	ρ	= density (kg/m^3)
	u	= gas velocity (m/s)
	E	= $e + ½u^2$ (m^2/s^2)
	e	= internal energy (m^2/s^2)
	p	= pressure (Pa)
	x	= coordinate (m)
	t	= time (s)

The vapour pressure of the flashing liquid, which is dependent on the evaporated liquid mass fraction, and the consequent vapour mass flow serve as boundary conditions for a time stepwise numerical solution of the equations through a Flux-Corrected Transport scheme (Boris, 1976). The exact formulation of the boundary conditions is dependent on the release scenario considered.

- If the rupture of the vessel and the subsequent liquid release are nearly instantaneous, the boundary condition for the gas dynamics of the expanding vapour and surrounding air consists in the vapour pressure of the superheated liquid. The liquid vapour pressure falls with the growing evaporated liquid mass fraction.
- If the failure of the vessel is not instantaneous, a steady release rate of liquid is assumed over some span of time. The vapour mass flow equals the flash fraction of the liquid mass release flow. The boundary condition for the gas dynamics consists in the steady vapour mass flow for the duration of the release time.

The numerical solution of the Euler equations results in the blast effects originating from the explosive evaporation of a superheated liquid dependent on the geometry of the numerical mesh. A one-dimensional approach may result in three basic solutions.

- A hemispherical mesh results in the blast effects of a BLEVE in free space at the earth's surface.
- A cylindrical mesh results in the blast effects of a BLEVE in the space between two parallel rigid planes.

- A plane mesh results in the blast effects of a BLEVE in a tube.

If a one-dimensional approach does not satisfy, a BLEVE can be simulated in a three-dimensional space. Any area within a three-dimensional numerical mesh can be simply provided with the flash-evaporating liquid boundary conditions.

4. Results

4.1 Acoustic modelling

To enable a quick quantitative exploration of the volume source analogue in half space, a linearly growing liquid release rate is assumed as a consequence of a cracking pressure vessel of a volume V of 50 m^3 propane with a temperature of 326 K. The consequent constant time derivative of the source strength results in a blast wave with a square shape and a duration equal to the release time. The overpressure is inversely proportional to the square of the release time.

Taking the properties of propane at the superheat limit into account: a flash fraction F of approximately 50% and a liquid to vapour expansion factor α of about 260, the blast overpressure can be calculated for various release times Δt from:

$$\frac{\Delta P}{P_o} = \frac{\gamma}{2\pi R.c_o^2} \cdot \frac{2V.F.\alpha}{(\Delta t)^2}$$

The blast overpressure varies inversely proportional to the distance from the source and has been tabulated for a distance of 10 m from the source in Table 1.

Table 1: Blast overpressure at 10 m distance dependent on release time for a 50 m^3 propane BLEVE in half space

Release time (s)	ΔP (kPa) R=10 m
0.05	106
0.1	27
0.2	6.7
0.5	1.1
1	0.3

Contrary to what occurs in free space, in a tunnel tube a steady release of superheated liquid does give rise to a blast effect. Now, the assumption of a steady release of 50 m^3 of propane of 326 K in a tunnel tube of 60 m^2 cross-sectional area results in a square wave of a duration equal to the release time. The blast overpressure is inversely proportional to the release time. The overpressure has been calculated for various release times and tabulated in Table 2:

$$\frac{\Delta P}{P_o} = \frac{\gamma}{2A.c_o} \cdot \frac{V.F.\alpha}{\Delta t}$$

Table 2: Blast overpressure dependent on release time for a 50 m^3 propane BLEVE in a tunnel tube with a cross-sectional area of 60 m^2

Release time (s)	ΔP (kPa)
0.1	228
0.2	114
0.5	46
1	23
2	11

The acoustic wave remains unaltered propagating down the one-dimensional tunnel tube.

Although many of these figures are well beyond the application range of the acoustic source approximation, they do demonstrate interesting qualitative features. They indicate, for instance, that the blast effect of a collapsing vessel of liquefied gas is highly dependent on the release time and thereby on the exact rupture mode of the vessel and the consequent liquid release development. They also suggest that if a nearly instantaneous disintegration of a pressure vessel can be avoided (by a tailored structural design, for instance), the blast effects from a BLEVE in a three-dimensional space will be minor. An important consequence is also that any experimental observations on BLEVE blast effects can only make any sense if the rupture mode of the vessels used and the consequent release scenario can be controlled and are fully repeatable.

4.2 Gas dynamic modelling

4.2.1 Validation

The simple acoustic volume source analogue for BLEVE-blast in half space indicated that BLEVE-blast is highly dependent on the exact release and evaporation rate of the liquefied gas. If the crack and the consequent release of a vessel take just a fraction of a second, the blast effects are already minor.

A BLEVE blast model can only match experimental data if the exact liquid release and evaporation rate upon the rupture of a vessel in an experiment can be used as input. This seems almost an impossibility in practice. Only in the case when a liquefied gas is heated beyond the superheat limit and the vessel is near-instantaneously disintegrated, will the liquid release rate be at its maximum and the experimental conditions may approximate the model assumptions.

Such an experiment is used to validate the gas dynamic modelling. Giesbrecht et al. (1981), report on blast observations from exploding vessels of liquefied propylene. The BLEVE's were conditioned by heating the propylene up to 50-80 ^{o}C and about 4 MPa, followed by a more or less instantaneous disintegration of the vessel by detonation of a small high-explosive charge. The test series comprised vessels of volumes varying from 0.226 to 1000 liter, containing 0.125 to 452 kg of propylene. Blast overpressures had been recorded at various scaled distances from the vessels. A striking feature of the data is that the largest vessel containing 452 kg propylene produced the relatively highest blast overpressure, probably because the rupture of the largest vessel approximated an instantaneous disintegration best of all. Therefore, only the data concerning the largest vessel are considered in the validation exercise of which Figure 1 shows the result.

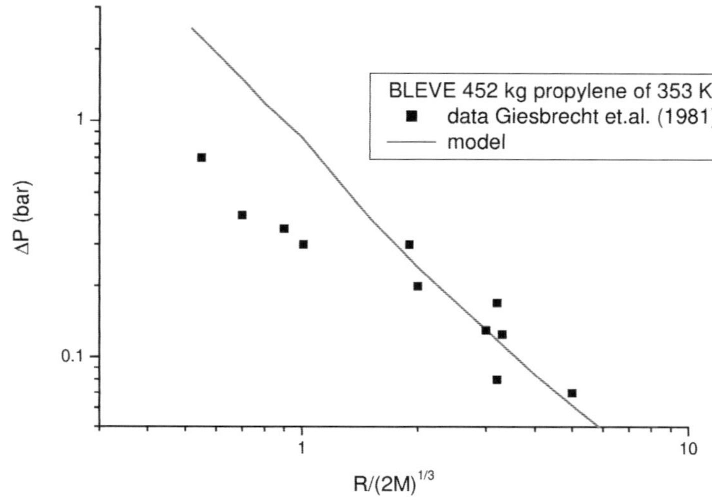

Figure 1: Observed and computed blast overpressures originating from an exploding vessel of 452 kg of propylene

Because the exploding vessel was positioned at some small height above ground level, the blast wave reflected at the earth's surface. The experimental blast peak overpressures in Figure 1 show how, with a growing distance from the vessel, the primary and reflected wave increasingly coincide. In order to take the earth's surface into account, following the representation by Giesbrecht et al. (1981), the data in Figure 1 have been scaled with twice the mass of propylene.

The experiment has been simulated by solving the Euler equations provided with the liquid propylene vapour pressure boundary conditions in a one-dimensional spherical mesh. Apart from the natural scatter inherent to blast experimentation, the result of this validation exercise in Figure 1 indicates that the gas dynamic model gives a realistic representation of this BLEVE blast experiment.

Comparison of the experimental data with outcome of the simple acoustic volume source under the assumption of a linearly growing release flow shows that the acoustic model and data match more or less for a release time of approximately 0.01 s.

4.2.2 Propane BLEVE blast charts for half space

The blast effects from an instantaneous disintegration of a pressure vessel of liquefied gas on the earth's surface are obtained by solving the Euler equations in a spherical mesh under the flashing liquid boundary conditions. The equations can be solved for any liquid properties and intial mass. Here the blast effects have been computed dependent on the distance from the source for propane at two initial temperatures:

- 326 K, the super heat limit temperature
- 300 K.

The blast effects have been recorded in terms of the blast peak overpressure and the wave's positive phase duration. The parameters for an approximately triangular wave shape have been represented dependent on the distance in the form of blast charts. The quantities in the blast charts have been scaled with the atmospheric pressure and the cube root of the initial liquid mass (509 kg/m^3). The scaling makes the charts applicable for any liquid mass of propane, flashing from the respective temperatures.

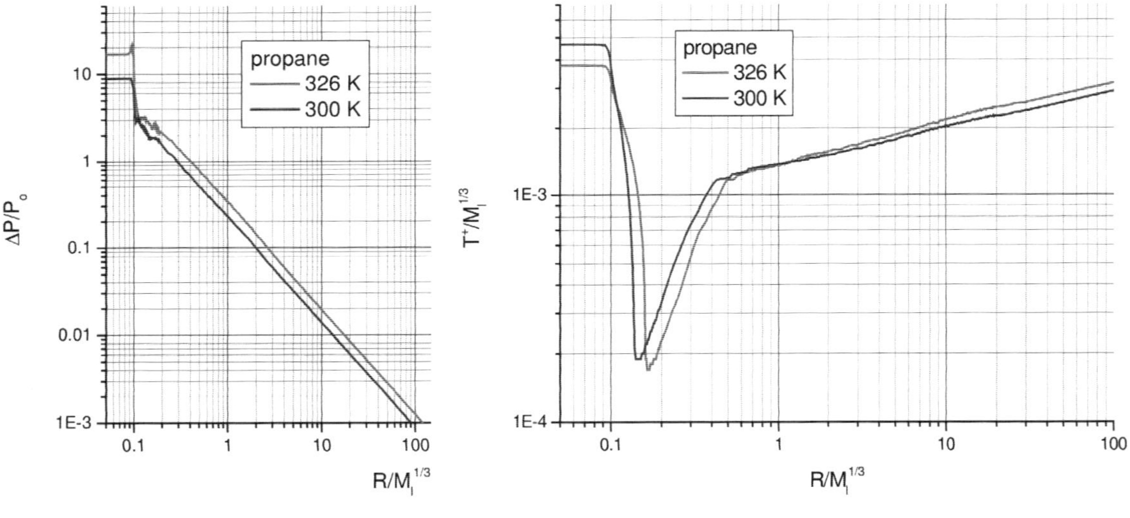

Figure 2 BLEVE blast charts for propane, flashing from 326 K and 300 K respectively
ΔP = overpressure (kPa); P_o = ambient pressure (kPa);
T^+ = positive phase duration (s); M_l = liquid mass (kg);
R = distance from source centre (m)

The hemispherical liquid source area extends up to a scaled distance of approximately $R/M_l^{1/3}$ = 0.1 m/kg$^{1/3}$. Close to the source area the development of the blast parameters shows a somewhat irregular behaviour, which is due to the violence of the gas dynamics. Directly outside the source, the overpressure falls from the initial vapour overpressure to a much lower initial blast overpressure, equal to the shock wave overpressure in a shock tube after membrane rupture. Driven by the vapour pressure of the liquid, the spherically diverging expansion flow velocity becomes supersonic developing typical shock formation. Such phenomena go hand in hand with substantial underpressures locally. In this area the blast wave is characterised by a very short duration. The blast positive phase duration strongly grows over the distance where the vapour is pushing the surrounding air away. Further away from the source in the ambient atmosphere, the blast parameters assume their more regular development.

Compilation of fully non-dimensionalised blast charts, generally applicable for any amount of any substance at any initial conditions, would require a quantitative measure for the expansion energy of a flashing liquid to scale with. This expansion energy is equal to the expansion energy of the vapour that expands on release from the liquid vapour pressure down to ambient pressure, integrated over the full liquid flash course. As the liquid temperature and consequently the liquid vapour pressure fall during the flash process, an analytical expression for the expansion energy could not be developed.

Although the gas dynamic blast wave decay is stronger than the 1/R acoustic wave decay, a local match with a linearly growing acoustic volume source analogue in half space is found around a scaled distance of $R/M^{1/3}$ = 1 $m/kg^{1/3}$ for a scaled release time of approximately $\Delta t/M^{1/3}$ = 0.002 $s/kg^{1/3}$.

4.2.3 A BLEVE in a tunnel

4.2.3.1 One-dimensional gas dynamic model

In urbanised areas, future road and rail infrastructure is increasingly projected underground or covered over. The consequences of possible incidents play an important role in the assessment of the risk involved in hazardous materials transport. A crash of a road tanker of 50 m^3 of propane in a tunnel leading to vessel rupture is a credible scenario in this respect.

In this demo, the vessel rupture is assumed not to be fully instantaneous but such that the consequence is a steady release of liquid over a short span of time. This release time is varied in the computational results below.

Apart from some numerical irregularities, Figure 3 gives a clear view of the potentials of the model. Close to the source, the gas dynamic model results in a square blast wave whose overpressure is dependent on the release time and whose duration is equal to the release time. With increasing distance from the source the wave shape is more and more affected by nonlinear gas dynamics. Propagating down the tube, the amplitude of the blast wave remains unaltered until the tail expansion wave catches up with the lead shock. From that instant on the blast wave amplitude starts to decrease due to the energy dissipation in the lead shock. The shorter the release time the higher the blast wave overpressure and the earlier the decay process sets in.

The initial blast wave overpressure is, of course, predominantly dependent on the release time of the liquid and has been tabulated for various release times in Table 3. It is good to see how well the gas dynamic solution approximates the acoustic volume source results in Table 2 for longer release times.

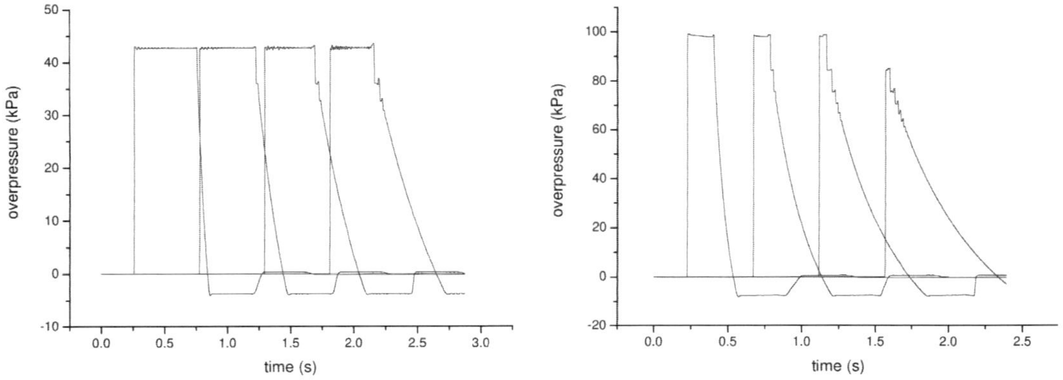

Figure 3 Blast wave at 100, 300, 500 and 700 m distance from a steady release of liquid propane with a temperature of 326 K from a 50 m^3 vessel in a tunnel tube of 60 m^2 cross-sectional area. Release time = 0.5 and 0.2 s respectively.

Table 3: Blast overpressures dependent on release time for 50 m^3 of propane of 326 K in a tunnel tube of 60 m^2 cross-sectional area

Release time (s)	ΔP (kPa)
0.1	180
0.2	98
0.5	43
1	22
2	11

4.2.3.2 Three-dimensional gas dynamic model

If the simple acoustic analogue and the one-dimensional gas dynamic modelling do not satisfy, a BLEVE can be simulated three-dimensionally. Any area within a three-dimensional mesh can be simply provided with the flash-evaporating liquid boundary conditions. Results of a numerical simulation of an LPG road tanker BLEVE in a tunnel have already been presented in a paper by Van den Berg et al. (2004).

Here, the potentials of this method are demonstrated by the simulation of a BLEVE due to an instantaneous rupture of an LPG road tanker in the forecourt of a petrol station. The flashing liquid with a temperature of 326 K was positioned at about 10 m distance from the refuelling area. To limit the size of the computation, the planes through the vessel's centre are considered planes of symmetry, bounding a mesh of 350×150×100 cells of 0.1×0.1×0.1 m^3.

An impression of the gas dynamics is given in Figure 4 showing the pressure distributions in one of the vertical planes of symmetry at three consecutive points in time. The pressure distributions have been visualised by a set of isobars as well as by a colour shading. Shock phenomena are visible where isobars accumulate. The pictures show, among other things, how a powerful blast wave propagates through the forecourt reflecting at the pylons and the refuelling units underneath the roof.

The blast overpressures in Figure 5 have been recorded in the centre of the roof both underneath and on top. The visualisations and the pressure records show that the confinement by the roof substantially slows down the blast overpressure decay compared to the decay in the free space on top of the roof. As a consequence the roof is struck by a blast load that will fully demolish the light-weight roof construction.

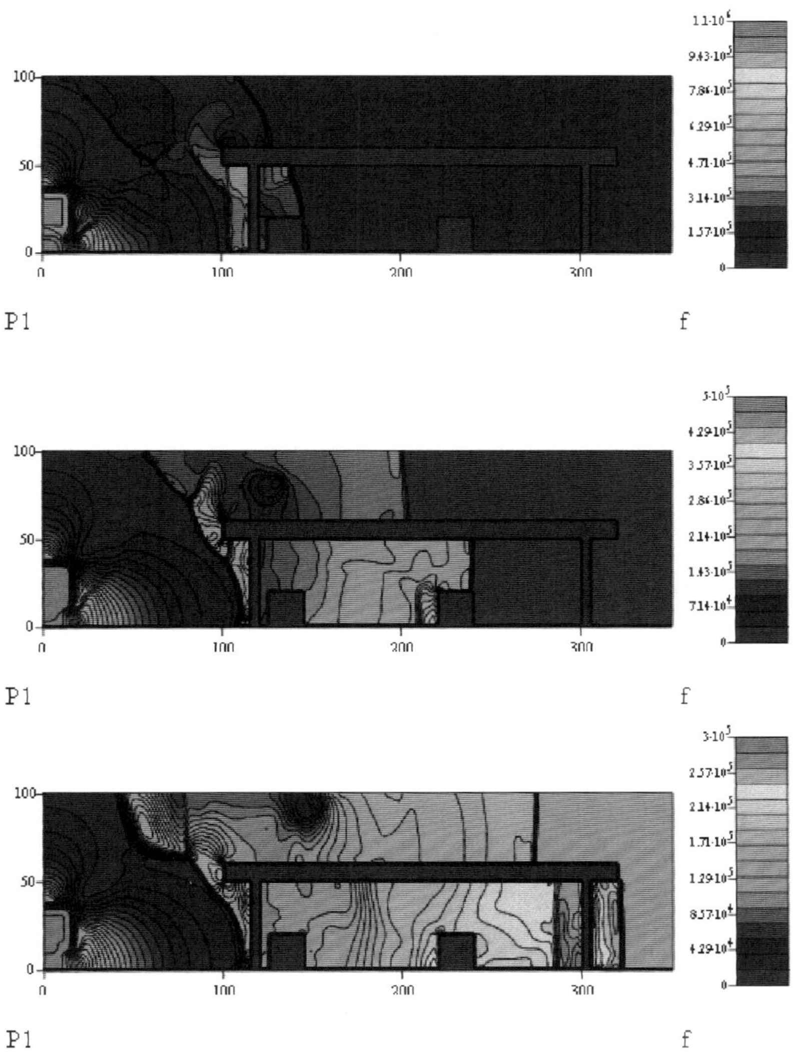

Figure 4: Pressure distributions (Pa) in a petrol station at three consecutive points in time, originating from a BLEVE of 50 m^3 of LPG.

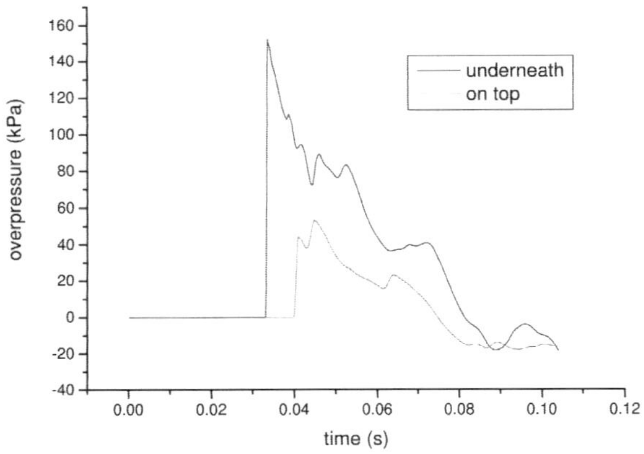

Figure 5: Pressure-time load records underneath and on top of the roof

5. Conclusion

A new method has been presented to calculate the blast effects originating from an exploding vessel of liquefied gas. Adequate blast calculation requires full knowledge of the blast source characteristics, i.e.: the release and consequent evaporation rate of a flashing liquid. A literature review showed that the conditions that allow explosive evaporation are not entirely clear. As the evaporation rate of a flashing liquid is unknown, the safe assumption that the evaporation rate is expansion-controlled has been adopted as the starting point in the modeling. In this way, blast effects from a BLEVE have been numerically computed by imposing the vapour pressure of flashing liquid as boundary condition for the gas dynamics of expansion. The modeling has been quantitatively explored just for liquefied propane.

In addition, it has been demonstrated that often an estimate of BLEVE blast effects can be made with very simple acoustic volume source expressions.

The modeling showed that the rupture of a pressure vessel containing a liquefied gas in free space develops blast of a significant strength only if the vessel nearly instantaneously disintegrates. Even if a rupture and the consequent release of a liquefied gas extend over just a short period of time, the blast effects are minor.

6. References

Birk, A. M. and Cunningham, M. H., (1994), "The Boiling Liquid Expanding Vapour Explosion", *J. Loss Prev. Process Ind.*,Vol.7, No.6, pp.474-480, 1994.

Boris J.P., (1976), "Flux-Corrected Transport Modules For Solving Generalized Continuity Equations," NRL Memorandum Report 3237, Naval Research Laboratory, Washington, D.C.

CCPS (1994), *Guidelines for Evaluating the Characteristics of Vapor Cloud Explosions, Flash Fires and BLEVE's*, American Institute of Chemical Engineers, Center for Chemical Process Safety, New York, 1994.

CPR 14E (Yellow Book) (1997), *Methods For The Calculation Of Physical Effects,* Committee for the Prevention of Disasters, ISSN: 0921-9633/2.10.0114/9110, Sdu Uitgevers, 1997

Giesbrecht, H., Hess, K., Leuckel, W. and Maurer, B., (1981), "Analysis Of Explosion Hazards On Spontaneous Release Of Inflammable Gases In The Atmosphere, Part 1: Propagation And Deflagration Of Vapour Clouds On The Basis Of Bursting Tests On Model Vessels", *Ger. Chem. Eng.,* Vol.4, pp.305-314, 1981.

Lighthill J., (1978), *Waves in Fluids*, Cambridge University Press, Cambridge, 1978.

Reid R.C. (1976), "Superheated Liquids," *Amer.Scientist,* Vol.64, pp.146-156, 1976.

Reid R.C. (1979), "Possible Mechanisms For Pressurized-Liquid Tank Explosions Or BLEVE's," *Science*, Vol.203, pp.3, 1979.

Strehlow, R.A. (1981), "Blast Wave From Deflagrative Explosions: An Acoustic Approach," 13th AIChE Loss Prevention Symposium, Philadelphia (PA), 1981.

Van den Berg, A. C., Van der Voort, M. M., Weerheijm, J. and Versloot, N. H. A. (2004), "Expansion-Controlled Evaporation – A Safe Approach To BLEVE Blast," *J.of Loss Prevention in the Process Industries*, Vol.17, No.6, pp.397-405, 2004.

Venart, J. E. S., Rutledge, G. A., Sumathipala, K. and Sollows, K. (1992), "To BLEVE Or Not To BLEVE: Anatomy Of A Boiling Liquid Expanding Vapour Explosion," Proceedings, paper no. 115, 7th Int. Symp. on Loss Prev. and Safety Promotion in the Process Ind., Toarmina, Italy, 1992

Venart, J. E. S., Sollows, K. F., Sumathipala, K., Rutledge, G.,A. and Jian, G.,A. (1993), "Boiling Liquid Compressed Bubble Explosions: Experiments/Models," *Gas-Liquid Flows*, Vol.165, pp.55-60, ASME, Fluids Engineering Division, 1993.

Paper TG002 - 5

Modeling the Consequences of Bursting Vessels Inside Enclosed Structures Using Wall Failure Criteria

David D. Herrmann, P.E.
DUPONT Engineering Technology
1007 Market Street
Wilmington, DE 19898
(302) 774-1766
david.d.herrmann@usa.dupont.com

Prepared for Presentation at
American Institute of Chemical Engineers
2005 Spring National Meeting
39th Annual Loss Prevention Symposium
Atlanta, Georgia
April 11-13, 2005

UNPUBLISHED

Modeling the Consequences of Bursting Vessels Inside Enclosed Structures Using Wall Failure Criteria

David D. Herrmann, P.E.
DUPONT Engineering Technology
1007 Market Street
Wilmington, DE 19898
(302) 774-1766
david.d.herrmann@usa.dupont.com

Abstract

It is desirable to have a better understanding of the consequences of potential explosion scenarios within a complex structure. When the magnitude of the explosion is such that the building is essentially destroyed, the consequences are relatively easy to ascertain, but frequently we expect that the event would result in some partial damage to the building, where some rooms have major damage, while others remain relatively unharmed. Methods for specifying failure criteria have been incorporated into certain software, however; the level of validation is minimal and techniques to evaluate these scenarios are not well established. Commercially available CFD codes for modeling of vessel bursts incorporate different approaches for dealing with the problem of removing solid surfaces (such as walls) from the calculation domain when a given failure criteria is reached. The ability to model the failure of walls within a structure has many potential benefits to improve our understanding of human and business risk. This paper discusses an example using FLACS® and CEBAM to discuss the pros and cons associated with these studies.

1. Background

When operational risk is to be determined, it is essential to understand both the consequences of an event, and the likelihood of that event. Often the business issues are significant. The consequences of a major explosion may include the health and well being of employees and potentially those off site, millions of dollars of investment, and credibility as a supplier that may be reflected in market share.

Many manufacturing structures are complex, consisting of areas that contain offices and control rooms that may be separated by several walls from areas where an explosion might occur.

In general, we try to avoid going into details to ascertain what portions of the building might survive, and use conservative assessments of consequences; however, such consequence analyses may drive the PHA team to make costly recommendations that are not required.

When an explosion impacts a complex structure such as an enclosed manufacturing building, it may destroy the entire structure or more frequently, a portion of the building may be damaged.

Other portions of the building remain intact enough to the point where the probability of survival is high, and equipment may be relatively undamaged.

When a blast wave hits a wall of a building, there is an effect on the wall, and an effect on the overall structure as the force is transmitted through the frame. If the wall fails, the force on the frame may be reduced by the failure, and a portion of the blast wave is deflected either away from, or to another part of the structure.

When a blast wave travels into a structure after failure of a wall, doors or windows, it often provides internal support against the external forces of the blast wave on the building. The effect of leakage pressure is hard to quantify in a complex structure.

Computational Fluid Dynamics (CFD) software provides us with the opportunity to sort out some of the complexities. This paper will focus on a relatively simple example, using two commercially available software tools.

2. Analysis Details

2.1 Structure and Event Description

The structure being studied in this paper consists of a series of explosion barricades that share a wall with a large manufacturing building. A single explosion barricade is identified as a 3-story structure of total dimensions: 14 ft wide x 32 ft tall x 32 ft long. The intermediate floors are located at 8 ft and 21 ft respectively. Referring to figure 1, each barricade has a frangible vent wall and roof that aim upwards and towards the viewer. The vent wall is constructed of fiberglass panels. The roof is lightweight stick frame, wood construction. A large horizontal vessel is located on the third floor (not shown in figure for proprietary reasons, but was incorporated into all modeling).

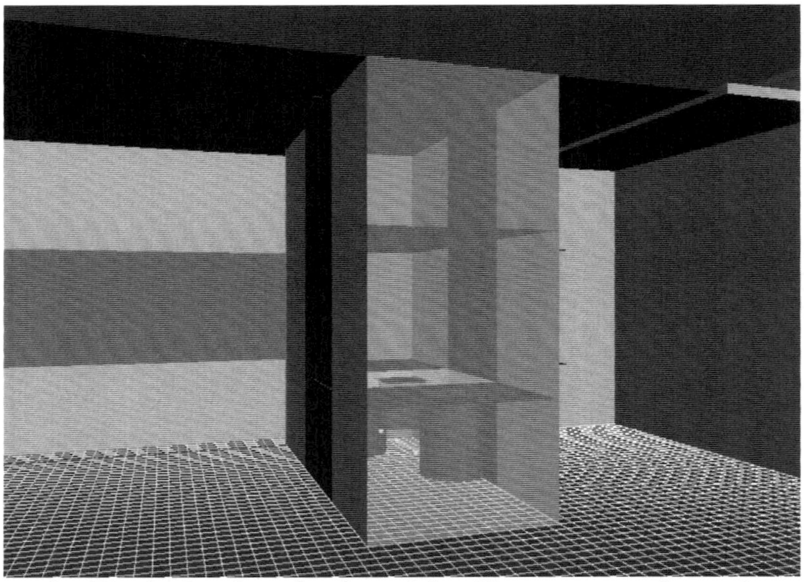

Figure 1 – Explosion barricades with adjacent manufacturing area.

The explosion event originates in a vessel on the ground floor located centrally between the side walls and 18 feet from the back wall that is shared with the manufacturing building. The burst pressure of the vessel is approximately 2000 psi, and the byproducts of the deflagration have a heat capacity ratio (C_p/C_v) at burst of approximately 1.1. The volume of the vessel is approximately 200 gallon. The manway of the vessel extends through the floor above the ground level. The vessel is also designed so that the bolts holding the manway will fail, directing the force of the explosion upward.

The intermediate floors are 4-inch reinforced concrete and will fail in the event of an explosion on the ground floor. The vertical walls are a heavier reinforced concrete design with metal plate used for protection from missiles.

In order to obtain a reasonable feel for the propagation of blast waves in a barricade where a portion of the heavy wall structure fails, the event was simulated with two different approaches used in 2 different CFD codes: FLACS® and CEBAM.

2.2 Uncertainties

There are a number of uncertainties in simulating the failure of walls within a structure. Some significant examples are listed below.

- Failure criteria for wall panels have been developed based on (1) estimates of material properties and (2) structural analysis techniques using single degree or 2 degree of freedom models. These models do not take full credit for the distribution of forces throughout the entire frame of the building or additional support from some members adjacent to the barricade being studied.
- The results may be very sensitive to the mode of failure. Some wall panels may fail at the connections and release as a solid piece. Some may fail due to shear and fragment, while others may fail in a ductile manner and either fragment or simply tear and peel back. The mass and shape of resulting fragments affect the amount of explosion energy that is translated to the drag forces that move missiles. Conceptually an explosion might more easily vent from a wall section where the wall becomes fragmented into small pieces versus one where the wall section is translated as a single rectangular object.
- Construction techniques may vary. If the construction of the structure was not carefully monitored to ensure construction per design, the response of a wall or connection to a load may not act in the way it was designed. Some of these discrepancies may be revealed from a visual inspection of the as-built structure.
- In currently available CFD blast modeling codes, walls either fail at some criteria, or remain in place. Permanent deformation of the structural components is not permitted, and this limitation could impact the results in certain cases.
- Wall failure may result from impact with missiles, rather than the blast wave itself. This paper does not deal with this uncertainty. In this example, the effect of missiles is assumed to be minimal.

2.3 CFD Code Comparison

In order to compare the two CFD codes, a simulation was run with each to replicate the scenario with a spherical release of pressurized gas at the ground floor with the assumption that all walls and the intermediate floors except for frangible panels remain firmly in place. This same scenario was documented in an earlier paper to the Center for Chemical Process Safety [1]. The hole in the floor above the vessel due to the manway is neglected in these simulations. In Figure 2 and Figure 3, the average pressure on the vertical wall surfaces of the ground floor is shown. The peak pressure on panel 3 and the back wall in Figure 2 was obtained through extrapolation due to the fact that CEBAM reports data for average pressure every 10 iterations of the program.

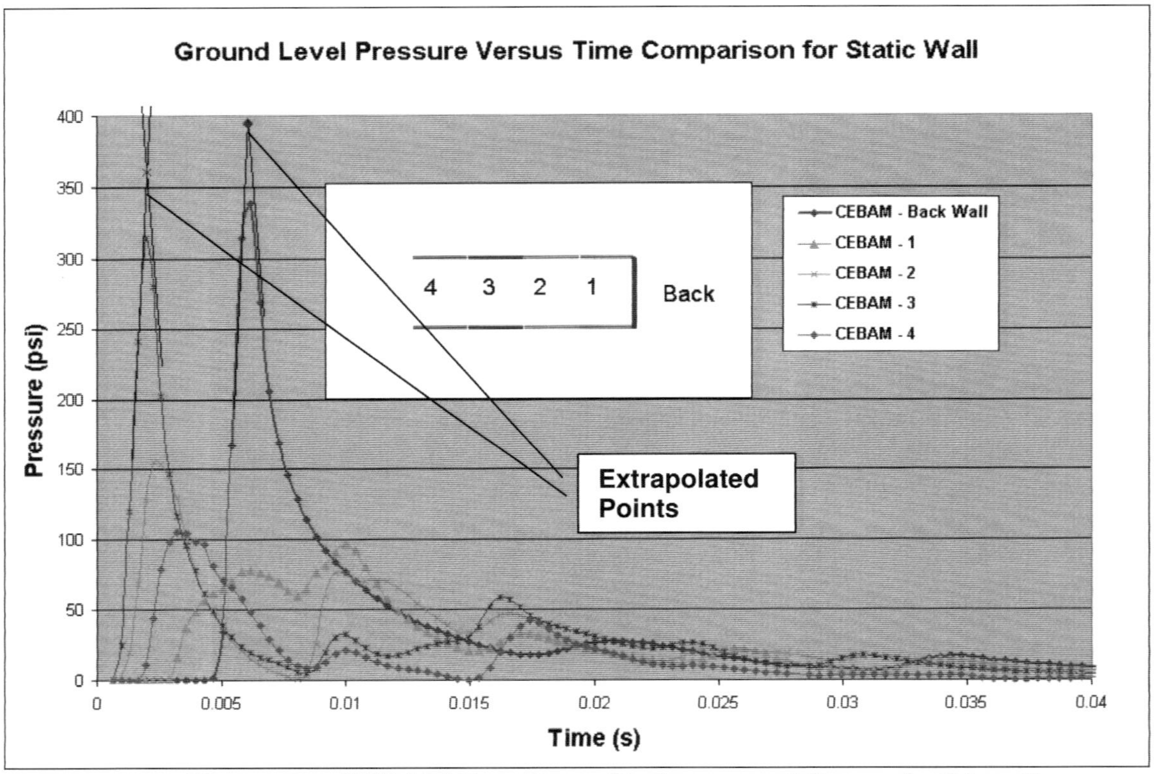

Figure 2 – CEBAM P vs T results for ground floor w/solid walls.

The comparison results with FLACS® are in Figure 3.

The impulse comparison is shown in Figure 4. The labels below the bars tell at what point in time the impulse calculation was stopped. This decision was based on input from structural engineers as to the portion of the curves that had value to the structural response of the walls.

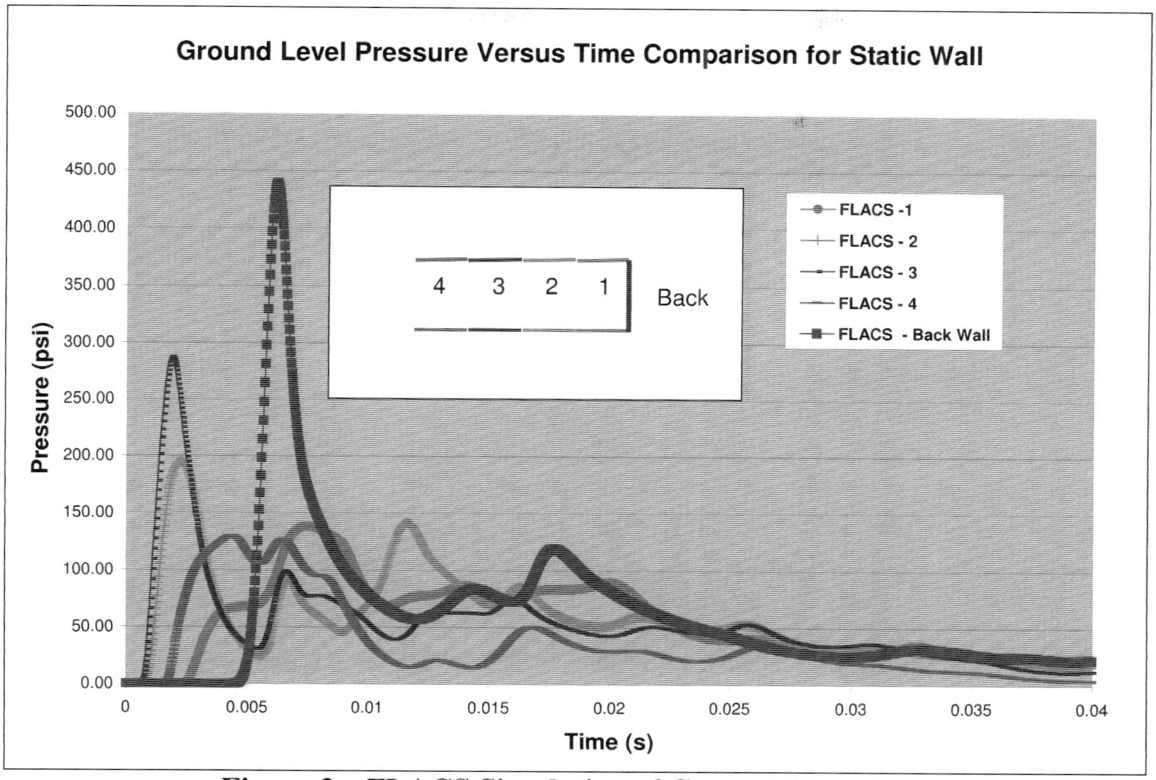

Figure 3 – FLACS Simulation of Ground Floor Wall Loads (non-moving walls & floors)

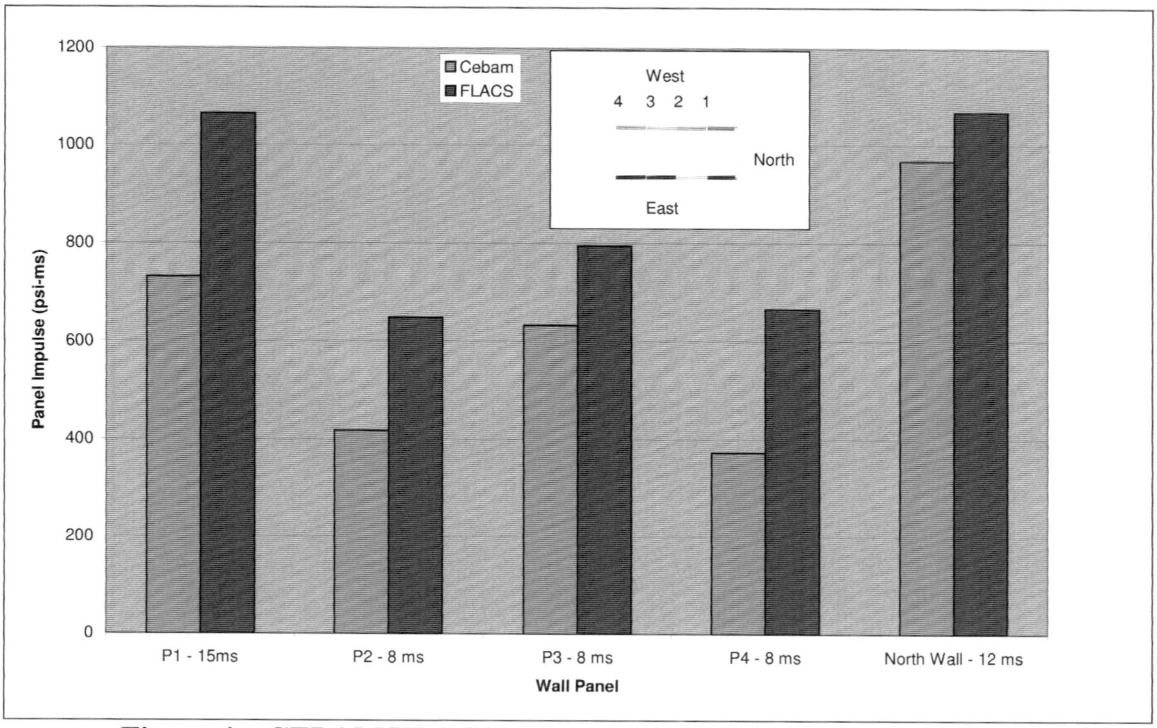

Figure 4 – CEBAM/FLACS Impulse Comparison for Solid Floor Scenario

The primary differences between the numbers likely result from the use of a constant heat capacity in CEBAM (input as 1.1), versus a variable heat capacity in FLACS. It is expected that a variable heat capacity relationship would result in somewhat higher numbers. Both programs track each other very well in terms of the relative shape of the pressure versus time profiles and onset of pressure rise at each panel.

It is important to note that FLACS has not undergone any rigorous validation testing in a vessel burst application. In terms of explosion calculations, differences of 20-30% are considered reasonable.

2.4 CEBAM

The CEBAM 3D model that was created to simulate the explosion is shown in Figure 5. The vessel is shown as a cylinder on the ground floor aimed up through a square hole in the second floor. The panel sections that will keep track of the pressure time history are shown as different shades and are separated by floor. The translucent panels on the vent opening towards the reader are the fiberglass panels currently in place.

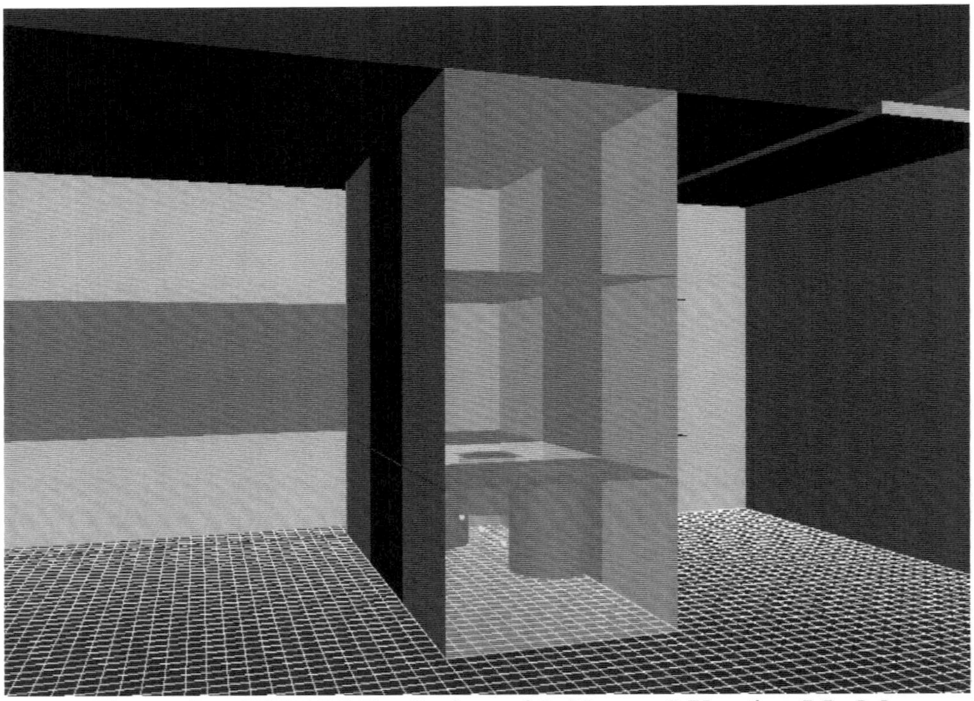

Figure 5 – CEBAM Explosion with Upward Venting Model

CEBAM (Computational Explosion and Blast Assessment Model) incorporates a method where wall sections are defined as rectangular panels of no thickness that have failure criteria assigned. Figure 6 is an example for one section of the barricade 2nd floor (nearest the back wall), where failure criteria of 1.4 psi and 50 psi-ms have been assigned as recommended through a structural assessment.

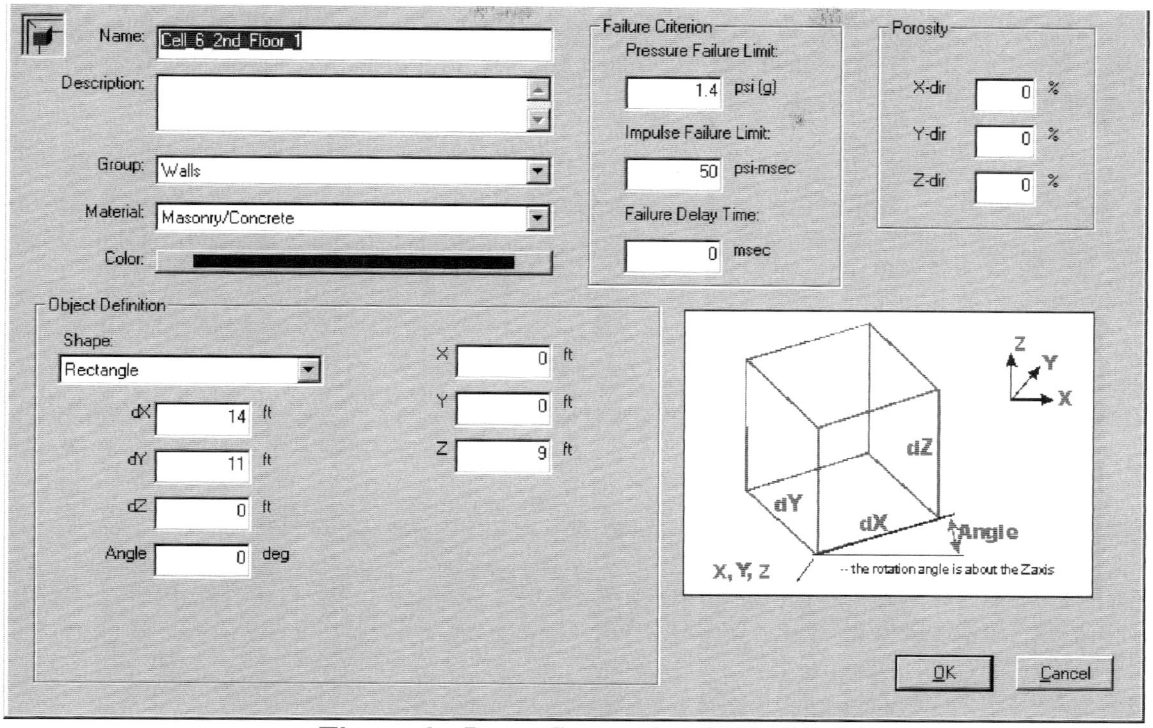

Figure 6 – Input Screen for CEBAM objects.

The model is based on the theory that a pressure versus impulse diagram like that shown in Figure 7 can represent the response of the wall to blast load.

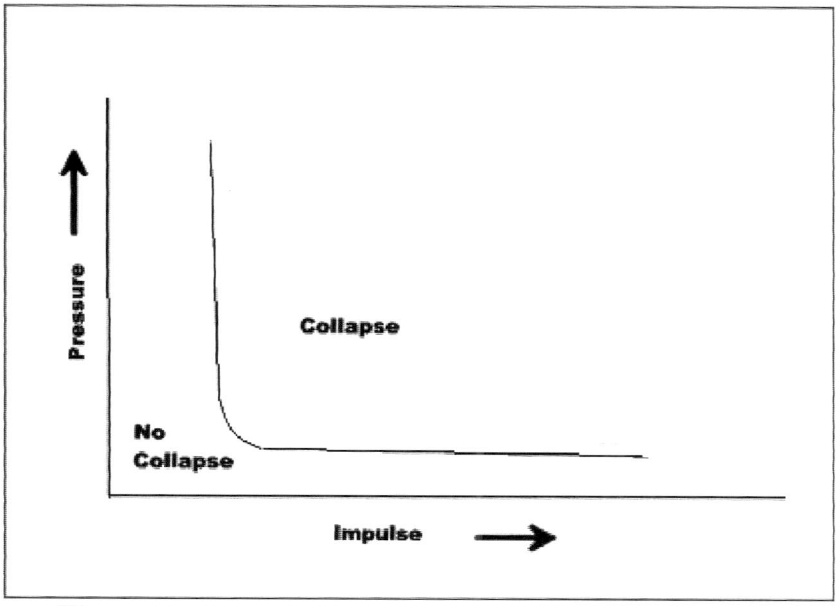

Figure 7 – Generic Pressure versus Impulse diagram.

The model treats a "curve" more simply as a right angle with a pressure and impulse threshold. In this case, the collapse criteria are placed into the model. When both the pressure and impulse criteria are met, the wall panel is eliminated from the computational domain completely. It is possible to specify a failure delay time in order to deal with any delay associated with overcoming the inertia of the structural component. This does not model generation of fragments, or the blast energy lost in displacing the fragments.

It is important to know the direction from which the blast is impinging on the floor as the resistance to blast is higher when the force is downward on the barricade floors, as opposed to a blast force from below. This may require some iteration of the model to establish the load directions.

2.5 FLACS

The FLACS CFD code was developed specifically for modeling flammable vapor/air explosions in semi-confined environments. Its use has recently been extended to model vessel bursts.

The FLACS model that was created to simulate the explosion is shown in Figure 8.

The vessel is shown as a cylinder on the ground floor (behind the cylinder nearest the vent wall) aimed up through a square hole in the second floor. The panels were specified the same size as that in the CEBAM model of Figure 5.

Figure 8 – FLACS Explosion with Upward Venting Model

FLACS has a validated model that is used to simulate the effect of a frangible panel. Model inputs are shown in Figure 9.

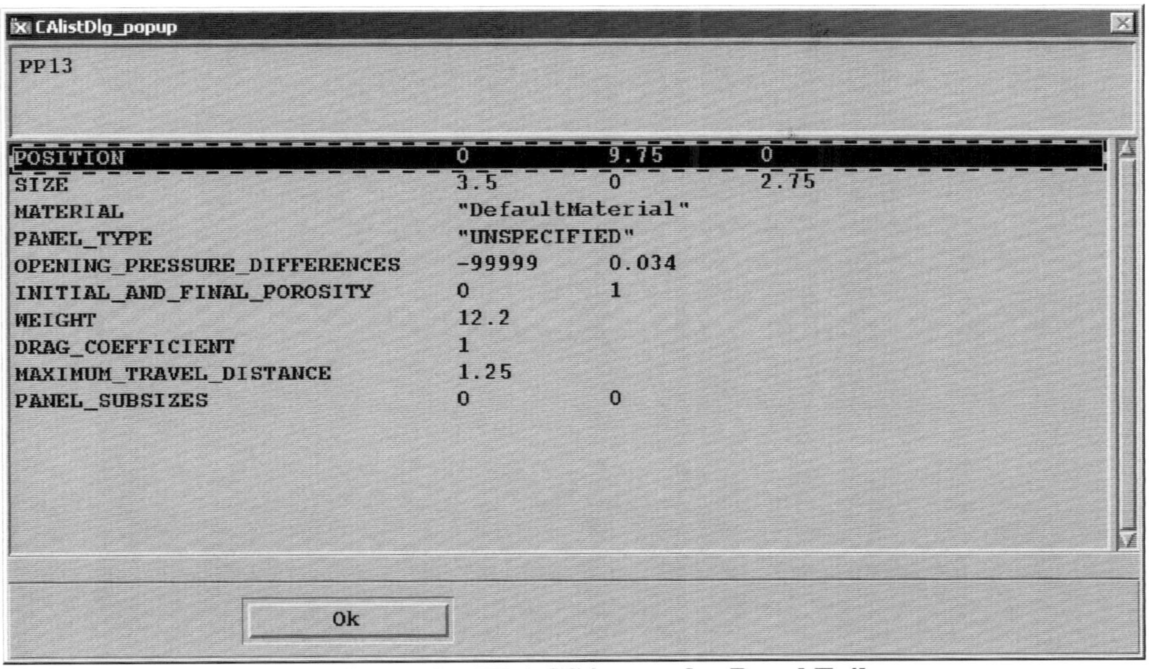

Figure 9 – FLACS inputs for Panel Failure

The theoretical basis for this is that lightweight panels are designed to fail at a given pressure. The user specified the failure pressure in the positive and negative direction (a very high number indicates failure in only one direction). Once the panel fails, the calculations for conservation of energy mass and momentum take into account the kinetic energy of the panel and loss of blast energy to drag forces on the panel. After the panel has moved away a specified distance, the panel effects are removed from the computational domain. In order to extend this methodology to heavy walls, we simply input a panel weight (kg/m^2) consistent with that of the wall, and input failure criteria in terms of pressure only.

3. Results Of Analysis Including Upward Venting

3.1 CEBAM

Two simulations were performed. The first used failure criteria supplied by a structural consultant for the floors and walls. No failure delay time was specified. The second simulation used a 5 ms failure delay time for the 2nd and 3rd floors, with no failure delay on the roof (due to the light material of construction).

The pressure and impulse failure criteria for both simulations were specified as follows.

Panel Type	Failure Criteria Pressure (psi)	Impulse (psi-ms)
Frangible Wall/Roof	0.5	10
2nd Floor	1.4	50
3rd Floor	1.4	50
Side Walls	20	165
Back Wall	250	615

Figure 10 – Pressure/Impulse Failure Criteria

3.1.1 Upward Venting with no Failure Delay

The results of the simulation are illustrated in Figure 11 through Figure 14. Note the significantly reduced pressures on the ground floor. In this simulation, the second floor did not fail. The plot plan shows the orientation of the wall panels on each figure. The naming convention identifies the specific wall panel. For example, F1E1 stands for "floor 1, East Side, wall panel 1.

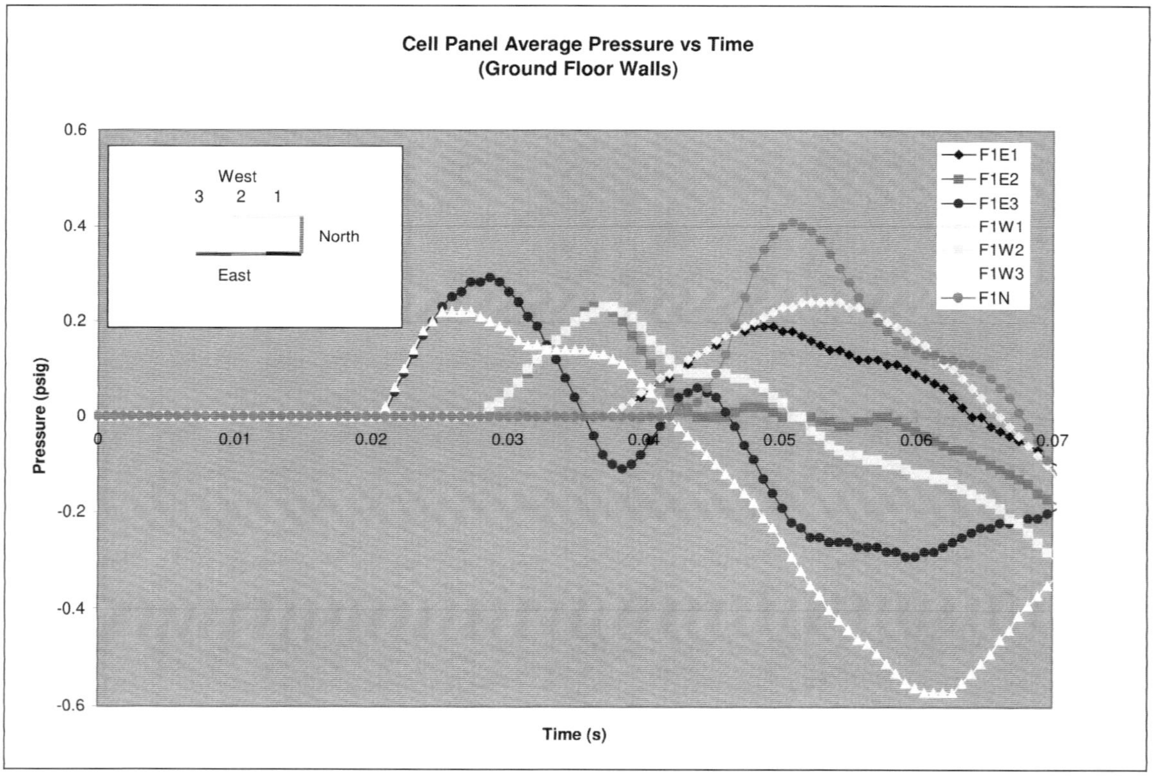

Figure 11 – Ground Floor Wall Panels - Pressure vs Time

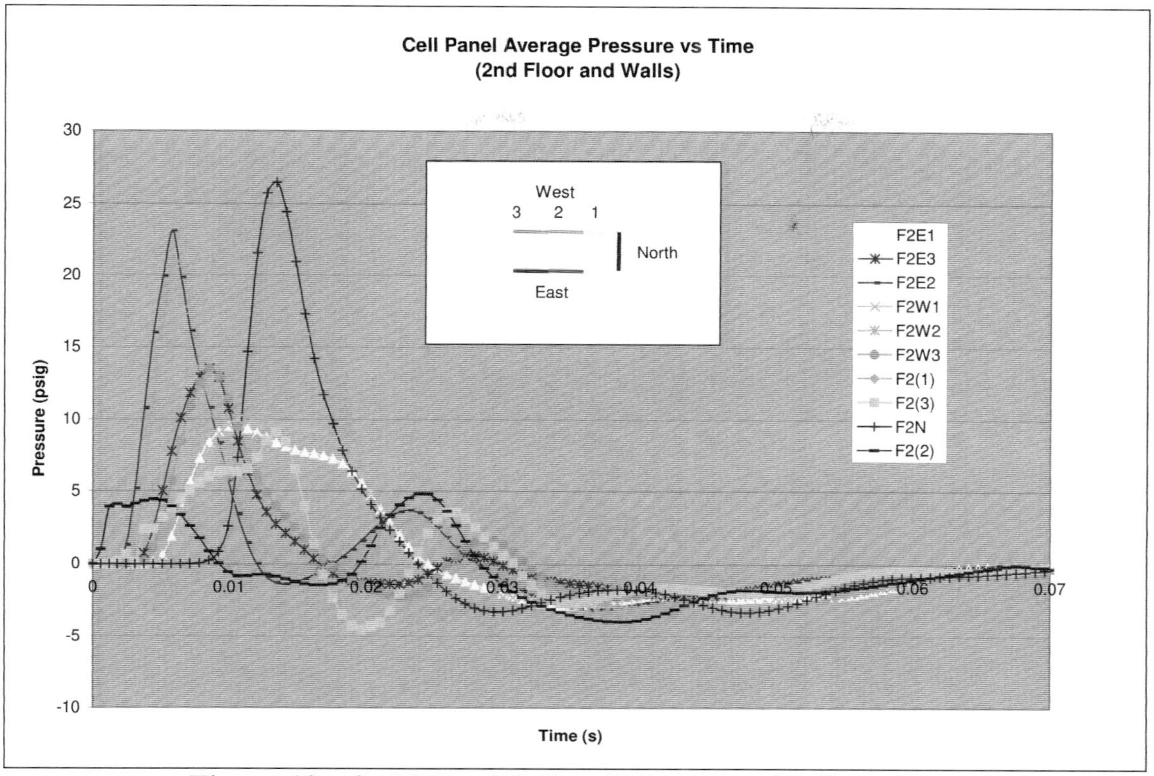

Figure 12 – 2nd Floor Wall and Floor Panels - Pressure vs Time

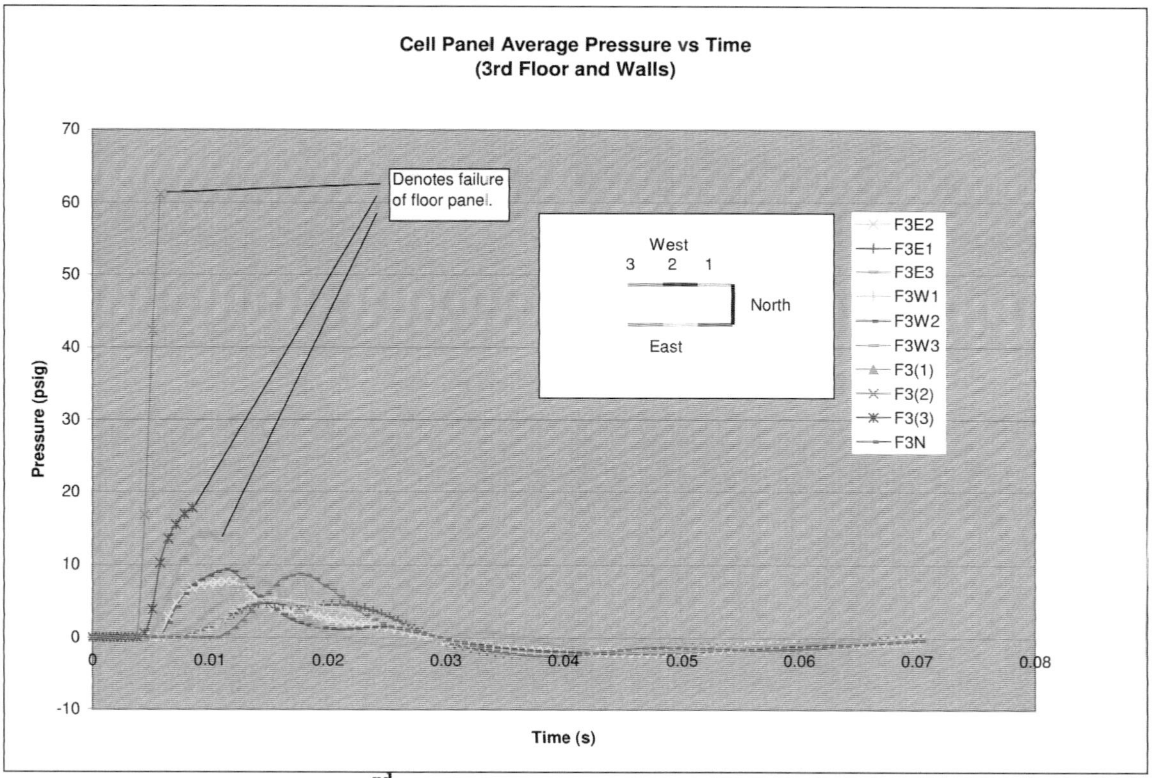

Figure 13 – 3rd Floor Wall and Floor Panels – Pressure vs Time

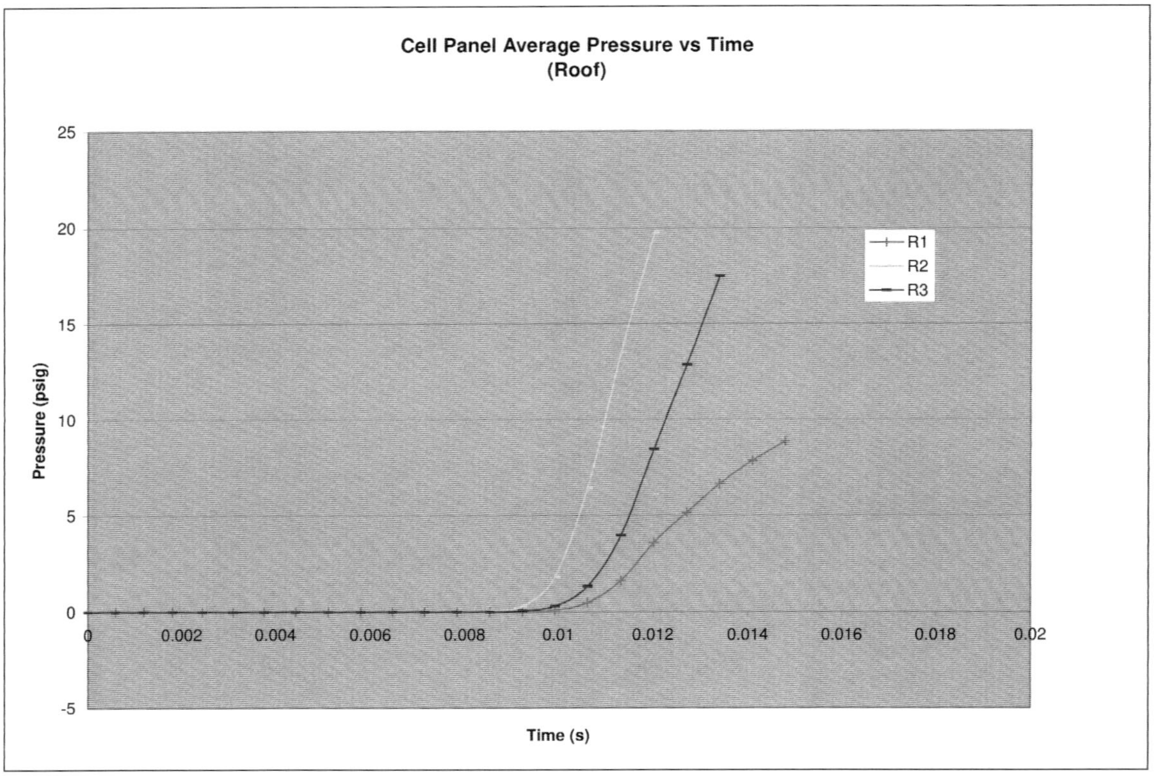

Figure 14 – Roof Panels – Pressure vs. Time

These results show a significant reduction in the loads on the barricade walls as compared to Figure 2 and Figure 3.

3.1.2 Upward Venting with 5-ms Delay Time

The results of the simulation with 5 ms Delay are shown in Figure 15 through Figure 18. The significant difference between this simulation and the previous is that several vertical wall panels on the 2nd floor failed. This means that the pressure would have ruptured that wall section sending fragments to the sides.

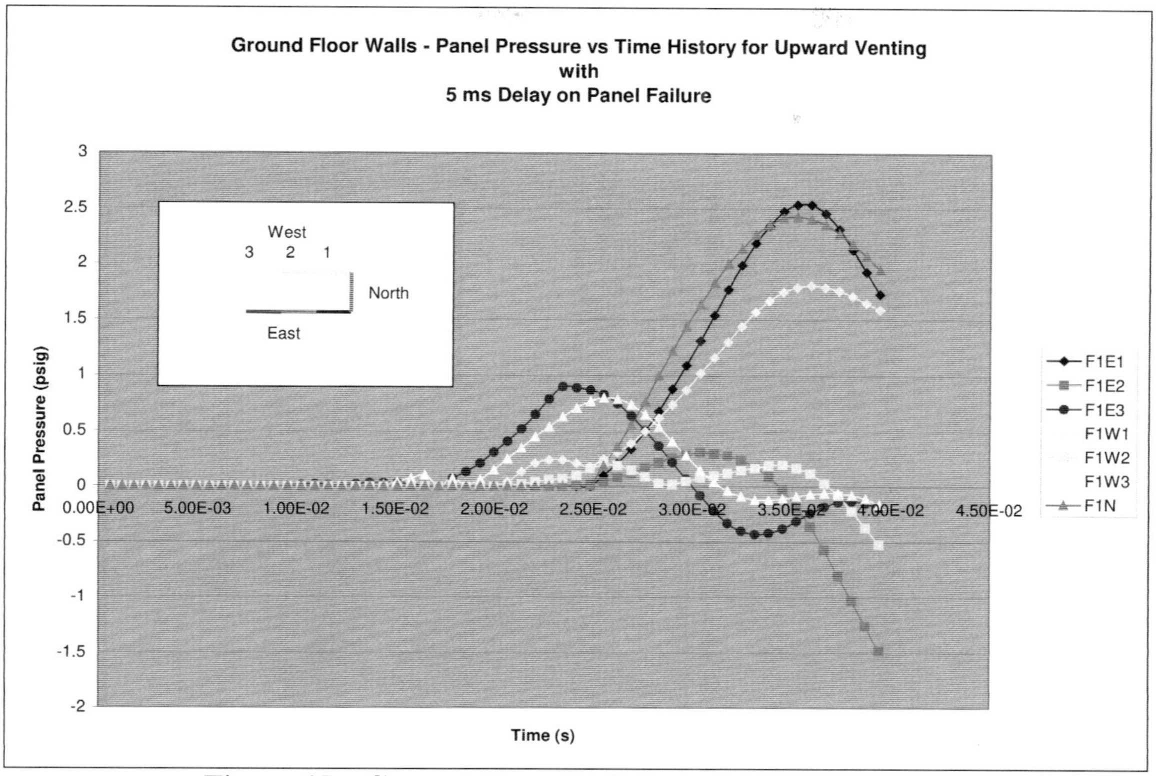

Figure 15 – Ground Floor Wall Pressure vs Time (5 ms delay)

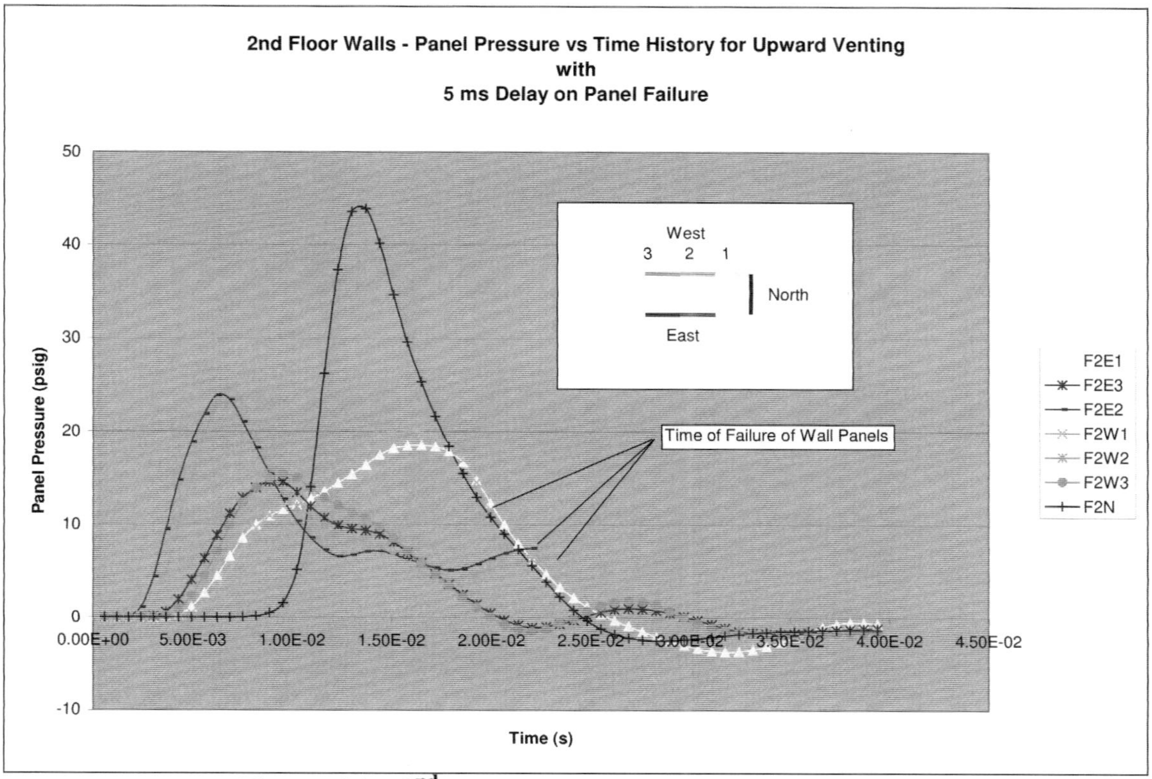

Figure 16 – 2nd Floor Wall Pressure vs Time (5 ms delay)

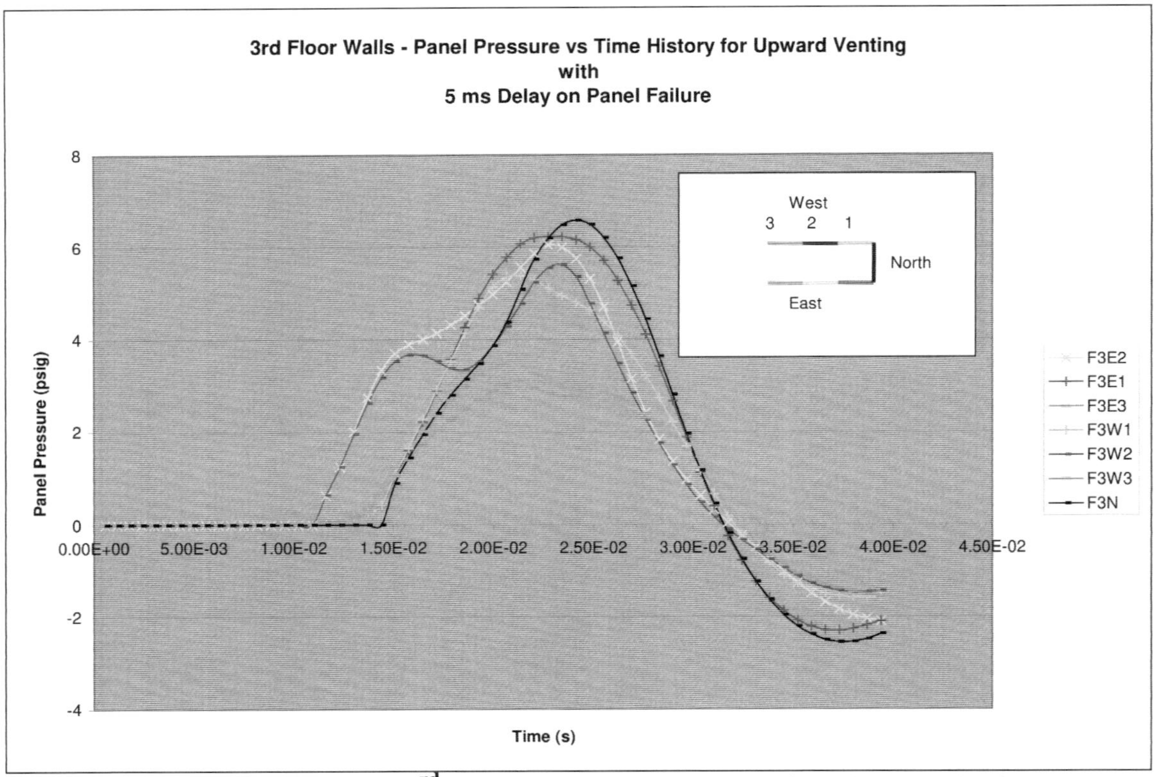

Figure 17 – 3rd Floor Wall Pressure vs Time (5 ms delay)

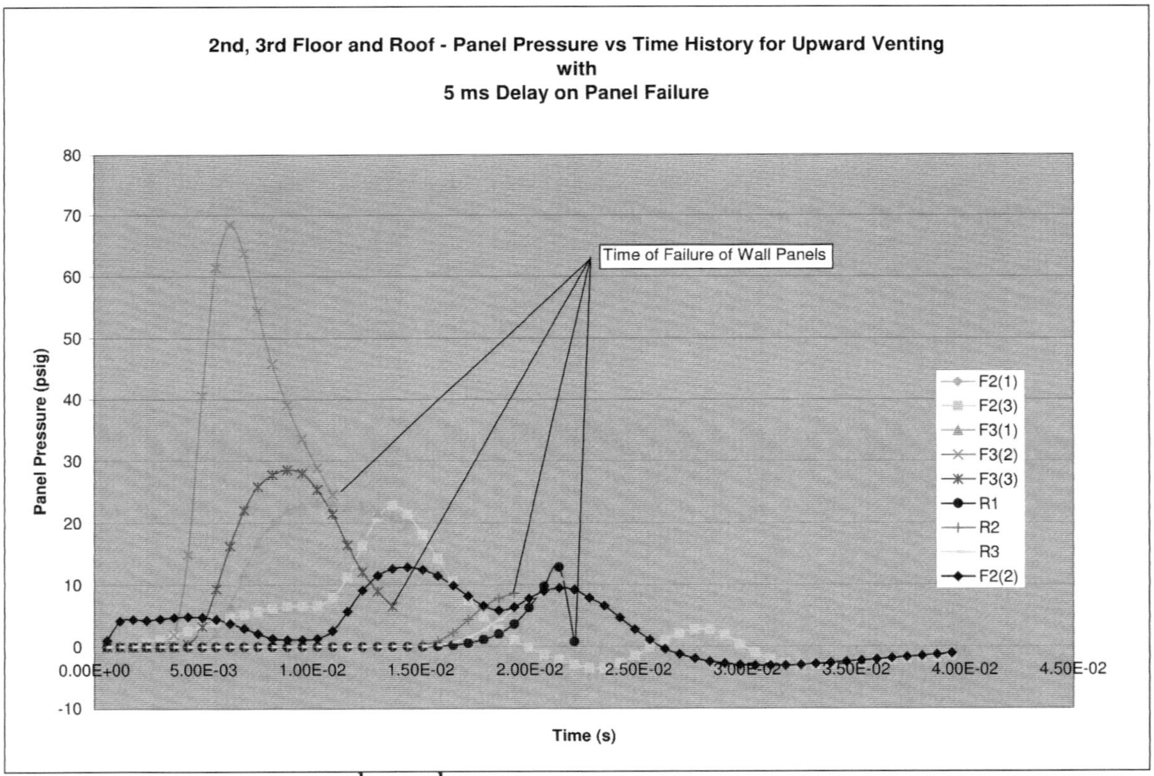

Figure 18 – 2nd & 3rd Floors and Roof Pressure vs Time (5 ms delay)

3.2 FLACS

Results of the simulation are illustrated in Figure 20 through Figure 22. In the FLACS simulation, the solid wall panels were not given failure criteria, while the intermediate floors were assigned failure criteria (Figure 19). This maximized the pressure on the back barricade wall.

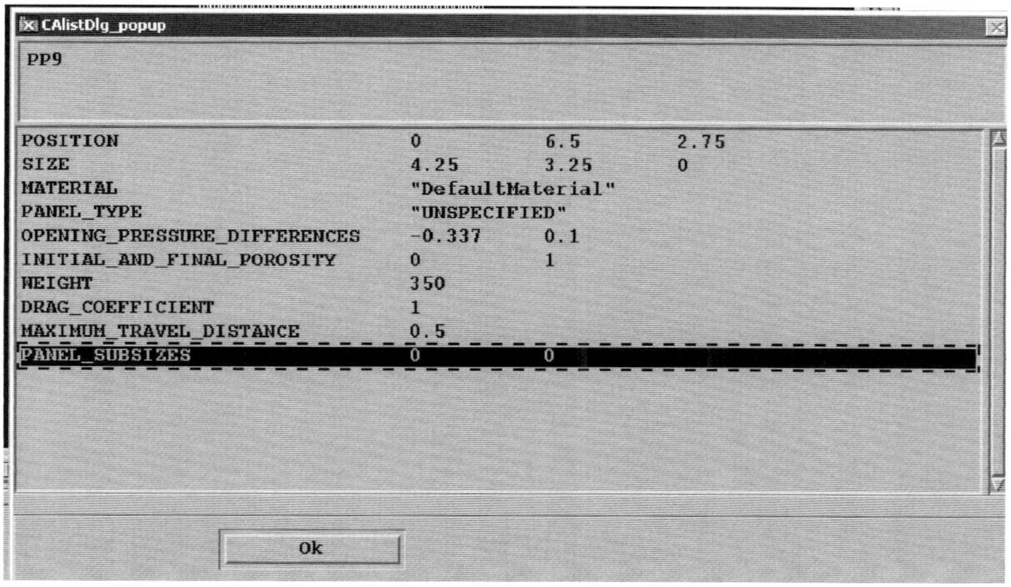

Figure 19 – Floor Failure Criteria for all Floors and Roof

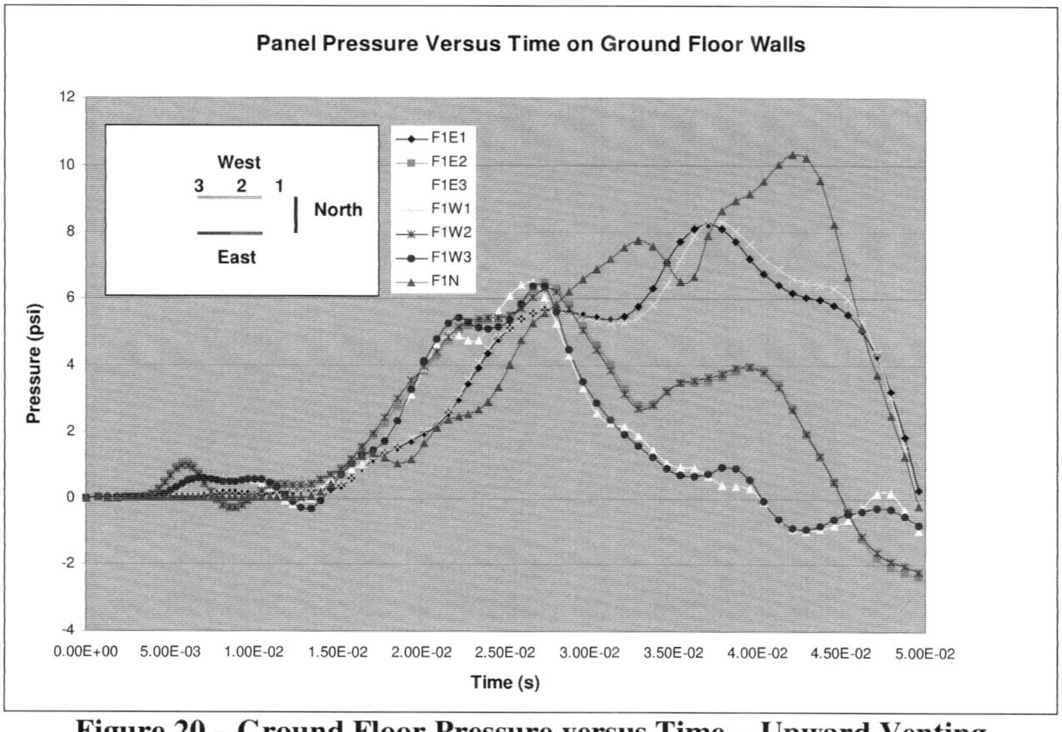

Figure 20 – Ground Floor Pressure versus Time - Upward Venting

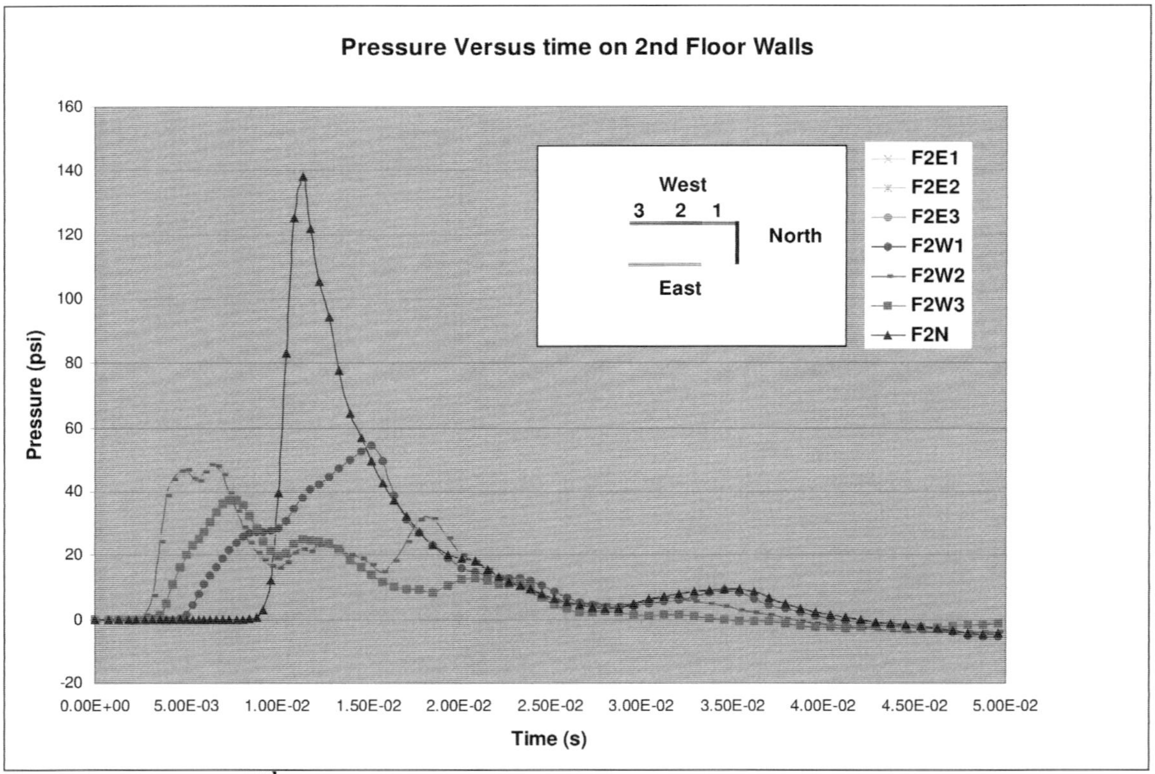

Figure 21 – 2nd Floor Wall Panel Pressure versus Time – Upward Venting

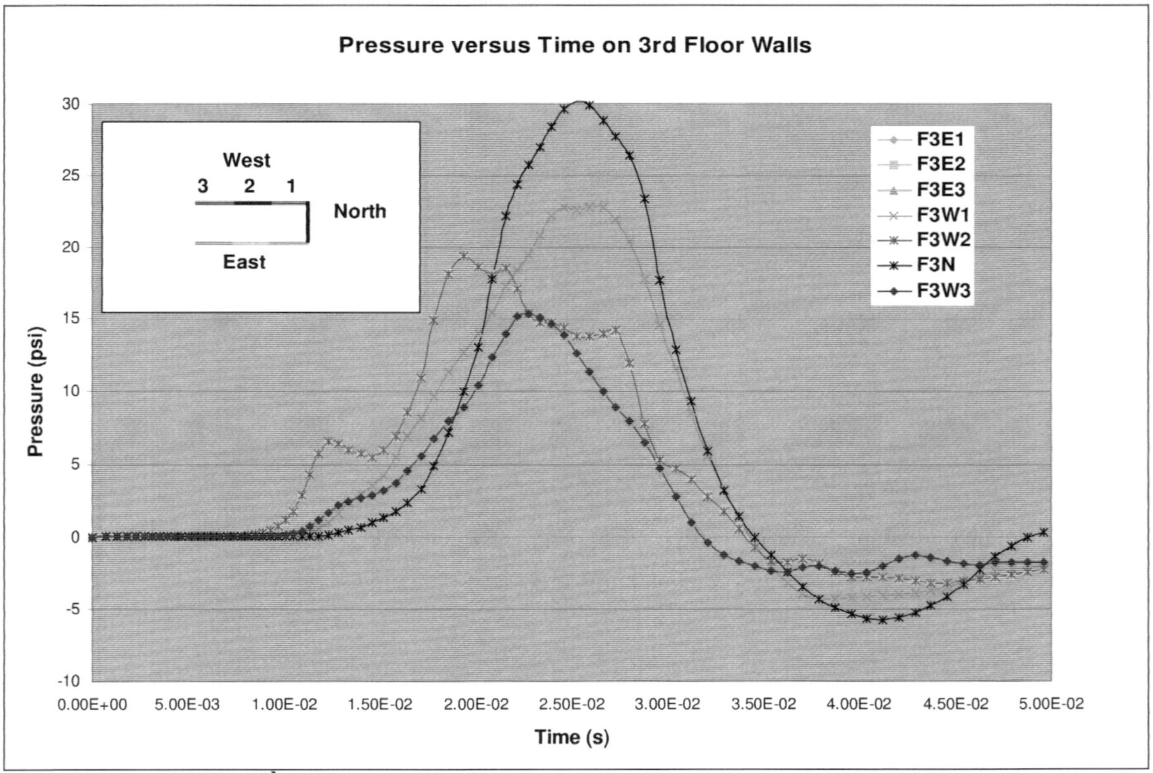

Figure 22 – 3rd Floor Wall Panel Pressure versus Time – Upward Venting

3.3 Comparison of Upward Venting Simulations

The results are compared in Figure 23. FLACS predicts higher blast loads overall. Even in simulations where the wall panels would not fail, failure of the connections of the steel members that tie the walls together is anticipated based on conversations with the structural engineer. This would result in loss of support for the back wall. The wall is supported somewhat from the opposite side by the building, and the blast will hold the wall to the building until the average pressure is negative, therefore; significant ingress of the blast wave into the manufacturing building is not anticipated.

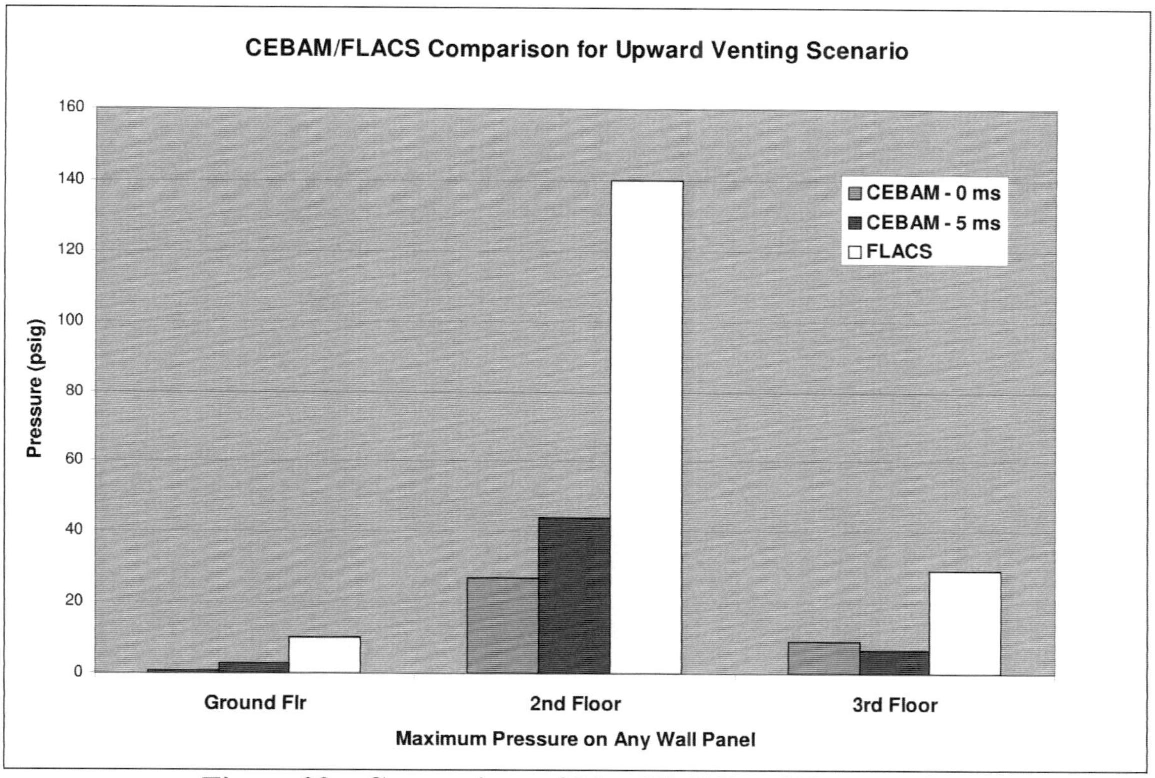

Figure 23 – Comparison of Max Wall Panel Pressure by Floor

4. Summary and Conclusions

In order to reduce the conservatism associated simplified blast models that assume that the blast wave will enter the manufacturing building, we have simulated the beneficial effect of failure of the concrete floors in the barricades to reduce the blast loads on the walls. The walls on the long sides are anticipated to fail and lose their connections. The back wall is expected to remain in one piece, but since the back wall will be supported from the opposite side against the building, significant ingress of the blast wave into the manufacturing building is not expected. As such, the estimated consequences to people and equipment inside the manufacturing building are significantly reduced in severity.

While we do not expect significant ingress of the blast wave, the building will still be shaken by the event. As such, damage and injury from falling (suspended) objects is possible.

When performing analysis such as this, some level of conservatism is still warranted due to the uncertainties discussed. The CFD analyst cannot perform this study in a vacuum. The analyst must collaborate with a competent structural engineer to get full advantage from the modeling techniques.

The CFD models have a long way to go to deliver the complexity desired to have high confidence in the results. In this case, each model has some desirable traits that could be incorporated in the other. In the extreme ideal case, a model that combines computational fluid dynamics with finite element analysis of structural components offers the greatest potential.

5. References

1. Herrmann, D., "Developing a Sound Basis for the Design of Vented Explosion Barricades in Chemical Processes," CCPS International Conference on Risk, Scottsdale, AZ, September 2003.

2. Clutter, J.; Whitney, M.; "Use Of Computational Modeling To Identify The Cause Of Vapor Cloud Explosion Incidents," *Journal of Loss Prevention in the Process Industries.* (14), 2001.

Paper TG002 - 6

Design Considerations In Dust Explosion Inerting And Suppression

Ashok G. Dastidar
Safety Consulting Engineers, Inc., Schaumburg, IL, USA 60173

John E. Going
Fike Corporation, Blue Springs, MO, USA 64015

Faisal I. Khan
Memorial University of Newfoundland, St. John's, NL, Canada A1B 3X5

Paul R. Amyotte
Dalhousie University, Halifax, NS, Canada B3J 2X4

Prepared for Presentation at
American Institute of Chemical Engineers
2005 Spring National Meeting
39th Annual Loss Prevention Symposium
Atlanta, Georgia
April 11-13, 2005

February, 2005

UNPUBLISHED

Design Considerations In Dust Explosion Inerting And Suppression

Ashok G. Dastidar
Safety Consulting Engineers, Inc., Schaumburg, IL, USA 60173

John E. Going
Fike Corporation, Blue Springs, MO, USA 64015

Faisal I. Khan
Memorial University of Newfoundland, St. John's, NL, Canada A1B 3X5

Paul R. Amyotte*
Dalhousie University, Halifax, NS, Canada B3J 2X4

*For correspondence: Phone (902-494-3976); Fax (902-420-7639); Email (paul.amyotte@dal.ca)

Abstract

This paper describes the use of inert dusts to reduce the risk of dust explosions through both prevention and mitigation schemes. A functional distinction is first made between inerting and suppression by explaining each term within the contexts of explosion prevention and explosion mitigation, respectively. The use of solid inertants is then described in terms of the various inhibitor and situation-specific parameters that can influence their effectiveness – e.g. inhibitor composition and scale of the process equipment to be protected, respectively. Practical guidance is provided on critical design considerations: explosion detection, suppressant delivery, the minimum inerting concentration (MIC), and the SEEP phenomenon (Suppressant Enhanced Explosion Parameter). Finally, the importance of inerting and suppression to industrial practice is emphasized by a discussion of their roles within the hierarchy of inherent, engineered and procedural safeguards.

1. Introduction

Risk is a measure of human injury, environmental damage, or economic loss in terms of both the incident likelihood and the magnitude of the loss or injury (CCPS, 2000). With respect to dust explosions, risk must therefore be assessed by considering the likelihood of the occurrence of such an explosion, as well as the severity of the consequences arising from a dust explosion. This naturally leads to first a consideration of the explosion hazard itself (e.g. Cashdollar, 2000), and then measures that can be taken for both prevention and mitigation in an effort to reduce the risk.

One of these risk reduction measures is the subject of the current paper – the use of solid inertants in the prevention and mitigation of dust explosions. Other techniques are available (e.g.

prevention: removal of ignition sources or inert gas blanketing, and mitigation: explosion relief venting or process isolation); there are several excellent resources in the literature that may be consulted on these matters (e.g. Bartknecht, 1989, Barton 2002, and Eckhoff, 2003). This paper covers the selected topic by referring primarily to the work of the authors and their research colleagues over the past several years (presented, in part, by Amyotte, 2004). Again, there are numerous other references available for further information (e.g. the texts by Bartknecht, Barton and Eckhoff referenced above, as well as others such as Cybulski, 1975). Additionally, the references in each of the citations in the current paper contain a wealth of information on the topic at hand.

The approach taken here is to describe the general use of solid inertants by referring to various parameters that can influence the effectiveness of these materials in preventing and mitigating dust explosions. These are factors related to both the inhibitor itself (composition and particle size) and the physical conditions under which the explosion occurs (co-presence of flammable gas, ignition energy, and scale of equipment). The latter part of the paper attempts to present practical guidance on critical design considerations: explosion detection, suppressant delivery, the minimum inerting concentration (MIC), the SEEP phenomenon (Suppressant Enhanced Explosion Parameter), and the role of inerting and suppression within a general loss prevention hierarchy. A distinction is first made between inerting as related to dust explosion prevention, and suppression as related to dust explosion mitigation.

2. Inerting And Suppression

In the preceding introductory section, reference was made to the issue of dust explosion *risk* and how this concept is composed of two components – likelihood of occurrence and severity of consequences. The term *inerting* arises when focusing on *preventing* the occurrence of a dust explosion. In the present context, use of an inert dust in sufficient amount can lead to removal of the heat necessary for combustion. This is somewhat analogous to explosion prevention by use of an inert gas (e.g. nitrogen) to ensure that process operation occurs at oxygen concentrations below the maximum permissible level.

On the other hand, the term *suppression* arises when focusing on *mitigating* the consequences of a dust explosion. The intent is the same as with inerting – to remove the heat necessary for sustained combustion and thus to limit the generation of destructive overpressures in an enclosed volume. In the case of suppression, however, the inert dust is injected into the just-ignited explosible dust/air mixture rather than being intimately premixed with the explosible dust prior to ignition (as in the case of inerting). This fundamental difference, as well as the linkage between inerting and suppression, is clearly shown in Figure 1 from Moore and Siwek (1992).

As described by Dastidar and Amyotte (2002a), Figure 1 is a pictorial representation of suppressant (or inertant) requirements to mitigate an explosion, coupled with suppressant delivery from high rate discharge canisters. The quantity of suppressant required to mitigate an explosion begins at some initial value at zero time, and increases rapidly as the explosion fireball increases in size. This initial amount of suppressant is dependent upon material properties and the nature of the ignition source. The upward curvature of the line is dependent upon the nature of flame propagation through the explosible dust in a given vessel geometry.

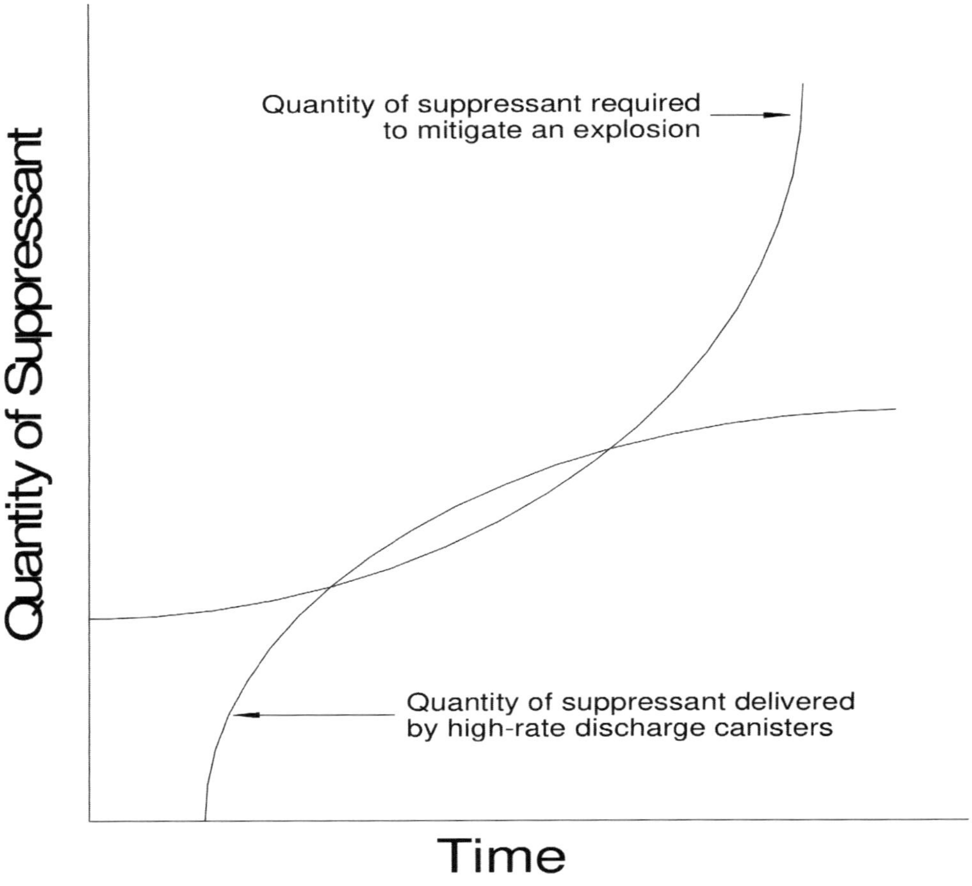

Figure 1. Suppressant (or inertant) requirement and delivery curves (from Moore and Siwek, 1992).

The delivery of the suppressant from the discharge canisters begins after an initial delay, which is dependent upon the explosion detection system. The curve then increases rapidly and begins to level off as the suppressant propellant is spent. For adequate suppression of an explosion, the suppressant delivery curve must cross and exceed the suppressant requirement curve. The nature and curvature of the suppressant delivery curve are dependent upon the design characteristics of the suppressant device (e.g. number and type of detectors, number and size of discharge canisters, and flow rate from the discharge canisters). A key factor in determining if the two curves cross is the intersection of the curve showing the quantity of suppressant required to mitigate an explosion, with the ordinate (i.e. the suppressant requirement at zero time). This value is the inerting level, or the suppressant requirement when the suppressant (or inertant) and the explosible dust are intimately mixed.

Whether the term *inertant* or *suppressant* is used, it is the same material – an inert (i.e. noncombustible) dust. The important difference is in the application of the material – be it *inerting* or *suppression.* This same distinction is clearly articulated by Eckhoff (2003) in his chapter on dust explosion research and development during the period 1990 – 2002, wherein he presents different sections entitled: *Inerting by Adding Noncombustible Dust*, and *Automatic Explosion Suppression.*

As noted by Eckhoff (2003), suppression is more widely applicable than inerting in the process industries. This is due to the need to intimately premix the explosible and inert dusts when the application is explosion inerting; this mixing can lead to unacceptable product contamination. Inerting is perhaps most prevalent in the underground coal mining industry where inert rock or stone dust (e.g. limestone or dolomite) is mixed with coal dust generated during the mining process. Other applications do exist (e.g. Mintz et al., 1996, as reported later), and it is important to look for non-traditional opportunities to employ the risk reduction potential of inert powders. As discussed later in the paper, it may be possible through inerting to process a bulk powder in a less hazardous form and thus take advantage of the inherent safety principle of moderation (Amyotte et al., 2003a). Additionally, partial inerting can be expected to significantly increase the minimum ignition energy and thus lessen the contamination or inert loading while at the same time reducing the dust explosion risk.

The discussion in the next section on factors affecting inhibitor effectiveness is based largely on inerting research, and much of that research has specifically involved the use of rock dust in the inerting of coal dust explosions. An attempt has been made wherever possible to describe the relevance of the findings to the field of explosion suppression.

3. Factors Affecting Inhibitor Effectiveness

This section describes the use of solid inertants by referring to various parameters that can influence the effectiveness of these materials in dust explosion prevention and mitigation. The parameters discussed are inertant composition, inertant particle size, co-presence of a flammable gas, applied ignition energy, and the scale of the equipment being protected. The first two of these factors are related to the inert material itself while the latter three are more specific to the industrial scenario.

Chatrathi and Going (2000) have also examined the importance of material (inertant or suppressant) factors by considering the influence of specific heat, decomposition temperature, heat of decomposition, particle radius, and surface area. In addition to these and the other parameters considered here, there are several properties of the *explosible* dust that are critical in determining the inerting level – explosible dust composition, particle size and concentration (Amyotte and Pegg, 1992). Inerting level is simply the amount of inhibitor required to inert the explosible dust/air mixture. This inhibitor amount can be expressed in a variety of ways:

- mass of added inert material,
- percentage of added inert material in the total solids mixture (which consists of the inertant plus the explosible dust),
- percentage of incombustibles in the total solids mixture; referred to as the total incombustible content, TIC, or
- ratio of mass of added inert material to mass of explosible dust; referred to as the inerting ratio, IR.

3.1 Inertant Composition

Within the framework of inertant composition, a number of factors are important: specific heat, heat of reaction (decomposition), decomposition temperature, and rate of decomposition. The first of these factors determines, in part, the overall heat capacity of the dust cloud, while the other three relate to the possibility of an endothermic decomposition reaction involving the inertant. These points are expanded upon in the following discussion from Amyotte and Pegg (1992).

As described by Amyotte et al. (1992), tests were conducted in a 26-L explosion chamber using three different rock dusts: limestone (98 weight % $CaCO_3$), dolomite (~56 weight % $CaCO_3$ and ~38 weight % $MgCO_3$) and magnesite (14 weight % $CaCO_3$ and 81 weight % $MgCO_3$), in an attempt to determine the relative importance of rock dust decomposition. The inertants had similar particle size distributions, with mass mean diameters of 25 μm, 30 μm and 28 μm, respectively. The explosible dust tested was coal dust from the Prince seam of the Cape Breton Development Corporation, Nova Scotia (mass mean diameter of 22 μm and volatile content of ~35 weight %). If an endothermic decomposition reaction were to play a role, one would have expected the increasing order of inerting effectiveness to be limestone, dolomite, magnesite. This ranking is based on the decomposition temperatures, which are 825 °C, 750 °C and 350 °C, respectively.

The results showed, however, that this order of effectiveness did not exist. The inerting levels for the three rock dusts were the same – 84 weight % rock dust (TIC of 86 weight % or IR of 5.3). The relative proportions of $CaCO_3$ and $MgCO_3$ in each rock dust did not influence the amount of inert material needed for explosion inerting. The dominant inerting mechanisms in these tests, therefore, was the increase in solids heat capacity due to the presence of rock dust. The specific heats of limestone, dolomite and magnesite do not vary greatly from ~0.9 kJ/kg·K, and the specific gravities are all about 3.0 or slightly less.

This lack of a thermal effect due to rock dust decomposition can be explained by the short residence time in the reaction zone. Hertzberg et al. (1984) have shown that the decomposition induction time for limestone can be a factor of ten slower than the devolatilization induction time for coal dust, and its half-life for the generation of carbon dioxide a factor of four slower than the volatiles generation half-life of coal. Assuming similar decomposition limitations exist for dolomite and magnesite, it is concluded that rock dust is simply too slow in decomposing relative to coal devolatilization (which is a critical step in the homogeneous flame propagation mechanism of coal dust). This occurrence has been interpreted as rock dust being "less than thermal" in inerting effectiveness (Hertzberg et al., 1984), meaning that the added thermal loading due to rock dust decomposition does not necessarily occur. The potential magnitude of this heat-of-reaction effect is, however, quite significant. Experiments by Choi et al. (1986) in a jet-stirred reactor system have shown that with an appropriate residence time and 100 % decomposition, the percentage of limestone required to inert can be nearly half of that needed when no decomposition occurs.

The preceding discussion has focused on the thermal influence of dust explosion inertants; as discussed shortly, other effects that are chemical in nature are also possible with some inhibitors.

To recap, inertants such as limestone act to inert or suppress dust explosions by "draining away" the heat necessary for combustion and therefore limiting flame propagation through the unburned fuel cloud. This thermal effect is due mostly to the high heat capacity of the inert material, but may also include contributions from inertant decomposition if conditions permit (e.g. if the residence time in the flame zone is sufficiently long). These decomposition effects would include the endothermic heat of reaction as well as dilution of the reaction zone by decomposition products such as carbon dioxide in the case of $CaCO_3$ calcination. Large amounts of these thermal inhibitors are typically required for effective inerting (recall the 86 weight % TIC from Amyotte et al., 1992; see also Sapko et al., 2000).

Knowledge of the manner in which thermal inhibitors act to prevent flame propagation can be applied to other, perhaps seemingly unrelated, applications. For example, in the electrical utility industry, incomplete combustion and subsequent contamination of a waste stream can pose a serious explosion hazard. This is the problem of *fuel carry-over*, as might arise in the case of contamination of fly ash with pulverized coal. The coal, if present in sufficient quantities in the mixture, can act as a fuel source for a potential explosion. Solutions to the risk posed by fuel carry-over can, however, be identified by recognizing that contamination of an inert dust (i.e. fly ash) with an explosible dust (i.e. coal dust) is simply explosion inerting "turned around". The underlying phenomena are the same whether one is attempting to thermally inert coal dust with fly ash or determine the percentage of coal dust that would render explosible a mixture of said coal dust and fly ash (Dastidar and Amyotte, 2002b and Amyotte et al., 2004).

Figures 2 and 3 from Amyotte et al. (2003b) and Amyotte et al. (1991a), respectively, show another example where knowledge of the behaviour of thermal inertants can be helpful in interpreting the results from an application other than explosion inerting. In the study by Amyotte et al. (2003b), the primary objective was to determine the degree of reduction in the dust explosion hazard that would be afforded by the substitution (wholly or partially) of petroleum coke for coal in the burner feed in fossil fuel-fired power plants.

Figure 2 presents the rate of pressure rise, $(dP/dt)_m$, data from Siwek 20-L vessel testing of blended fuels of coal and petroleum coke; the rate of pressure rise displays a general decrease from essentially the onset of petroleum coke admixture. This trend is analogous to that in Figure 3, which shows the reduction in rate of pressure rise brought about by the admixture of dolomite to a typical coal dust from a Nova Scotia mine (Amyotte et al., 1991a). Petroleum coke, although obviously not totally inert, does have a significantly lower volatile content than the coals and therefore exerts an "inerting-like" influence when mixed as a fuel with coal dust. This analogy also applies to explosion overpressure results; as described by Amyotte et al. (2003b), admixture of either dolomite or petroleum coke with coal dust results in a lowering of explosion overpressure only at relatively high additive percentages.

As previously mentioned, some inertants have the ability to exert a chemical influence in addition to the physical mechanism of thermal energy absorption. These inhibitors chemically participate in the combustion sequence by acting to terminate chain branching reactions via free radical capture, thus providing kinetic interference in the process of flame propagation. Chatrathi and Going (2000) describe Halon 1301 (CF_3Br) as an example of an inertant that chemically participates in the combustion reaction, as contrasted with the previously described rock dust, which is primarily a physical (i.e. thermal) inertant. The papers by Swift (1988) and Moore (1996) also provide good overviews of various inertant (suppressant) agents and their effectiveness; of particular note is Moore's description of the chemical industry's efforts in the mid-nineties to secure halon alternatives due to the displacement of the these inhibitors because

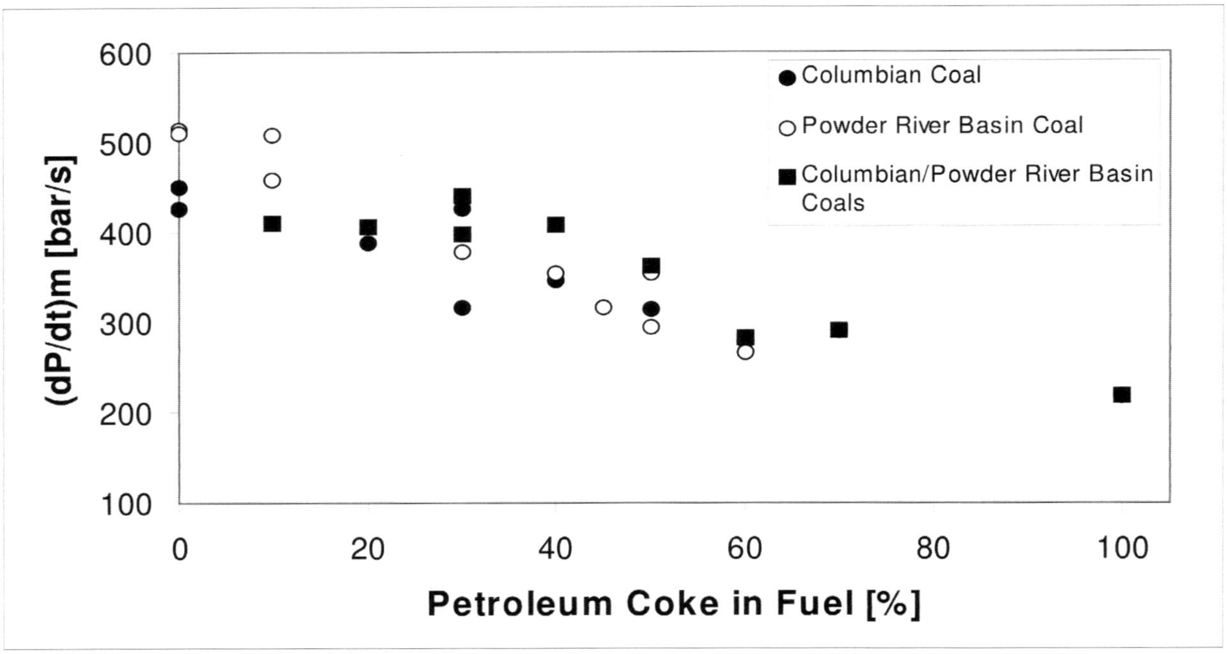

Figure 2. Reduction of rate of pressure rise achieved by using blended fuels of coal and petroleum coke (from Amyotte et al., 2003b).

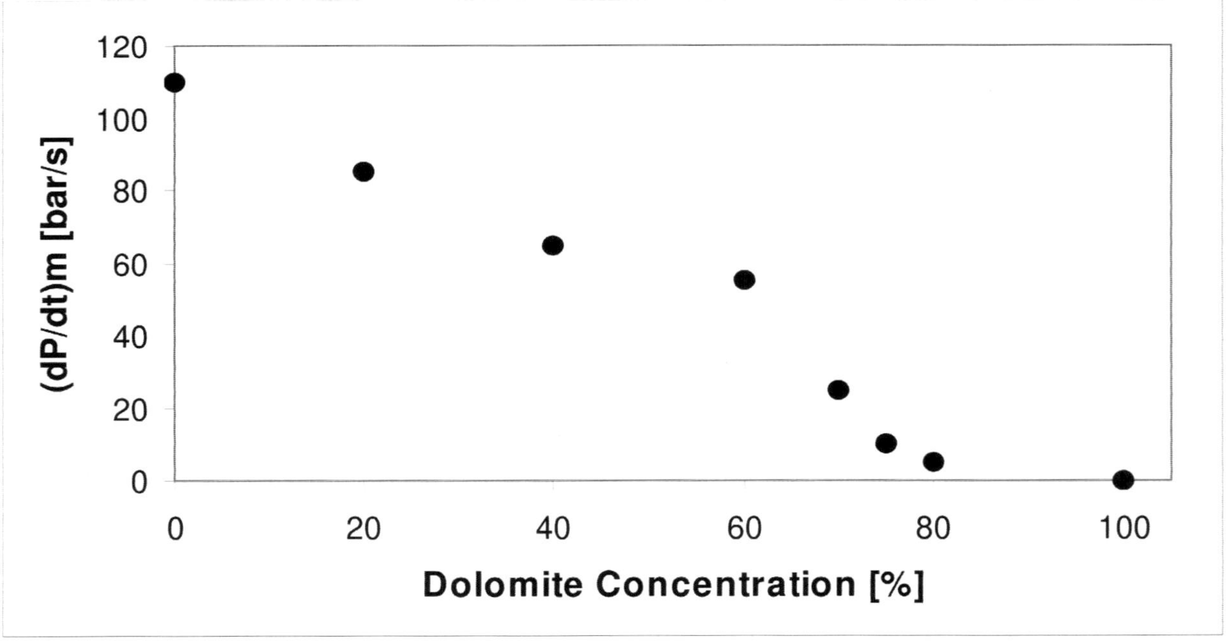

Figure 3. Influence of admixed dolomite (rock dust) on rate of pressure rise of a coal dust (from Amyotte et al., 1991a).

of their ozone-depleting effects.

The inertants monoammonium phosphate ($NH_4H_2PO_4$ or MAP) and sodium bicarbonate ($NaHCO_3$ or SBC) are thought to act as explosion inhibitors by both physical and chemical means (Chatrathi and Going, 2000). These materials have been tested extensively by the current authors in the Fike 1-m^3 chamber (Dastidar et al., 1998, Dastidar et al., 1999, Chatrathi and Going, 2000) and a Siwek 20-L chamber (Dastidar and Amyotte, 2002a). The explosible dusts tested were aluminum, anthraquinone and polyethylene (with cornstarch and Pittsburgh pulverized coal, and the inertant limestone, to a lesser degree). The primary purpose in these tests was to investigate a flammability parameter known as the minimum inerting concentration, or MIC (as described in subsequent sections of the current paper).

Of interest here is Figure 4 (1-m^3 inerting testing) from Dastidar et al. (1999) and Chatrathi and Going (2000). The data points represent the explosibility limit of the given fuel/inertant mixtures. They have been interpolated by averaging the highest inertant concentration tested that produced an explosion with the lowest inertant concentration tested that did not produce an explosion for a given fuel concentration. The area to the left of each curve in Figure 4 represents the explosible area for the fuel/inertant mixture. The area to the right of each curve represents the nonexplosible area; here there is sufficient inertant to prevent an explosion. The "nose" of each curve (or envelope) represents the least amount of inertant that would prevent an explosion regardless of fuel concentration – i.e. the aforementioned MIC.

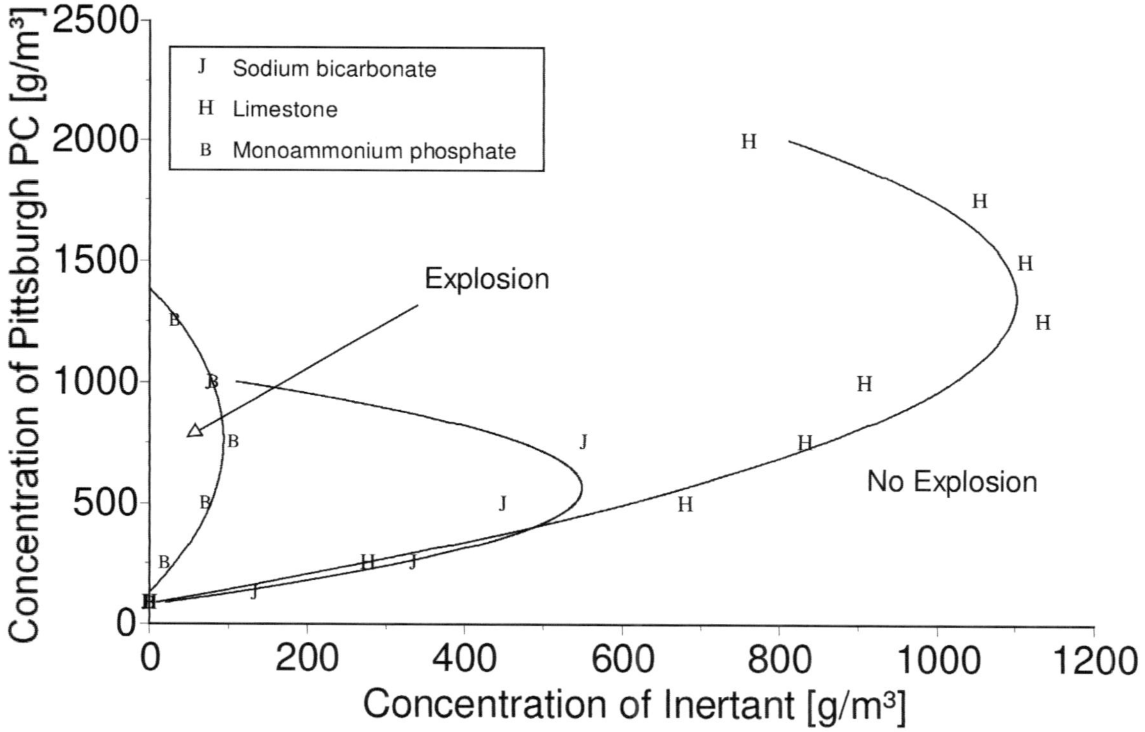

Figure 4. Inerting envelope for Pittsburgh pulverized coal dust with sodium bicarbonate, monoammonium phosphate and limestone as inertants; experiments conducted in a spherical 1-m^3 chamber using two 5-kJ ignitors as ignition source (from Dastidar et al., 1999, and Chatrathi and Going, 2000).

Figure 4 clearly illustrates that for inerting of Pittsburgh pulverized coal (PPC), limestone is the least effective inhibitor and MAP is the most effective inhibitor. This occurrence is explainable by the inability of limestone to decompose in the rapidly advancing flamefront (Chatrathi and Going, 2000). Additionally, the chemical inhibition properties of MAP and SBC would be expected to enhance their effectiveness over limestone. It should be noted that the superior performance of MAP over SBC illustrated in Figure 4 did not hold for all combinations of fuel and inertant tested (Dastidar et al., 1998, Dastidar et al., 1999, Chatrathi and Going, 2000, and Dastidar and Amyotte, 2002a). This means that, as would be expected, an inertant's effectiveness is dependent on the composition of the explosible dust. Further, as demonstrated by Amrogowicz and Kordylewski (1991), such effectiveness may also be dependent on whether the application is *inerting* or *suppression*. They found that for their suite of explosible dusts (melamine, wood dust, wheat flour and coal dust), MAP was more effective than SBC for *inerting*, whereas SBC performed as well or better (depending on inhibitor concentration) than MAP for *suppression*.

3.2 Inertant Particle Size

Particle size considerations are a key element of dust explosion testing, research, and data interpretation and application to industry. The effects of a reduction in explosible dust particle size are numerous and are well-established; these include, for example, an increase in the maximum explosion pressure – P_{max}, a potentially significant increase in the maximum rate of pressure rise – $(dP/dt)_{max}$, a lowering of the minimum ignition energy – MIE, and a lowering of the minimum explosible concentration – MEC (Amyotte et al., 1991b, and Cashdollar, 2000). Decreasing the mass mean diameter of an explosible dust by reducing particle size throughout the entire size distribution can also significantly increase the inerting level (Amyotte et al., 1991a).

With respect to *inertant* particle size, a general statement can be made that finer sizes of a given inertant are more effective at explosion inerting and suppression than coarser size fractions. There are limitations to this statement; ultrafine inertants may tend to agglomerate, thus negating the intended benefit of a reduction in inhibitor particle size, and extremely coarse sizes may show little incremental inerting effectiveness. Overall, however, it is well-established that a reduction in inertant particle diameter leads to enhanced inerting performance. This has been demonstrated by Hertzberg et al. (1984) for MAP and by the current authors in work involving fly ash (Dastidar et al., 2002b and Amyotte et al., 2004).

A preliminary study on the effect of dolomite particle size on coal dust inerting requirements (Amyotte et al., 1992), prompted later more comprehensive investigations (Amyotte et al., 1995, and Dastidar et al., 1997). In the follow-up study by Amyotte et al. (1995), tests were conducted using a 26-L spherical chamber; the test matrix incorporated two size fractions of coal (fine and coarse) and several broad and narrow size distributions of limestone. The use of broad limestone fractions ensured practical applicability of the work, while the narrow sizes enabled more fundamental conclusions to be drawn from the research results.

Figure 5 from Amyotte et al. (1995) shows the inerting levels as a function of limestone mass mean diameter for the narrow-size limestone tests. The data trends in Figure 5 were similar to those determined for the broad-size limestone tests, thus confirming that the fine (< 20 μm) particles in a broad size distribution of rock dust make the most significant contribution to explosion inerting. Also illustrated by Figure 5 is the existence of a mean particle diameter above which the inerting level remains essentially constant. This " characteristic diameter" is about 65 μm for the fine coal and 80 μm for the coarse coal.

The findings described in the preceding paragraph permit the conclusion that a broad particle size distribution of rock dust can be thought of as a series of narrow size distributions, each of which makes a contribution to the inerting level. This contribution will be independent of rock dust particle size down to some characteristic diameter (re. Figure 5). Rock dust particles smaller than this characteristic diameter can then influence the inerting level depending on their relative abundance. The underlying phenomena responsible for this behaviour can be understood by recalling that rock dusts (limestone, dolomite, etc.) inhibit dust flame propagation by thermal means. A reduction in rock dust particle size leads to a corresponding increase in rock dust surface area, which in turn leads to greater radiant heat absorption. The possibility also exists that smaller rock dust particles are able to decompose fast enough that inertant decomposition is not a rate-limiting step in relation to coal dust devolatilization.

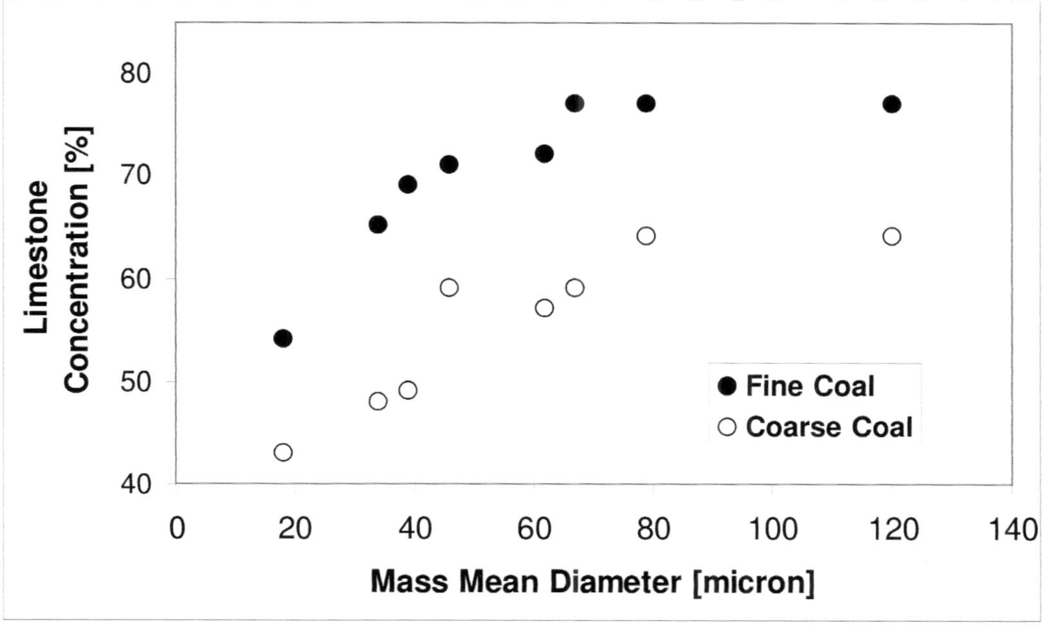

Figure 5 Effect of limestone particle size on inerting level; narrow sizes of limestone (from Amyotte et al., 1995).

3.3 Co-Presence of Flammable Gas

Hybrid mixtures of flammable gas and explosible dust are of particular concern in industry – perhaps no more so than in the underground coal mining industry where methane usually co-exists with coal dust (Amyotte and Pegg, 1993). Similar to the effects of particle size, the influence of the co-presence of a flammable gas on the explosibility parameters of a fuel dust alone is well-established (Amyotte et al., 1991b and Cashdollar, 2000). With respect to mixtures of fuel dust and inertant, the co-presence of a flammable gas leads to an increase in the inerting level. This again is well-established (Amyotte et al., 1991a and Amyotte et al., 1995).

The magnitude of this increase in inerting level can be significant. For example, in a series of 26-L tests conducted by Amyotte et al. (1991a), the inerting level for one of the coal dusts examined was raised from 84 weight % dolomite to 89 weight % dolomite by the addition of 2 volume % methane to the oxidizing atmosphere. While the dolomite percentage increase is only 6 % of its value when no methane is present (i.e. 84 x 1.06 = 89), the increase in actual mass of dolomite for a given mass of coal dust is 54 %. This is caused by the exponential relationship observed when rock dust mass is plotted as a function of rock dust percentage. The increase in mass can be quite large for seemingly modest increases in percentage.

With respect to the coal mining industry, the above discussion indicates the extreme inefficiency of rock dust as applied to methane. Considering that it is essentially gas-phase flame propagation which must be arrested during a coal dust explosion, the results also help to explain why large amounts of rock dust are required even in the absence of methane. On a more general note, these data demonstrate that inerting levels based on the presence of an explosible dust alone can be entirely inadequate when applied to hybrid mixtures of solid and gaseous fuels.

3.4 Ignition Energy

Although discussed in separate sections in this paper, ignition energy and equipment scale are inextricably linked in dust explosibility testing. The size and geometry of the test equipment play a major role in the choice of ignition source strength. This is the case not only for determination of the inerting level, but also for other parameters such as the MEC (Chawla et al. (1996) and LOC – limiting oxygen concentration (Going et al., 2000). These are all limit flame conditions and, as measured in standard test equipment, are strongly dependent on the applied ignition energy. The issue is essentially one of avoiding a phenomenon known as *overdriving*.

Overdriving of an explosion occurs when the ignition source (and as a result the flame produced by the ignition source) is large compared to the vessel volume. This has two consequences as described by Dastidar et al. (2001). First, it changes the initial test conditions of the dust cloud (increasing the temperature and pressure), thereby possibly making a nonexplosible dust explosible. Second, it can also result in burning of the dust within the ignitor flame, but with no real propagation beyond the ignition source. Since the ignition source volume is a large fraction of the test volume, the dust appears to explode and as a result the overpressure and rate of pressure rise are overestimated. The clear implication of these considerations is that the inerting level – and MEC and LOC – are not fundamental parameters of a particular material (at least as they are measured in standard dust explosibility test equipment).

The effect of overdriving is illustrated quantitatively by Figure 6 from Dastidar et al. (1997). In these Siwek 20-L vessel tests, mine-size coal dust was inerted with rock dust (dolomite) using ignition energies ranging from 0.25 kJ – 20 kJ and a fixed coal dust concentration of 500 g/m^3. As shown in Figure 6, the explosion overpressure, P_m, decreased for all ignition energies tested as rock dust concentration was increased. When enough rock dust was added, the mixture was totally inerted and explosions were no longer possible (as evidenced by the explosion overpressure dropping below the explosion criterion of P_m = 1 bar(g)). A wide range of inerting levels exists in Figure 6 – from just under 50 weight % dolomite at an ignition energy of 0.25 kJ to just over 80 weight % dolomite at an ignition energy of 20 kJ. The selection of the inerting level to be used for industrial applications depends entirely on the size and geometry of the equipment to be protected. In other words, choosing the "correct" inerting level requires knowledge of inerting levels measured in chambers larger than laboratory scale (i.e. > 20 L). This point is expanded on in the next section of the paper in which the work of Dastidar et al. (2001) is briefly discussed.

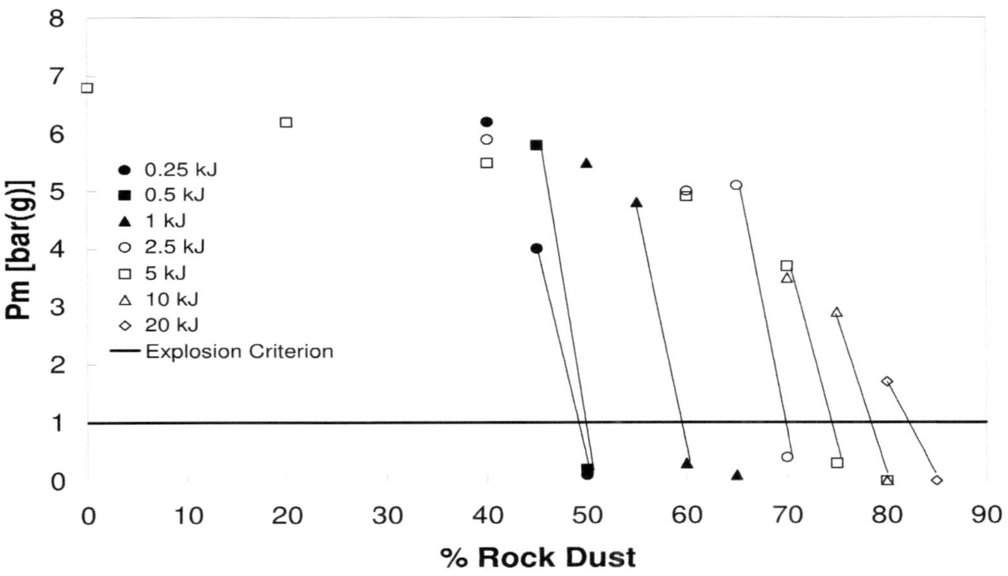

Figure 6. Effect of ignition energy on the inerting level of mine-size coal dust inerted with rock dust (from Dastidar et al., 1997).

Before moving to an examination of equipment scale, Figure 7 from Dastidar and Amyotte (2002a) is presented as an alternate representation of the effect of ignition energy on inerting level. As previously mentioned, the study by Dastidar and Amyotte (2002a) involved Siwek 20-L vessel testing and a comparison with the Fike 1-m^3 vessel tests of Dastidar et al. (1998) and Dastidar et al. (1999). Of interest here are the 20-L chamber results for cornstarch inerted with SBC; the data are presented as inerting or flammability envelopes in a manner similar to that shown in Figure 4. Figure 7 illustrates that decreasing the ignition energy in the 20-L chamber from 5 kJ to 1 kJ lowered the MIC from ~1500 g/m^3 to ~500 g/m^3. (Recall that the MIC, or minimum inerting concentration, is the least amount of inertant that would prevent an explosion regardless of fuel concentration. In other words, the MIC is simply an alternate way of expressing the inerting level for a given explosible powder.)

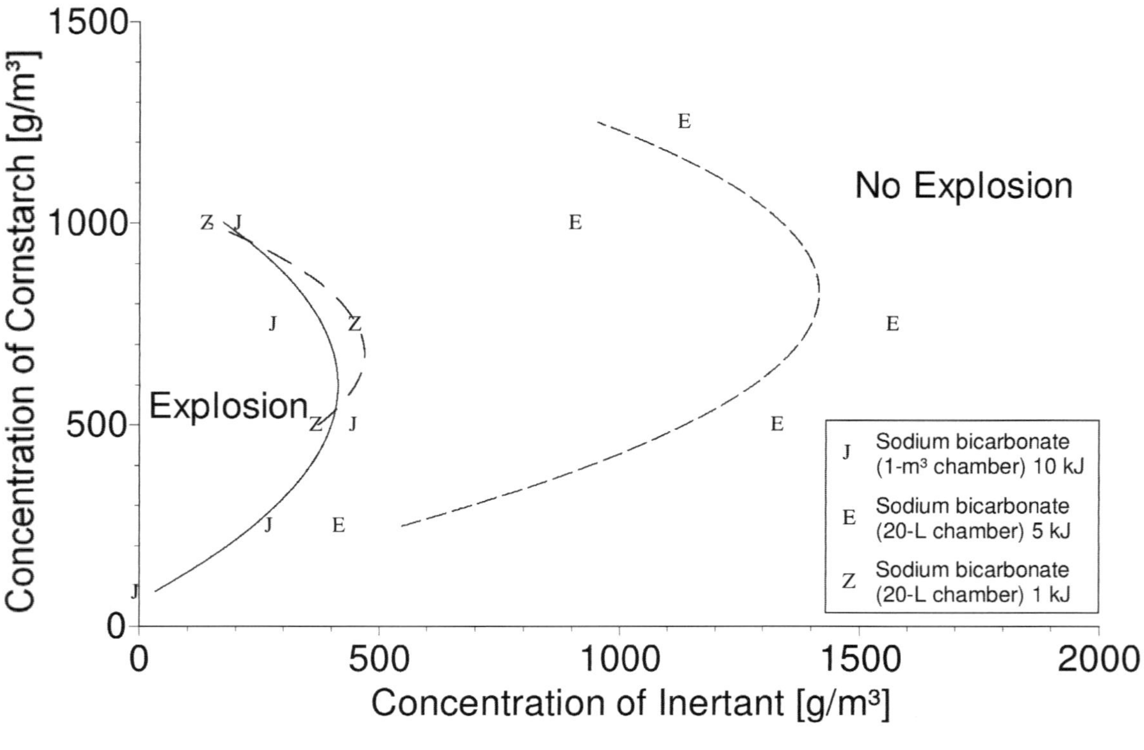

Figure 7. Effect of ignition energy (and vessel volume) on the inerting requirements of cornstarch inerted with SBC (from Dastidar and

3.5 Equipment Scale

The importance of vessel size and geometry is well-established in the field of dust explosion testing and research when investigating the parameters of maximum explosion pressure and maximum rate of pressure rise. A minimum volume of 20 L and spherical or near-spherical geometry (coupled with a 10-kJ ignition energy) are essential for measuring P_{max} and $(dP/dt)_{max}$ values that are applicable to industrial-scale equipment. With respect to the limit flame conditions of MEC and LOC, similar size and geometry considerations apply. The desirability of making MEC and LOC comparisons between laboratory-scale chambers (20 L) and intermediate-scale chambers (1 m^3) has been illustrated by the work of Chawla et al. (1996) and Going et al. (2000), respectively. These comparisons are necessary for specification of an appropriate ignition energy that yields values of MEC and LOC applicable to industrial scenarios.

Precisely the same considerations apply to the determination of inerting levels that have relevance beyond the particular test equipment in which they were measured. Referring back to Figure 7, it can be seen that inerting results from the 1-m^3 chamber are approximated by the 20-L values only when the ignition energy in the 20-L chamber is decreased to 1 kJ. This is an energy level in the 20-L chamber that would not be expected to overdrive the fuel/inertant system, just as the higher 10-kJ ignition energy does not overdrive the larger 1-m^3 volume. The conclusion, therefore, is that cornstarch/MAP inerting levels measured under conditions of a 5-kJ ignition energy and 20-L test volume would be unrealistically high if applied to an enclosure volume of 1 m^3.

Figure 8 from Dastidar and Amyotte (2002a) shows inerting envelopes for Pittsburgh pulverized coal inerted with monoammonium phosphate in volumes of 1 m^3 (10-kJ ignition energy) and 20 L (ignition energies of 0.5, 1, 2.5 and 5 kJ). Clearly, decreasing the ignition energy from 5 kJ results in 20-L inerting levels approaching those measured in the 1-m³ chamber. While this was the case for the carbonaceous dusts examined by Dastidar and Amyotte (2002a), aluminum dust did not require a reduction in the 20-L ignition energy from 5 kJ to produce similar results in both chambers. This occurrence clearly demonstrates that the ignition energy required to produce inerting results in the 20-L chamber similar to a 1-m³ volume is dependant on the explosible material being tested.

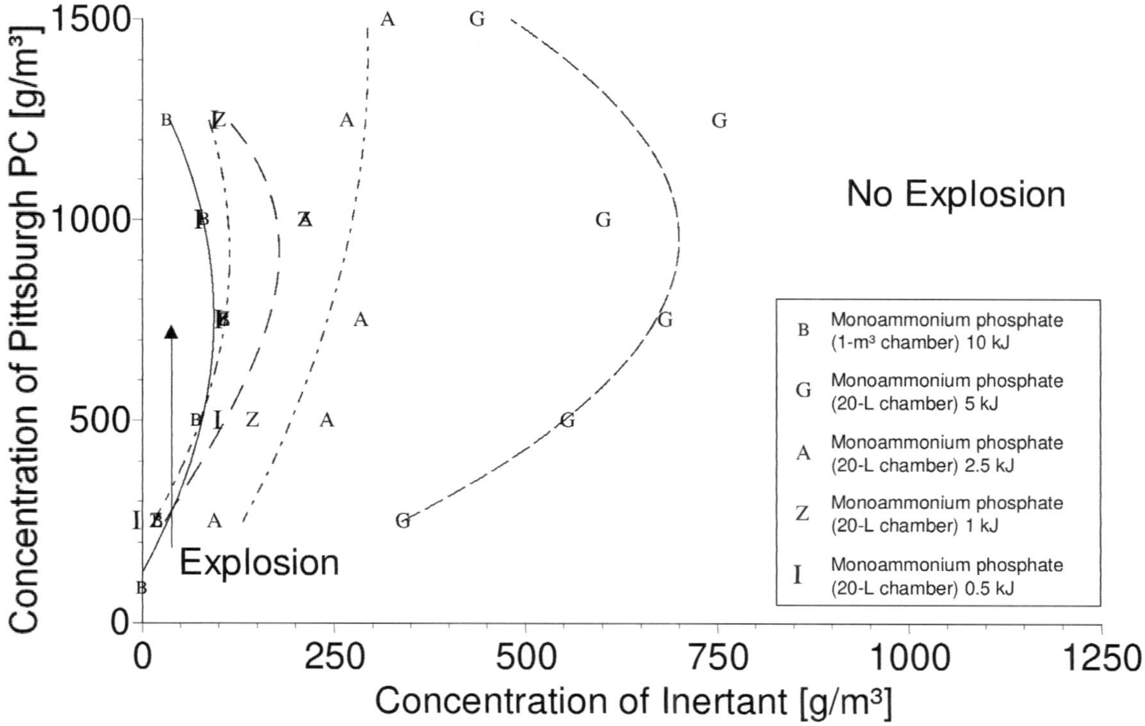

Figure 8. Comparison of 1-m^3 and 20-L chamber inerting curves at different ignition energies for Pittsburgh pulverized coal inerted with MAP (from Dastidar and Amyotte, 2002a).

In a companion study, Dastidar et al. (2001) investigated coal dust explosion inerting with limestone in 20-L and 1-m^3 volumes. Key to the interpretation of results from this work is the availability of inerting level data for coal dust/limestone mixtures from full-scale mine tests.

There is a large body of evidence that an ignition energy of 5 kJ is required in either the Siwek 20-L chamber (Dastidar et al., 2001) or the NIOSH 20-L chamber (Cashdollar, 2000) to match coal dust/limestone inerting levels determined in the NIOSH Lake Lynn experimental mine (Sapko et al., 2000). Interestingly though, Dastidar et al. (2001) showed that coal dust/limestone inerting levels in the Siwek 20-L chamber (over an ignition energy range of 1 – 10 kJ) were consistently higher than those measured in the Fike 1-m^3 chamber (over an ignition energy range of 5 – 30 kJ). As mentioned previously, the higher ignition energies would be expected to overdrive the 20-L chamber but not the 1-m^3 chamber for these mixtures of explosible dust and inertant at the limit flame condition represented by the inerting level.

When reviewing this dissimilitude among 20-L, 1-m^3 and mine-scale coal dust/rock dust inerting levels, it appears that energy density, in conjunction with vessel geometry, must be considered. For comparison to vessels with high L/D ratios where flame acceleration due to pressure piling can occur (e.g. a mine gallery), a high energy density is required in the 20-L chamber to produce similar inerting results. In chambers with lower L/D ratios where the propagating flamefront in the burning dust cloud will reach the vessel walls at approximately the same time (making allowances for buoyancy and turbulence factors), a lower ignition energy density is required.

As a final note to this section, it must be remembered that mine-scale experiments replicate a specific industrial scenario. Inerting levels determined in an experimental mine are therefore entirely representative of explosion inerting requirements in that scenario. Attempts to match those inerting levels through manipulation of test variables (e.g. ignition energy) in 20-L testing, are attempts to replicate the full-scale mine scenario. The fact that flame acceleration and overdriving may be involved in the physics of mine-scale and laboratory-scale testing, respectively, becomes somewhat irrelevant. This is analogous to a situation where the scale of the equipment to be protected through inerting or suppression in fact matches that of the test equipment. For example, if the industrial scale is of the order of 20 L (perhaps in the processing of high-value added materials), and massive ignition energies are possible, then overdriven inerting levels would be of relevance.

4. Design Considerations

The objective of this section is to provide practical guidance for several design considerations related to inerting and suppression. The topics covered include: explosion detection, suppressant delivery, the minimum inerting concentration (MIC), the SEEP phenomenon (Suppressant Enhanced Explosion Parameter), and the role of inerting and suppression within a general loss prevention hierarchy.

The phenomena of detection and suppression involve a series of events that occur in both serial and parallel fashions. The overall event is seldom greater than a few hundred milliseconds in duration and is controlled by kinetic factors as much as by thermodynamics. The general order of events is: ignition followed by deflagration growth or propagation and development of a detectable pressure or flame, physical detection, signal processing to generate a fire or activation signal, activation of a discharge container, propulsion of the suppressant agent toward the flame, mixing of the agent and fuel, and finally interruption/termination of the flame propagation by physical or chemical means. The last three steps will occur in a parallel fashion. The key steps of detection and agent delivery are now expanded upon.

4.1 Explosion Detection

A successful ignition event leads to flame propagation or growth, which is most readily characterized by pressure due to heat generation and radiation. Both parameters can be utilized for detection, although pressure is most commonly used. Pressure is used for detection by specifying a specific pressure level or a specific rate of pressure rise. These are often referred to as static and dynamic detection and each method has its pros and cons. It should be noted that in either case, a finite amount of time is required for the pressure requirement to be met. This permits flame growth in a vessel or propagation down a duct during that time period. This time to detection is affected by the vessel volume, K_{St} (size-normalized rate of pressure rise), vessel obstructions, and detection set points. A low-K_{St} dust explosion in a very large vessel can take considerable time to reach a detectable level. Recalling from the cubic expression that $dP/dt = K_{St}/V^{1/3}$ (or $dt = dP \cdot V^{1/3}/K_{St}$), it is readily seen that for a fixed dP or activation pressure, the time to detection increases with an increase in volume and a decrease in K_{St}.

The use of low static set points can provide faster response, but often at the expense of activations from process pressure variations. Rate detection is used then to differentiate between process pressure changes and actual deflagrations. The set point now is given in terms of a dP/dt (e.g. 20 psi/s) that is characteristic of a deflagration and outside of expected process variations. While high static or rate set points will provide stability, they will also delay the onset of suppression and lead to reduced suppression pressures, P_{red}, that may exceed the process design intentions.

The actual detection of pressure, whether static or rate, requires stable, high-speed hardware that is also rugged in nature. The Fike CEREx detector, for example, uses a ceramic membrane that is able to withstand full deflagration pressures, particle impaction, and cleaning with no damage or loss of calibration. Ceramic membranes are also more corrosion and vibration resistant than metal versions. These detectors operate on a piezoresistive basis and collect pressure data at a rate of 4 kHz. A high sampling rate allows millisecond level signal averaging to reduce the effect of electrical spikes. Modern detection and control systems are microprocessor based and are able to make logical decisions from multiple detectors, respond to static and/or rate criteria, and store pre-and post-activation pressure data.

A flame is also characterized by its emission spectrum or radiation. Radiation is released in the ultraviolet, visible and infrared regions and can be used for detection. In the absence of any obstruction, radiation detection can be extremely fast and is limited only by the sensitivity of the detection device. However, physical obstruction can be a major obstacle in the use of optical detection. Physical obstruction can arise due to partitions or components inside a process that interfere with the light path from source to detection. The combustible dust in a vessel will also significantly interfere with radiation and limit the applicability of such detection to short distances. Detection in ductwork with multiple sensors, however, is a viable application of optical detection and is complementary to pressure detection.

4.2 Suppressant Delivery

As previously defined, suppression is the process of providing protection by controlling the development of pressure and flame. It accomplishes this by extinguishment of the flame and limiting of the pressure to ~5 % of the maximum permissible value. Key variables are detection (see previous section), choice of agent, and delivery of agent. The agent predominantly in use today is sodium bicarbonate with limited use of water or monoammonium phosphate (hence our emphasis on the inerting/suppression properties of SBC and MAP).

Because suppression is reactive, it is acknowledged that pressure will develop before complete extinguishment. This pressure, P_{red} or Total Suppressed Pressure (TSP), results in fact from two sources – combustion and suppressant delivery. From the time of ignition, pressure will increase until the activation pressure, P_1, is reached (see Figure 9). This pressure is set on the detector/control panel. At this point, the panel has released a fire signal which will result in the opening of the high rate discharge (HRD) container and the beginning of the discharge. During this time period, the pressure has continued to grow (P_2). Agent is propelled into the vessel/flame and commences to mix and interfere with the combustion. The time required will be related to how far the agent must travel (~vessel volume), and how fast the agent reaches the combustion area (dependent on HRD characteristics of discharge velocity and pattern). Pressure grows again in this step (P_3), and is often the major contributor to the overall pressure. The final contribution to pressure will be the propelling gas from the HRD, which is typically nitrogen at

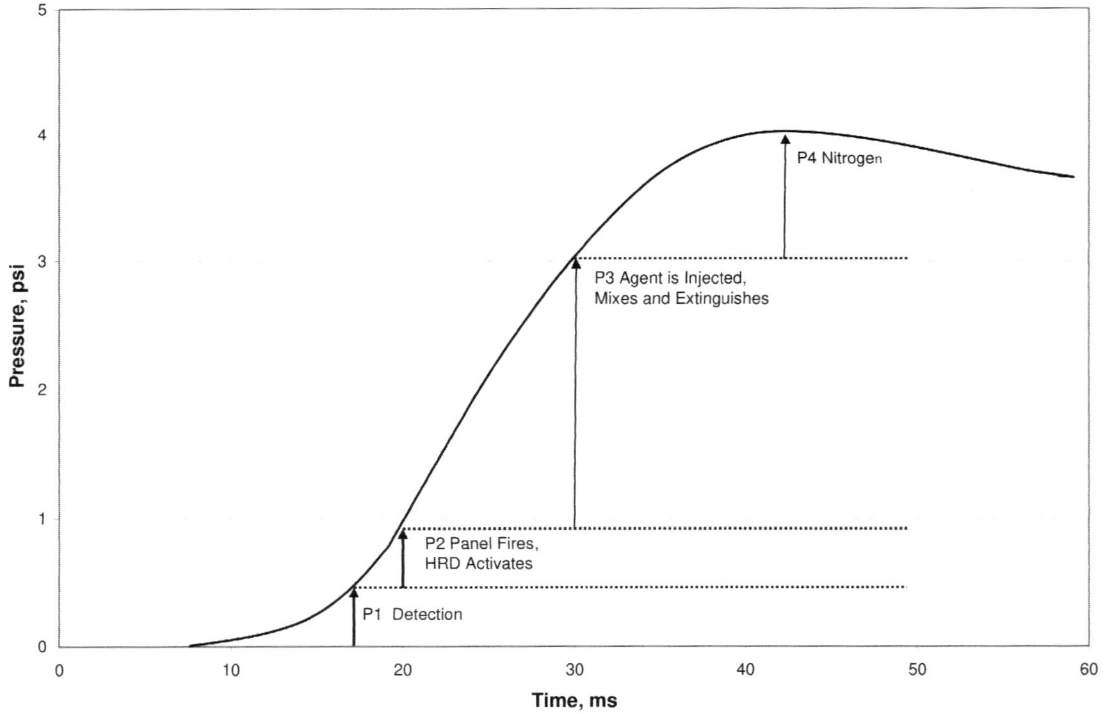

Figure 9. Growth and contributions to total suppressed pressure (TSP).

~60 bar. In a normal application, this value, P_4, can be between 1 and 1.5 psi. The TSP is the sum of the individual contributions: $P_1 + P_2 + P_3 + P_4$ (see Figure 9).

Figure 10 provides a graphical illustration of the discharge characteristics of a 30-L HRD container with a 4-in. nozzle. The times shown on the profiles represent the values in milliseconds from the first appearance of agent at the nozzle. The sequence of curves shows the arrival of agent at 10 ft in ~50 ms after firing. The velocity past 6 ft is in excess of 300 ft/s (200 mph), which is a requirement for effective suppression. It should be noted that the distribution is three-dimensional, and effectively reaches all parts of the protected vessel. HRD containers with large nozzles (e.g. 6 in.) are able to reach greater distances with greater velocities.

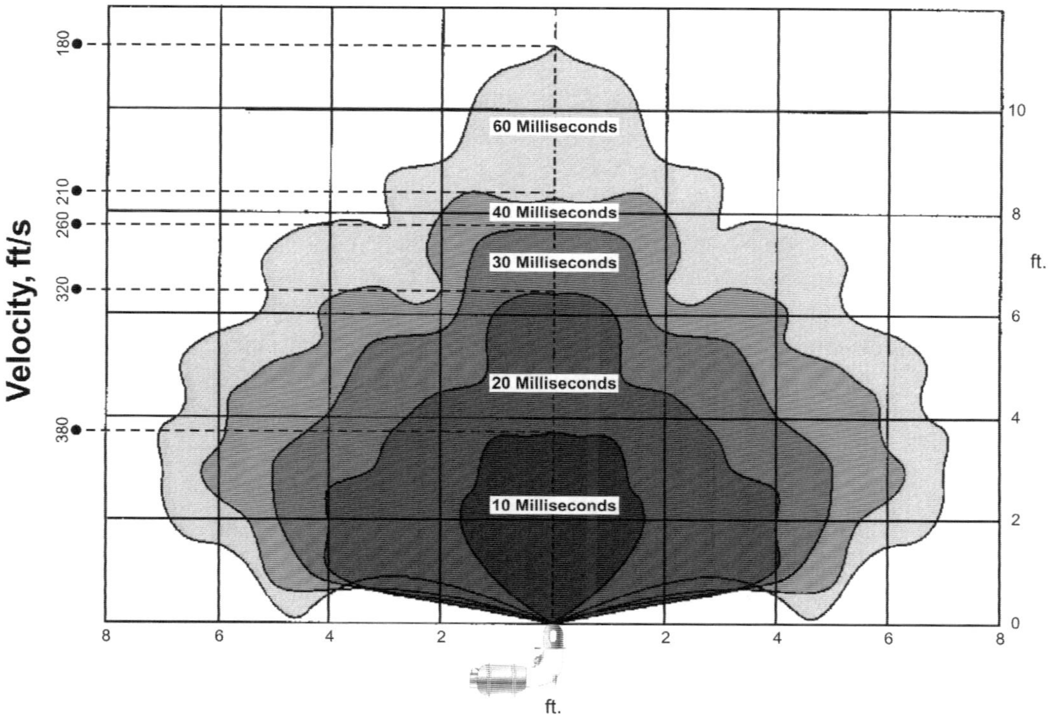

Figure 10 30-L HRD agent distribution pattern and velocity.

Protection design is generally driven by the process vessel design strength. This, in turn, affects the activation pressure and the HRD container selection (number, location and size). From Figures 9 and 10, the importance of speed in each step is obvious. Final suppression system design often involves a compromise. The use of several smaller containers rather than a few larger containers will shorten the time to suppression, reducing P_3 and TSP. Additional containers beyond this point will have little effect on P_3, but will increase the nitrogen contribution P_4, and therefore TSP.

Thus far in the discussion in this section, agent concentration has been assumed to be sufficient (reference Figure 1). This concentration is expressed as an equilibrium term, often as lb_m/m^3 or

kg/m^3, and has its primary basis in testing. Insufficient agent would obviously fail or only partially suppress (or worse, as described in the subsequent section on SEEP). Excess agent would require additional containers and would in fact lead to increased TSP. A critical suppressant mass has been identified as the mass required to absorb the excess energy content of the fireball. The limited information available suggests that the practical requirement for an agent is ~1.7 – 1.9 kg/m^3. This concentration has been shown to successfully suppress conventional fuels. The minimum suppressant concentration also has a linkage to the minimum inerting concentration previously introduced – although this relationship has not been definitively established, partly because of the experimental challenges as discussed in the following section.

4.3 Minimum Inerting Concentration

The MIC, or minimum inerting concentration, provides a useful conceptual link between the technologies of inerting and suppression. As described previously, an MIC value is determined by measuring the full flammability curve for a given fuel/inertant mixture. Again, this approach differs from suppression testing in that with MIC evaluation, the fuel and inertant are premixed and collectively injected into the test vessel. The ordinate for the upper curve in Figure 1 (quantity of suppressant required) is an inerting level; this ordinate value can therefore be taken to also represent the MIC provided the inerting level is independent of fuel quantity and is measured with an appropriate ignition energy. The bridge between inerting and suppression is made by considering the design features required to minimize the difference between the MIC and the actual amount of suppressant required in a given application. Figure 1, and the discussion in the previous sections, illustrate that focusing on explosion detection and suppressant delivery are highly beneficial in achieving this goal.

The current research has highlighted the fact that the MIC may be thought of as the amount of suppressant required in the ideal case of instantaneous detection and suppressant injection. For some combinations of fuel dust and inertant, however, it is not possible to measure the MIC – at least in volumes of 20 L (Dastidar and Amyotte, 2002a) and 1 m^3 (Dastidar et al., 1998, Dastidar et al., 1999, and Chatrathi and Going, 2000). In these cases (e.g. aluminum inerted with MAP or SBC), the flammability curve does not bend back to form an envelope with a "nose" as shown in Figures 4, 7 and 8. Here, larger test volumes may be required, or alternatively an effective MIC can be specified; this is the concentration of inertant required at the highest fuel dust concentration tested.

Either the measured or the effective MIC could possibly be used to assist engineers in the design of a suitable suppression strategy, particularly when it is difficult to estimate the suspended fuel dust concentration present in process equipment. This might be during normal process operation or more likely in the event of an upset condition. A fixed number of high rate discharge canisters could be attached to the process unit to deliver a fixed concentration of inertant using the MIC as a design basis, and accounting for other design factors such as detector response and injector efficiency. At both low and high fuel concentrations there would be enough inertant to mitigate an explosion. Thus, if the concentration of explosible dust were to increase or decrease in the unit, there would always be enough inertant for effective suppressant.

4.4 MIC Testing

Limited testing has recently been conducted with the goals of improving our understanding of the relationship between the minimum suppressant agent concentration and the MIC, and of screening new fuels for suppression application. As noted previously, an agent concentration of 1.7 – 1.9 kg/m^3 has acceptance as an effective suppressant concentration. Inerting of various fuels was tested at this level and compared to suppression experience. The inerting tests were done in a 20-L vessel with a 1-kJ ignition source to avoid overdriving (see previous discussion of this phenomenon). Table 1 contains the collected results using sodium bicarbonate as the inerting and suppressing agent.

Table 1 Summary of recent inerting and suppression testing.

Fuel	**Concentration [g/m^3]**	**K_{St} [bar·m/s]**		**Inerting at 1.8 kg/m^3, 1 kJ**	**Suppression[3] at 1.8 kg/m^3**	**MIC[4] [kg/m^3]**
		5 kJ	**1 kJ**			
Coal	750	nd[1]	nd[1]	nd[1]	Yes	0.55
Magnesium Stearate	250	372	305	Yes[2]	Yes	nd[1]
Lycopodium	750	124	89	Yes[2]	Yes	nd[1]
Yellow Toner	250	432	301	Yes[2]	Yes	nd[1]
Cornstarch	750	266	218	Yes[2]	Yes	0.63

[1]nd = not determined.
[2]Combustion pressure was no greater than that due to the ignitor alone.
[3]Suppression testing conducted in a 1-m^3 vessel.
[4]See Chatrathi and Going (2000).

The data are incomplete at this point, yet do suggest that inerting testing of a new fuel can predict the suppressibility of that fuel. Thus, and assuming that agent delivery parameters are equal, the need for actual demonstration of suppression can possibly be limited to non-conventional fuels or conditions.

4.5 Suppressant Enhanced Explosion Parameter

The importance of ensuring that sufficient inertant is present during an explosion scenario is demonstrated by Figure 11 from the 1-m^3 testing conducted by Dastidar et al. (1999). Figure 11 illustrates a phenomenon termed SEEP – Suppressant Enhanced Explosion Parameter – where the explosion overpressure, for the case of insufficient inertant in the fuel/inertant mixture, is enhanced to levels greater than that for the pure dust. This occurrence can be attributed to decomposition of the inertant by heat from the combustion of the fuel. For the example shown in Figure 11, the hot-burning aluminum powder is thought to bring about at least partial decomposition of the SBC and hence generation of carbon dioxide. This underscores the importance of a proper suppression design that ensures sufficient agent is introduced. A somewhat analogous situation investigated by Pegg at al. (1997) is the decomposition of sodium azide (NaN_3) and subsequent generation of nitrogen; this leads to overpressures much greater than those possible due solely to the combustion of sodium azide. (The authors are unaware of any similar situation with conventional hydrocarbon fuels.)

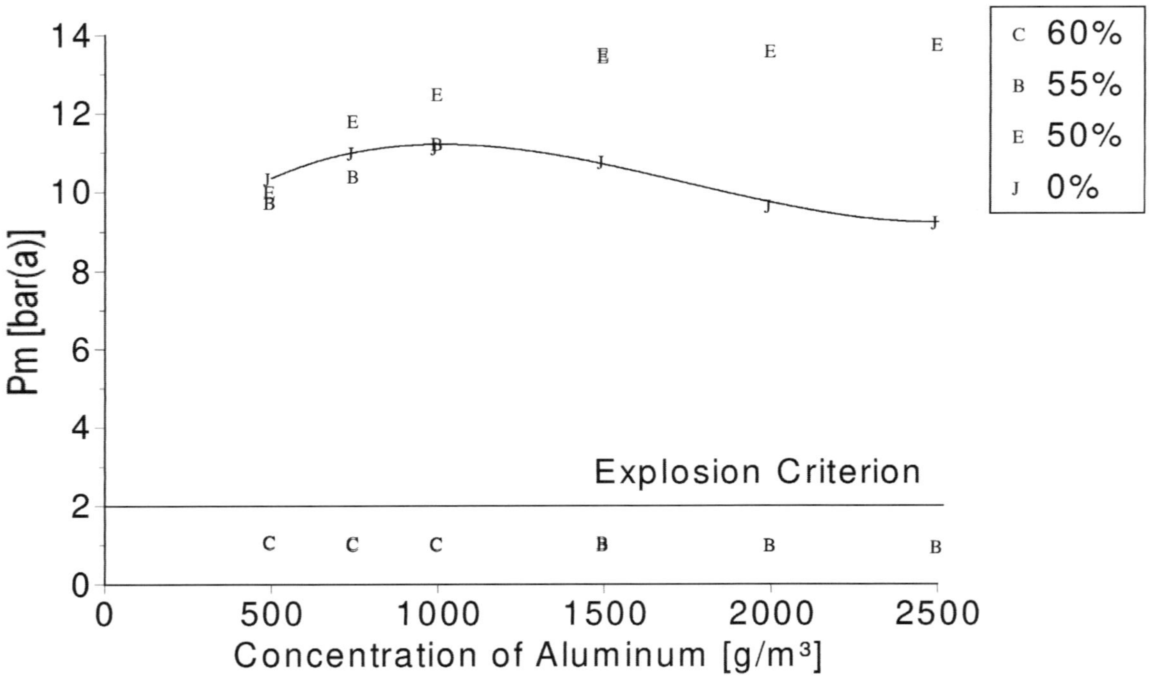

Figure 11. Explosion overpressure plotted against aluminum concentration for different amounts of SBC in mixture (from Dastidar et al., 1999).

Inerting and suppression are therefore applications in which "a little is not good enough". As commented by Moore (1996), the consequence of insufficient deployment of a suppressant agent is a failed suppression in which the resulting overpressure is greater than for the unsuppressed explosion. He uses the suppression example of halon molecule dissociation contributing to the overall system pressure to explain the SEEP phenomenon. Other inerting examples where SEEP has been observed are aluminum/MAP, polyethylene/MAP and polyethylene/SBC (Dastidar et al., 1999), and coal dust/fly ash and petroleum coke dust/fly ash (Amyotte et al., 2004).

4.6 Loss Prevention Hierarchy

To conclude this section, reference is made to recent work by the authors in the field of inherent safety – a topic area best understood by examining its role in the overall setting of industrial safety. Loss prevention in processing industries is generally achieved in three ways (Etowa et al., 2002): i) engineered safety, ii) procedural safety, and iii) inherent safety. Engineered, or add-on, safety involves the addition of safety devices to a process unit; these devices do not perform any fundamental operation, but are designed to act when a process upset occurs. Engineered features can be further categorized into passive devices (e.g. an explosion relief vent) and active devices (e.g. an explosion suppression system). Procedural safety measures, or administrative controls, utilize safe work practices and procedures to reduce risk. An example here would be the implementation of a rock dusting program aimed at inerting explosible coal dust in an underground mine.

On the other hand, inherent safety uses the inherent properties of a material or process to eliminate or reduce the hazard. The fundamental difference between inherent safety and the other two categories is that inherent safety seeks to remove the hazard at the source as opposed to accepting the hazard and looking to prevent its occurrence or mitigate the effects. A degree of inherent safety can be achieved by consideration of the four basic principles shown in Table 2 from Goraya et al. (2004). These principles and other inherent safety concepts are further described in the review paper by Khan and Amyotte (2003). The papers by Amyotte and Khan (2002) and Amyotte et al. (2003a) are attempts to more visibly demonstrate the integration of inherent safety into the field of dust explosion prevention and mitigation.

Table 2 Inherent safety principles (from Goraya et al., 2004).

Principle	**Description**
Minimization	Use smaller quantities of hazardous materials when the use of such materials cannot be avoided. Perform a hazardous procedure as few times as possible when the procedure is unavoidable.
Substitution	Replace a substance with a less hazardous material or processing route with one that does not involve hazardous material. Replace a hazardous procedure with one that is less hazardous.
Moderation	Use hazardous materials in their least hazardous forms or identify processing options that involve less severe processing conditions.
Simplification	Design processes, processing equipment, and procedures to eliminate opportunities for errors by eliminating excessive use of add-on (engineered) safety features and protective devices.

Of relevance here is the specific application of inherent safety to the design of both inerting/suppression systems and industrial plant. Inerting, although carried out through a procedural program, clearly achieves the goal of *moderation* – that of using a hazardous material in its least hazardous form (see Table 2). The paper by Mintz et al. (1996) illustrates how this principle, combined with *substitution* of a processing route, led to an inherently safer process for the manufacture of refractory materials. As described in their work, the preparation of powdered metals (e.g. aluminum) for refractory use is an application where the inert dust (e.g. magnesite) is present in the final product. It was determined that the inertant could be introduced at an earlier stage of the process in amounts sufficient for inerting of the explosible aluminum throughout virtually the entire manufacturing sequence.

Another recent example from a proprietary client report illustrates the potential use of inert powders to enhance the inherent safety of a processing facility. The objective in this study was to determine the inerting level for an explosible dust and an admixed inertant. In practice, the inertant is introduced into the manufacturing process for both quality control and explosion prevention purposes. It was established in a series of baseline tests that the explosible dust was similar to a typical bituminous coal dust in terms of explosion overpressure and rate of pressure rise. A literature review provided some confidence that the additive was similar to rock dust in terms of its effectiveness as an explosion inhibitor. This enabled the inerting tests to be optimized so as to provide data with a minimum number of tests (which was a requirement of this work). Additionally, the use of a 5-kJ ignition energy gave reasonable confidence that the measured inerting level was a conservative value given the low L/D ratio of the industrial equipment. Key to the successful implementation of the inerting protocol was the fact that the inerting level could be achieved without degradation of product quality (thus employing the principle of explosible dust *minimization*).

Another example of the *substitution* principle is given by Amyotte et al. (2003b) in their investigation of petroleum coke as a partial replacement for pulverized coal in the feed to utility boilers. The lowering of the explosion overpressure and rate of pressure rise brought about by petroleum coke substitution could potentially lead to a reduction in venting and suppression requirements. Similar possibilities exist through application of the inherent safety principle of *simplification*, wherein process equipment is built strong enough to withstand (or partially withstand) explosion overpressures. Further, *minimization* of fuel loadings (if possible) can lessen both inerting and suppression requirements. Although the situation will seldom be as straightforward as presented here – with operational and cost considerations playing a key role in the selection of prevention and mitigation strategies – the hierarchical approach described by Amyotte et al. (2003a) highlights the importance of inherent safety considerations early in the design of process plant.

5. Concluding Remarks

This review of solid inertants and their use in dust explosion prevention and mitigation has focused primarily on the research of the authors and their colleagues. As commented earlier, there is a significant body of work in the literature that should be consulted to provide further detail and fill in the gaps. Here, the purpose was to clearly delineate between inerting (i.e. explosion prevention) and suppression (i.e. explosion mitigation). Regardless of the particular application, however, it is important to gain an understanding of the properties of solid inertants that impact on their effectiveness. These properties include specific heat, heat and rate of decomposition, decomposition temperature and particle size, in addition to the inertant composition itself (which will determine whether inerting occurs through thermal or chemical means or a combination thereof). Important physical parameters to consider are the co-presence of a flammable gas, ignition energy and equipment scale (both size and geometry).

The latter part of the paper has focused on the practical application of the results of research on solid inertants. Explosion detection and suppressant delivery considerations were addressed. The potential relevance of the MIC for design engineers was emphasized, as well as the importance of understanding the effects of adding insufficient inertant (with respect to SEEP). Inherent safety considerations were raised as being important for the design of both explosion inerting/ suppression systems and industrial plant.

6. Acknowledgements

The authors gratefully acknowledge the financial support provided by the Natural Sciences and Engineering Research Council of Canada, and other agencies and companies that funded the research reported in this paper.

7. References

[1] Amrogowicz, J. and Kordylewski, W. (1991). *Effectiveness of Dust Explosion Suppression by Carbonates and Phosphates*, Combustion and Flame, **85**, 520-522.

[2] Amyotte, P.R. (2004). *Solid Inertants and Their Use in Dust Explosion Prevention and Mitigation*, Proceedings of 5th International Symposium on Hazards, Prevention and Mitigation of Industrial Explosions, Krakow, Poland, October 10-14, 2004, pp. 111-135.

[3] Amyotte, P.R. and Khan, F.I. (2002). *An Inherent Safety Framework for Dust Explosion Prevention and Mitigation,* Journal of Physics IV – Proceedings, **12**, Pr7-189 – Pr7-196.

[4] Amyotte, P.R. and Pegg, M.J. (1992). *Dust Explosion Prevention by Addition of Thermal Inhibitors*, Plant/Operations Progress, **11**, 166-173.

[5] Amyotte, P.R. and Pegg, M.J. (1993). *Explosion Hazards in Underground Coal Mines*, Toxicological and Environmental Chemistry, **40**, 189-199.

[6] Amyotte, P.R., Khan, F.I. and Dastidar, A.G. (2003a). *Reduce Dust Explosions the Inherently Safer Way*, Chemical Engineering Progress, **98**, 36-43.

[7] Amyotte, P.R., Basu, A. and Khan, F.I. (2003b). *Reduction of Dust Explosion Hazard by Fuel Substitution in Power Plants*, Process Safety and Environmental Protection (Transactions of the Institution of Chemical Engineers, Part B), **81**, 457-462.

[8] Amyotte, P.R., Mintz, K.J. and Pegg, M.J. (1992). *Effectiveness of Various Rock Dusts as Agents of Coal Dust Inerting*, Journal of Loss Prevention in the Process Industries, **5**, 196-199.

[9] Amyotte, P.R., Mintz, K.J and Pegg, M.J. (1995). *Effect of Rock Dust Particle Size on Suppression of Coal Dust Explosions*, Process Safety and Environmental Protection (Transactions of the Institution of Chemical Engineers, Part B), **73**, 89-100.

[10] Amyotte, P.R., Mintz, K.J., Pegg, M.J., Sun, Y.H. and Wilkie, K.I. (1991a). *Effects of Methane Admixture, Particle Size and Volatile Content on the Dolomite Inerting Requirements of Coal Dust*, Journal of Hazardous Materials, **27**, 187-203.

[11] Amyotte, P.R., Mintz, K.J., Pegg, M.J., Sun, Y.H. and Wilkie, K.I. (1991b). *Laboratory Investigation of the Dust Explosibility Characteristics of Three Nova Scotia Coals*, Journal of Loss Prevention in the Process Industries, **4**, 102-109.

[12] Amyotte, P.R., Khan, F.I., Basu, A., Dastidar, A.G., Dumeah, R.K. and Erving, W.L. (2004). *Explosibility Parameters for Mixtures of Pulverized Fuel and Ash*, Proceedings of 5th International Symposium on Hazards, Prevention and Mitigation of Industrial Explosions, Krakow, Poland, October 10-14, 2004, pp. 71-80.

[13] Bartknecht, W. (1989). *Dust Explosions: Course, Prevention, Protection*, Springer-Verlag, Berlin, Germany.

[14] Barton, J., Editor. (2002). *Dust Explosion Prevention and Protection*, Institution of Chemical Engineers, Rugby , UK.

[15] Cashdollar, K.L. (2000). *Overview of Dust Explosibility Characteristics*, Journal of Loss Prevention in the Process Industries, **13**, 183-199.

[16] CCPS (Center for Chemical Process Safety) (2000). *Guidelines for Chemical Process Quantitative Risk Analysis*, 2nd edition, American Institute of Chemical Engineers, New York, USA.

[17] Chatrathi, K. and Going, J. (2000). *Dust Deflagration Extinction*, Process Safety Progress, **19**, 146-153.

[18] Chawla, N., Amyotte, P.R. and Pegg, M.J. (1996). *A Comparison of Experimental Methods to Determine the Minimum Explosible Concentration of Dusts*, Fuel, **75**, 654-658.

[19] Choi, T., Rahimian, S. and Essenhigh, R.H. (1986). *Studies in Coal Dust Explosions: Influence of Additives on Extinction of High Intensity Coal Dust Flames*, 21st Symposium (International) on Combustion, The Combustion Institute, Pittsburgh, USA, 345-355.

[20] Cybulski, W. (1975). *Coal Dust Explosions and Their Suppression* (English translation from Polish), Foreign Scientific Department of the National Centre for Scientific Technical and Economic Information, Warsaw, Poland.

[21] Dastidar, A. and Amyotte, P. (2002a). *Determination of Minimum Inerting Concentrations for Combustible Dusts in a Laboratory-Scale Chamber*, Process Safety and Environmental Protection (Transactions of the Institution of Chemical Engineers, Part B), **80**, 287-297.

[22] Dastidar, A.G. and Amyotte, P.R. (2002b). *Explosibility Boundaries for Fly Ash/Pulverized Fuel Mixtures*, Journal of Hazardous Materials, **92**, 115-126.

[23] Dastidar, A.G., Amyotte, P.R. and Pegg, M.J. (1997). *Factors Influencing the Suppression of Coal Dust Explosions*, Fuel, **76**, 663-670.

[24] Dastidar, A.G., Amyotte, P.R., Going, J. and Chatrathi, K. (1998). *Scaling of Dust Explosion Inerting*, Archivum Combustionis, **18**, 21-45.

[25] Dastidar, A.G., Amyotte, P.R., Going, J. and Chatrathi, K. (1999). *Flammability Limits of Dusts - Minimum Inerting Concentrations*, Process Safety Progress,**18**, 56-63.

[26] Dastidar, A.G., Amyotte, P.R., Going, J. and Chatrathi, K. (2001). *Inerting of Coal Dust Explosions in Laboratory- and Intermediate-Scale Chambers*, Fuel, **80**, 1593-1602.

[27] Eckhoff, R.K. (2003). *Dust Explosions in the Process Industries*, 3rd edition, Gulf Professional Publishing/Elsevier, Amsterdam, The Netherlands.

[28] Etowa, C.B., Amyotte, P.R., Pegg, M.J. and Khan, F.I. (2002). *Quantification of the Inherent Safety Aspects of the Dow Indices*, Journal of Loss Prevention in the Process Industries, **15**, 477-487.

[29] Going, J.E., Chatrathi, K. and Cashdollar, K.L. (2000). *Flammability Limit Measurements for Dusts in 20-L and 1-m^3 Vessels*, Journal of Loss Prevention in the Process Industries, **13**, 209-219.

[30] Goraya, A., Amyotte, P.R. and Khan, F.I. (2004). *An Inherent Safety-Based Incident Investigation Methodology*, Process Safety Progress, in press.

[31] Hertzberg, M., Cashdollar, K.L., Zlochower, I. and Ng, D.L. (1984). *Inhibition and Extinction of Explosions in Heterogeneous Mixtures*, 20th Symposium (international) on Combustion, The Combustion Institute, Pittsburgh, USA, 1691-1700.

[32] Khan, F.I. and Amyotte, P.R. (2003). *How to Make Inherent Safety Practice a Reality*, Canadian Journal of Chemical Engineering, **81**, 2-16.

[33] Mintz, K.J., Bray, M.J., Zuliani, D.J., Amyotte, P.R. and Pegg, M.J. (1996). *Inerting of Fine Metallic Powders*, Journal of Loss Prevention in the Process Industries, **9**, 77-80.

[34] Moore, P.E. (1996). *Suppressants for the Control of Industrial Explosions*, Journal of Loss Prevention in the Process Industries, **9**, 119-123.

[35] Moore, P.E. and Siwek, R. (1992). *New Developments in Explosion Suppression*, VDI Berichte Nr. 975, 481-505.

[36] Pegg, M.J., Amyotte, P.R., Lightfoot, P.D. and Lee, M.C. (1997). *Dust Explosibility Characteristics of Azide-Based Gas Generants*, Journal of Loss Prevention in the Process Industries, **10**, 101-111.

[37] Sapko, M.J., Weiss, E.S., Cashdollar, K.L. and Zlochower, I.A. (2000). *Experimental Mine and Laboratory Dust Explosion Research at NIOSH*, Journal of Loss Prevention in the Process Industries, **13**, 229-242.

[38] Swift, I. (1988). *Design of Deflagration Protection Systems*, Journal of Loss Prevention in the Process Industries, **1**, 5-15.

Paper TG002 - 7

Barrier Spacing to Mitigate Explosion Propagation in Ducts

Joseph A. Senecal, and Henry Garzia
Kidde-Fenwal, Inc.
400 Main Street
Ashland, MA 01721

Prepared for Presentation at
American Institute of Chemical Engineers
2005 Spring National Meeting
39th Annual Loss Prevention Symposium
Atlanta, Georgia
April 11-13, 2005

UNPUBLISHED

Barrier Spacing to Mitigate Explosion Propagation in Ducts

Joseph A. Senecal, and Henry Garzia
Kidde-Fenwal, Inc.
400 Main Street
Ashland, MA 01721

Abstract

This paper reviews the principles relating to process analysis for the placement of deflagration isolation barriers in ducts separating process locations. The approach developed by Moore and Spring for making a quantitative analysis of the process volume in which a primary deflagration may originate, the characteristics of the combustible materials, the features of the deflagration isolation system and the duct on which it is mounted is reviewed in brief. Features of the resulting software design tool are noted and examples of field applications are given.

1. Introduction

The methods of explosion protection of bins, silos and similar structures by vent panels or explosion detection and suppression systems is well established. Guidance for these practices is given in NFPA standards 68 (*Guide for Venting of Deflagrations*) and 69 (*Standard on Explosion Prevention Systems*). However, the design guidance for such systems is predicated on certain assumptions about the worst case characteristics of an underlying deflagration event, namely the maximum pressure (P_{MAX}) and normalized rate of pressure rise (K_{ST}) associated with the fuel-air mixture. Recent analysis suggests that the validity of such assumptions is in jeopardy when vessels are interconnected by ducts, a common occurrence. Deflagration progress in a two vessel system is complex. Initial flame propagation and pressure rise in the deflagration *source* vessel, say A, will cause flow into the duct connection to Vessel B. Such flow may very likely include a flame front. Duct isolation, by a rapid acting valve or chemical barrier, will only be successful if the distance of the barrier from the source vessel A is sufficient to assure adequate time from the ignition event to allow for deflagration detection, actuation of the barrier (closure in the case of a valve; discharge in the case of a chemical agent barrier). Such protection, however, can be maximized only if the location of an isolation barrier on a duct is specified with detailed quantitative consideration of the underlying fuel and system parameters that govern the flame propagation, flame detection and valve response characteristics. Failure to isolate a duct-bound deflagration may result in passage of the deflagration event into the *target* vessel, vessel B, initiating therein a deflagration that is much stronger than would be anticipated based on the fuel explosibility properties (P_{MAX} and K_{ST}) alone. This situation would likely put vessel B at greater risk than expected. This paper discusses a new design tool for placement of explosion isolation barriers which considers the relevant design parameters, demonstrates how the model was validated, and presents several examples of application of the model to actual solids processing systems.

2. Protection of Interconnected Process Vessels by Isolation Barriers

Processing combustible materials in closed or semi-closed vessels has an associated risk of ignition and deflagration of the internal flammable atmosphere with the further risk of a explosion. Characterization of such process explosion risks and mitigation strategies are generally well defined in the body of existing fire protection standards. NFPA 68 provides design guidance on minimum requirements for deflagration vents to mitigate deflagration pressure development within tolerable values. NFPA 69 provides minimum design guidance for explosion prevention in process systems by several means particularly by use of explosion suppression systems. Both venting and suppression have a long history of successful application in the protection of isolated process vessels. This paper focuses on the special considerations associated with interconnected process vessels. In this case, a deflagration event initiated in one vessel (the *source* vessel) can propagate through the connecting duct to the second vessel (the *target* vessel) thereby propagating the deflagration event. Several strategies are identified in NFPA 69 that can be employed to isolate the source vessel and the target vessel, i.e., to mitigate the propagation of a deflagration from one vessel to the other. Two isolation practices in common use create barriers employing high-speed valves or nonflammable chemical isolation blocks within the duct. What has been lacking in the available guidance on deflagration isolation barrier systems is quantification of the deflagration propagating event. That is, how do deflagrations propagate from within a source vessel to the interconnecting duct and thence down the duct toward the target vessel? Isolation barrier systems are actuated upon detection of the deflagration event. Once actuated they require a finite amount of time to fully and effectively deploy. Given an understanding of the details of event propagation and barrier function, what criteria should be used to locate an isolation barrier along the length of a duct to assure a high probability that the deflagration event will in fact be intercepted? These issues have been addressed by Kidde Research in a recent paper (Moore and Spring). This paper reviews the principle finding of that work and provides examples of how the resulting recommendations on isolation barrier placement have been employed in practice in cases involving flammable gas and combustible dust hazards.

3. Problem Statement

Simply stated process vessels connected by a duct are at risk of much more severe consequences due to a deflagration event than isolated vessels. Consider the simple representation shown in Fig. 1. Illustrated is the simple case of a two-vessel system with an interconnecting duct.

Studies of the consequences associated with deflagration propagation from one process vessel to another have been reported by several authors. See in particular Lunn et al and Holbrow et al. The results of these studies show that the deflagration effects achieved in the "target" vessel can be substantially more severe than for the case of a simple deflagration event. The more severe results arise from three principle causes:

1. Deflagration discharge from a duct gives rise to a much more energetic ignition source than employed in standardized deflagration testing per ASTM E1226;

2. Discharge from the duct into the target vessel gives rise to elevated turbulence in the combustible atmosphere therein thereby accelerating flame propagation;

3. Rapid discharge from the source vessel can raise the initial pressure in the target vessel prior to ignition which can contribute to both higher P_{MAX} and $(dP/dt)_{MAX}$ events in the target vessel.

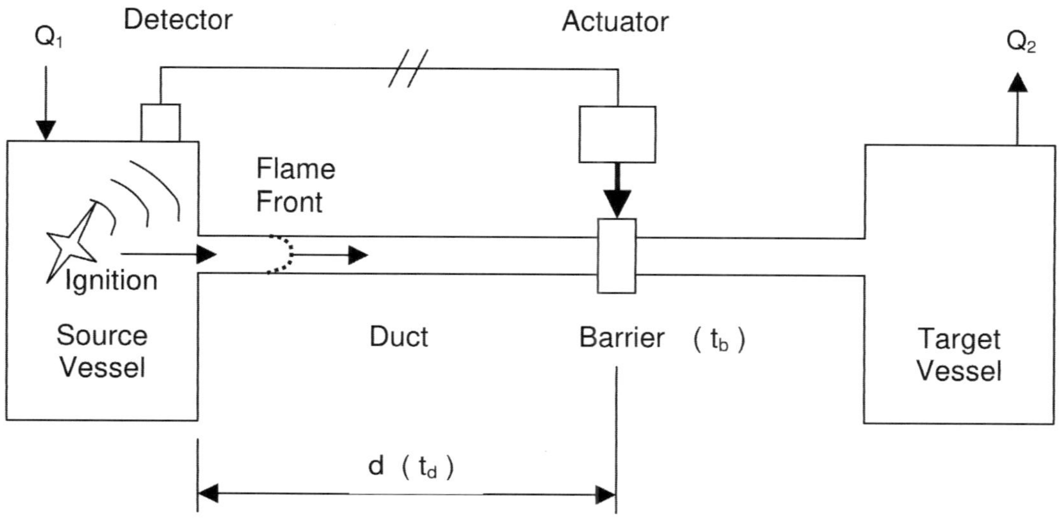

Figure 1. Representation of interconnected process vessels at risk for deflagration transmission through an interconnecting duct.

For the foregoing reasons it is important that means intended to mitigate communication of a deflagration from one vessel to another be designed and installed in consideration of recent analysis to assure high probability of successful performance.

4. Principles of Deflagration Propagation between Vessels

The time of progress of a deflagration flame front from its point of origin to a duct barrier position can be modeled by considering separately each mode of flame propagation and its associated effect on flame travel within the source vessel and duct.

Deflagration detection. Regardless of the means of initiation of a deflagration event the explosion protection process for an "active" system begins at the time of detection of this event. For closed or semi-closed process vessels the most common mode of deflagration detection is by a determination that the pressure in the process space has risen above a predetermined alarm value. Alternatively, the deflagration alarm may be related to detection of a rate of pressure rise that exceeds normal process characteristics. With "smart" pressure sensors deflagration alarm may be related to both pressure rise and rate of pressure rise criteria. Optical flame detection is also used to detect the onset of a deflagration. Optical means are well suited to detection of flames in ducts where the path optical length is limited. In any case, the time after ignition at which the deflagration event is detected is designated as t_A for the purposes of analysis.

Deflagration passage to the duct entrance. Passage of a deflagration flame front from its point of origin to the entrance of the duct connecting the "source vessel" to the "target vessel" requires an amount of time related to the flame speed, or, conversely, the equivalent K_{ST} in the case of combustible dust systems. The time after ignition at which the deflagration flame enters the duct is designated t_E.

Deflagration passage from duct entrance to barrier position. Once a deflagration flame front has entered a duct it propagates down the duct in the direction of the barrier. The time of transport of the flame front from the duct entrance to the barrier position is designated t_D.

Barrier function time. Once the deflagration event has been detected an alarm is sent to the barrier actuator setting in motion the closure of the barrier valve, or in the case of a chemical isolation system, the discharge of chemical agent. The time from deflagration detection for the barrier to be fully deployed is designated t_B.

The success criteria for a barrier isolation system is as follows:

$$t_A + t_B \leq t_E + t_D \qquad \text{Eq. 1}$$

Central to using this model to design barrier systems is the necessity to understand the functionality and limits on the values of the four characteristic time elements t_A, t_E, t_D, and t_B.

The magnitude of t_A will depend on the type of detection device employed and the flame speed developed in the combustible atmosphere. The common choices of deflagration detection device are based on measurement of pressure rise, rate of pressure rise, a combination pressure and rate of pressure rise, or optical flame sensing. Pressure detectors are normally mounted on the wall of a process vessel whereas optical flame detectors are normally mounted at or near the entrance to a duct. The time to response of a pressure detector can be predicted using established theory[1] and is related primarily to the vessel size and the combustion properties of the fuel-air mixture therein. The time of response of an optical flame detector will be related to t_E which will depend on the location within a vessel that ignition occurs. The magnitude of t_E can also be predicted using the same theory as is used to predict vessel pressure rise.

Flame front velocity in duct in the duct is modeled as

$$V_{FF} = V_{CONV} + V_P + S(x) \qquad \text{Eq. 2}$$

Each contributing component of the flame front velocity can be understood in terms of conventional processes.

[1] See Zabetakis, p. 15.

V_{CONV} is the initial pneumatic conveying velocity prevailing in the duct prior to the ignition event.

V_P is the velocity contribution due to the source vessel pressure, *P(t)*, which increases in time after ignition. A suitable model of the function V_P is taken as

$$V_P = a\left(\frac{P}{\rho_a(1+P/2)}\right)^b \qquad \text{Eq. 3}$$

where *P* is the post-ignition pressure rise, ρ_a is the air density at pre-ignition process conditions, and *a* and *b* are constants determined by the best fit of experimental data.

S(x) is the instantaneous flame front speed relative to the air. *S(x)* development in pipes has been found to increase with distance as follows

$$S = jS_f\left(1+(d/D)^k\right) \qquad \text{Eq. 4}$$

where S_f is the flame velocity developed in the source vessel, *D* is the duct diameter, *d* is the barrier distance from the source vessel, *j* and *k* are constants determined from best fit of the experimental data.

The estimated flame transit time from duct entrance to barrier position is t_{FF} and is modeled as

$$t_{FF} = \frac{L}{V_{FF}} \qquad \text{Eq. 5}$$

where *L* is the distance to the barrier from the duct entrance.

The flame propagation model constants were derived from data developed by the FSA[2] in vessels of several sizes (0.5, 1.0. 4.25, and 9.4 m^3) at their test facility in Kappelrodeck, Germany . Vent ducts in the range of 100 to 400 mm diameter were affixed to the vessels along with the requisite number of pressure and flame sensors to obtain fundamental event data. Explosion tests were carried out using dried corn starch (P_{MAX} = 9 bar g; K_{ST} = 190 bar-m/s) under ISO 6814 test conditions and with quiescent propane (P_{MAX} = 8 bar g ; K_G = 100 bar-m/s).

Tests were carried out under several conditions of explosion severity. The flame front arrival times at several positions in the vessel's attached vent duct were determined and used to determine the best values of the constants in equations (3) and (4). Details on validation test data and how they were used to obtain best fit constants for the equations can be found in the source paper of Moore and Spring. Figure 2, below, is reproduced from Moore and Spring to illustrate the results achieved in estimating flame front arrival time for dust explosions in tests carried out in a 2.4 m^3 vessel connected to a 0.4 m diameter duct at the test facility of the Swiss Safety Institute in Zegligen, Switzerland.

---

[2] *Forschungsgesellshaft für angewandte Systemsicherheit und Arbeitsmedizin.*

The Moore and Spring paper addresses different modes of detection – optical and pressure – and the implications of using either alone or both together relative to "residual risk", i.e., the probability that an isolation barrier system may not be successful. Use of deflagration pressure sensing alone on the source vessel has risk related to two causes: ignition close to the duct entrance and "sluggish" deflagration events (low K value). In either case there is some probability that the flame front may enter the duct prior to detection. Use of optical flame sensing alone at the duct entrance has risk associated with ignition far from the duct in which case the flame speed at the duct entrance may exceed assumptions employed in the model. The authors concluded that use of both optical and pressure detection offers the greatest probability of isolation success.

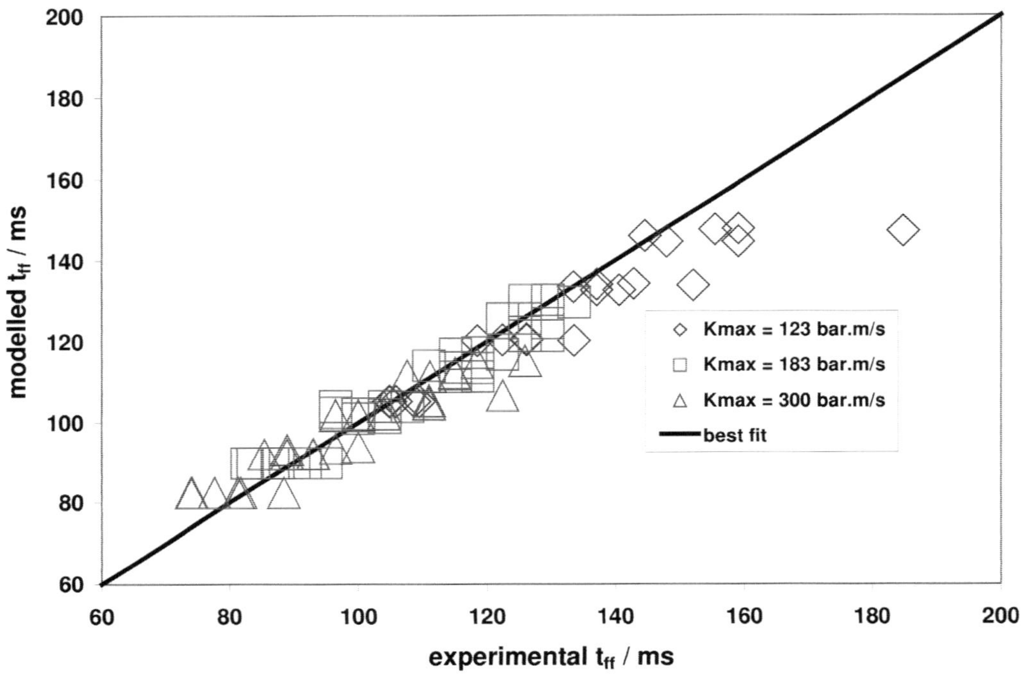

Figure 2. Model and experimental results for dust explosions carried out in a 2.4 m^3 vessel. Figure 4 from Moore and Spring.

5. Smart IS™ Barrier Location Tool

The complex analysis of modeling flame front progress in the vessel-duct system and comparison with the barrier function time parameter is carried out with a software design tool, called *Smart IS*™. During the period December 2002 – June 2003, validation tests of barrier placement designs based on *Smart IS*™ were undertaken by FSA at their test facility as part of the ATEX approvals of Kidde explosion protection systems and hardware (Wingerden).

Data are entered into the design tool by means of a simple Windows style screen, illustrated in Fig. 3. The data entry screen allows for input of the principal features of the hazard such as primary vessel dimensions, volume, duct dimensions, background air velocity, initial pressure and temperature, and explosibility properties of the combustible process material. Barrier system features include the relevant characteristics of an isolation valve (valve size) or extinguisher (container size, agent quantity, charging pressure, elbow or no-elbow mounting style) or detection system features and set point. The output of the design tool is a clear statement of requirements on placement of an isolation barrier. An example of the input and output data for *Smart IS*™ is given in the two application examples shown below.

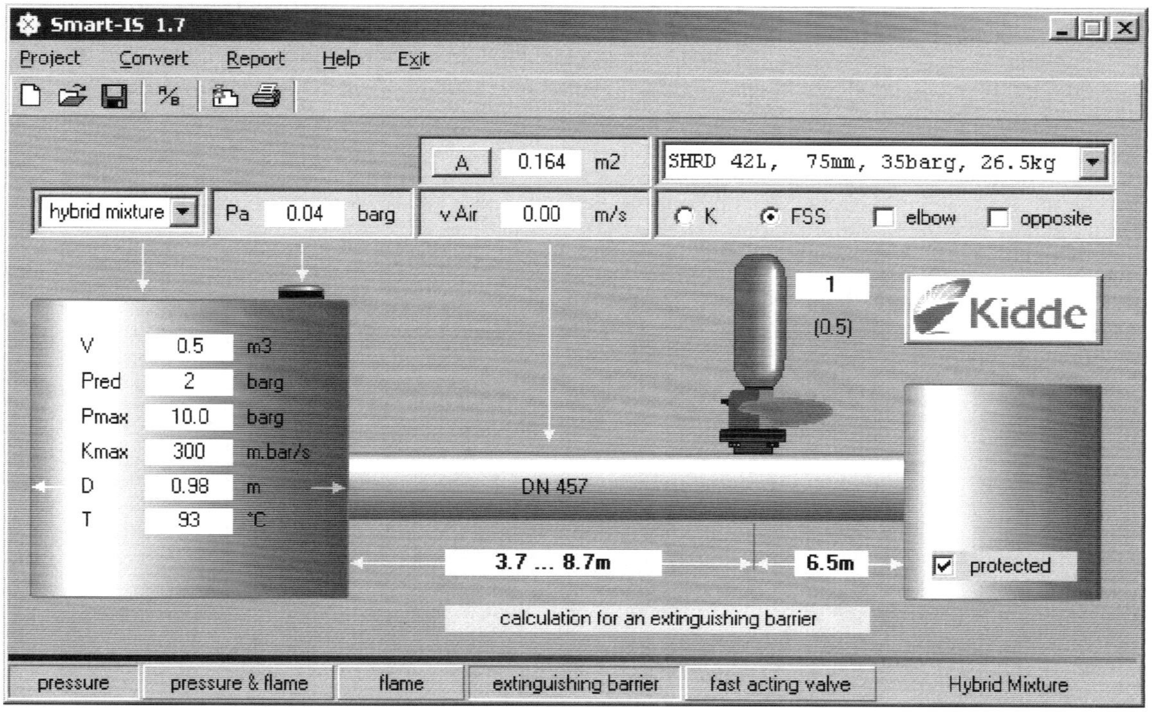

Figure 3. Typical *Smart IS*™ Screen shot showing case of suppression barrier with pressure detection employed on the source vessel.

- Minimum distance from the source vessel. This distance is computed from the algorithms described above and assures that there is a high probability that the isolation valve will be fully closed or chemical isolation barrier is discharged.

- Maximum distance from the source vessel. The prescribed maximum barrier distance is set to the lesser of either 5 m beyond the minimum barrier distance or a distance estimated, conservatively, to be less than that which would give rise to DDT (deflagration-to-detonation transition).

- Downstream safety distance. For mechanical isolation valves this value is zero. For chemical isolation systems this is the minimum distance prescribed to assure complete quenching by barrier agent of flame passed from the source vessel.

6. Application examples

Two examples of the application of *Smart IS*™ to determine isolation barrier location are given, one for a combustible dust application and one for a hybrid (combustible dust and vapor) application.

Example 1 – Isolation of duct conveying a combustible dust discharged from a pin mill.

Figure 4 illustrates an actual field example of the application of the isolation of a pin mill pulverizer system to mitigate deflagration propagation down stream. Pulverizing mills are known sources of ignition. Metal contaminants in the feed can lead to generation of metal sparks having sufficient energy to ignite the primary process dust.

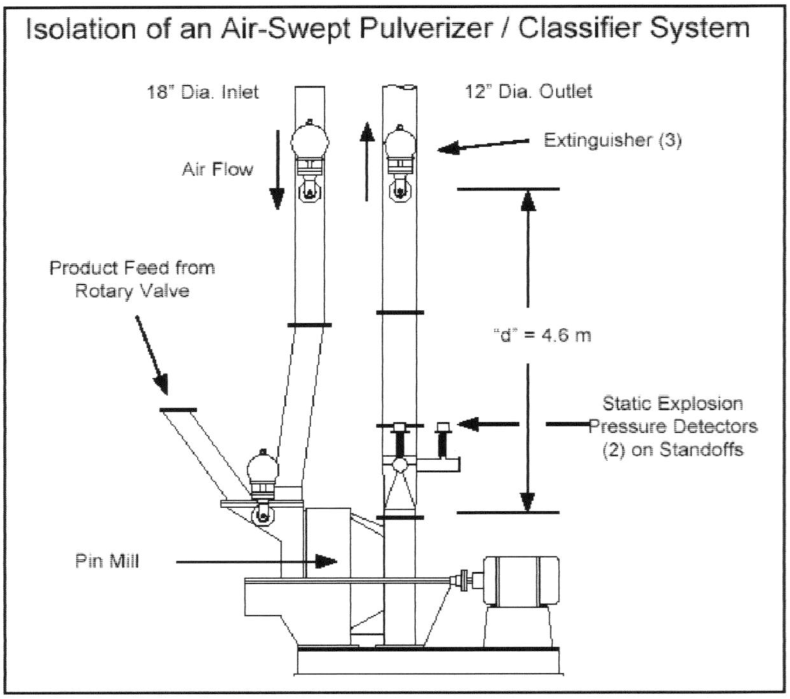

Figure 4. Layout of pin mill, product feed, air supply and return ducts, pressure detectors, and suppression barrier extinguishers.

The pin mill of Example No. 1 processes a dust that was determined to have the following explosibility properties: P_{MAX} = 8.6 bar, K_{ST} = 274 bar-m/s as determined by testing in accordance with ASTM E 1226. The primary mill volume was fitted with an explosion suppression system designed to limit P_{RED} to 1.2 bar gauge. A suppression isolation barrier was employed with pressure detection (actuation pressure of 0.07 bar) at the mill outlet. Isolation standoffs were used with the pressure detectors to reduce the likelihood that mill vibration would lead to an unintended system actuation. The complete set of input data and results of the *Smart IS*™ analysis are shown in Table 1. The isolation barrier system of this example was commissioned in 1996, before the rigorous implementation of *Smart IS*™ but in full conformance with its requirements. This system has since actuated twice to perform its isolation function, once in 2003 and again in 2004. Post-event analyses revealed that the cause in both events was ignition due to tramp metal entering the feed hopper. The value, therefore, of the isolation system has been demonstrated in successfully mitigating propagation of flame to downstream locations.

Example 2 – Isolation of a supply duct to a fan moving a flammable hybrid dust-vapor mixture.

Figure 5, illustrates an actual field example of the application of a chemical isolation barrier in a dust - vapor transfer duct where the ignition risk is at the exhauster fan. The vapor in question is isopropyl alcohol (IPOH). A distinguishing feature of this example is the use of an electronic sensor to measure pressure and rate-of-pressure-rise as indicators of a deflagration event. The dP/dt sensing functionality reduces the risk of system actuation due to abrupt, but relatively slow, pressure swings on operation of the fan. Note that barrier position limits specified to the installer is between the minimum distance of 3.7 m and a maximum distance of 8.7 m. Complete details of the input data and resulting barrier characteristics are given in Table 2.

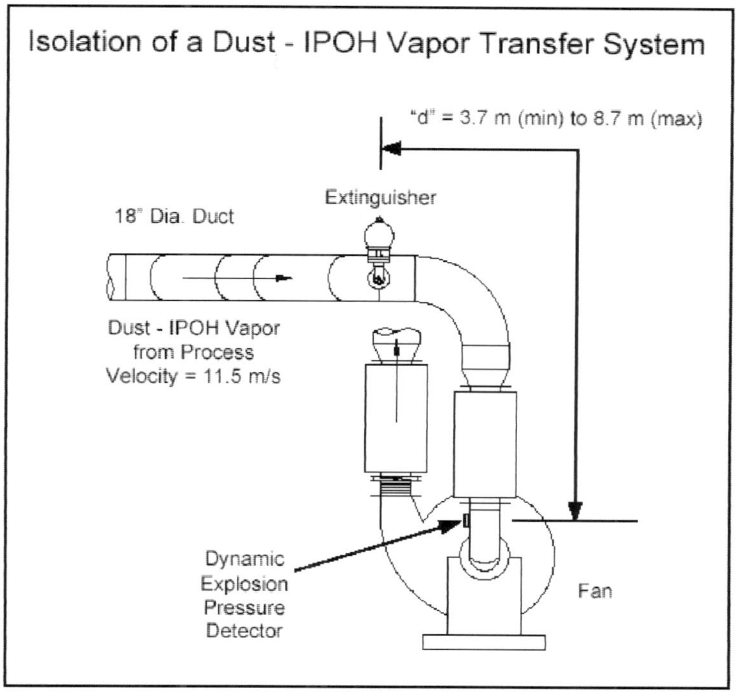

Figure 5. Dust-vapor transfer duct with isolation barrier upstream of the exhauster fan.

7. Summary

Placement deflagration isolation barriers in chemical process transfer ducts, whether using a mechanical valve or a chemical suppression unit, must be carried out with the support of a detailed analysis of material and process conditions that govern the progress of the worst case flame front development. Additionally, the system analysis must consider the operating characteristics of the means of deflagration detection and the isolation barrier actuation time. The work of Moore and Spring provides a basis for relating fuel combustion properties, process vessel and duct dimensions, method of deflagration detection, barrier operation characteristics into a holistic model for use in specifying limits on barrier location. Barriers located too close to the duct entrance are at risk of allowing flame passage before being fully actuated. Barriers located to far from the duct entrance are at risk of allowing a duct deflagration to accelerate ("run up") to detonation conditions that would place severe stresses on system components. Kidde has codified the deflagration isolation model into a software tool called *Smart IS*™, now used in all isolation barrier applications. Examples of the field use of the barrier design tool were given for combustible dust and dust-vapor hybrid applications.

8. References

[1] Moore, P. and D. Spring, "Design of Explosion Isolation Barriers," *Proceedings of Hazards XVIII: Process Safety – Sharing Best Practice*, IChemE NW Branch, UMIST, Manchester, UK, November 22-25, 2004.

[2] Lunn, G.A., P. Holbrow, S. Andrews and J. Gummer, "Dust Explosions in Totally Interconnected Vessel Systems," *J. Loss Prevention in the Process Industries*, 9(1) pp. 45-58 (1996).

[3] Holbrow, P., S. Andrews and G.A. Lunn, "Dust Explosions in Interconnected Vessels," *J. Loss Prevention in the Process Industries*, 9(1) pp. 91-103 (1996).

[4] NFPA 68 *Guide for Venting of Deflagrations*, Chapter 5 "Fundamentals of Venting of Deflagrations," National Fire Protection Association, Quincy, MA (2002).

[4] NFPA 69 *Standard on Explosion Prevention Systems*, National Fire Protection Association, Quincy, MA (2000).

[5] NFPA 654, Standard for the Prevention of Fire and Dust Explosions from the Manufacturing, Processing, and Handling of Combustible Particulate Solids, National Fire Protection Association, Quincy, MA (2000).

[6] Zabetakis, Michael G., *Flammability Characteristics of Combustible Gases and Vapors,"* Bulletin 627, Bureau of Mines [now the NIOSH Pittsburgh Research Center], U.S. Department of the Interior (1965).

[7] ASTM E 1226, "Standard Test Method for Pressure and Rate of Pressure Rise for Combustible Dusts," ASTM International.

[8] Wingerden, K. v., 2004, Erfahrungen bei der ATEX-Zertifizierung von Explosionsschutzprodukten, Technische berwachung, 45: 18–21.

Table 1
Input Data & Results for Example No. 1
Application: 18 Inch Mill Air Inlet

Equipment:	Source Vessel Volume	V_{vessel}	= 1.4
	Reduced Explosion Pressure	P_{red}	= 1.2
	Initial Pressure	P_i	= 0.00
	Duct Area	A_{duct}	= 0.164
	Duct Diameter	D_{duct}	= 457
	Duct Height	H_{duct}	=
	Duct Width	W_{duct}	=
	Process Air Velocity	V_{air}	= 0.00
	Air Temperature	T_{air}	= 35
	Source Vessel Diameter	D_{vessel}	= 1.4
	Downstream Protection		= yes
Fuel:	**Dust**		
	Explosion Overpressure	P_{max}	= 8.6
	Explosion Rate Constant	K_{max}	= 274
Sensor:	**Pressure**		
	Detection Pressure	P_a	= 0.07
Barrier:	**Extinguishing**	FSS	
	Type of HRD's	SHRD 18L, 75mm, 35barg, 11.56kg	
	Opposite	no	
	Elbow	yes	
Results:	**Calculation for an extinguishing barrier**		
	Number of HRD's	n_{HRD}	= 1
	Min. Distance - Vessel to Barrier	d_{min}	= 4.6
	Max. Distance - Vessel to Barrier	d_{max}	= 9.6
	Downstream Safety Distance	d_{safety}	= 5.0
	Mass of Suppressant Required	M_{supp}	= 6.1
	Max. Pressure in Duct and at Barrier	P_b	= 1.20

Table 2
Input Data & Results for Example No. 2
Application: 18 Inch Fan Inlet Duct

Equipment:	Source Vessel Volume	V_{vessel}	= 0.5	m3
	Reduced Explosion Pressure	P_{red}	= 2.0	bar(g)
	Initial Pressure	P_i	= 0.00	bar(g)
	Duct Area	A_{duct}	= 0.164	m2
	Duct Diameter	D_{duct}	= 457	mm
	Duct Height	H_{duct}	=	mm
	Duct Width	W_{duct}	=	mm
	Process Air Velocity	V_{air}	= 0.00	m/s
	Air Temperature	T_{air}	= 93	°C
	Source Vessel Diameter	D_{vessel}	= 1.0	m
	Downstream Protection		= yes	
Fuel:	**Hybrid mixture**			
	Explosion Overpressure	P_{max}	= 10.0	bar(g)
	Explosion Rate Constant	K_{max}	= 300	bar·m/s
Sensor:	**Pressure**			
	Detection Pressure	P_a	= 0.04	bar(g)
Barrier:	**Extinguishing**	FSS		
	Type of HRD's	SHRD 42L, 75mm, 35barg, 26.5kg		
	Opposite	no		
	Elbow	no		
Results:	**Calculation for an extinguishing barrier**			
	Number of HRD's	n_{HRD}	= 1	
	Min. Distance - Vessel to Barrier	d_{min}	= 3.7	m
	Max. Distance - Vessel to Barrier	d_{max}	= 8.7	m
	Downstream Safety Distance	d_{safety}	= 6.5	m
	Mass of Suppressant Required	M_{supp}	= 9.8	kg
	Max. Pressure in Duct and at Barrier	P_b	= 2.00	bar(g)

Containment of Pharmaceutical Compounds: A Risk Based Approach

Brian Dean Moore
Eli Lilly and Company
Lilly Corporate Center
Indianapolis, IN 46285 U.S.A.
Phone: 317-433-0950
moorebr@lilly.com

Prepared for Presentation at
American Institute of Chemical Engineers
2005 Spring National Meeting
39th Annual Loss Prevention Symposium
Atlanta, Georgia
April 11-13, 2005

UNPUBLISHED

Containment of Pharmaceutical Compounds: A Risk Based Approach

Brian Dean Moore
Eli Lilly and Company
Lilly Corporate Center
Indianapolis, IN 46285 U.S.A.
Phone: 317-433-0950
moorebr@lilly.com

Abstract

One of FDA's initiatives is to improve the regulatory and quality control systems for pharmaceutical products, using a science-based, risk management approach. With this initiative in mind, a methodology, based on Failure Mode and Effect Analysis (FMEA), was developed to conduct containment capability assessments. This methodology was chosen because it allows identified risks to be quantified and subsequently prioritized.

The identified containment capability risk is quantified by scoring each risk on three (3) aspects: severity, probability, and detection. These aspects are multiplied together to obtain a Risk Prioritization Number (RPN). Based on the RPN, the identified risks can be prioritized and action taken to eliminate or mitigate the identified risk. After corrective actions have been completed, the identified risks are re-scored, prioritized, and corrective action implemented. The process is repeated until all identified risks have RPN's below an agreed upon level.

This methodology is applicable to all areas of pharmaceutical material production, including laboratory and clinical trial material manufacturing. An example utilizing the methodology and an analysis of a theoretical case will also be presented.

1. Safe Handling of Potent Pharmaceutical Compounds

1.1 Containment of Pharmaceutical Compounds

Pharmaceutical compounds have become more biologically active in recent years, leading to considerable lowering of the occupational health and safety guidelines. Safe handling, or specifically, containment of pharmaceutical compounds presents unique challenges. The overall objective for these compounds is to contain the material, as closely as feasible, to its potential emission sources. This objective translates, in practice, as primarily within the process equipment and secondarily within the process rooms. Failing this objective, product emissions and subsequent product migration have the potential to affect Health and Safety (personnel exposure), Quality (product exposure), Environmental impact (outdoor exposure) and operations impact (financial exposure). The containment objective, therefore, consists of isolating material from other material, people and the surrounding environment.

In August 2002, the U.S. Food and Drug Administration (FDA) announced that it was undertaking a new initiative to improve its regulatory and quality control systems for pharmaceutical products. One of the stated goals of this initiative is to focus the FDA's current Good Manufacturing Practices (cGMP) requirements on a science-based, risk management approach. A gap was perceived as to how to implement a risk management approach to qualitative containment capability assessments. The methodology presented in this paper was developed to address this gap. [1]

1.1.1 What Is Containment?

Fundamentally, manufacturing consists of transforming raw materials into finished goods. Energy and labor are two additional inputs to the manufacturing function. The transformation of raw materials during the manufacturing process must be performed safely. Personnel, the surrounding environment, and often, adjacent products must be protected. Containment can be defined, at a very basic level, as consisting of the isolation of the raw materials, and subsequently the finished goods, from personnel, the environment and adjacent products. Simply stated, containment is the equipment, procedures and operational mindset that prevent unwanted materials from getting to unwanted locations. Using this definition of containment, it is reasonable to think of containment as a risk reduction activity.

1.1.2 Containment Historical Background

There have been five influences on containment during the past half century. In relative chronological order, the influences are nuclear industry, asbestos abatement, semi-conductor manufacturing, sterile manufacturing, and aquatic studies. The nuclear and asbestos abatement industries primarily focused their containment activities on personnel protection; whereas, semi-conductor and sterile manufacturing focused their containment activities on product protection. Containment activities that focused on environmental protection were one of the outcomes of aquatic studies. This history shows an industry's response to each of the objectives for containing material at the source. Today, containment in the pharmaceutical industry takes a more synergistic approach. The underlying philosophy of containment is the protection of personnel, product, and environment. When these three elements are adequately protected, the risks of adverse financial consequences are less likely. The process defined within this document defines a methodology to rate and prioritize the risks associated with improper or inadequate containment. A fundamental assumption that is inherent in the methodology presented is that any material emission is undesirable.

1.2 Risk

For the purposes of this document and the methodology presented, risk is defined as a function of hazard and exposure.

$$\textbf{Risk} = f\,\textbf{(Hazard, Exposure)} \quad \textit{Equation 1}$$

The hazard is inherent in the material. Toxicological effects and particle size are two examples of the hazards inherent in a material. Exposure is a response to a hazard and is measurable and controllable. To help understand the difference between hazard and exposure, here is an example. Great white sharks are inherently dangerous. They present themselves as unique hazards to swimmers. However, many people have been literally inches from a shark. What protected them from this inherent hazard was the plate glass window at the aquarium. The aquarium controlled the exposure, thereby lessening the risk.

The focus for containment is to provide for the four key stakeholders (quality, health & safety, environmental and operations) effective means of eliminating or controlling material emissions that may ultimately result in exposure. Before discussing aspects of risk mitigation and examples of safe handling of pharmaceutical compounds, focus should be given to the methodology.

1.3 Case Examples of Poor Containment

Examples of the consequences of poor containment are abundant in the media. Here are a few recent examples.

CNN.com reported in 2003 on the unintended consequences of drugs, hormones, steroids and personal care products, such as soaps and perfumes, in U.S. rivers. Early analysis of the data collected indicates that low levels of these materials may delay the development of aquatic and amphibious species. The U.S. Geological Survey is currently studying this issue to determine how the buildup over time could affect a species. Sewage treatment plants are not equipped to filter out any of the hundreds of different prescription drugs that are present in wastewater. Effective containment attempts to eliminate or minimize contaminated water into waste streams.

A search of the National Institute for Occupational Safety and Health website resulted in numerous examples of personnel exposure. In one example, workers at a specialty metal processing facility were being exposed to heavy metals. Exposure occurred during the processing and laboratory analysis of metal billets. In another example, a health hazard existed from occupational exposures during textile flocking. Dust generated during the flocking process was being inhaled, resulting in cases of lung disease being diagnosed. In both examples, the root cause was attributable to poor containment.

The Food and Drug Administration focuses on assuring the strength, purity and quality of drug products. With this focus in mind, FDA inspects drug manufacturing facilities to ensure that the risk of contamination is being properly managed. When a manufacturer cannot adequately demonstrate that this risk is being properly managed, FDA is often very firm with its response. As an example, an inspection by the FDA of a drug manufacturer in 1999 resulted in a warning letter being issued. The warning letter specifically mentioned a lack of adequate separation and containment that created a potential for cross contamination.[2, 3]

2. Risk Based Approach Methodology

One of FDA's initiatives is to improve the regulatory and quality control systems for pharmaceutical products, using a science-based, risk management approach. With this initiative in mind, a methodology, based on Failure Mode and Effect Analysis (FMEA), has been developed to conduct containment capability assessments. This methodology has been chosen because it allows identified risks to be rated and subsequently prioritized.

The risk is rated by scoring each risk on three (3) aspects: severity, probability, and detection. These aspects are multiplied together to obtain a Risk Prioritization Number (RPN). Based on the RPN, an individual risk can be prioritized and action taken to eliminate or mitigate the risk. After corrective actions have been completed, the identified risks are subsequently re-scored and, prioritized for follow-up corrective actions. The process is repeated until all identified risks have RPN's below an agreed upon level, indicating an acceptable level of risk.

This methodology is a tool to assist in the analysis of identified emission risks that may lead to exposure risks due to poor containment.

After emission risks have been identified, the methodology aids in documenting recommendations and actions that will reduce risk.[4, 5]

2.1 History of Failure Mode and Effect Analysis (FMEA)

The methodology that provides the basis for Failure Mode and Effect Analysis (FMEA) was formally introduced in the late 1940's. In the 1960's, FMEA was effectively used during the early years of the space program. Automotive manufacturers, in the late 1970's, adopted and revised the basic methodology to address safety, regulatory and production improvements. Today, FMEA is used in a wide variety of industries from semi-conductor manufacturing to consumer product design to telephone support operations.

FMEA methodology has gone through a transformation of sorts recently as a focus on severity and occurrence has replaced Risk Prioritization Number (RPN) driven activity. However, the fundamental approach of the FMEA methodology to rating risks, prioritizing risks and presenting recommendations to manage the risks remains the basis for the methodology for managing risks associated with containment of pharmaceutical compounds.

It is important to note that FMEA is not a design process; rather, it is a complement to the product or process design activity and aids in the definition of what characterizes a robust and capable design.[4, 5]

2.2 Key Definitions

The key to successful implementation of the methodology is common terminology consistently applied. Key terms are defined below.

Risk is defined as vulnerability to loss or harm. The loss or harm may be personnel, equipment related, or financial in nature. For the purposes of this document and the methodology presented, risk is defined as a function of hazard and exposure.

Emission Severity is defined as to the degree to which a potential emission may have adverse effects on product quality, personnel health or environmental compliance. *Severity* as it relates to containment is defined as the amount of material emissions, the frequency of material emissions and the location of material emission. Potency of the material emitted is not a consideration when assessing the severity of a material emission. The severity scale is from 1 to 10, with 1 being low severity and 10 being high severity. Identified risks having a severity of 10 will require that the risk be addressed.

Emission Probability is defined as the likelihood of occurrence. The probability scale is from 1 to 10, with 1 being low probability and 10 being high probability.

Emission Detection is defined as the ease with which a material emission is observed. Method of detection may impact the ease of material emission observation. Methods may include visual, analytical or a combination of both. The detection scale is from 1 to 10, with 1 being easy to detect and 10 being difficult to detect.

Risk Prioritization Number (RPN) is defined as the product of the Severity (S), the Probability (P) and the Detection (D). RPN = (S) x (P) x (D). RPN's above a pre-determined threshold require that the risk identified be addressed.

Risk Mitigation is defined as actions taken to reduce the risk identified. Risk mitigation actions may include equipment or process design modification, implementation of training programs, and implementation of procedural and administrative controls.

Risk Reduction Matrix is defined as a matrix that shows the identified risks, their associated RPN, recommendations to reduce the identified risks and their associated costs and time to implement.

Risk Threshold is defined as a "line drawn in the sand" above which corrective action is required. The risk threshold is defined by operating management based partly on how much risk they are willing to assume. Other factors that may influence the setting of the risk threshold include material potency, the percent of potent material in a formulation, frequency of processing (e.g., one lot per year versus one lot per day). Conceptually, a more hazardous pharmaceutical compound should have a lower risk threshold or RPN.

2.3 Risk Assessment Steps

<u>Step 1</u>. Define the specific risk threshold for the specific process being assessed. Inputs used to set the risk threshold may include the level of overall risk deemed acceptable, the material's physical and chemical properties, toxicological effects of the material, percentage of material in a formulation and frequency of processing. The risk threshold is typically set by operating management.

<u>Step 2.</u> Conduct the containment capability assessment as outlined in an accepted protocol. Identify and document all risks observed.

<u>Step 3</u>. Assess each identified risk in terms of severity, likelihood and impact. Calculate the risk prioritization number (RPN).

<u>Step 4.</u> Rank order each risk based on the RPN from highest to lowest RPN. Highlight those RPN's that are greater than the risk threshold or have a severity of 10.

Step 5. Define recommendation(s) for each risk identified. Recalculate the RPN assuming each recommendation has been completely implemented.

Step 6. Prepare a qualitative assessment report. Include as appendices the table of initial RPN's and the table of recalculated RPN's. This report should be issued in draft form to allow the key stakeholders (quality, health & safety, environmental and operations) to comment on the risks identified, the scoring, and the recommendations. Highlight in the report's executive summary those risks whose initial RPN's are greater than the risk threshold or have a severity rating of 10. Highlight in the report's executive summary the recommendations that reduce all identified risks; specifically, the recommendations that address the risks having an RPN greater than the risk threshold or a severity rating of 10. The report should include the aspects that were included in setting the risk threshold and the rationale for the scoring of each risk. Rationales for scoring may be cost, risk to supply chain, potential harm to manufacturing personnel, potential harm to patients or a combination of factors.

Step 7. Prepare a risk reduction matrix that identifies the risk, its associated cost, the cost to mitigate the risk and the resulting RPN. Include in the risk reduction matrix, opportunities for obtaining synergies with multiple recommendations and partial implementation of recommendations. This iterative process of evaluating and obtaining a RPN is repeated until all identified risks have RPN's below an agreed upon level, indicating acceptable risk.

2.4 Rating Scales

The rating scales for the severity, probability and detection risk factors are shown in the following three (3) tables.

Table 1. Emission Severity Rating Scale

Rating	Description	Definition
10	Extremely High	Material emission renders the facility inoperable or unfit for use until cleaning or decontamination is complete. Cleaning and decontamination may be ineffective.
9	Very High	Material migrates to locations that are unable to be cleaned.
8	High	Material emission requires the shutdown and cleaning of entire process, process room and building.
7	Moderately High	Material migrates to locations that are difficult to clean.
6	Moderate	Material emission requires the shutdown and cleaning of entire process, process room and adjacent areas.
5	Low	Material emission requires the shutdown and cleaning of entire process and process room.
4	Very Low	Material emission requires the shutdown and cleaning of entire process.
3	Minor	Material emission requires cleaning of entire process room.
2	Very Minor	Material emission requires the cleaning of the exterior of process equipment.
1	None	Material is completely contained during the entire process. Any material egress is contained within the process or failsafe equipment.

Emission probability is the likelihood that a particular failure will result in material emissions or subsequent material migration. The rating scale for emission probability is non-linear.

Table 2. Emission Probability Rating Scale

Rating	Description	Definition
10	Extremely High	More than one occurrence per day.
9	Very High	One occurrence every three to four days.
8	High: Repeated Failures	One occurrence per week or a probability of 5 occurrences in 100 operations.
7	Moderately High	One occurrence every month.
6	Moderate: Occasional Failures	One occurrence every three months
5	Moderately Low	One occurrence every six months to one year.
4	Low: Relatively Few Failures	One occurrence per year.
3	Very Low	One occurrence every one to three years.
2	Negligible	One occurrence every three to five years.
1	Remote: Failure is unlikely	One occurrence in greater than five years.

The ability of the current controls to detect material emissions or subsequent material migration is a measure of emission detection. The descriptions and definitions in the table are not quantitative; rather, subjective, relative terms are used.

Table 3. Emission Detection Rating Scale

Rating	Description	Definition
10	Absolute Uncertainty	Material emission is not visually obvious and facility monitoring is not employed.
9	Very Remote	Material emission is visually obvious and facility monitoring is not employed.
8	Remote	Material emission is not visually obvious but is detectable only during special exposure monitoring studies.
7	Very Low	Material emission is not visually obvious but is detectable only during special facility monitoring studies.
6	Low Certainty	Material emission is not visually obvious and a greater than one month response is needed between facility monitoring and response.
5	Moderate	Material emission is visually obvious and a greater than one month response is needed between facility monitoring and response.
4	Moderately High	Material emission is not visually obvious and a less than one-month response is needed between facility monitoring and response.
3	High Certainty	Material emission is visually obvious and a less than one-month response is needed between facility monitoring and response.
2	Very High	Material emission is not immediately visually obvious and there is continuous facility monitoring with immediate response.
1	Almost Certain	Material emission is immediately visually obvious and there is continuous facility monitoring with immediate response.

2.5 *Risk Assessment Data Collection and Reporting*

2.5.1 Data Collection and Reporting

A qualitative containment capability assessment is focused on performing risk-based evaluations that pertain to key tasks and unit operations that involve the handling, sampling, and processing of pharmaceutical compounds and their respective raw materials. Below, is a breakdown of the aspects of the qualitative assessment reference criteria that are used during the risk assessment. The evaluation of the aspects is the basis of the scoring, leading to the calculation of RPN's for individual risks.

- Architectural Design - The physical layout of the processing rooms, adjacent airlocks, and corridors that support the facility's strategy for isolating and containing emissions at the source and/or within the defined containment zones.
- Process Flows - The movement of personnel, equipment, product, material, and waste through the facility. This evaluation will also focus on how the potential mixing of clean and contaminated personnel, equipment, material and waste are managed.
- Equipment - The design, functionality, and level of source containment provided for each piece of processing equipment, including operator accessibility for maintenance (both routine and during shutdown) and cleaning activities.
- Culture/Mindset - The overall awareness and mindset of management and production, cleaning, and maintenance operators for how easily product emissions can occur, how widespread these emissions can become, and the impact this potential release can have on both product cross-contamination and operator exposure risks.
- Decontamination - The procedures and techniques for the decontamination of equipment, personnel, processing rooms and adjacencies. Emphasis will be placed on decontamination methods.
- Procedures and Training - Operator techniques for waste handling, equipment and room cleaning, gowning and degowning, and personnel protective equipment decontamination. Also included will be procedures and training for emergency responders, emergency egress and waste handling. Emphasis will be placed on open operations, degowning/suit-showering procedures, and product migration.
- Future Regulatory Requirements - From a regulatory risk-based perspective, what are the "perceived" gaps? In essence, are there regulatory trends that may impact the process or facility design and operation? An example may be the need to define a process's critical step.

The analysis of the aspects of the qualitative assessment is achieved through: a review of relevant project documentation; a review of material, equipment and personnel flows; and interviews with process owners, quality and health and safety representatives. Specific questions that probe the various aspects of containment should be prepared in advance to facilitate the assessment.

Recommendations will be made that will attempt to eliminate or mitigate the risks. The format of the data collected is shown below.

Table 4. Risk Assessment Data Collection and Reporting

Item No.	Unit Operation	Potential Material Migration Vector	Potential Effect(s) of Material Emission	Severity	Potential Cause(s) of Material Emission	Probability	Current Controls	Detection	RPN	Recommended Action
1										
2										
...										
n										

The recommended action to address emission or migration risk should also include an assessment of unintended consequences.

Table 5. Risk Reduction Matrix

Risk Reduction Matrix				
Initial RPN	**Identified Risk**	**Cost of Associated Risk**	**Cost to Mitigate Risk**	**Resulting RPN**

It is important that it is clear what has been included in the costs shown in the risk reduction matrix. Costs of associated risk may include loss of business opportunity, replacement cost and supply chain disruption. Facility design cost, commissioning and qualification cost, and training cost are examples of costs to mitigate risk that may be included.

2.5.2 Risk Quantification Summarization

During a qualitative containment capability assessment, it is recommended that the identified risks that are above the designated risk threshold or have a high severity rating should be summarized. Table 6 is an example of a matrix form of Severity versus Probability, in terms of low, medium and high risk. The individual risks could also be further categorized by Health and Safety impact (personnel exposure), Quality impact (product exposure) and Environmental impact (outdoor exposure) aspects with separate matrices for each, if appropriate.

Table 6. Risk Quantification Summarization Table

		PROBABILITY		
SEVERITY	**High** (8-10)	Medium	High	Immediate
	Medium (4-7)	Low	Medium	High
	Low (1-3)	Low	Low	Medium
		Low (1-3)	**Medium** (4-7)	**High** (8-10)

Immediate	Indicates an Immediate Risk Hazard
High	Indicates a High Risk Hazard
Medium	Indicates a Medium Risk Hazard
Low	Indicates a Low Risk Hazard

2.6 Managing Risk

2.6.1 Risk Management

There are inherent risks associated with the handling of pharmaceutical compounds. These risks may be to personnel health, product quality, environmental and operational efficiency. These risks, however, can be successfully dealt with through the implementation of a risk management plan. The four key elements of a successful risk management plan include.

1. Maintaining a risk-conscious perspective - In other words, having an understanding and appreciation for the risk management requirements imposed by stakeholders (Quality, Health & Safety, Environmental, Operations).
2. Risk identification - This entails reviewing the process or operation and identifying what can go wrong. This is accomplished through the rating of the identified risks, i.e., developing RPN's.
3. Risk analysis - This matches an identified risk with the potential business impact. A comparison of a risk's RPN with the risk threshold identifies those risks that may have the largest business impact.
4. Contingency planning - The development and maintenance of plans to eliminate or mitigate the identified risks.

This document focuses on aspects of risk analysis. Aspects of contingency planning will also be presented. The rating and prioritization of identified risks is a tool that operating management may use during risk analysis. The RPN and risk threshold will aid operating management with assessing the potential business impact of emissions risk.

2.6.2 Containment Controls Used to Assist in Mitigating Risk

The containment controls discussed below are aspects of a contingency plan for addressing the emissions risk.

2.6.3 Process Controls

Process controls consist of the specific actions to optimize the process that can be implemented to minimize the risk of material emission and migration during routine and upset processing conditions. Process optimization may also include the elimination or substitution of a hazardous material.

2.6.4 Engineering Controls

Engineering controls, as they pertain to containment, have the ability to physically eliminate or reduce emission risks, through facility, equipment and process equipment modifications. These controls will usually focus on a facility's architectural layout, process equipment design, air handling pressurization and filtration. It should be noted that engineering controls, alone, will not achieve long-term, sustainable containment capability, because the best engineering controls can be put into a state of non-control based on poor procedural practices.

Within the engineering controls, there is a hierarchy of controls. These are briefly discussed below.

- Automation/Robotics is the engineered removal of personnel interaction with the process or material.
- Isolators/Closed Transfers are engineered systems that "control" personnel interaction with the process.
- Laminar Flow is the engineered airflows with appropriate capture velocities. The intent of these airflows is to entrain any material emissions and filter the material from the air stream before the air is exhausted, either back into the room or to the environment.
- Local Exhaust is non-engineered ventilation that only provides dilution of the emitted material in air-through-air volume air exchanges.
- Open Operations are engineered processes that are open to their surroundings.

The higher an engineering control is on the list, the interaction by personnel with the process or material becomes further removed and containment at the source is better achieved.

2.6.5 Procedural Controls

Procedural controls focus on the specific training, procedural and work practice controls that can be utilized to address containment risk during routine and upset conditions. Most of these controls focus on training curriculums and procedural requirements that are designed to help reinforce safe handling techniques for processing, cleaning, maintenance and other activities that address the risk material emissions and migration.

2.6.6 Administrative Controls

Addressing the risk of material emission and migration and the potential for subsequent exposure through production scheduling, shift rotation and limiting or restricting access to areas where exposure could occur are examples of administrative controls. Another aspect of administrative controls is the use of personnel protective equipment (PPE). PPE should be seen as the last measure, and should only be required if engineering solutions are not technically feasible or financially reasonable to achieve adequate containment of the process.

2.6.7 Response Profile

Engineering controls, procedural controls and administrative controls are all responses to maximizing the containment and minimizing the risk of emissions and their consequences. Experience has proven that engineering controls alone will not maximize the overall containment of emissions because of operator interaction, but engineering controls in conjunction with both procedural and administrative controls has been proven effective in minimizing long-term risk. In fact, a typical overall containment response will include a combination of engineering, procedural and administrative controls. Moreover, experience has shown that putting the initial focus on engineering controls typically yields a more robust overall response to containment challenges. This minimizes the remainder of the total containment challenge for the less repeatable administrative and procedural responses to address, as correct adherence to such becomes more problematic in the long term.

It should be understood by operating management that they are in the key position to provide the lead for managing the emissions risks that may be identified. As stated above, engineering solutions are preferred. However, it should be noted that alternative recommendations are available, yet they rely more heavily on procedural controls. Procedural controls, by their very nature, rely more heavily on human intervention and action. This human intervention and action is often not as repeatable over the long term nor are they as effective as engineering controls.

3. Benefits of Risk Based Methodology

The methodology presented here is applicable to all areas of pharmaceutical material production, including laboratory and clinical trial material manufacturing.

3.1 Process Knowledge is Improved by a Risk Analysis Approach

The risk assessment methodology presented here provides a systematic way of examining all the ways in which material emission and migration may occur within a process. Part of the risk assessment exercise involves an estimation of the potential impact each potential emission or migration may have on the overall system.

3.2 Risk Based Approach Exposes Key Process Steps

Through the risk assessment methodology, process-operating ranges can be identified under which a process can be run without causing unacceptable material emissions. This methodology should drive a company to make decisions and understand what is important about their processes.

The next step is to implement the appropriate manufacturing and process control technologies.

A final step is to ensure that a sound quality system infrastructure exists so that " a plant operation can reliably manage any process you give them" by controlling and limiting operator error.[6]

4. Managing Risk

A key element for successfully managing the risks associated with the handling of potent pharmaceutical compounds is contingency planning. Contingency planning consists of controlling the risk. Standard methods include:

- Prevent the release of material
- Minimize the impact of released material as much as possible
- Detect the release of material quickly
- Respond appropriately and quickly
- Recover as best as possible

These standard methods attempt to deal with risks early.

4.1 Costs to Consider during Risk Management

Several costs need to be considered when developing a risk management plan to address safe handling of pharmaceutical compounds. The costs associated with the risk and potential adverse impacts must be compared with the costs associated with the risk management plan.

4.1.1 Cost of Disaster

The costs associated with a poorly contained process can be broken into two broad categories, tangible or costs that are relatively easy to quantify and intangible or costs that are difficult to quantify.

Tangible costs:

- Cost of clean-up
- Cost of fines
- Closer scrutiny by regulatory agencies
- Loss of product
- Loss of raw materials

Intangible costs

- Disruption of the production schedule
- Reduced amount of product in the marketplace, potential reduction in sales for a given sales period
- Loss of production facility use during clean-up
- Negative publicity
- Cost of litigation

4.1.2 Cost of Risk Management Plan

The costs associated with implementing a risk management plan to address the safe handling of pharmaceutical compounds may be broken into two categories, the costs associated with mitigating the emission risk and the costs associated with implementing measures to address the sources of emissions. Costs to consider when mitigating the emission risk include:

- Cost of engineering design
- Cost of creating standard operating procedures and training materials
- Cost of training employees on the new procedures
- Cost of training employee on the new equipment
- Cost of equipment upkeep

Costs to consider when implementing measures to address the sources of emissions include:

- Cost of new equipment design
- Cost of new equipment fabrication and installation
- Cost of production disruption during installation of new equipment

4.1.3 Costs of Installation, Operation, and Maintenance

In determining, which control approach makes sense, the costs of initial installation and ongoing operation and maintenance needs to be evaluated against the level of protection that various risk elimination or mitigation options could offer.

4.1.4 Cost of Disaster (Assumed Risk) vs. Cost of Mitigation & Implementation

Management has to decide what level of risk they are willing to assume, based on the costs involved to mitigate the risk. In essence, what is the cost of implementing corrective actions to address risk versus the cost of a "bad" outcome? To make an informed decision all the risks, corrective actions and costs need to be clearly defined.

Initially, the cost of disaster is greater than the cost of mitigation. This is what is driving the need to address the risks identified.

Cost of Disaster > Cost of Mitigation *Equation 2*

However, after corrective actions have been taken, the cost of disaster is less than the cost of mitigation. The point, at which the inequality in Equation 2 changes, is the risk threshold.

$$\textbf{Cost of Disaster} < \textbf{Cost of Mitigation} \qquad \textit{Equation 3}$$

Unfortunately, the amount of risk that an operation is willing to accept is frequently dictated or bounded by financial constraints.

It is an informed management decision as to how and when the risks are assumed and the cost of mitigating the risks are amortized.

When risk mitigation activities are presented to management, the activities should be presented in such a way as to capitalize on synergies. Often, implementing a corrective action to address one risk will also address other risks. For example, an engineering solution that addresses a gap in a facility's HVAC containment capabilities may address risks associated with product cross-contamination and personnel exposure.

5. Summary

5.1 Challenges

Containing pharmaceutical compounds present unique challenges. These include:

- Managing the differing requirements of Health and Safety, Quality, Environmental and Operations
- Potency of pharmaceutical compounds is increasing
- Containment Capability has not been a priority for equipment manufacturers
- Pharmaceutical compounds are being developed at a record pace

5.2 Conclusion

The importance of effective risk management has significantly increased over the past several years. Safe handling of pharmaceutical compounds requires special attention. However, the fundamental concepts of risk management apply whether the process involves a pharmaceutical compound or a common industrial chemical.

The methodology shown in this document has been used successfully to rate and prioritize the risks associated with the safe handling of pharmaceutical compounds and subsequently develop an effective risk management plan.

6. References

[1] Announcement- Pharmaceutical cGMPs Initiative, www.fda.gov/oc/guidance/announce (accessed Oct. 16, 2003)

[2] Kolpin, Dana W., "Pharmaceuticals, Hormones, and Other Organic Wastewater Contaminants in U.S. Streams 1999-2000: A National Reconnaissance," *Environmental Science & Technology*, Volume 36, NO. 6, 2002

[3] "Frogs, Fish And Pharmaceuticals, A Troubling Brew" www.cnn.com/2003/TECH/science/11/14/coolsc.frogs.fish (accessed Sept. 3, 2004)

[4] *The Basics of FMEA*, Productivity Inc., ISBN: 0-527-76320-9, 1996, Service@productivityinc.com,

[5] "FMECA," www.fmeca.com (accessed June 17, 2004)

[6] "The Gold Sheet" Newsletter Published by F-D-C Reports, Inc. Vol. 37, No. 10 Oct 2003, ISSN 1530-6194/03

Additional references not cite:

[7] "Guidance for Industry Q7A Good Mfg Practice Guidance for Active Pharmaceutical Ingredients (FDA)," Aug 2001, www.fda.gov/cder/guidance/4286fnl.htm (accessed Oct. 26, 2004

[8] "Compliance Program Guidance Manual Program (FDA)," 7356.002, Form FDA 2438, 30 June 2001,www.fda.gov/ora/cpgm/default.htm#drugs (accessed Oct. 26, 2004)

[9] "GMPs for the New Millennium: A Risk Based Approach," www.fda.gov/oc/guidance/gmp (accessed Oct. 16, 2003)

[10] "Pharmaceutical cGMPs for the 21st Century - A Risk-Based Approach (Final Report FDA)," Sept. 2004, www.fda.gov/cder/gmp/gmp2004/GMP_finalreport2004 (accessed Oct. 27, 2004)

[11] "Pharmaceutical cGMP Initiative Questions and Answers," www.fda.gov/oc/guidance/qsas (accessed Oct. 16, 2003)

[12] *Containment in the Pharmaceutical Industry: Drugs and the Pharmaceutical Sciences*, Volume 108, James P. Wood, Ed., Marcel Dekker, Inc. ISBN: 0-8247-0397-9, 2001

[13] Hirst, Nigel; Brocklebank, Mike; Ryder, Martyn; *Containment Systems, A Design Guide*, Gulf Professional Publishing, ISBN: 0-7506-76124, 2002

[14] "FDA News,"U.S. Food and Drug Administration Press Release, P02-288/21/2002 http://www.fda.gov/bbs/topics/NEWS/2002/NEW00829.html (accessed Oct. 26, 2004)

[15] "Health Hazard Evaluation Report," National Institute for Occupational Safety and Health, 97-0141-2819, www.cdc.gov/niosh/hhe/select.asp, (accessed Sept. 3, 2004)

[16] "Health Hazard Evaluation Report," National Institute for Occupational Safety and Health, 96-0093-2685, www.cdc.gov/niosh/hhe/select.asp, (accessed Sept. 3, 2004)

Low Temperature Oxidation Of Methanol In The Presence Of Sponge Nickel Leads To Filter Overpressure Incident

E. Mark Davis
Pfizer Global Research and Development
Holland MI
Currently: Eli Lilly & Co. Chemical Hazards Lab,
DC 4813 Indianapolis, IN 46285
davisem@lilly.com

Prepared for Presentation at
American Institute of Chemical Engineers
2005 Spring National Meeting
39th Annual Loss Prevention Symposium
Atlanta, Georgia
April 11-13, 2005

UNPUBLISHED

Low Temperature Oxidation Of Methanol In The Presence Of Sponge Nickel Leads To Filter Overpressure Incident.

E. Mark Davis
Pfizer Global Research and Development
Holland MI
Currently: Eli Lilly & Co. Chemical Hazards Lab,
DC 4813 Indianapolis, IN 46285
davisem@lilly.com

Abstract

The chemical investigation of an A-7000 sponge nickel / methanol / anhydrous ammonia hydrogenation incident lead to the discovery of a non-condensable gas producing thermal instability of methanol in the presence of sponge nickel at 60°C, significantly below the temperatures previously reported in the literature. This previously unrecognized process hazard is of concern because these hydrogenations are done in sealed systems often at temperatures within the decomposition range.

1. Introduction

The chemical investigation of an A-7000 sponge nickel / methanol / anhydrous ammonia hydrogenation incident lead to the discovery of a non-condensable gas producing thermal instability of methanol in the presence of sponge nickel at 60°C, significantly below the temperatures previously reported in the literature[1]. This previously unrecognized process hazard is of concern because these hydrogenations are done in sealed systems at temperatures within the decomposition range. It is likely that the long time to maximum rate at the 50 °C process temperature (compared to the production scale process time) and the very low initial heat output have conspired until now, to hide this side reaction within the heat of reaction of the hydrogenation.[2]

On February 22, 2001 at about 4:00 pm, an isolated pressure plate filter heated up, overpressured and ruptured a chemical hose attached to the discharge port. There were no injuries and no fire outside of the filter. Damage was limited to the chemical hose and pressure stress to the filter plates and housing.

The incident lot was a routine sponge nickel - methanol and anhydrous ammonia reduction to produce a high volume pharmaceutical intermediate. At the end of the hydrogenation, the 50°C

---

[1] Mizuno, K., et. al. Chem. Lett., 1986, 1969 ; Takashi, K., Kobayashi,H., Takezawa,N.,Chem. Lett., 1985, 759.
[2] An incident in a $1 m^3$ hydrogenator is described where under 100 bar hydrogen, a methanol - Raney Nickel system unexpectedly heated up from 130°C to 160°C with a pressure rise from100 bar to 190 bar. Klais, O., "Hazards from Pressure," IChemE Symp. Ser. No. 102, 25-36, Oxford, Pergamon, 1987

batch is filtered through two large pressure plate filters. Two filters are required to isolate the large volume of sponge nickel catalyst from the process stream. The pressure plate filters are set up in parallel. After the first is loaded, the valving is switched to isolate it while the remainder of the batch is sent to the second filter. Both filters are then rinsed with water to remove any remaining product and flammable solvent prior to removal and disposal of the spent sponge nickel catalyst cake. In this incident, the first pressure plate filter had been loaded, blown free of liquid with nitrogen and isolated with manual valves while the remainder of the batch was being filtered through the second pressure plate filter. Thirty to forty minutes after this first filter was isolated, the chemical hose on the discharge side of the isolated filter ruptured, accompanied by a loud bang and pressure release which discharged smoke or vapor with an acrid odor into the building.

The area was immediately evacuated. Emergency responders found the pressure plate filter very hot. The radiant heat could be felt through their protective equipment. The filter was flushed with a large volume of water until cool.

After opening the filter the top 3 or 4 plates (perforated metal trays lined with filter paper and loaded with 1 to 2 inches thick of nickel catalyst) showed signs of exposure to high heat. The ordinarily wet and flexible filter paper was found to be stiff and brittle with a slightly charred appearance.

The 10 foot long 1.5 inch diameter chemical hose, rated for 200 psig working pressure (1000 psi burst strength), ruptured in three places. The center of the hose completely separated. Ruptures also occurred near the Kamlok® connections at both ends; however the end fittings did not separate from the hose.

In this incident the first and most obvious suspect was a catalyst fire; however no source of sufficient oxygen could be found. Plant maintenance and gas delivery records were reviewed for possible contamination sources. None were found. The plant nitrogen was tested in numerous areas and assayed for oxygen content; this too found minimal oxygen contamination. The involved operators and supervisors, all highly experienced, were interviewed in detail as to procedures that were used on this process and in particular this batch. Special attention was paid to having them relate anything they felt was unusual, new or different about this batch or the plant environment. All signs pointed to a routine batch.

A quick calculation of the pressure at the burst point, based on a rough guess of the loaded pressure plate filter's internal volume, a 1.5 inch outlet and the estimated length of exhaust time as reported by the startled operators, gave an extremely rough estimate of >800psi. This was consistent with the burst strength of a used chemical hose.

In a parallel investigation, samples from the incident batch, the next batch (already in process when the incident occurred), and some "typical" archived samples from previous batches were delivered to the R&D Process Hazards Laboratory of Pfizer Inc in Holland, MI for thermal stability evaluation.

A detailed thermal hazard study using Differential Scanning Calorimetry (DSC) and Accelerating Rate Calorimetry (ARC) was conducted on all samples. An exothermic decomposition was found in the unwashed catalyst cake of the next batch but not in a washed catalyst cake sample from the incident batch. No other samples showed any thermal instability. DSC and ARC studies on samples made from combinations of various known components of the process wet catalyst cake mixture, found an exothermic reaction between virgin A-7000 sponge nickel (as removed from the supplier's drum and not exposed to hydrogen) and aqueous methanol which does not require any other of the process components to occur. Virgin Sponge Nickel is produced by leaching the aluminum from finely ground Ni-Al alloy. This produces hydrogen, some of which is retained on the catalyst as shipped.[3] This methanol consuming exothermic reaction reaches ARC onset (≥0.02°C/min) between 120°C and 140°C and produces large volumes of gaseous byproducts.

Reactions between sponge nickel and alcohols can be found in the literature. Hydrogen under high pressure (100 bar) and temperature (>110°C) reduces methanol to methane and water in the presence of sponge Ni catalyst with generation of "large amounts of heat and pressure"[4]. Another known reaction is the dehydration of alcohols to carbonyls at high temperatures[5]. Dehydrogenation of methanol produces formaldehyde, methylformate, hydrogen and CO_2 in the presence of water (figure 1) or alternately formaldehyde, hydrogen and CO without the participation of water (figure 2). [6]

Figure 1

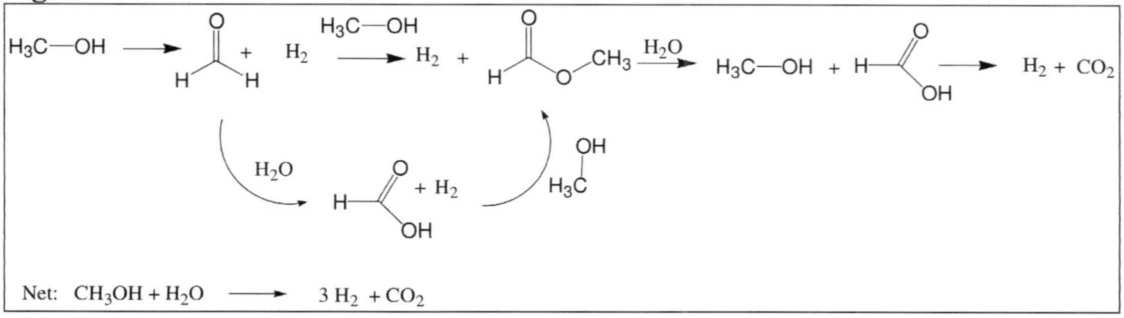

Figure 2

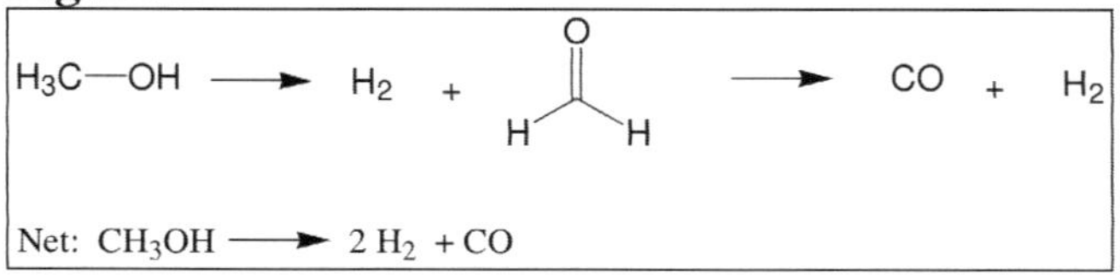

Either reaction path produces hydrogen which could be fuel for the previously mentioned reduction of methanol to methane. Formaldehyde and reaction products of formaldehyde may also explain the acrid smell reported by eye witnesses. Direct injection gas phase mass

[3] Fieser and Fieser, Reagents for Organic Synthesis I, 1967 Wiley & Sons inc, 723-731

[4] Klais, O., Hazards from Pressure, IChE Symp. Ser. No. 102, 25-36, Oxford, Pergamon, 1987

[5] Mizuno, K., et. al. Chem. Lett., 1986, 1969 ; Takashi, K., Kobayashi,H., Takezawa,N.,Chem. Lett., 1985, 759.

[6] Takagi, K., Morikawa, Y., Ikawa, T., Chem Lett., 1985, 527

spectrometry of the off-gases from ARC experiments strongly indicated the presence of CO, CO_2, and H_2, as well as other gases which may have included methane and formaldehyde.

2. Equipment and Methods

DSC was performed on Mettler model DSC 821^e calorimeters. Data were collected and analyzed using Mettler STAR^e ® software. All DSC samples were run in high pressure gold plated steel 40μl crucibles sealed under nitrogen or normal atmospheric air as specified.

ARC was performed on a Colombia Scientific Instruments "classic" ARC with an A. D. Little "New" ARC upgrade. Our ARC was modified for sampling off gases by the addition of a T in the pressure transducer line which leads to a valve outside the clam shell via fine diameter stainless steel tubing. Appropriate calibrations were performed prior to HWS (heat, wait, search) or isothermal runs. The Heat-Wait-Search method is where the calorimeter is heated to a specified temperature, waits for a specified time based on the sensitivity selected, then searches for an exotherm as defined by the sensitivity. If an exotherm is found, the calorimeter switches to exotherm mode where the calorimeter temperature tracks the sample temperature. If no exotherm is found, the calorimeter heats up by the step temperature to begin another cycle.

Mass spectrometry was performed using a Pfeiffer Vacuum OmniStar™ GSD 300 O mass spectrometer and analyzed using Blazers QuadStar™ 422 software. The sample was collected from the ARC in a 1L polyethylene bag which allowed for multiple runs from one sample.

3. Results

3.1 DSC:

The DSC results are representative of a screen of the different process samples from a typical batch. Sponge nickel catalyst was run as a blank and found to have no exotherms to 350°C. Figure 3, DSC #1, is the unwashed process catalyst cake sealed under nitrogen. It shows a 183.7 j/g exotherm at 232.5°C. Figure 3, DSC #2, is the same sample sealed under air. It shows a slightly larger heat of decomposition at roughly the same onset. Figure 3, DSC #3, the washed filter cake did not show any thermal decomposition to 350°C however adding back a little methanol restored the thermal decomposition (Figure 3, DSC #4). This lead to the ARC study of sponge nickel and methanol.

Figure 3[7]

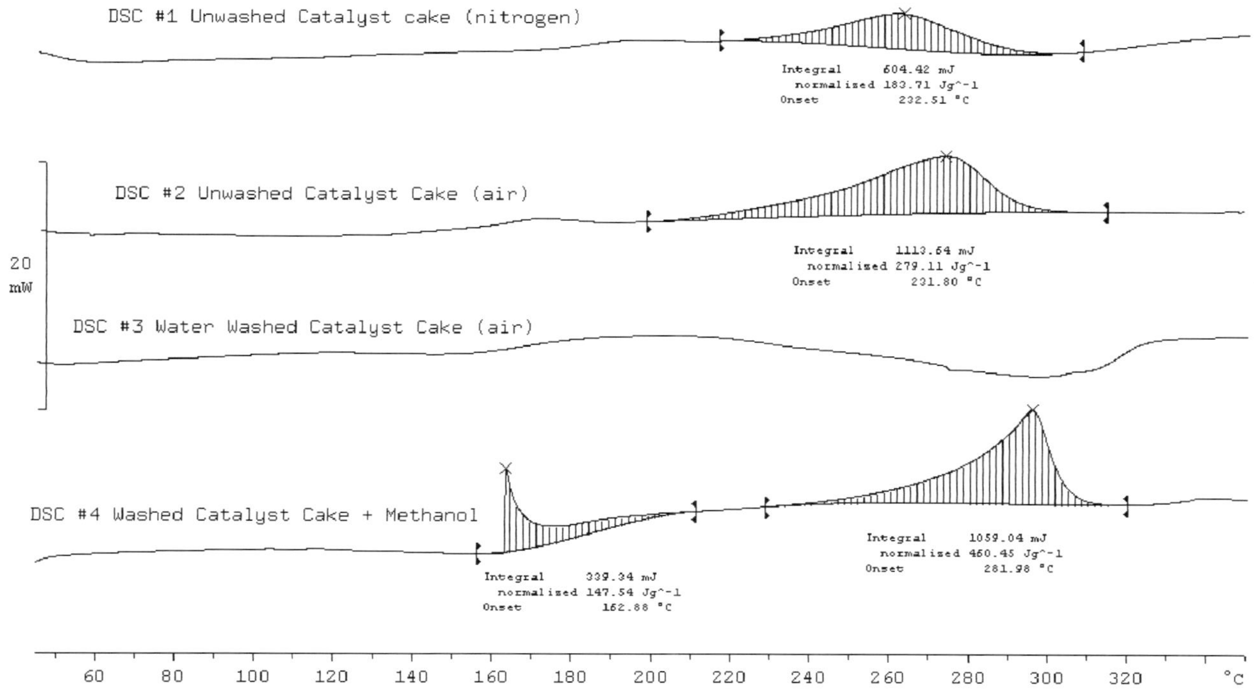

Table 1 (see figure 3)

DSC#	Program[8]	Sample Description	Results
1	35-350@10	Lab filtered catalyst cake of typical reduction mixture (unwashed). prepared and sealed under nitrogen	exotherm 183.7j/g@232.5°C
2	35-350@10	Lab filtered catalyst cake of typical reduction mixture (unwashed). sealed under air	exotherm 279.1j/g@231.8°C
3	35-350@10	Lab filtered catalyst cake of typical reduction mixture water washed 2X.	no exotherm
4	35-350@10	Lab filtered catalyst cake of typical reduction mixture water washed 2X. plus methanol	exotherm 147.5j/g@162.9°C exotherm 450.5j/g@282.0°C

[7] The DSC baselines are relatively poor due the high phi factor produced by running samples of <3mg in cells averaging about 630mg. The small sample size was needed to keep the samples from bursting the cells at high temperature. During this investigation, a cell containing 7.18mg of pure methanol burst at 350.9°C.

[8] DSC programs are described in the following format: T_i-T_f °C at R DPM, where T_i is the starting temperature, T_f is the final temperature, R is the rate in DPM (°C per minute).

3.2 ARC:

ARCs were run on various samples of A-7000 catalyst and amounts of methanol.

Table 2

Fig.#	Program[9]	Sample	Interpretation of Results
4	HWS 50-450° step=5° sens. 0.02°/min phi=2.34	2.16g A-7000 sponge Ni (damp) 0.37g bulk methanol Nitrogen inerted	single self sustaining 748 j/g (MeOH) exotherm onset 151°C. ΔT_{ad} = 69.75°C Final weight of bomb indicated a weight loss of 0.37g to non-condensable gases.
5	Isothermal 60°C sens. 0.02°/min phi=1.66	2.31g A-7000 sponge Ni (damp) 0.62g bulk methanol Atm= air	Temperature rises below exotherm threshold until 120°C (~115 hrs). Self heats to ~215°C
6	Isothermal 60°C sens. 0.02°/min phi=1.66	3.60g A-7000 sponge Ni (damp) 0.64g bulk methanol Atm= air	Manually induced negative drift. Despite isothermal conditions pressure steadily increases over duration of experiment. this indicates that the reaction is taking place at temperatures as low as 59-60°C.

Attempts to replicate the actual incident conditions in the ARC were thwarted by the basic design differences between an ARC bomb and a plate filter. In the filter, the liquids settle in the bottom with the driest solids occurring in the upper most plates - not coincidentally where the most heat damage occurred during the incident. The opposite occurs in an ARC bomb where the solids settle to the bottom and become "protected" from the atmosphere by a layer of liquid. This basic difference effects the calorimetry measurements too as the dryer the nickel cake the lower the heat capacity. A lower heat capacity in turn causes a more rapid ΔT and a larger ΔT_{ad} (adiabatic heat rise of the system) per given unit of heat of reaction. This could explain the variance in both onset temperature (where ΔT>0.02°C/min) and calculated ΔT_{ad} between multiple runs. This difference in settling effect also virtually eliminates the possibility of oxygen induced nickel hot spots initiating the thermal decomposition in the ARC as would be quite plausible in the catalyst cake at the top of a pressure plate filter. Thus, little difference was seen between ARC runs padded with air, 6% O_2 in nitrogen, 100% nitrogen or argon. Of interest were the gaseous by-products of the ARC runs. Mass Spec. of gas phase products indicated with a high degree of confidence the following: Ar (the inertion gas) CO, CO_2, H_2, and NH_3. Also, to a lesser degree of confidence methane, formaldehyde and NO. Of interest are the nitrogen products which indicate the inefficiencies of inerting an unstirred slurry. The other gasses are those expected from an oxidation or reduction of methanol. In most ARC runs the measured weight loss at the end of the run was roughly equal to the weight of the methanol charged to the

[9] ARC programs are described in the following format: HWS T_i-T_f (T_{exo}) °C X°C steps y°C/min sensitivity Or Isothermal T_i°C y°C/min sensitivity, where HWS is heat , wait, search mode, T_i = starting temperature, T_f = final HWS temperature, T_{exo} = final exotherm mode temperature, X= step size in °C and y is the lower exotherm detection limit.

bomb. For this reason all heats of reaction are presented in joules per gram of charged methanol. However these must be calculated based on an assumed cp (2.5 j/°C) for the bomb contents.

The first two isothermal ARC runs (Figures 4 and 5) at 60°C indicate the presence of an exothermic reaction which produces non-condensable gases and is not dependant in the atmosphere present (air vs. Nitrogen). In Figure 6, a third run was done, where the calibration data was manually manipulated to induce a small negative drift. This was done to eliminate all possibility that the temperature rise seen in the first isothermal ARC (Figure 5) was an artifact of the instrument. The pressure rise resulting from the production of gas and resulting pressure rise is clearly visible in spite of the negative temperature drift which kept the calorimeter at 59-60°C. The final pressure at 23°C is about 29 psi above the starting pressure at 23°C proving a non-condensable gas producing reaction has occurred. The weight loss in the bomb was 0.12g or 19% of the methanol added.

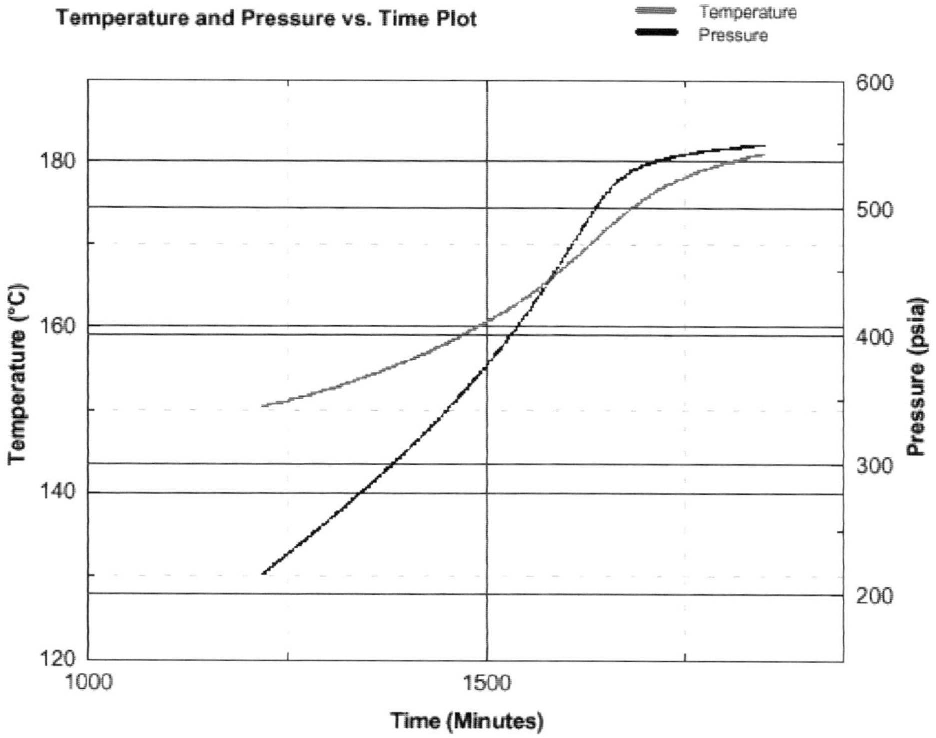

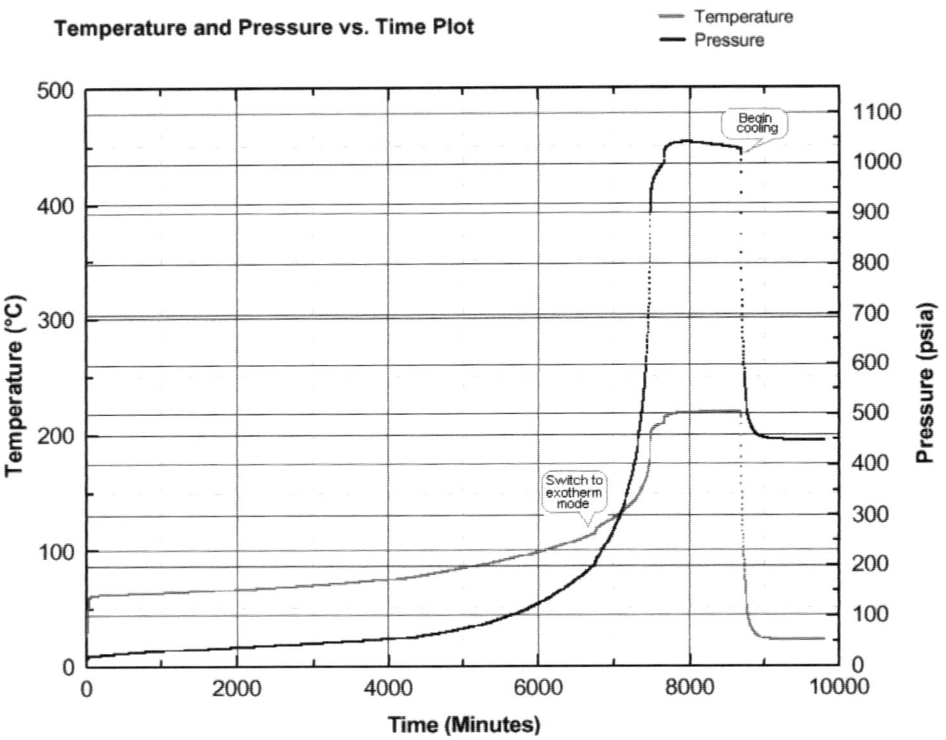

Temperature and Pressure vs. Time Plot
Temperature
Pressure
Begin cooling
Switch to exotherm mode
Temperature (°C)
Pressure (psia)
Time (Minutes)

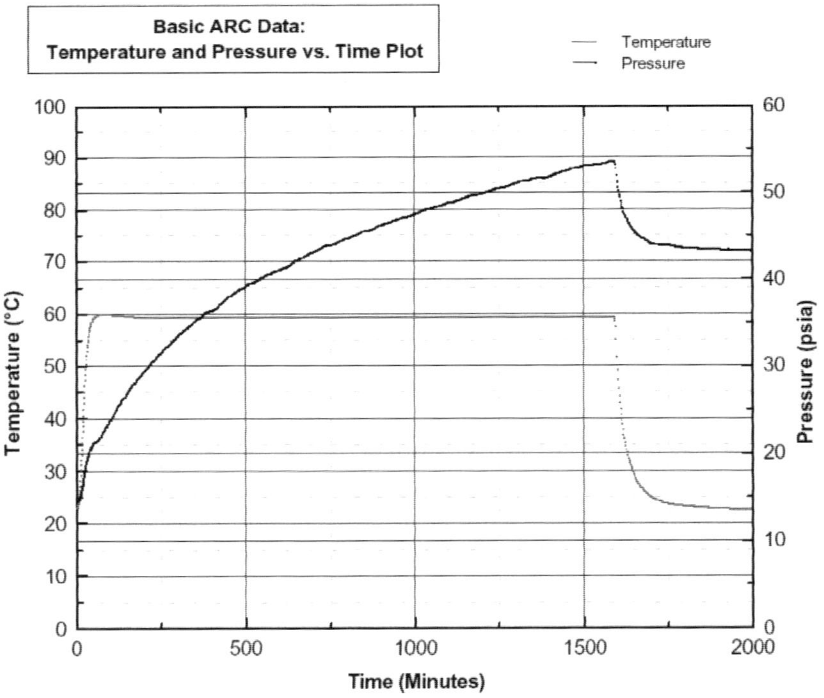

Basic ARC Data:
Temperature and Pressure vs. Time Plot
Temperature
Pressure
Temperature (°C)
Pressure (psia)
Time (Minutes)

4. Conclusions and Recommendations

Based on the test data and eye-witness accounts, the incident was likely caused by a thermally induced catalytic oxidation of methanol by the A-7000 sponge nickel which resulted in overpressure of the pressure plate filter. Air was introduced into the top of the pressure plate filter - probably through the shaft seal because a vacuum was produced upon cooling of the hot sealed methanolic system. The filter was pressure checked prior to use but not vacuum checked. Sealing for vacuum uses a different side of the O-ring and seat grove than pressure sealing. The introduced air caused some portion of the sponge nickel on the upper plates where the catalyst was driest, to begin auto-oxidizing creating a heat source. Once initiated, the methanol oxidation became self sustaining even though the inertion was restored by consumption of all available oxygen. It is plausible that through this mechanism, the upper three plates could reach temperatures high enough to explain the brittle filter papers and radiant heat noted by emergency workers. These temperatures along with the gas produced by the sponge nickel-aqueous methanol reaction could explain the increase in system pressure which ruptured the chemical hose. The evidence points to a slow pressure buildup to structural failure, not an internal fire, deflagration, or explosion. It is unknown if other types of sponge nickel behave in a similar manner.

It is recommended that all methanolic sponge nickel catalyst filter cakes be promptly water washed (in-situ) to remove residual methanol. No self sustaining exotherms were found in water washed catalyst cakes.

It is further recommended that all sponge nickel hydrogenation cakes containing alcohols be assessed for this potential exothermic reaction and appropriate actions taken on a case by case basis.

It seems unlikely that this reaction would cause problems during the reduction phase of a routine batch. Reductions in general, tend to be fairly exothermic already, thus heat removal is necessarily occurring already through the application of coolant to the vessel jacket. It is conceivable that if jacket cooling were interrupted, heat and pressure could build despite the standard emergency purge of hydrogen from the reactor to stop the intended chemistry. It is advised that any nickel reduction left under these conditions should be monitored for pressure and heat buildup during the loss of cooling upset period. The effect of a loss of agitation should also be considered in light of the buildup of metal catalyst at the bottom of the tank leading to a significant local heat capacity decrease which could lead to rapid local temperature rises.

A Method To Determine If A Flowing Suspension Might Cause Electrostatic Damage To The Glass / TFE Liner Of Steel Equipment

Daniel Muzzio* and Michael Toth
Merck & Co, Inc.
One Merck Drive, Whitehouse Station, New Jersey 08889

* Corresponding author Tel: 732-594-2540, daniel_muzzio@merck.com

Prepared for Presentation at
American Institute of Chemical Engineers
2005 Spring National Meeting
39th Annual Loss Prevention Symposium
Atlanta, Georgia
April 11-13, 2005

UNPUBLISHED

A Method To Determine If A Flowing Suspension Might Cause Electrostatic Damage To The Glass/ TFE Liner Of Steel Equipment

Daniel Muzzio* and Michael Toth
Merck & Co, Inc.
One Merck Drive, Whitehouse Station, New Jersey 08889

* Corresponding author Tel: 732-594-2540, daniel_muzzio@merck.com

Abstract

Electrostatic damage in the form of pin-holes was observed after certain suspensions/slurry streams were processed in glass or Teflon lined steel equipment. An investigation was performed to determine what electrostatic properties of the slurry combined with physical handling conditions of the stream are useful in predicting liner damage. The electrostatic properties measured were conductivity, charge relaxation times, and slurry chargeabillty. Slurry chargeability and the methods used to determine it are discussed in detail and experimental results are compared with results published by other scientists.

1. Introduction

Electrostatic damage in the form of pin-holes was observed after certain suspension/slurry streams were processed in glass or Teflon lined steel equipment. An investigation was performed to determine what electrostatic properties of the slurry combined with physical handling conditions of the stream are useful in predicting liner damage.

To determine the electrostatic properties of the slurry, the conductivity of the liquid phase was measured using a tera-ohm meter and a liquid conductivity cell. Next the volume resistivity (R_v) of the dry solids was measured using a tera-ohm meter and parallel plate electrodes cell (BS5958, 1991). Conductivity is the inverse of resistivity. The slurry/suspension was then constituted from these parts and the conductivity was measured using the same method as for the liquid. Two general observations were made; the conductivity of the slurry essentially remained the same as the liquid phase or it was much more conductive. The second observation is important because it justifies the need to measure the electrostatic properties of the slurry not just its individual parts.

The increase in conductivity of the slurry versus the clear liquid is attributed to ionic species coming off the solid material. These ionic species could be the material itself or some impurities. The ionic species create bridges that allow the electrical charge to be carried between electrodes, hence increasing the conductivity of the en tire system.

An observation that was expected, but never observed, was that the addition of a highly resistive solid could lower the conductivity of a liquid. This never occurred in any of the 50 different systems tested, even though most of the solid materials had very high volume resistivities. There is concern that the measurement method itself was preventing this observation from being seen. Conductivity/resistivity measurements show the fastest route for charge migration available, not the slowest or average. Basically, a tiny part of the slurry could allow the electricity to flow, hence not being representative of the true conductivity of the entire slurry.

To dispel the concern about the accuracy of the conductivity measurements, a method for estimating the conductivity of the slurries was used, based on measurement of the charge relaxation time. This is done by using a voltage corona discharge to deposit a charge onto the slurry. An electrostatic field meter is then used to measure how quickly the deposited charge migrates away by measuring the decreasing surface potential. Relaxation time is defined as the time from an initial peak surface potential to some specified fraction of it. If the material is ohmic, the migration of the charge is logarithmic and is usually independent of the amount of charge, 1/e (~37%) is a convenient time fraction used for this comparison.

The charge relaxation time was measured for all the different systems that had conductivity data, but no difference in electrostatic behavior was observed. That is, the charge relaxation time and conductivity of the slurry system gave similar results (Britton 1999). No new insight was gained.

The preceding testing all provide the initial charge to the sample. A measure of the process stream's ability to generate charge is also needed to properly analyze the issue. The use of high speed stirring to charge-up the slurry proved to be effective with the added advantage of including a physical handling condition, slurry velocity, as a measurable variable. This led to the development of the slurry chargeability test.

2. Apparatus

The slurry chargeability test is performed by placing approximately 250 ml of the slurry sample into a nitrogen blanketed, 3 liter glass round bottom four neck flask fitted with a variable high speed mechanical stirrer. A glass shaft fitted with a PTFE 14 cm x 2.6 cm blade is connected to the stirrer. This blade is just able to fit at the bottom of the flask and allows stable stirring at 1100 RPM.

A field meter, JCI 140 static monitor, is positioned 4 cm from the flask wall, just above the upper surface of the unstirred liquid. The field meter sensitivity varies with distance from the surface being measured. The meter is calibrated to directly read surface potential at a distance of 10 cm. When positioned at 4 cm the reading on the meter is 60% higher than if it were at the 10 cm distance (Chubb June 2003). The field meter is connected, via an analog to digital converter, to a computer. A commercial software program is configured to log data every second. The program takes as many readings as possible (about 1 per millisecond) during each data period, and then saves the average of the readings. A plot of surface potential in kilovolts (kV) versus time in seconds is generated.

2.1 Field Meter

A field meter is well suited to measuring an electric charge being generated by the movement of slurries, because a field meter is able to measure the resulting electric field from a distance and without contact. Between any charged bodies there is a force of attraction or repulsion, this force at any point is called the electric field, *E*. The field meter responds to the strength of the electrostatic field into which it has been placed. Essentially a field meter can be thought of as placing a test charge into an electric field and measuring the force exerted on that charge by the electric field. (Pratt 1997)

Field meters operate on the principle of electrical induction. There are two basic types of field meters: induction probe and field mill. The instrument used in this paper is the field mill type, which is mechanically more complex, but also more sensitive, has a longer zero stability, and does not require 'zeroing out' before each measurement. Measurements of the electric field can be made to a few V/m with response time to a few milliseconds. This instrument displays the observed surface potential in kilovolts by sensing the resulting electric field at the meter's aperture. (Chubb Feb. 2003 and Secker 1975).

Some limitations of using a field meter are that the meter itself distorts the field lines, also the measurement being taken should ideally be of a plane surface. These measurement distortions will make an absolute value of the observed surface potential questionable; however, as long as the experimental set-up is constant, a relative field strength or surface potential is useful (Glor 1988). In this investigation slurries streams with unknown charge generation tendencies, will be compared to streams of known chargeability so consistent relative surface potential reading is all that are needed.

2.2 The Model System

The slurry systems examined throughout most of this paper consist of glass beads at a concentration of 40 kg/m^3 in different liquids. The liquids are water and isopropyl alcohol (IPA) which are conductive: hexane, toluene, and cyclohexane, which are highly non-conductive: as well as different mixtures of these liquids. The glass beads, which are the solid component of the slurry, are either of uniform size of 106 microns, or a range of sizes between 150-212 microns. The glass beads are either used as received and are non-resistive ($R_v = 1 \times 10^7$ ohm-m) or are acid washed before use and are highly resistive ($R_v = 1 \times 10^{12}$ ohm-m).

3. Results

In Figure 1, the surface potential generated over time of two different slurry systems is observed using the slurry chargeability set-up. It is clear from the plots that the test set-up can distinguish between a system that does build up an electrical charge and one that does not. At 1200 seconds, after stirring at 1100 RPM is started, only the cyclohexane slurry begins to accumulate a charge. This is seen by both an increase in the measured surface potential which peaks at approximately –9 kV and an oscillation range of 6 kV. The surface potential oscillation instantly stops as the surface potential slowly weakens towards zero when the stirring is stopped at 1750 seconds. Essentially no surface potential or oscillations of the surface potential are observed for the stirring of the water slurry, indicating no electrostatic charge accumulation. The only difference between these two slurry systems is the liquid used to make the slurry. Cyclohexane which is nonconductive (0.2 pS/m) is used in the one system, while water which is conductive (1,000,000 pS/m) is used in the other slurry.

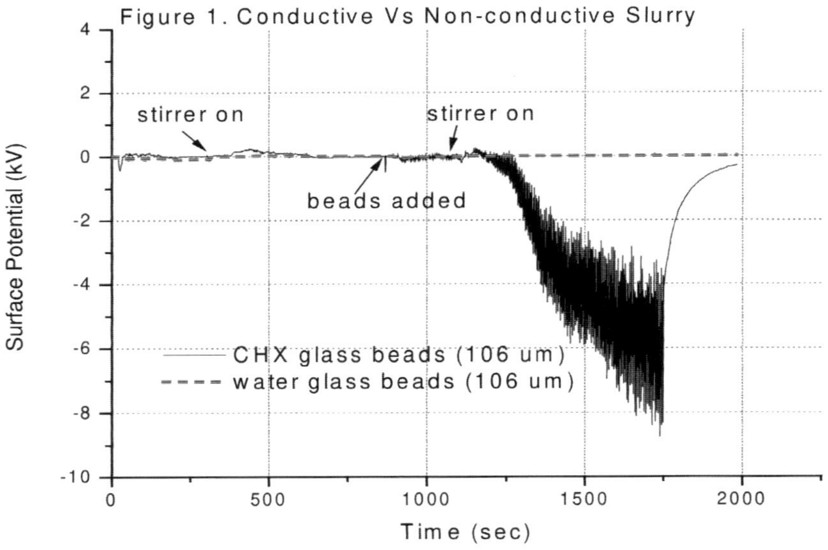

Again in Figure 1 looking at the period between 950 – 1200 seconds, which is before stirring occurs but after the glass beads are added, no significant surface potential and hence charge is observed, so charge generation requires stirring. The relationship between charge generation and stirrer speed was also investigated. Expressed in terms of tip speed, no significant charge generation was observed below a tip speed of 5 meters per second for any slurry system, including highly non-conductive ones. Tip speed expressed as meters per second (m/s) is equal to length of stirrer blade in meters times the rotation speed of the stirrer in 1/seconds times π. Using a 3000 ml round bottom flask with a 14 cm long blade the stirring rate needs to be above 690 rpm to reach this tip speed. It was observed that increasing the stirrer above this speed increased the surface potential. Velocity gradients caused by the high speed stirrer between the two phases is what generates the electrostatic charge as observed by the surface potential. (Vos *et al* 1974)

Another important observation from Figure 1 is that stirring the liquid without the 106 microns size glass beads does not accumulate significant static charge. This is seen by looking at the time between 315 and 730 seconds. In this period, cyclohexane without glass beads is stirred at 1100 rpm, but the maximum observed surface potential during this period is only 0.22 kV with an oscillation range of 0.06 kV. So the strength of the surface potential observed, the oscillation range, and hence the charge generated by stirring the cyclohexane slurry is much larger than stirring pure cyclohexane. The presence of a solids phase with the liquid is responsible for this increase in charge generation. The charge generation takes place at the particle-liquid, at the particle-stirrer blade, and at the particle-vessel wall interfaces.

The conductivity of the liquid is an important property in evaluating whether the slurry will generate/retain static charge. As previously stated cyclohexane is highly non-conductive.

Adding a polar solvent e.g., isopropyl alcohol (IPA), to the cyclohexane increases the conductivity (see Table 1).

Table 1. Electrical Conductivity of isopropyl alcohol in cyclohexane

v/v IPA in Cyclohexane	Conductivity pS/m
0.0 %	0.2
9.0 %	5.5
13.0%	67
18.7 %	1956
25.0%	13300
28.6%	37500

Figure 2. Chargeability Plot Increasing Conductivity

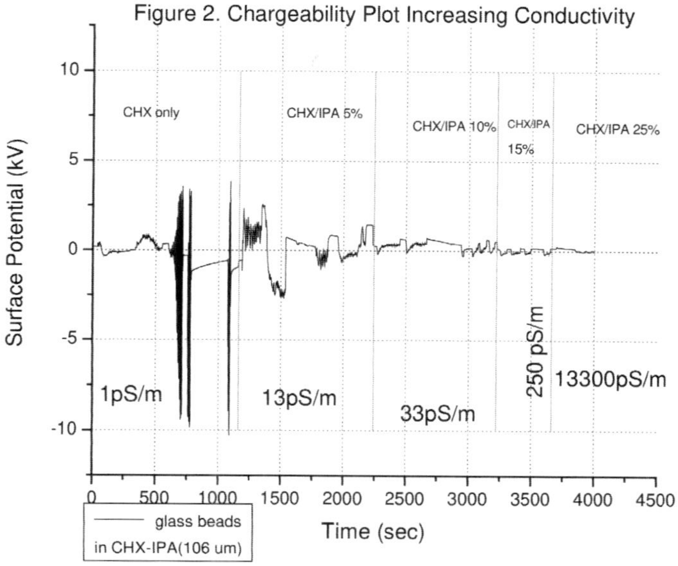

Figure 2 is the observed surface potential generated over time of cyclohexane/IPA glass bead slurry with increasing amounts of IPA which increases the system's conductivity. From this plot, two trends stand out. First, as the conductivity of the liquid increases the peak surface potential weakens to essentially zero when the measured conductivity is about 13,000 pS/m. The second trend is that the oscillation of the surface potential dampens towards zero as conductivity increases. Both these trends are related and expected because as slurry becomes more electrically conductive the speed at which the generated charges can migrate and hence neutralize any charge imbalance created by the stirring increases.

Up until now, the experiments have shown that the slurry chargeability test can distinguish between systems that will generate a static charge and those that will not. The characteristic of systems that generate a static charge are highly nonconductive liquid containing a solid phase. The conductivity of the slurry system is important not just the conductivity of the liquid. Both these properties can easily be determined without using a slurry chargeability test set-up. The slurry systems tested and results shown in Figure 3 cannot be so easily differentiated. The only difference between the two slurries shown in this figure is the size of the glass beads used in the make-up of the slurry system. The measured electrical properties of each system and their individual parts are essentially the same (see table 2).

In Figure 3 after stirring at 1100 RPM has started the slurry made with 106 microns glass beads has a measured maximum surface potential of over –10 kV and an oscillation range greater than 14 kV, where the slurry containing the 150-212 microns glass beads has peak surface potential of 0.035 kV and essentially no observable oscillation.

Table 2. Conductivity and Charge Relaxation times for Cyclohexane Slurries		
Sample	Conductivity (pS/m)	Relaxation time (msec)
Cyclohexane pure (CHX)	0.2	160
Glass beads (acid washed) 106 um	0.5	90
Glass beads (acid washed) 150-212 um	0.3	90
Slurry CHX-glass beads 106 um	33-140	90 to 120
Slurry CHX-glass beads 150-212 um	20-120	90 to 190

Figure 3.Chargeability Plot for Slurries with Same Conductivity and Relaxation time

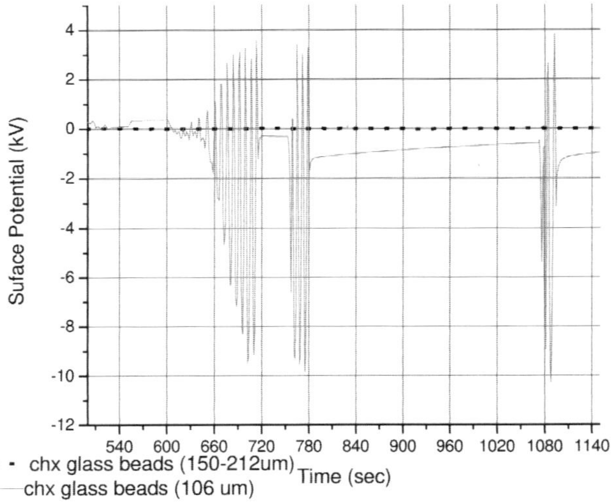

From all these results it has been demonstrated that the important parameters for charge accumulation in slurries are speed of stirring, slurry system conductivity, and particle size. Charging may be related to velocity gradient between particles and the liquid due to fluctuations caused by turbulence and the settling velocity of the particle due to gravity. Smaller particles with slower settling velocities allow the particle to float through the liquid minimizing any velocity gradient and hence decreasing any charging rate. The larger particles may remain motionless or stuck at the bottom of the vessel. Additionally, the surface area to volume ratio for the larger particles is smaller and charging is probably proportional to the total area of the exposed particle surface (Vos *et al*, 1974).

Figure 4 is the observed surface potential generated over time of actual process streams with the only difference between each stream being the average particle size of the material in the solid phase. The liquid phase is a 50/50 volume to volume mix of toluene and heptane with measured electric conductivity for the mixture of 3 pS/m and the solid material is a single organic compound with a measure R_v of $1x10^{14}$ ohm-m.(>0.001 pS/m) Each slurry's measured conductivity is essentially the same as the clear mixture. A similar stream, as the 38.9 micron size stream, caused pin-hole damage in the glass-lining of a vessel which was assumed to be caused by static discharges.

In this example, the particle size of the solid material is reduced initially from 38.9 microns to 14.1 microns but the tendency of the slurry to generate charge does not appear to change. The measured peak surface potential are 5.8 kV and 4.5 kV with an associated oscillation range of 6 kV and 7.5 kV for the 38.9 microns average particle size slurry and 14.1 microns average particle size slurry, respectively. Reducing the average particle size of the solids further to 7.9 microns the charging tendency greatly decreases as measured by surface potential (-2.5 kV/m) and oscillation range (0.3 kV/m).

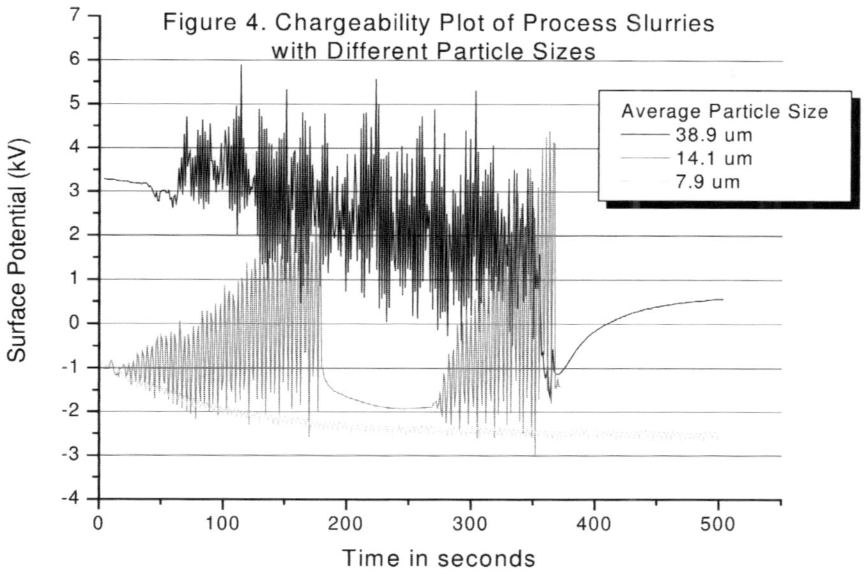

It was reported by other researchers (Maurer 1997) and demonstrated here that slurries made with conductive or nonconductive organic or inorganic particles can generate large electrostatic charges. The properties that determine how particles will interact with the velocity gradient (size, shape, and density of the particle) are more important then the chemical make-up of the particle. This agrees with other works pertaining to electrostatic charging of liquids and solids rubbing against solid walls, which states that charging is strongly dependent upon the relative velocities and not so dependent upon the chemical composition of the material.

All the findings from this investigation are consistent with the charge generation mechanism known as double-layer charging. Charge separation occurs at any liquid interface, be it liquid-solid, liquid-gas or even liquid-liquid. Ions of one polarity are preferentially adsorbed at these interfaces. The adsorbed ions attract oppositely charged ions within the liquid forming a layer near the interface. This layer is thin for conductive liquids and becomes more diffused as the conductivity of the liquid decreases due to reduced ion mobility. As the liquid flows relative to the interface, ions in the more diffuse layer are sweep away. There is a charge imbalance created, causing the electrical potential to increase. The charges generated tend to move toward an interface to recombine with charges of opposite polarity and this process is called charge relaxation. When the rate of charge generation exceeds the rate of charge relaxation, charge accumulates.(Kong 2002)

4. Conclusion

Electrostatic damage to the glass/TFE lining of steel equipment occurs in the chemical and pharmaceutical industries. Ignitions caused by these discharges are usually avoided by inerting the atmosphere in the process equipment. Still, equipment is damaged and extraneous materials may be introduced which can contaminate the product being produced. Replacement or repairs of the damaged equipment along with rejection or additional processing of the contaminated product can be very costly.

An important parameter for understanding when a process stream might cause electrostatic discharges is its ability to become charged while it is moving. In this paper, a simple inexpensive bench-top set-up has been described and testing performed that allows the chargeability of slurries to be determined. Our next step is to use the slurry chargeability test to examine streams which have documented histories of causing liner damage and determine if this testing can be used to exactly predict when a stream will electrostatically damage liners. Future questions to answer are whether the observed oscillations are simply artifacts of the test set-up or are they some fundamental physical phenomenon, and to determine quantitatively the relationship between the observed surface potential and published breakdown voltage for liner material used in plant equipment.

5. Literature Cited

British Standard 5958 Part 1 Control of undesirable static electricity, Appendix A, 1991.

Britton, L. G., *Avoiding Static Ignition Hazards in Chemical Operations*, AIChE, pp 98-101, 1999.

Chubb, J. N., User Manual for Static Monitor JCI 140, UM140 issue 17, section 3.2, Sensitivity, June 2003.

Chubb, J. N., Introduction to Electrostatics, pp 15-17, February 2003. Paper may be read at this website www.jci.co.uk/Electrostatics/Wshop-03.2.pdf

Glor, M., *Electrostatic Hazards in Powder Handling*, John Wiley & Sons Inc, pp 109-111, 1988.

Kong, D. Unpublished communication, 2002.

Maurer, B., "Damage To Enamel On Stirred Apparatus Triggered By Static Electricity Discharges Of Material Being Stirred," *Journal of Electrostatics,* 40 & 41, pp 517-522, 1997.

Pratt, T. H., *Electrostatic Ignitions of Fires and Explosions*, Burgoyne Inc, Marietta, Georgia, p 91, 1997.

Secker, P.E., "The Design of Simple Instruments for Measurement of Charge on Insulating Surfaces," *Journal of Electrostatics*, 1, pp 27-36, 1975.

Vos, B., Douwes, C., Ramackers, L. and Van de Weed, J. M., "Electrostatic charging of suspensions during agitation," In: Buschman, C.H. (ed) *Proceedings Of The First International Symposium On Loss Prevention And Safety Promotion In The Process Industries*, Published by Elsevier, Paper No.34, pp 381-384, 1974.

Paper TG001 - 4

A Safe and Practical Procedure to Prepare Ethyl Diazoacetate

Steve S. Y. Wang, Atul S. Kotnis, James H. Simpson,
William Merkl, Daniel Domina, San Kiang
Process Research and Development
Bristol-Myers Squibb Pharmaceutical Research Institute
One Squibb Drive, P.O. Box 191
New Brunswick, NJ 08903-0191

Prepared for Presentation at
American Institute of Chemical Engineers
2005 Spring National Meeting
39th Annual Loss Prevention Symposium
Atlanta, Georgia
April 11-13, 2005

UNPUBLISHED

A Safe and Practical Procedure to Prepare Ethyl Diazoacetate

Steve S. Y. Wang, Atul S. Kotnis, James H. Simpson,
William Merkl, Daniel Domina, San Kiang
Process Research and Development
Bristol-Myers Squibb Pharmaceutical Research Institute
One Squibb Drive, P.O. Box 191
New Brunswick, NJ 08903-0191

Abstract:

Ethyl diazoacetate (EDA) is a versatile reagent. It is commonly used as a source for metal-mediated cyclopropanations. Bulk quantities of EDA are not available primarily due to the risks associated with transporting this highly dangerous substance. EDA was synthesized through a diazotization reaction of ethyl glycine hydrochloride with sodium nitrite. The ability to control the initiation of the diazotization and the rate of this potentially explosive reaction was critical for large-scale preparation. A safe and practical procedure for the efficient preparation of EDA in toluene on a bulk scale has been developed. The process incorporates extensive thermo-chemical safety studies related to the preparation and handling of this potentially carcinogenic and explosive reagent. In addition, safety neutralization procedures for the destruction of unused EDA and waste streams were also developed for large scale processing.

1. Introduction

In connection with one of our projects, the melatonin agonist, the need arose to prepare and use large quantities of ethyl diazoacetate (EDA) for an asymmetric cyclopropanation. EDA is a versatile reagent and is commonly used as a carbene source for metal-mediated cyclopropanations. [1] EDA has also found application in the synthesis of furans[2], in ring expansions of cyclic ketones[3], in homologation of esters (including synthesis of propargylic esters) via reaction with alkyl boranes[4], and in additions to 1,4 Quinones.[5] Although EDA can be purchased for laboratory scale work, bulk quantities are not commercially available. EDA is highly toxic, a possible carcinogen[6], and is known to be prone to developing sensitivities to metals such as Cu, Sb and Zn.[7] It is an extremely explosive material and decomposes to emit toxic fumes of NO_X.[8] As a result, EDA must be handled with *extreme caution.* Therefore, a safe, practical process to prepare bulk quantities of EDA in-house was needed.

EDA was synthesized on a one-mole scale according to the literature procedure[9] and was obtained in the expected 83-88% yield. In this process, ethyl glycine hydrochloride is diazotized with sodium nitrite and sulfuric acid at -5 °C in methylene chloride (see **Scheme 1**). Upon addition of the acid, the reaction becomes highly exothermic and is complete in about 10 minutes, as indicated by the decrease in the heat evolved and consumption of the oxidant.[10] Diazotization occurs in the aqueous phase and the EDA is extracted into methylene chloride, thus protecting it from excess acid, which promotes decomposition. The reactor for this process needs to be well-vented because gaseous side-products NO, NO_2, and N_2 are formed. The EDA is isolated, after a cold sodium bicarbonate solution wash, by removal of the solvent from the organic phase at 20 mm Hg at a maximum pot temperature of 35 °C. The purified EDA was stored cold and protected from light.

Scheme 1

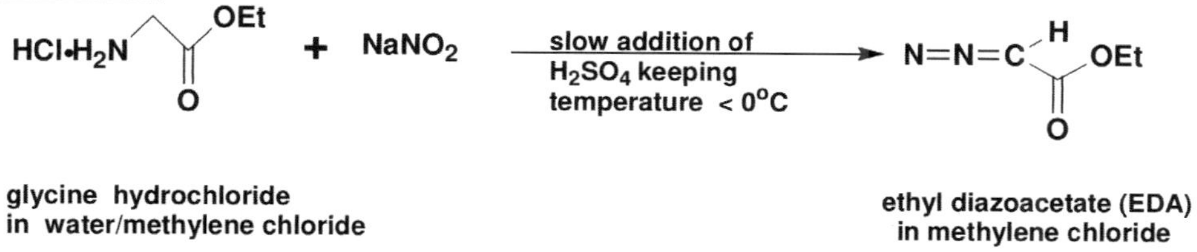

2. Thermochemistry and Reactive Hazard Evaluation

2.1 Heat of Reaction During the Formation of EDA

In order to establish parameters for safe operation of this process on scale, thermochemical safety studies of the reaction and product were performed. The reaction to produce EDA from glycine hydrochloride were performed in RC1 and CRC 90 heat flow calorimeters to evaluate the exothermicity associated with different process conditions (variation of the acid, pH and solvent ratio). The heat of the reaction was found to be ca. 280 cal/gm of glycine hydrochloride with an adiabatic temperature rise of ca. 33 °C (sodium tetraborate/ aqueous phosphoric acid/ toluene solution system). The heat evolution rate can be controlled by active cooling and addition rate of the aqueous phosphoric acid.

2.2 EDA Thermal Stability and Flammability Hazards

Laboratory scale thermal stability tests were undertaken on two levels of sensitivity for development projects in Bristol-Myers Squibb. Screening testing was conducted with Reactive System Screening Tool (RSST)[11], which has the ability to clearly identify exothermic activity along with any gas evolution. RSST studies were performed on a 5.0 mL sample of EDA (90 wt%) in methylene chloride (10 wt%). The temperature was ramped at 0.5 °C/minute to 160 °C under 300 psig of nitrogen. The self-heating rate of EDA was observed starting at 85 °C and rapidly accelerated above 100 °C with a sudden pressure surge. The RSST cell was completely shattered during this event. Due to the explosive nature of EDA and the strong exothermic activity observed in RSST experiment, an adiabatic calorimeter study for further quantifying the thermal run-away reaction was performed. The thermal data from an Accelerating Rate Calorimetry (ARC) study (performed in a stainless steel bomb with 2.15 g of EDA 90 wt% solution, phi = 3.1) were as follows:

Observed initial reaction temperature	80 °C
Maximum self-heat rate	> 500 °C/min.
Maximum pressure	1976 psig
Calculated adiabatic temperature rise	528 °C
Decomposition energy estimation	240 cal./g
Time to maximum rate at 100 °C	25 minutes

EDA has been reported to explode upon heating at a temperature > 50 °C. EDA containing <10% methylene chloride was tested and shown to be insensitive to mechanical impact from our Drop Dart Tester (Bureau of Explosives Machine). The detonation testing done by Clark[12] showed benign results for a 14:1 EDA and aldehyde mixture. However, friction testing in our study (BAM Friction Apparatus) showed positive results at a force above 40 Newtons.

2.3 Catalytic Effects of Metals on EDA Decomposition

To further define the safe process limits, we evaluated the impact on reaction streams of various metals. Solid copper sulfate was added as a 2% concentration to 5 mL of EDA in a Carius Tube Apparatus (13). As the sample temperature neared 40 °C there was a rapid run-away reaction which produced enough pressure to shatter the Carius tube (maximum recorded pressure was 25.4 bar). On the other hand, a similar experiment with rust (iron oxide) showed no evidence of self-heating. Our results are in agreement with similar thermochemical studies disclosed by a Monsanto group.[14, 15]

3. Optimized Diazotization Process For Scaling Up

With the knowledge in hand, we set about to determine conditions under which EDA could be prepared, handled and employed in our desired reaction with assurance that there were minimal (preferably, no) potential hazards to our workers and facilities. At the outset, we established two goals:

- The solvent should be changed from methylene chloride to one lighter than water to minimize the possibility of worker exposure during the aqueous workup of EDA-rich solutions. In selecting the solvent, stringent criteria for the levels of residue water, acid and unreacted glycine would need to be met.
- The reaction must be run in a manner such that there is absolute control of the exotherm and gas evolution during the diazotization reaction.

Our initial efforts focused on finding a more suitable solvent than methylene chloride. MTBE and toluene were viewed as potential replacements. MTBE proved unsuitable in the subsequent chemistry, due to incomplete cyclopropanation reactions. EDA prepared in toluene was comparable in yield and quality to that prepared in methylene chloride. Toluene was thus selected for further study.

In addition to meeting our criteria for a solvent with minimal potential for worker exposure, toluene has a greater heat capacity than methylene chloride, thus minimizing the possibility of a runaway reaction. In the context of the subsequent cyclopropanation, toluene also was superior.

While the vinyl dihydrobenzofuran could be isolated, our preference was to telescope the cyclopropanation reaction into the preparation of vinyl dihydrobenzofuran, which was isolated and stored in MTBE at ~ 5°C for over 6 weeks. MTBE was then replaced by toluene prior to cyclopropanation with EDA.[16]

With the issue of the solvent resolved, we turned our attention to the preparation of EDA under controlled conditions. As indicated above, the initial two-phase mixture of sodium nitrite and ethyl glycine hydrochoride in water/toluene (or methylene chloride) is acidic enough to permit slow diazotization before the addition of mineral acid. **The ability to control the initiation of the diazotization and the rate of this potentially explosive reaction was critical for large-scale preparation.** While slow addition of the mineral acid provided improved control, the presence of the background reaction was not acceptable.

Our first approach to a more controlled reaction was to replace the acid used from sulfuric acid to a buffered mixture of acetic acid and sodium acetate at a pH ~ 4.5. Under these conditions, nitrous acid is slowly generated, thus carrying out the diazotization (and the resulting exotherm) at a much slower rate. However, the solubility of acetic acid in methylene chloride caused substantial decomposition of the EDA, thereby lowering the overall yield of the reaction. Based on a patent report (17), sodium borate decahydrate (18) was added to toluene-overlayed aqueous solution of ethyl glycine hydrochloride and sodium nitrite to form a 3.58 M solution. The pH of this resulting solution was ~ 5.2, and at this pH the diazotization reaction was not completely suppressed. Additional sodium borate decahydrate was added until the pH remained constant at > 6.0, resulting in suppression of further diazotization. Dropwise addition of sulfuric acid then led to a completely controlled diazotization. When acid addition was stopped, the pH rose to ca. 6.0 virtually immediately and diazotization ceased. For tactical reasons (more controlled exotherm and benefits during the work-up of the reaction) sulfuric acid was later replaced with 2% phosphoric acid.

3.1 Heat Tranfer Consideration upon Scale up

Based on the heat removal capacity of the reactors to be used, we established 50 watts/L as the maximum heat of the reaction for a safe controlled process on-scale. Simulations of this reaction in a scaled up batch (125 kg of Glycine.HCl / 500 gallon Pfaudler reactor) were performed using Vesselpak (19) by calculating the addition of 2 wt/wt % phosphoric acid over time periods of 1, 1.5 and 2 hours for examining the worse case scenario and desired operational conditions. **Effects of phosphoric acid charging rate on the temperature profile are shown in the graph below right.**

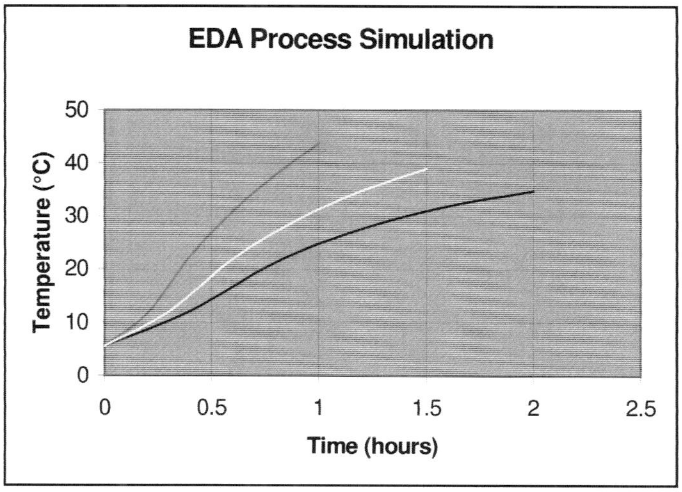

In addition to confirming the expected control of the exotherm by slower addition of the acid, these heat flow calorimeter studies demonstrated that a 2 hrs addition period generated 37 watts/L, well within the equipment cooling capacity.

In further optimizing the reaction, it was found that, under these conditions only 1.05 equivalents of sodium nitrite was required for complete reaction. The change (from 1.5 equivalents) led to a substantial reduction in evolution of the byproduct gases NO and NO_2 without decreasing the yield or quality of the DEA. To provide greater ease in processing, the temperature range of the reaction was broadened from < 0 to 5 to 20 °C (1). Addition of the acid was done at a controlled rate such that the reaction temperature was maintained between 5 to 20 °C. Under these conditions, the pH only needed to be lowered to 4.5 for complete conversion. The milder pH conditions also decreased the amount of decomposition. The overall yield of EDA by this process was > 90 mole % and was determined by GC analysis of the toluene solution (see **Scheme 2**).

Scheme 2

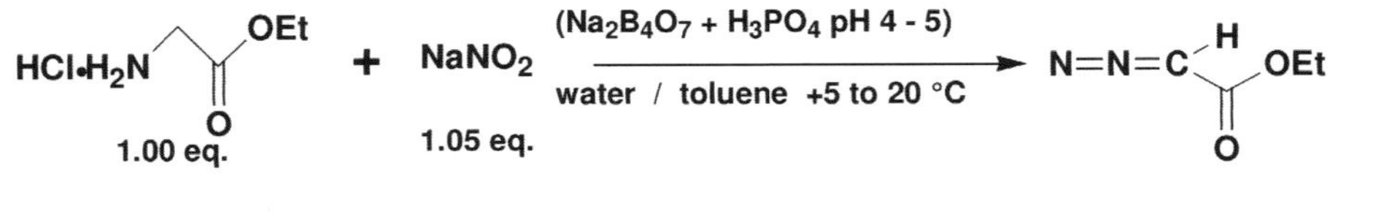

4. Neutralization of Unused EDA and Waste Streams

We required a process for the safe destruction of solutions of EDA should a process upset occur. In addition, the procedure would be needed for treating reaction streams after the cyclopropanation, since the use of excess EDA was anticipated. Here, our earlier observations about the instability of EDA in the presence of acetic acid proved of value. Slow addition of the EDA in the toluene solution to a 10 fold molar excess of 50 wt/wt% acetic acid in water provided a controlled generation of heat and gas. The experiment was carried out in a 250 ml beaker. A 25 ml solution of EDA in toluene was charged into 10 equivalents of acetic acid/water solution in 20 minutes. The chart below shows the temperature, rate of evolution and total volume of gas vs time for neutralization of EDA with acetic acid.

EDA Neutralization

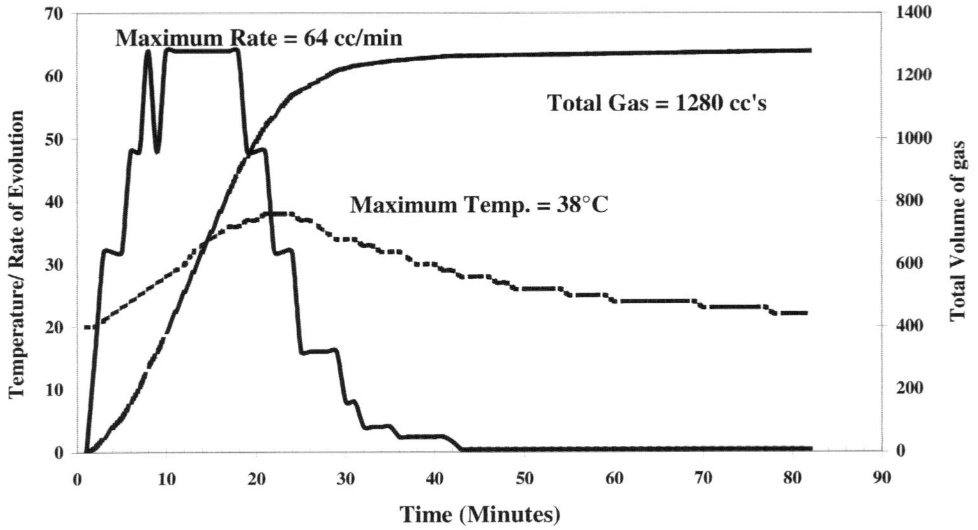

The aqueous waste streams generated in this process were transferred to holding tanks, sampled, and analyzed for EDA and /or sodium nitrite content. These streams were neutralized with 20 wt/wt% phosphoric acid by charging the aqueous waste to the phosphoric acid at a rate such that temperature and rate of gas evolution were controlled. The sodium bicarbonate waste streams were assayed for EDA and shown to be free of this material. RSST (heated at 1.0 °C/ minute to 120 °C; under 300 psig nitrogen) showed no self-heating or gas evolution. These, and the neutralized aqueous waste steams, were placed in plastic lined drums and disposed of as hazardous waste.

5. In Summary

A safe and practical procedure to prepare EDA in toluene on a bulk scale has been developed. It has been successfully telescoped "as is" in an asymmetric cylclopropanation of styrenes.

6. References and Endnotes

[1] (a) Katsuki, T.; Irie, R.; Uchida, T., Niimi, T., *Adv. Synth. Catal.* 2001, 343, 79-88. (b) Colacot, T. J., *Proc. Indian Acad. Sci.* (Chem. Sci), 2000 112, 197-207. (c) Yamada, T.; Ikeno, T., P40 at 17th International Symposium on Synthesis in Organic Chemistry, Oxford, UK

[2] Storm, D.L.; Spencer, T.A.; *Terahedron Lett.* 1865, (**1967**).

[3] (a) Warnhoff, E. W.; Tai, W.T., Can. J. Chem. **42,** 1333 (**1964**) (b) Mock, W.L.; Hartman, M.E., *Am. Soc.* **92**, 5767 (**1970**) (c) Majumdar, S.P.; Liu, H.J., *Syn. Commun.* **5**, 125 (**1975**).

[4] (a) Linke, S.; Hooz, J.; *Am. Soc.* 90, 6891 (**1968**) (b) Hooz, J.; Layton, R.B., *Can. J. Chem.* 1105 (**1972**).

[5] Fieser, L.F.; Peters, M.A.; *J. Am Chem. Soc.* **53**, 4080 (**1931**).

[6] (a) Lewis, R. J., in *Sax's Dangerous Properties of Industrial Material*, 9th Edition. (b) *Ullmann's Encyclopedia of Industrial Chemistry*, Vol. A8.

[7] (a) Noyori, R., Nozaki, H.; Takaya, H.; Moriiuti, S.; *Tetrahedron* 24, 3655 (1968). (b) Huisgen, R., Binsch, G.; Ghosez, L.; *Chem. Ber.* 97, 2628 (1964) (c) Arnold, Z.; *Chem. Commun.* 299 (1967). (d) Hammond, G.S., Kopecky, K.R., Couch, M. M., Cowan, D. O.; *J. Org. Chem.* 29, 1922 (1964).

[8] *Bretherick's Handbook of Reactive Chemical Hazards*, 6th Ed. (1999)

[9] Searle, N.E., *Org. Syntheses Coll. Vol.* **4,** 424-426.

[10] A positive starch iodide test indicates the presence of excess nitrous acid and thus completion of the reaction.

[11] Center for Chemical Process Safety, *Guidelines for Pressure Relief and Effluent Handling System*, CCPS, AIChE, New York (1998).

[12] Clark, J.D.; Shah, A. S.; Peterson, J. C.; Patelis, L.; Kersten, R.J.; Heemskerk, A. H.; *Thermochemica Acta* 2002, 386, 73.

[13] *Guidelines for Chemical Reaction Hazard Evaluation,* The Association of the British Pharmaceutical Industry, London (1989).

[14] Clark, J.D.; Shah, A. S.; Peterson, J. C.; Grogan, M.; Camden, S., *Thermochemica Acta* 2001, 367, 75.

[15] Clark, J.D.; Shah, A. S.; Peterson, J. C.; *Thermochemica Acta* 2002, 392, 77.

[16] Deshpande, R.; Yang M.; Kuehner, D.; Rao, M.; Grosso, J.; *Org. Proc. Dev.*, manuscript in preparation.

[17] Takeda, N.; Ishii, S.; Sumitomo Chemicals Co.; Ltd. Patent JP48103514, 12/25/93. CA.: 80:132837.

[17] Anhydrous sodium borate is very difficult to dissolve in water, and so the decahydrate form of it was used to minimize the amount of water.

[18] Vesselpak software program developed by Madison Technical Software Inc. The simulation was carried out under assumptions (1) 70% tank fill in Pfaudler reactor (2) 80 RPM (3) jacket syltherm flowrate 120LPM with inlet temperature of 10°C (4) heat conductivity and capacity of 100% toluene (5) calculated heat transfer coefficient for the vessel system, 143.8 W/sqM°K.

Pfizer's Global Approach For Reducing Risks In Hydrogenation Operations

Cigdem Karayigitoglu
Pfizer Inc., Corporate EHS
235 East 42nd St., MS 219-9-19
New York, NY 10017-5755
Cigdem.F.Karayigitoglu@Pfizer.com

John Holroyde
Pfizer Inc., Pfizer Global Manufacturing Ramsgate Road,
Sandwich, UK, CT13 9NJ Tel: 44 1304 646320 John.Holroyde@Pfizer.com

Prepared for Presentation at
American Institute of Chemical Engineers
2005 Spring National Meeting
39th Annual Loss Prevention Symposium
Atlanta, Georgia
April 11-13, 2005

UNPUBLISHED

Pfizer's Global Approach For Reducing Risks In Hydrogenation Operations

Cigdem Karayigitoglu
Pfizer Inc., Corporate EHS
235 East 42nd St., MS 219-9-19
New York, NY 10017-5755
Cigdem.F.Karayigitoglu@Pfizer.com

John Holroyde
Pfizer Inc., Pfizer Global Manufacturing
Ramsgate Road, Sandwich, UK, CT13 9NJ
Tel: 44 1304 646320
John.Holroyde@Pfizer.com

Abstract

Pfizer API manufacturing consists of sixteen sites around the world and represents a large manufacturing capability. This network has expanded in recent years through mergers. This community of manufacturing sites represents a wide range of practical knowledge for conducting safe API manufacturing. Taking advantage of this wealth of experience, Pfizer manufacturing operations performed a global survey of hydrogenation current practices. From the information gathered, "Hydrogen and Catalyst Handling Good Practice" guide was developed which covered the facility design and operational practices. The team that developed the "Good Practice" guide is also involved in overseeing the implementation of the recommended practices at all manufacturing sites conducting hydrogenations.

1. Introduction

In 1997, global teams were set up which meet regularly to review all aspects of the global business. The Drug Substance Management team (DSMT) was an early example of this, and its earliest statement of aims included the sharing of Best Practices across the plant network. At that time, the plant network consisted of just four API (Active Pharmaceutical Ingredient) Manufacturing sites. With the acquisition of Warner Lambert, and later Pharmacia, this network expanded to 16 API sites. The expansion of the network and the development of the business have changed the scope of the Best Practice sharing within the team, but the basic principle remained unchanged. Since 2002 evaluation of Hydrogenation Good Practices across the network has been one of the top priorities for this team.

2. Development of Pfizer Best Practice Program

In 1997, Pfizer US manufacturing and Pfizer International manufacturing organizations were unified to become Pfizer Global Manufacturing. Prior to that reorganization, the two organizations had a separate reporting structure and this did not encourage the sharing of best practices between the two groups. However, once the two groups merged, the identification of Best Practices across network was seen to be a significant benefit to the organization. Best Practice teams were established to look at a number of key areas.

The Best Practice teams looked at two facets of the business.

- Product teams looked at the manufacturing processes which were being carried out on more than one site.
- Equipment teams looked at setting guidelines for the standards Pfizer wanted to incorporate in the design and operation of its plants.

The teams developed an "overall best practices" by combining the individual best approaches practiced by each site. The scope of the teams was limited when it came to significant changes to facilities. The Best Practice documents generally provided more direction on designing new installations than guidance on retrofitting an established facility. The Best Practice teams faced a challenge in managing the implementation of the best practice across different sites.

3. Hydrogen Unit Operations Good Practice

One of the new Best Practice teams which formed was the Hydrogenation Best Practice team. The team drew up a comprehensive document covering the design of hydrogenation facilities.

Prior to the formation of the Best Practice teams, Pfizer had a hydrogen usage guidance document based on NASA Glenn Safety Manual. In 2002 "Pfizer Global Manufacturing Hydrogen Unit Operations Good Practice" guide was issued. The intent of the document was to be used as a reference during the design and installation of new unit operations, the retrofitting of particular unit operations, and the review of currently installed unit operations involving the use of hydrogen.

This guide included a discussion about the hazards of hydrogen and provided guidance on design and operation of the hydrogen systems. The main topics in the guide were

- General mechanical design concepts: This section provided guidance on process vessels and storage systems with references to the NASA Glenn Safety Manual and ASME Boiler and Pressure Vessel Code.
- Storage and delivery: This section listed key considerations for hydrogen storage and delivery systems, and provided references to codes and standards such as NFPA 50A Appendix C "Standard for Gaseous Hydrogen Systems at Consumer Sites", BCGA Code of Practice CP8, 8/1986, ISSN 0260 4809 "The Safe Storage of Gaseous Hydrogen in Seamless Cylinders and Similar Containers", OSHA 29 CFR 1910.103 "Hydrogen", ASME Boiler and Pressure Vessel Code Section VIII and ANSI B31.1-1969 for Pressure Piping.
- Hydrogenation process: This section provided a short discussion on the process safety issues such as the effect of mixing and hydrogen gas pressure on the reaction rates, catalyst handling, emergency relief design and hydrogen venting.
- Module design: This section covered work place safety, structural issues, ventilation, building emergency pressure relief and electrical system requirements with references to the codes and standards.
- Standard operating procedures: This section discussed requirements for personnel such as training on the properties of hydrogen, process equipment, protective equipments, etc.
- Emergency procedures: This section gave guidance on emergency procedures for general considerations such as area and personnel precautions, supply piping leaks, hydrogen gas leaks from cylinders and accumulated combustible gas mixture, and fire fighting techniques.

The gap analysis against this document was completed by all of the sites in the network, and gap closure programs were put in place.

4. Changes in Best Practice Approach: "Minimum" and "Contemporary" Standards

The Pharmacia acquisition in 2003 brought about an even more significant expansion in terms of the number of sites in the network. There are now 16 API sites in the network, ranging from older completely manual plants to state-of-the-art fully automated units. The diversity of the network made it more difficult to get a vision of a single Best Practice outcome.

In particular there was a need to clarify where recommendations from a Best Practice team represented what Pfizer consider to be absolute needs for safe operation, and where the recommendations are more aspirational – i.e. what would be expected when the new facilities were built.

It was decided that the recommendations would be categorized as being either "Minimum" or "Contemporary" standard.

It was not the expectation that the sites should only have to reach minimum standards, but the limitations on the ability to reach contemporary standards on all facilities in the short term had to be recognized.

The goal was to establish a uniform baseline safety for all sites. The sites with older facilities could not always implement the latest design features, but would be required to achieve an equivalent standard by the development of their existing installations.

5. Hydrogen and Catalyst Handling Good Practice

The new Best Practice model was the basis for the revised Hydrogenation Good Practice guide. Shortly before developing the new guide, an incident occurred at one of the sites. We used the investigation of this incident as an opportunity to review the revised Good Practice document. We had to question whether the document was not recognising a significant feature which needed to be included to ensure safe operation. The main observation made by the team was that the best practice document adequately covered the design requirements for safe operation. However, we also found that to be comprehensive, the document needed to include more guidance on operational standards. One example of this was in the way that we discharged spent catalyst from filters. Discharging of spent catalyst has been one of the most hazardous operations.

The common issues with catalyst discharging were

- Catalyst drying out
- Insufficient water washing (residual solvent content)
- Air exposure

Development of catalyst handling good practices has been more challenging. Preventing incidents in this area requires a more detailed approach.

The new document combines the hydrogenation good practices and the catalyst handling good practices, and contains two main sections: Hydrogenation and Catalyst Facility, and Operational Practices

Hydrogenation and Catalyst Facility: This section covers the design concepts such as hydrogen storage and piping, layout of hydrogenation and catalyst handling module, process vessels and storage systems, hydrogen feed, other vessel inlets, reactor emergency relief, hydrogen venting, detection and fire protection, control room, structure, ventilation, building emergency relief, electrical equipment classifications, catalyst preparation and charging, and spent catalyst separation.

Operational Practices: This section covers safety review, automation, procedures and training, change control, access to hazardous areas, tools and equipment, maintenance, catalyst charging, catalyst filter operation, and hydrogenator operations.

The revised Good Practice guide is now being reviewed by sites in the network. In this paper we share some of the contents of this guide, which is in the final stage of its development.

Examples from "Hydrogen and Catalyst Handling Good Practice" Guide

The following recommendations are abstracted from hydrogen and catalyst facility, and operational practices sections of the good practice guide.

The recommendations are categorized as "Minimum" (M) and "Contemporary" (C) standards.

Hydrogenation and Catalyst Facility

Hydrogen Feed

- Deliver the gas via a dedicated supply line equipped with flow measurement and a means of shutting down the Hydrogen supply in the event of excess flow (C)
- Equip the Hydrogen supply line with automated shutoff valve(s) interlocked with Hydrogen detection systems in the reaction module. Other conditions (e.g. over-pressurization, high temperature) may also activate the shutdown function as highlighted during the Process Hazards Analysis of the process (M)

Other Vessel Inlets

- To prevent migration of Hydrogen to other plant vessels or systems, equip all inlets to the hydrogenator system with double-block and bleed valves, regardless of the pressure of the source stream. Collect the bleed valve vents to a single outlet and equip this vent network with a Nitrogen bleed flow to prevent collection of Hydrogen (C)

- All the inlets to the Hydrogenators must be equipped with double isolation valves and a pressure test must be carried out between batches to ensure that the valves are not leaking. This will reduce the risk of migration of Hydrogen to the other plant vessels (M)
- All the inlet and outlets must be interlocked with the Hydrogen supply under normal operation (C)

Hydrogen Venting

- The vents from Hydrogenator and ancillary equipment should not be tied into the main plant header, to prevent the risk of migration of hydrogen and catalyst to other systems of the plant (M)
- Maintain a constant Nitrogen bleed flow into the vent piping downstream of the last control valve – preventing air migration into the vent piping thus preventing a flammable Hydrogen mixture forming in the vent line during the venting operation (M)
- A temperature probe must be installed at the end of the vent pipe to detect flame and shut down the venting (C)
- If a flame arrestor is used, adequate control measures need to be in place to prevent fouling of the tight apertures and address the variant composition of the fuel (M)
- If hydrogenator is vented through a scrubber or condenser, a program must be in-place to prevent the risk of blockages and accumulation of pyrophoric catalyst (M)

Catalyst Preparation and Charging
Selected list of recommendations

- Catalyst must be prepared in a separate tank and then gravity-transferred to the hydrogenator. This enables the catalyst to be handled in a separate area away from the hydrogenator. The slurry tank should be located directly above the hydrogenator to minimize the risk of line blockages during transfer. There must be triple block and bleed arrangements (C)
- There must be a dedicated Nitrogen purge line for the hydrogenator and catalyst slurry tank. Hydrogen must not be tied-in to the Nitrogen purge line (M)
- If it is required to open the manway hatch for any purpose, there must be a mechanism (mechanical/software sequence/others) to verify to the operator, that the hydrogenator/slurry tank is inert before opening the hatch. Permanent H2 and O2 analyzers have been installed on some sites to satisfy this requirement (M)

Spent Catalyst Separation
Selected list of recommendations
Here experience has taught us to be more prescriptive in the way our facilities are operated, and most of the standards we require are designated as minimum standards.

- Self-discharging (e.g. Schenk or Funda filters) are the preferred means of spent catalyst isolation. If these are not available, the area around filter during discharge must be fully contained (M)

- A contained system should be provided, in order to prevent spillages and leakages of catalyst during discharge (M)
- In-line filters must have water available to flush the housing internally and externally in all circumstances (normal as well as emergency). In the event that the primary filter blocks, a mechanism must be available to flush the filter housing with water. Piping design must allow for this (M)
- In-line filter housing should be fitted with drain and/or vent points to remove any residual liquid before change out (M)
- Pumps for catalyst filtration must be designed to avoid dry running operation. The pump installation should not allow for build up of liquid in the base plate as a results of leaks of the seal (M)

Operational Practices

Catalyst Filter Operations

Selected list of recommendations

- A pressure test should be conducted prior to every batch to identify leaks (M)
- Primary catalyst filter should be inerted to bring down the O2 concentration below 2% (M)
- The spent/waste catalyst must be water washed to remove residual solvent prior to discharge. The solvent-water ratio in the catalyst, at the point of discharge, must be low enough not to support combustion in the event of catalyst sparking. This ensures that the catalyst is safe to discharge (M)
- The effectiveness of water wash may be affected by channeling of water through the catalyst cake, cracking of the cake, different flow rates of water and uneven distribution of cake on the filter. Measure the solvent-water ratio in the discharged catalyst to ensure the effectiveness of water wash. This should be repeated for any changes to the process or any new equipment (M)
- Catalyst should be discharged as wet slurry. This eliminates risk of fire as a consequence of the pyrophoric nature of the catalyst (C)
- The containers used for the storage of spent catalyst must have a vent to allow for release of residual Hydrogen, which can be entrained within the spent catalyst (M)
- If the filter is in the same area as the hydrogenator then the discharge must not take place while hydrogenation is in progress (M)
- A nitrogen atmosphere should be maintained at all times in the filter between batches of a campaign. This maintains an inert atmosphere when the filter is not in use but may still contain residual catalyst (M)
- A mechanism should be in place to monitor the filter performance, without opening the filter, which will trigger the changing of the filter. Typically, ΔP is a system that many sites use (C)

- In-line filter elements must be adequately water washed and wetted before change out. In-line filters should not be Nitrogen blown to remove residual water. Residual water should be drained from the filter before change out (M)
- Spent in-line filters must immediately be immersed in water for disposal (M)

Picture 1: A wet catalyst discharge arrangement as an example of good practice

6. Summary

This paper gives an overview of Pfizer's Hydrogenation and Catalyst Handling Good Practice guide including its development and implementation. The focus of the earlier hydrogenation good practice guide was more towards the design of the facilities. While this format was good for the design of the "state of the art" new facilities, it did not always provide the necessary guidance for existing operations in plants which were developed at different times to different guidelines. It became necessary to categorize the recommendations to the minimum and contemporary standards where the sites were at least required to meet the minimum standards and work towards the contemporary ones. We also found it necessary to add more guidance on operational practice to the guide. To make better use of good practice guide the document is converted to a shorter and more user-friendly format with a focus on operational details especially in catalyst handling operations.

Effectively Managing Change

Shane Thornton
Ciba® Expert Services
Ciba Specialty Chemicals
P.O. Box 113
McIntosh, AL 36653
(251) 436-3772
shane.thornton@cibasc.com

Prepared for Presentation at
American Institute of Chemical Engineers
2005 Spring National Meeting
39th Annual Loss Prevention Symposium
Atlanta, Georgia
April 11-13, 2005

UNPUBLISHED

Effectively Managing Change

Shane Thornton
Ciba® Expert Services
Ciba Specialty Chemicals
P.O. Box 113
McIntosh, AL 36653
(251) 436-3772
shane.thornton@cibasc.com

Abstract

Pharmaceuticals are one of the most regulated industries in the United States. Efficient regulatory compliance during product development, process introduction, and commercial production is essential to ensure minimum time to market, with the most cost effective compliance and production strategy. One of the critical regulatory requirements, compliments of 29 CFR 1910.119, is "Management of Change," or MOC. "Paper-based" MOC systems are quite inefficient and can easily bottleneck process change authorization and assignment completion. With paper systems, there is also an increased risk of changes being implemented without proper documentation or authorization by management. In contrast, an electronic management of change system will allow the archival of required information, and the assignment of relevant tasks such as updating operating procedures, process hazard analysis, training, and so on. An electronic management of change system should notify management of recommended process changes, and provide a mechanism for review and approval of the change. The use of an electronic management of change system not only enhances the existing business practice, but also dramatically increases the efficiency and time management of personnel.

1. Introduction

The Occupational Safety and Health Administration (OSHA) developed the Process Safety Management (PSM) standard (29 CFR 1910.119) as a direct result of several catastrophic incidents in the mid 1980's through the early 1990's that resulted in approximately 2,050 deaths and 300 injuries. More specifically, OSHA determined that there were no specific regulations in place for preventing or minimizing the consequences of catastrophic releases of toxic, reactive, flammable, or explosive chemicals.[1] The OSHA standard defines the chemicals covered in a PSM process. Appendix A of 29 CFR 1910.119 lists specific toxic and reactive chemicals and threshold quantities, above which regulatory coverage is invoked. In addition, flammable liquids and gases, as defined in 29 CFR 1910.1200(c), are covered in quantities in excess of 10,000 pounds. However, experience in both the specialty chemical and pharmaceutical industries has shown that the elements of the PSM standard define a minimum "best practice" even for chemical processes that are not "covered processes" as defined above.

---

[1] 29CFR §1910.119 "Purpose"

The PSM standard consists of the following fourteen elements: employee participation, process safety information, process hazard analysis, operating procedures, training, contractors, pre-startup safety review (PSSR), mechanical integrity, hot work permit, management of change (MOC), incident investigation, emergency planning and response, compliance audits, and trade secrets. Eleven of the fourteen elements were developed to ensure that "covered processes" are thoroughly reviewed and that measures are in place to minimize the probability and severity of the hazards, to manage and protect employees and contractors, and to manage the continued integrity of processing equipment. The three exceptions, incident investigation, MOC and PSSR were developed to ensure that the inherent safety measures initially established for the covered processes were maintained as the process evolved over the years.

Establishing an effective incident investigation system is required to ensure that close calls, minor incidents, and major incidents are adequately investigated and that proper corrective measures are implemented to prevent reoccurrence. Similarly, the MOC and PSSR systems provide a review and approval structure for improving safety, quality, and the overall cost efficiencies of the process. For any change not considered a "Replacement in Kind," the MOC/PSSR systems must be initiated, with a definitive scope of change, expected benefits, technical basis, and impact on health and safety of affected personnel. Once approved, all of the pertinent process safety information, process hazards analysis, training requirements, impact on operating procedures, and other critical elements are reviewed to ensure the proposed change will not negatively impact operations. Once the analysis is complete, final management approval is required to facilitate implementation. Typical changes can range from the substitution of a reaction reagent source to changes in the material of construction of process piping, each having a potential impact on failures in the process.

2. Electronic Management of Change Systems

Currently, "paper-based" systems are used extensively to facilitate documentation and approval of a change. Unfortunately, these systems are inherently inefficient and are themselves difficult to modify. In today's competitive environment, change is no longer the exception but more "the rule". Managers are challenged with balancing optimal performance of the process with reduced headcount, complying with environmental health and safety guidelines, and maintaining the integrity of process equipment while at the same time striving for increased production targets. While trying to manage multiple priorities, it is inevitable that administrative tasks suffer and documentation goes lacking. In addition, the MOC work processes are also constantly evolving to ensure optimal performance (faster – better – cheaper). When these changes in system complexity are coupled with the changes in pharmaceutical manufacturing processes, "paper-based" systems quickly become unmanageable.

Currently, there are several electronic management of change systems available. These systems range in price and performance. The most common systems are "modules" associated with existing maintenance or quality systems. These systems appear to be cost effective, but the architecture is not easily adaptable to customer-specific needs. Not only is customization problematic, but most systems do not integrate tracking and notification features that are essential to managing change. On the other end of the spectrum, more comprehensive and costly systems are available. Unfortunately, it is often difficult to customize these systems in order for them to integrate into existing business work processes. In fact, we faced this dilemma as an organization. Consequently, our model identified the following key elements:

- Documentation
- Archiving
- Tracking
- Approval/rejection
- Customizable business process
- Reporting
- Notification
- Cost effectiveness

Our goal was to enhance the existing business practice, to increase the effectiveness and efficiency of change management, and to improve time management of personnel. We also sought a system that was configurable in order to adapt to changing work processes. The solution that I am describing is a system that is engineered and built by "practitioners" who are not only governed by this regulation but have the burden of complying in a cost effective manner.

3. System Configuration

So you may ask, where do we start? To provide an effective mechanism for tracking MOC's, a numbering system should be utilized. This is important later in the process when trying to retrieve archived MOC's efficiently. The example in **Figure 1** demonstrates a system that utilizes the process unit information, a date, and a three digit numbering sequence to identify each MOC. This method is a way to associate change with a process unit and to provide a mechanism to easily retrieve unit specific changes as a part of a process hazards analysis five-year review for example.

Figure 1: MOC Numbering Scheme

SS = Site Code	YYYY-Mon-DD = Year, Month & Day
DD= Division Code	Seq = Three Digit Sequence Number
UU = Unit Code	

SS-DD-UU-YYYY-Mon-DD-Seq, for example MC-AD-AO-1999-Nov-10-002

To ensure the information is secure and approval can be recorded electronically, each user should have a valid login and password assigned to access the application. This is the basic framework for establishing roles in a system and configuring user profiles. Once the user profile is created, the user can be restricted with regard to the type of information they can view, print, and record. Certain features should only be available to users assigned to a specific Role within the system. **Figure 2** is an example of account types and permissions.

Figure 2: MOC User Profiles

Role	Description
Originator	Basic level of access to the system. Can initiate a MOC and forward to the Approver for review/approval.
Approver/ Owner	Approves/Rejects all MOC's. Manages MOC to completion.
Technical Representative (Tech Rep)	Works in conjunction with the approver to affirm the initial approval of an MOC request and assists in managing the MOC to completion.
Environmental	Can be assigned Environmental Review
Safety	Can be assigned Safety Review

4. Application Workflow Example

There are many approaches for establishing business rules for a basic MOC workflow. The most basic rule is to have a strategy that provides a method for documenting change management, approvals, comments, attachments, and task completion and that provides a means to track and to notify users of these events. Once a MOC is created, the approver is notified via email. The approver would then review the MOC and approve or reject the change. If the MOC is approved, a Technical Representative is notified via email to review the technical specification of the change and to document approval. Once the Approver and Tech Rep have both approved the MOC, the next step is to assign Responsible Parties (RP) for the variety of tasks that may need to be completed prior to implementation of the requested change. As assignments are made, each RP is notified via email of their assignment. If it is determined as a part of the review that it is not feasible to implement the change, all parties associated with the change are notified via e-mail of the rejection, regardless of whether they have completed their assignment prior to the rejection or not. An effective Management of Change system will complement your business practice in place to manage change. **Figure 3** illustrates the basic workflow approach.

Figure 3

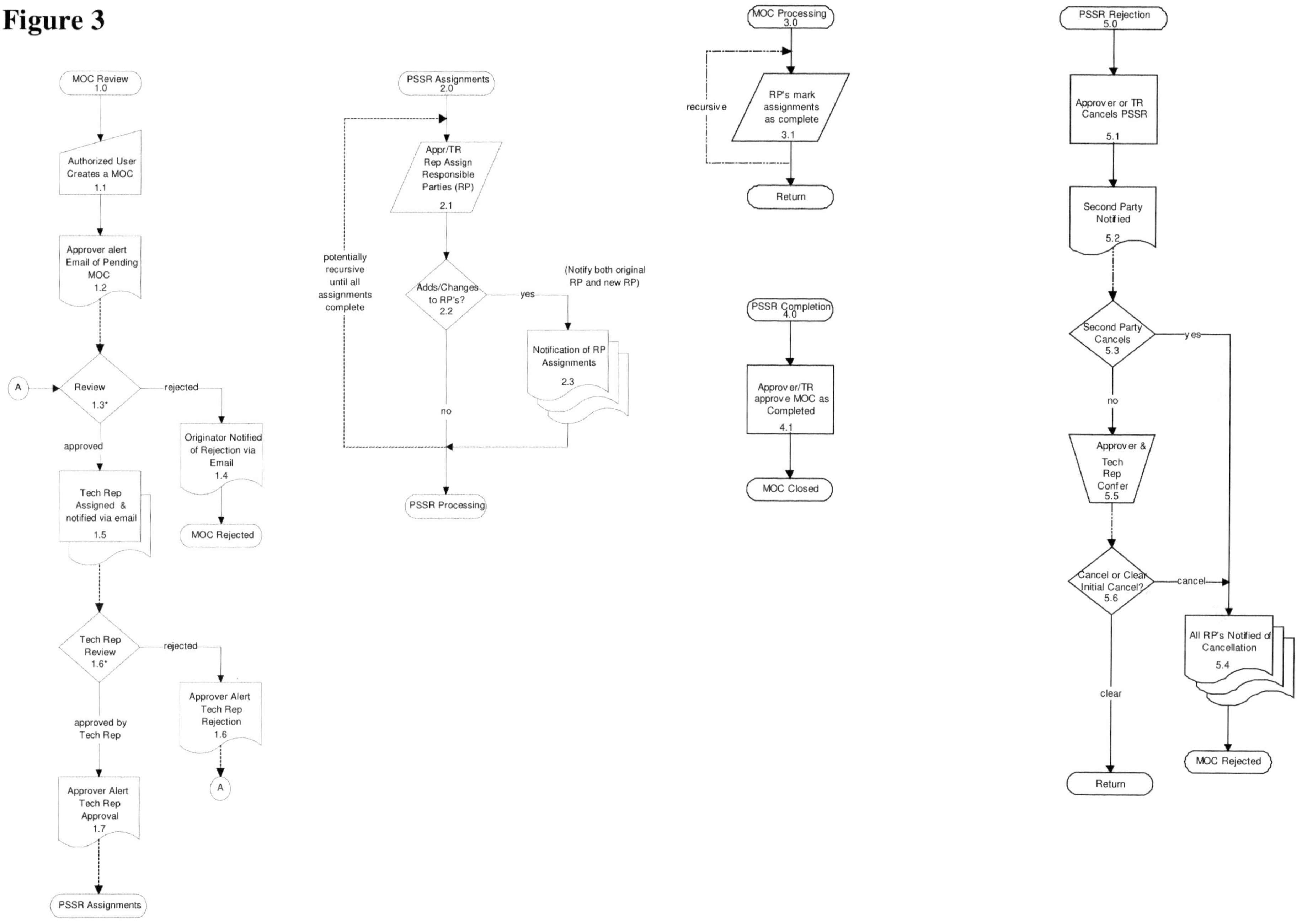

MOC Review 1.0
Authorized User Creates a MOC 1.1
Approver alert Email of Pending MOC 1.2
Review 1.3*
A
rejected
approved
Originator Notified of Rejection via Email 1.4
MOC Rejected
Tech Rep Assigned & notified via email 1.5
Tech Rep Review 1.6*
rejected
Approver Alert Tech Rep Rejection 1.6
A
approved by Tech Rep
Approver Alert Tech Rep Approval 1.7
PSSR Assignments
PSSR Assignments 2.0
Appr/TR Rep Assign Responsible Parties (RP) 2.1
potentially recursive until all assignments complete
Adds/Changes to RP's? 2.2
yes
(Notify both original RP and new RP)
Notification of RP Assignments 2.3
no
PSSR Processing
MOC Processing 3.0
recursive
RP's mark assignments as complete 3.1
Return
PSSR Completion 4.0
Approver/TR approve MOC as Completed 4.1
MOC Closed
PSSR Rejection 5.0
Approver or TR Cancels PSSR 5.1
Second Party Notified 5.2
Second Party Cancels 5.3
yes
no
Approver & Tech Rep Confer 5.5
Cancel or Clear Initial Cancel? 5.6
cancel
clear
All RP's Notified of Cancellation 5.4
MOC Rejected
Return

5. Conclusion

The MOC requirements of 29 CFR 1910.119 dictate that all changes are evaluated to determine if the modification is considered a "replacement in kind" or a change that should be further evaluated for potential impact to the safety of the process. Additionally, the standard requires that a review of the proposed change's impact on the process, operating procedures, process hazards analysis, and a review of the change with affected personnel be completed before management gives final approval for implementation/start-up of the change. However, an effective management of change system will include the following non-regulated elements, and should be considered when purchasing or developing a system:

- Customizability
- Efficiency
- User Friendliness
- Historical Archiving
- Pre-Screen Changes
- Data Retrieval
- Electronic Notification
- Comment Capabilities
- Task Management
- Underlying Business Process
- Prioritize Changes
- Reporting Functions

In reality, a "paper-based" system can be used to manage this process, but this approach typically results in incomplete documentation and the time-consuming process of trying to manage the paper work, track assignments, and effectively manage a change from start to finish. This coupled with the increasing demands of safely and efficiently managing our processes, cost, personnel, and equipment often results in incomplete documentation and ultimately an increased probability of a catastrophic failure leading to down time and increased vulnerability to scrutiny by regulatory agencies. The most effective management of change process utilizes an electronic system that compliments your existing change process and provides management with the tools to efficiently manage change, approvals, assignments, status, and documentation throughout the process and ultimately maintaining the inherent safety of your covered processes.

Back to the Basics In Dust Explosions

David C. Kirby
Baker Engineering and Risk Consultants, Inc.
3330 Oakwell Court, Suite 100
San Antonio, TX 78218
(210) 824-5960

Prepared for Presentation at
American Institute of Chemical Engineers
2005 Spring National Meeting
39th Annual Loss Prevention Symposium
Atlanta, Georgia
April 11-13, 2005

February, 2005

UNPUBLISHED

Back to the Basics In Dust Explosions

David C. Kirby
Baker Engineering and Risk Consultants, Inc.
3330 Oakwell Court, Suite 100
San Antonio, TX 78218
(210) 824-5960

Abstract

A potential dust explosion hazard exists any time a combustible dust is stored, handled or processed. Once that fact is recognized, steps can be taken to protect the workers and property at the plant where such operations are conducted. The purpose of this paper is to provide a basic overview of key dust explosion safety considerations. Qualitative guidance is provided on housekeeping and equipment design versus electrical classification.

A dust explosion can occur when a combustible dust is suspended in air at a suitable concentration in the presence of an ignition source. As little as a few hundredths of an inch of combustible dust accumulation, when dispersed in air, can support the propagation of a deflagration. An initiating event is required to disperse the dust accumulation. The resulting dust-air deflagration can develop damaging overpressures depending on the degree of confinement and congestion present.

1. Introduction

The purpose of this paper is to provide a basic overview of key dust explosion safety considerations. An "explosion" within the context of this paper refers to rapid combustion resulting in a pressure rise and the actual or potential damage to the enclosure within which the explosion occurs or the surrounding area. This paper does not deal with the dust explosion safety considerations unique to the grain or metals processing industry.

The focus of the paper is on secondary dust explosions. The dust explosions that have caused widespread loss of life and property damage are almost always secondary explosions. A secondary dust explosion is one that is triggered by some other initiating event. The most common example of a secondary dust explosion is the dispersion and ignition of accumulated dust within a building due to a primary explosion. The primary explosion may be a flammable gas or combustible dust explosion within an enclosed space or piece of equipment. The rupture of a piece of equipment from an internal dust explosion can also disperse unburned dust into a building and directly initiate a secondary dust explosion. A secondary dust explosion can also be caused by mechanically dispersing accumulated dust in the presence of an ignition source. The initiating event in this case is the local dust explosion involving the mechanically dispersed dust, and the larger secondary dust explosion involves the remainder of the dust accumulation within the building. Very thin accumulations of dusts within confined volumes can be both dispersed and ignited by electric ignition sources.

2. Background

Not much has really changed in the basics of dust explosion safety in the past 45 years. The first paragraph of the Combustible Dust chapter of the first edition (1959) of the Factory Mutual Handbook Of Industrial Loss Prevention states:

"Dust explosions are possible wherever materials that give off combustible dusts are stored, handled, or produced. Basic safeguards are:

1. *Prevent accumulations of dust outside equipment. Use dust tight equipment as much as possible. Prevent escape of dust from open hoppers and transfer points by providing local dust-collection systems. Maintain high housekeeping standards.*
2. *Eliminate ignition sources.*
3. *Isolate operations from important plant areas by distance or by construction.*
4. *Provide explosion vents on equipment and buildings.*
5. *Provide adequate fire protection.*
6. *Educate employees in the hazards of combustible dusts and in the necessary precautions."*

This guidance still captures the most important dust explosion safety considerations. Each of these items is covered separately in the following sections. It should be recognized that dust accumulation prevention and housekeeping (item 1 on the list above) is the most important consideration with respect to dust explosions. Preventing the accumulation of combustible dusts virtually eliminates the potential for any significant secondary dust explosion. This item is discussed below in connection with recognizing and assessing the level of a dust explosion hazard.

3. Hazard Recognition and Assessment

There are three basic steps involved in recognizing and assessing a dust explosion hazard:

1. Determine if the dust is combustible,
2. Determine if a dust explosion hazard exists, and
3. Assess the potential consequences.

Each of these is discussed separately below.

3.1 Determine if the Dust is Combustible

The first step is to determine if the dust of interest is combustible. If the dust is combustible, then it can cause an explosion if it is dispersed in air with an ignition source present. The Material Safety Data Sheet (MSDS) will generally identify if a material is a combustible dust. Generally, if the solid is combustible in air, then a dust generated from the material is a combustible dust.

NFPA 68[1] defines a dust in this regard as a material with a particle size less than 420 microns (40 mesh). Particle sizes greater than 420 microns could contribute to a dust explosion, but there is a very low probability that a dust explosion would be initiated if all the dust particles were this large. Furthermore, it would be difficult to generate damaging overpressures with particles this large. However, a dust explosion hazard may still be present even if a very large fraction of the particles are larger than 420 microns. A dust explosion hazard can develop in this case if the larger particles can be reduced in size during handling (e.g., pneumatic conveying) or if the small particles are separated (i.e., classified) during handling or processing. For example, dust accumulations in elevated sections of the plant tend to be the smaller size fractions since these particles are more readily transported via air currents within the plant. Similarly, smaller particles will tend to accumulate in a dust collector and create a dust explosion hazard. Hence, a dust explosion hazard can be readily created even if a very large fraction of the material being stored, handled, or produced is larger than 420 microns.

Dusts can be divided into two groups: dusts that are able to ignite and propagate flame (Group A) and dusts that do not propagate flame (Group B). The ASTM E 1226 20-liter test bomb with varying strength igniters is that standard test employed in the USA. The material is not considered to be combustible if tests are negative using two 5 kJ pyrotechnic igniters. The British recommend the "vertical tube" apparatus as a first screening test. If the dust is shown to be combustible with the 10 kJ ignition source, it is sometimes beneficial to further quantify the minimum ignition energy (MIE). If the MIE is greater than 100 J, there is a low probability of ignition from common ignition sources such as static electricity, frictional sparks, or overheated electrical equipment. The dust is "ignition sensitive" and very susceptible to static sparks and weak ignition sources if the MIE is less than 10 mJ. Hence, the MIE can be aid in estimating the probability of a dust explosion. It should be understood that a high MIE dust could still produce a catastrophic explosion.

3.2 Determine if a Dust Explosion Hazard Exists

A determination should be made if an actual dust explosion hazard exists once it has been determined that a combustible dust is present. A dust explosion hazard exists if enough of the dust is available to support a dust explosion.

The Minimum Explosive Concentration (MEC), defined per the new ASTM 1515 Test Methodology, is the concentration of dust in a uniform dispersion with air that will double the initial pressure within the test apparatus when ignited. The final pressure would therefore be approximately 15 psig if the explosion were initiated at atmospheric pressure. This may seem non-conservative since 15 psig is a significant pressure. However, once the contribution from the ignition source is accounted for, this yields roughly the dust concentration required to support propagation of the flame through a large enclosure. Tests conducted with the old Modified Hartman Apparatus generally give more conservative values (lower MEC values). In some cases, this is because the dust within the apparatus was not uniformly distributed. In some of Hartman tests, a "go" was defined by the rupture of a piece of paper at the top of the apparatus, whereas in some tests a plastic lid was placed on the apparatus, and in later tests the equipment was evacuated and pressure measured with an aim of achieving 15 psig.

It is instructive to consider how little dust is required to achieve the MEC. The MEC for coal dust[2] is about 60 g/m^3 (0.06 oz/ft^3). The coal mass required to establish this concentration within an enclosure with a 5 m by 5 m cross section and a height of 3 m room would be 4.5 kg (9.9 lb_m). The density of solid coal is about 1.3 g/cm^3. Assuming a layer density equivalent to 50% of the solid, the layer volume required to achieve the MEC within this enclosure would be just under 7 liters (1.8 gallons), corresponding to a uniform layer thickness of 0.3 mm (0.01 in.) across the floor. The optimum concentration of coal dust[2] is about 200 g/m^3, so that the layer thickness required to yield the optimum concentration would be roughly three times deeper (i.e., 0.9 mm or 0.03 in.). Hence, the thickness of the dust layer required to achieve the MEC is quite small. The presence of even a relatively thin layer of combustible dust over any significant area therefore represents a dust explosion hazard.

As another example, consider a layer of fine polyethylene dust within a 1 m x 1 m duct. The MEC for this dust is in the range of 50 g/m^3 and the density of a dust layer would be roughly 0.4 g/cm^3. The dust mass required to establish the MEC concentration within a 1-meter long section of duct would be 50 grams. The corresponding uniform layer thickness across the duct bottom would only be 0.1 mm (0.005 in.). This is a mere dusting that can cling to the walls and ceiling of the duct. The presence of larger dust accumulations at duct bends and intersections could support the development of significant pressures and result in duct rupture. A damaging secondary explosion could then occur if dust accumulations exist outside of the duct.

It may be helpful to consider the flammable gas volume required to give the same amount of energy that would be released via combustion of a dust layer. This comparison is valid since both combustion mechanisms take place at similar energy release (i.e., as opposed to high explosives such as TNT). For example, a square foot layer of the polyethylene dust discussed above (18,000 Btu/lb_m) with a depth of 1/8 of an inch would yield roughly the same combustion energy as 2 ft^3 of propane (i.e., 50 ft^3 of propane-air mixture at the stoichiometric concentration).

It is true that a combustible dust, even if present in sufficient quantity to support combustion, must be dispersed in air or another oxidant and ignited in order to cause a dust explosion. However, dispersion can occur via many routes and ignition sources can never be completely eliminated. Furthermore, an ignition source can disperse and ignite the dust at the same time. Some years ago, the Bureau of Mines conducted MEC tests on lycopodium in the Modified Hartman Apparatus in a series of "round robin" testing at the request of the International Electrochemical Committee (IEC).[3] In one test, the experimenter dispersed 1 gram of lycopodium in the 2.7-liter test vessel, passed a spark through the airborne dust, and failed to get ignition. He then allowed the dust to settle on the surfaces of the vessel, and a day later passed a spark of the same energy, achieving ignition. They found the results to be repeatable. Others have reported similar observations.[4] Hence, it should be assumed that a dust explosion hazard exists if enough of the dust is available to support a dust explosion.

3.3 Assess the Potential Consequences

Risk is calculated as the *consequences* from an incident multiplied by *probability* (frequency) of its occurrence. Risk cannot be properly managed without an understanding of the potential consequences. The resources devoted to prevention and/or mitigation of an event can be limited if the potential consequences are small. Conversely, if potential consequences are large, a major prevention and/or mitigation effort may be warranted.

A combustible dust-air mixture at the optimum dust concentration will generate a maximum pressure (P_{max}) of about 10 times the initial pressure if totally confined and will generate a fireball about 10 times the initial mixture volume if not totally confined. There are exceptions to these rules of thumb. Metal dusts can produce a P_{max} more than 10 times the initial pressure and, as a class, organic dusts generally produce a P_{max} of 7 or 8 times the initial pressure. A dust-air mixture with a concentration above the optimum concentration in a confined volume is oxidant-limited, and P_{max} decreases since the extra dust acts as a heat sink; note that it takes an enormous amount of dust to raise the mixture concentration above the upper explosive limit. Without confinement, unburned dust can be expelled beyond the fireball and combust, increasing the fireball volume. The maximum explosion pressure developed with un-vented interconnected or elongated equipment can exceed P_{max} due pressure piling and, in extreme cases, due to a deflagration to detonation transition. However, the values of 10 times the initial pressure with complete confinement and 10 times the initial volume without complete confinement are good rules of thumb to develop an appreciation of the potential consequences from a dust explosion.

An explosion can obviously occur if only a portion of the enclosure contains a combustible dust-air mixture (i.e., a "partial volume deflagration"), but the maximum explosion pressure is a function of the fuel loading in this case. A rule of thumb for consequence assessment with a partial volume deflagration is that there is a linear relationship between pressure generation and fuel loading. For example, if P_{max} is 100 psig and 2% of the enclosure volume contains an optimum dust-air mixture, then the maximum explosion pressure would be approximately 2 psig. It should be understood that this loading is quasi-static (i.e., of long duration) and several psig of internal pressure is sufficient to severely damage structures employing conventional construction. Equipping the enclosure with explosion vents can limit the explosion pressure such that the structure is not significantly damaged. However, the presence of congestion and partial confinement in the vicinity of the dust cloud will act to accelerate the flame and produce a blast wave yielding locally elevated loads (i.e., in the same fashion as a vapor cloud explosion) even in the average pressure within the enclosure is low. While a dust accumulation covering a few square feet within in a large enclosure is only a flash-fire hazard (i.e., no blast pressure), the explosion intensity increases as the dust accumulation area increases with respect to the enclosure volume. The flow of gas towards explosion vents or other openings will act to disperse additional dust. Large areas of dust accumulation, even in large structures, can result in severe dust explosion releasing large amounts of energy at a high rate and causing significant damage.

4. Separation / Isolation

The NFPA 654 standard states[5]: "*When separation is used to limit the fire or dust explosion hazardous area, the hazardous area shall include areas where dust accumulations exceed 1/32 inch or where dust clouds of a hazardous concentration exists.*" The 1/32-inch is based on an optimum mixture calculated for a room-size enclosure with a dust having a settled bulk density of 75 lb_m/ft^3. Settled bulk layer density varies widely between types of dust, ranging from about 15 lbs/ft^3 (or even less for some materials such as lint or flock) to about 75 lb_m/ft^3, and NFPA

654 allows the 1/32-inch value to be adjusted for the layer density. Considering fine polyethylene dust as an example (bulk layer density of 25 lb_m/ft^3), the hazardous area would be defined as those locations where dust accumulations exceed 3/32 of an inch.

A hazardous area defined this manner would yield an optimum dust concentration for a typical room-size enclosure if the dust were uniformly dispersed and mixed with air. This is obviously not a conservative value. The NFPA 654 standard hence implicitly recognizes that dust accumulations will usually not be uniformly distributed within an enclosure, and that the dust will probably not be dispersed completely by an initiating event. The NFPA 654 appendix suggests that if more than 5% of the building area has a hazardous accumulation, precautions should be taken to address the dust explosion hazard. Five percent is roughly equivalent to about the area of overhead bar joists and beams in a typical building. Accumulations in elevated spaces can be knocked down and readily dispersed by an initiating event are hence are more hazardous than accumulations at floor level, which are much more difficult to disperse.

This emphasizes the extreme importance of preventing dust accumulation by good design practice and good housekeeping in preventing a catastrophic incident. Good design practices include enclosing and sealing open bar joists and beams, or putting lightweight shields on beams, etc. that are at an angle greater than the angle of repose of the dust. The NFPA 654 standard specifies that: "*spaces inaccessible to housekeeping shall be sealed in order to prevent dust accumulation.*" Good design practice also includes the use of dust tight equipment as much as possible, and preventing escape of dust from open hoppers and transfer points by providing local dust-collection systems.

The NFPA 654 standard also specifies that hazardous areas include those where a dust cloud of hazardous concentration exists. This applies primarily to equipment interior spaces. Dust clouds within the combustible concentration range are dense, and visibility through such a cloud is limited to a few feet. A rule of thumb is that if a 100-watt light bulb is visible at 10 feet, then the dust concentration in the cloud is below the MEC.

With respect to separation and/or isolation of hazardous areas, the NFPA 654 standard states: "*The required separation distance between the hazardous and surrounding exposures shall be determined by an engineering evaluation that addresses the properties of the materials, the type of operation, the amount of material likely to be present outside the process equipment, the building design, and the nature of surrounding exposures. In no case should the separation distance be less than 30 feet.*" Operations involving combustible dust should therefore be separated or isolated from important plant areas by distance or by construction. In dusty operations, both separation distance and construction (a separate building) are good practice. A better approach is to house the operation in an enclosure without walls (or roof, if possible) and provide a separation distance of at least 50 feet from facilities to be protected. The potential for an unconfined dust explosion to generate damaging blast loads beyond the explosion site is low. Where the process exposes adjacent lines or facilities, or where dust can escape from the process, it is critical to employ good housekeeping along with a combination of separation distance (between hazardous areas) and construction (walls) to isolate the operation. It should be recognized that walls constructed to isolate a hazardous area, if they unable to withstand the explosion loads developed due a dust explosion within that area, could actually increase the severity of the event.

Often overlooked with respect to hazardous area isolation are adjacent areas where fugitive dusts from processing or handling operations can accumulate over a long period of time. Examples of these areas include suspended ceilings, mezzanines, above large ducts, and in difficult-to-reach areas behind large pieces of equipment close to walls or the ceiling. Suspended ceilings should be sealed, cleaned regularly or eliminated. Mezzanines should be sealed off from dust-generating equipment or cleaned regularly. Another consideration sometimes overlooked is that exhaust ventilation ducts for dust collection systems and air streams carrying fugitive dust should have adequate velocities to prevent the build-up of combustible dust within the ducts.

5. Explosion Venting/Protection

The direction given by the 1954 FM Handbook to "*Provide explosion <u>vents</u> on equipment and buildings*" should be expanded to "*Provide explosion <u>protection</u> on equipment and buildings*". The NFPA standard on explosion prevention systems[6] (i.e., NFPA 69) gives a range of options for protecting against explosions, one of which is to provide explosion venting. Others include oxidant control (i.e., inerting), fuel control and deflagration (explosion) pressure containment. It is seldom practical to control dust accumulations within equipment to a safe level (less than 25% of the MEC) as the means of explosion protection. The need for providing explosion venting on a building is a function of housekeeping, with strict housekeeping reducing the need for venting. However, even with excellent housekeeping, it is always a good practice to employ lightweight damage-limiting construction so that the entire building acts as an explosion vent if a dust explosion occurs.

6. Eliminate Ignition Sources

Protection against primary explosions within equipment that can result in ejection of a fireball into a building is probably the most important potential ignition source that can be eliminated. The NFPA 69 standard should be followed to protect against equipment explosions. If explosion venting is selected as the protection scheme, then venting to the building exterior should be assured. Special consideration should be given to explosion isolation between pieces of equipment so that the explosion venting can properly function.

Other ignition sources that should be controlled include cutting and welding, smoking and hot surfaces. Hot surfaces should be controlled to below the lower of 80% of the dust layer ignition temperature (expressed in °C) or 165°C. Where hot surfaces (e.g., compressors, uninsulated high pressure steam piping, heaters, shrink-wrap resistance heaters, etc.) cannot be avoided, it is imperative that such ignition sources be isolated by good housekeeping or barriers. Electrostatic and stray-current electrical potential ignition sources should be eliminated by bonding and grounding.

Electrical classification (i.e., minimizing the potential for electrical equipment to act as an ignition source) is a function of housekeeping. For those areas defined by NFPA 654 as "hazardous areas" (see previous discussion) the decision is clear-cut. The area must be classified, per Article 502 or Article 503 of NFPA 70, *National Electrical Code* (NEC) as applicable. Per the NEC, the classification is either Division I, if the accumulation exists continuously or frequently, or Division II if the area has infrequent accumulations (e.g., infrequent fugitive emissions from equipment). A hazardous area containing conductive dust must be classified as Division I. The classification area must extend to a solid wall or, per NFPA 654, at least 30 feet beyond the hazardous operation if space is being used to separate classified and unclassified areas. Presumably, if (1) the dust is non-conductive, (2) the hazardous area is Division I, and (3) the space separation is dividing classified zones, then a Division II area of at least 30 ft. would exist around the Division I operation. Although NFPA 654 is not clear on this issue, an exception would be a point source hazard, (e.g. filling a small open-topped container from a closed piping system with excellent housekeeping throughout the surrounding area). In this case an evaluation could be used to determine the extent of the Division II area, following guidelines of Table 1.

Table 1 provides a qualitative guideline for housekeeping verses electrical classification for a typical Group G plastic dust; this table is a proposed appendix for the 2006 edition of NFPA 654. The term "dust-tight" was used in this table, rather than something similar to "sealed to minimize penetration of dust", since "dust-tight" is an NEC term. The NEC defined dust-tight as "*constructed so that dust will not enter the enclosure under specified test conditions*". The NEC states: "*Where specifically permitted in Articles 501 through 503 general-purpose equipment or equipment in general-purpose enclosures shall be permitted to be installed in Division 2 locations if the equipment does not constitute a source of ignition under normal operating conditions.*" Class II is more restrictive than Class I in this regard. The NEC states that, for Class II, Division 2: "*Wireways, fittings, and boxes in which tapes, joints, or terminal connections are made shall be designed to minimize entrance of dust, be provided with covers to prevent escape of sparks, and shall have no openings through which sparks or burning material might escape. Enclosures for fuses, switches, circuit breakers, and motor controllers, including pushbuttons, relays, and similar devices shall be dust-tight.*"

Specific types of enclosures are used to house electrical equipment in a hazardous atmosphere. Class II locations require the use of dust ignition-proof enclosures (NEMA 9) that exclude all dust from the interior of the enclosure, and at the same time are designed such that enclosed heat-generating devices will not cause external surfaces to reach temperatures capable of igniting or discoloring dust on the enclosure or igniting dust-air mixtures in the surrounding atmosphere. An option that can be used to allow the use of conventional electrical equipment enclosures in hazardous areas is to create an enclosure that is less hazardous (or non-hazardous) by means of dry air or nitrogen purging and pressurization systems. For Class II locations types X, Y, and Z, pressurized equipment and intrinsically safe electrical equipment are applicable to both Division 1 and Division 2 locations.

Table 1. Guidance for Area Electrical Classification (typical Group G plastic dust)

Depth of Dust Accumulation (inches)	Frequency of Dust Accumulation	Housekeeping Requirement	Area Electrical Classification
Negligible [(a)]	N/A	N/A	Unclassified (General Purpose)
Negligible to <1/32 [(b)]	Infrequent [(c)]	Cleanup during same shift	Unclassified (General Purpose)
Negligible to <1/32 [(b)]	Continuous/frequent [(d)]	Clean as necessary to maintain an average accumulation <1/64 in. [(e)]	Unclassified, however electrical enclosures should be dust-tight [(f), (g)]
1/32 to 1/8	Infrequent [(c)]	Cleanup during same shift	Unclassified, however electrical enclosures should be dust-tight [(f), (g)]
1/32 to 1/8	Continuous/frequent [(d)]	Clean as necessary to maintain an average accumulation < 1/16 in.	Class II, Division 2
> 1/8	Infrequent [(c)]	Immediately shutdown and clean	Class II, Division 2
> 1/8	Continuous/frequent [(d)]	At frequency appropriate to minimize accumulation	Class II, Division 1

(a) Surface color just discernible under the dust layer.
(b) 1/32 inch is approximately the thickness of a typical paper clip.
(c) Infrequent - Episodic release of dust occurring not more than about 2 or 3 times per year
(d) Continuous/frequent - Episodic release of dust occurring more than about 3 times per year or continuous release resulting in stated accumulation occurring in approximately a 24 hour period.
(e) It has been observed that a thickness of about 1/64 inch of a low-density dust is sufficient to yield a small puffy cloud with each footstep.
(f) For example, National Electrical Manufacturers Association (NEMA) 12 or better. Note: Ordinary equipment that does not produce heat, such as junction boxes, can be significantly sealed against dust penetration by the use of silicone-type caulking. This can be considered in areas where fugitive dust is released at a slow rate and tends to accumulate over a long period of time.
(g) Guidance to be applied for existing facilities. For new facilities, it is recommended that the electrical classification be at least Class II, Division 2.

Certain areas can be considered as unclassified areas if experience has shown that the release of ignitable dust suspensions is so infrequent that area classification is not necessary. For example, where combustible dusts are processed, stored, or handled, it is not necessary to classify areas where:

- Materials are stored in sealed containers (e.g., bags, drums, or fiber packs on pallets or racks).
- Materials are handled within well-maintained closed piping systems.
- Closed tanks are used for storage and handling.

- Dust removal systems prevent dust clouds and dust layer accumulations that make surface colors indiscernible.

Note that Table 1 covers situations in which fugitive combustible dusts accumulate over a long period of time in normally unclassified areas and in other areas of the plant adjacent to electrically classified areas. Electrical installations in these areas will normally be ordinary (i.e., NEMA 1). Ordinary enclosures may have knockout plugs or other penetrations not sealed against dust. Periodic housekeeping can keep the general area clean, but over a period of time dust can penetrate the ordinary electrical enclosures. In case of a spark within the electrical enclosure, flames can shoot out of the enclosure and become a source of ignition for fire or explosion in the surrounding area. The explosion cannot propagate away from the enclosure if housekeeping is adequate (i.e., a secondary dust explosion will not occur), but, as noted above, a very small accumulation of combustible dust can trigger a secondary explosion. It may therefore be prudent to blow the dust out of such enclosures and securely seal the enclosure with silicone sealant to minimize dust penetration. NFPA 654 discourages blowing dust deposits, but the standard allows this procedure when the following requirements are met:

1. Area and equipment is vacuumed prior to blowdown,
2. Electrical power and other sources of ignition is shut down or removed from the area,
3. Only low gauge pressure (15 psi) steam or compressed air is used, and
4. There are no hot surfaces in the area capable of igniting a dust cloud or layer.

7. Fire Protection

The general topic of fire protection is beyond the scope of this paper. However, automatic sprinkler protection should obviously be installed in facilities with the potential for dust explosions, both to address fires that may result from dust explosions (if the sprinkler system remains functional) and to minimize the potential for a fire to cause an initiating event that could trigger a secondary dust explosion. Although it is unlikely that sprinkler activation could disperse dust deposits and trigger a dust explosion, dust explosions can result from the dispersal of burning dust deposits by hose streams.

8. Conclusions

The hazard of a dust explosion must first be recognized and assessed in order manage the associated risk. Measures can then be taken to reduce both the likelihood and severity of a dust explosion. Both the likelihood and severity of a dust explosion can be controlled by proper facility design and management practices.

A coal dust layer as thin as 0.01 inches (0.3 mm) is capable of propagating an explosion within a typical enclosure. A coal dust layer with an average thickness of 0.03 inches (0.9 mm) covering the floor of a typical enclosure is sufficient to produce a uniform dust cloud at the optimum concentration. The accumulation of 1/32 of an inch of combustible dust constitutes a hazardous area per NFPA 654. Elevated dust accumulations are more likely to be dispersed than accumulations on the floor. A combustible dust accumulation of 1/8-inch deep on the bar joists and beams of a typical building provides enough fuel, if dispersed and ignited, to cause a potentially damaging dust explosion.

Hence, while none of the fundamental dust explosion safety considerations can be ignored, it should be clearly recognized that the most important elements of dust explosion safety are prevention of dust accumulation and good housekeeping. Any time a combustible dust accumulation exists which can produce a concentration above the MEC, a dust explosion hazard exists. It is not possible to completely eliminate all potential initiating events that can trigger a secondary dust explosion. Without good housekeeping, the potential for a dust explosion capable of causing widespread plant damage and loss of life will always be present.

9. References

[1] Guide for Venting of Deflagrations, NFPA 68, 2002 Edition, National Fire Protection Association, Quincy, MA.

[2] Sapko, M.J., E.S. Weiss, K.L. Cashdollar and I.A. Zlochower, Experimental Mine and Laboratory Dust Explosion Research at NIOSH.

[3] Kubala, T.A., F.J. Perzak and E.L. Litchfield (1981) Electric Ignition of Lycopodium Powder in a Modified Hartman Apparatus, Bureau of Mines Report of Investigations RI 8600.

[4] Hughes, J.F. and A.W. Bright (1979) *Electrostatic Hazards Associated with Powder Handling in Silo Installations*, IEEE Trans. On Industry Applications, vol. IA-15, No. 1, pp. 100-103.

[5] Standard for the Prevention of Fire and Dust Explosions from the Manufacturing, Processing, and Handling of Combustible Particulate Solids, NFPA 654, 2000 Edition, National Fire Protection Association, Quincy, MA.

[6] Standard on Explosion Prevention Systems, NFPA 69, 2002 Edition, National Fire Protection Association, Quincy, MA.

Paper TG004 - 2

Dust Explosion Scenarios and Case Histories in the CCPS Guidelines for Safe Handling of Powders and Bulk Solids

Robert Zalosh
Worcester Polytechnic Institute, Worcester, MA.

Stanley Grossel
Process Safety and Design, Inc.

Russell Kahn
Syngenta Crop Protection, Inc

Daniel Sliva
Center for Chemical Process Safety

Prepared for Presentation at
American Institute of Chemical Engineers
2005 Spring National Meeting
39th Annual Loss Prevention Symposium
Atlanta, Georgia
April 11-13, 2005

UNPUBLISHED

Dust Explosion Scenarios and Case Histories in the CCPS Guidelines for Safe Handling of Powders and Bulk Solids

Robert Zalosh
Worcester Polytechnic Institute, Worcester, MA.

Stanley Grossel
Process Safety and Design, Inc.

Russell Kahn
Syngenta Crop Protection, Inc

Daniel Sliva
Center for Chemical Process Safety

Abstract

The new CCPS book, entitled *Guidelines for Safe Handling of Powders and Bulk Solids*, discusses several different types of accident scenarios, including scenarios that can lead to a combustible dust explosion. The dust explosion scenarios are divided into the following three categories: Explosions within Process Equipment, Explosion Propagation into Interconnected Equipment, and Secondary Dust Explosions in Buildings. Examples of process equipment explosion scenarios include dryer explosions involving particulate clouds or hybrid vapor-dust mixtures ignited by overheated (often smoldering) dust layers, blender explosions ignited by either electrostatic discharge or frictional heat from a failed bearing, and dust collector explosions ignited either by sparks from tramp metal impacting against the collector wall, by a smoldering nest from upstream equipment, or by an electrostatic discharge. Explosion propagation into interconnected equipment often occurs via either pneumatic or conveyor transport ducting. Secondary explosions in buildings occur when the primary explosion causes a breach of equipment or a weak enclosure, and the associated blast wave creates a suspended dust cloud that is ignited by the vented flame from the primary explosion or other local ignition source. Case histories illustrating these common dust explosion scenarios are described in this paper, along with a brief summary appropriate protection measures described in the CCPS guidelines book for the specific equipment and explosion scenarios.

1. Introduction

The new Center for Chemical Process Safety *Guidelines for Safe Handling of Powders and Bulk Solids*[1] includes a great deal of material on dust explosion scenarios, case histories, and methods for specific assessments of dust explosion hazards. Excerpts of some of this material are presented in this paper, categorized as Explosions within Process Equipment, Explosion Propagation to Interconnected Process Equipment, and Secondary Dust Explosions in Buildings. This categorization scheme often parallels the escalation of some of the more devastating dust

explosions, which originate in one piece of equipment, propagate to other equipment, and finally emerge from the equipment to produce a massive deflagration in a process building.
In many dust explosions the originating event is either a smoldering fire or an exothermic decomposition of particulates. The new CCPS *Guidelines* book[1] includes detailed background and descriptions of particulate fires and decompositions, but those discussions and descriptions are beyond the scope of this paper.

2. Explosions Within Process Equipment

Dust explosions occur in process equipment when there is a particulate cloud with a concentration between its Minimum and Upper Explosive Concentrations (MEC & UEC) which eventually finds an ignition source. The types of process equipment that routinely have combustible dust concentrations in at least a portion of the equipment volume include blenders, dryers, dust collectors, and grinders/pulverizers. Table 1 lists the number and percent of dust explosions that have been reported to occur in these and other equipment in surveys performed in the U.S.[2,3], U.K. and Germany[4] during the time periods indicated.

Table 1. Equipment Involved in Dust Explosions

Equipment	U.S. (FM,1985-95) + (IRI,1975-2001)		U.K. (1979-1988) (HSE)		Germany (1965-85) (Eckhoff, 1997)	
	# Incidents	%	# Incidents	%	# Incidents	%
Dust Collectors	156	42	55	18	73	17
Grinders/Pulverizers	35	9	51	17	56	13
Silos/Bunkers	27	7	19	6	86	20
Conveying System*	32	9	33	11	43	10
Dryer/Oven	22	6	43	14	34	8
Mixers/Blenders	> 12	> 3	7	2	20	5
Other or Unknown	84	23	95	31	114	27
Total	372	100	303	100	426	100

* Conveying systems include conveyors, ducts, and elevators.

2.1 Dust Collector Explosion Scenarios

Three possible reasons for the high occurrence of dust collector explosions are 1) they are almost omnipresent in particulate handling facilities, 2) they inherently concentrate the smaller particles which are easier to ignite than the mostly larger particles in other equipment, and 3) they often employ pulse jet cleaning, which by nature, periodically generates dust clouds inside the collector. Data from the German compilation of dust explosions indicate that the most frequent ignition sources have been mechanical sparks (41%), smoldering nests (11%), electrostatic discharges (10%), and mechanical heating via friction (7%).

Matsuda and Yamaguma[5] describe a tantalum dust collector explosion that they attribute to an electrostatic discharge in the collector. The small, corral shaped tantalum particles have a high resistivity, and become electrostatically charged by rubbing against the collector wall. Apparently an electrostatic discharge from the charged particles triggered the explosion in the 5-m high, 1.5-m diameter bag type collector.

The following account of an explosion in a baghouse dust collector used to collect a pharmaceutical product from a hammer mill/flash drying operation illustrates how frictional heating in upstream equipment can produce a smoldering nest ignition source for a dust collector explosion. The impact hammer mill had been operating for approximately 10 minutes when the operator heard unusual grinding sounds coming from inside the mill. He immediately shut down the mill just as an explosion occurred within the dust collector, located inside the building on the second floor. The pressure wave caused the explosion vent (a hinged panel) of the dust collector to open, and the explosion products and unburned powder were directed outside the building via a vent duct. However, a screen had been securely fastened at the end of the duct to prevent birds from entering, and as the vent panel swung upward and outward, it struck the screen and opened no further. It is estimated that the screen prevented the explosion vent panel from opening to no more than 50 percent of the vent area.

With the vent partially obstructed, the access door to the dust collector failed under pressure and released a dust cloud into the building, which ignited. The flame front went through the vent duct and followed the dust cloud through the access door, resulting in a fireball at both locations. Also, on the first floor, a fireball was seen exiting the vicinity of the rotary valve outlet at the bottom of the dust collector, which feeds a sifter. There was no secondary explosion on the first or second floors. However, windows were blown out at both floors. The ensuing fire in the dust collector engulfed the wool filter bags (which were burned up) and the remaining powder in the collector hopper, but the fire was quickly extinguished by the automatic sprinkler system inside the dust collector.

A subsequent investigation of the incident revealed that a carbon steel bolt from the inside of the feeder (which feeds wet powder to the hammer mill/flash dryer) had become loose and fell into the hammer mill. The bolt became trapped inside the 3600 rpm mill, where it became heated to above the autoignition temperature of the powder. The hot metal ignited some of the powder in the mill, which was pneumatically conveyed into the dust collector. In the collector, a dust cloud created by the blow ring (pulse jet), was ignited by the hot powder conveyed in from the hammer mill. An inspection of the feeder revealed that six 3/8-inch carbon steel bolts were missing.

Tyldesley[6] described a double fatality explosion that occurred while a pipefitter was trying to clear out a clogged rotary valve at the bottom of a cyclone dust collector at an aluminum powder packaging facility. This was likely an example of an impact spark or mechanical heating ignition event.

The CCPS *Guidelines* book[1] provides some dust collector explosion protection recommendations that are common for all types of collectors, and separate listings of recommended explosion protection measures for cyclone collectors, electrostatic precipitators, and fabric filter baghouses. Perhaps the most important common protection measure is effective explosion isolation on the ducting between the collector and the process equipment it is serving. Effective isolation devices include chemical suppression systems, diverter valves, and mechanical isolation valves. High-speed infrared detectors interlocked to shutdown the duct air flow to the collector, and possibly to initiate a water spray discharge, are also recommended, as are certified spark/ember suppression systems. Another general protection measure is installation of either explosion venting or an explosion suppression system for most dust collection applications.

In the case of cyclones, additional high priority preventive measures are: 1) thorough bonding and grounding and avoiding low conductivity coatings on walls and internal surfaces, 2) regular inspections and cleaning to minimize accumulations on cyclone walls, and 3) infrared or temperature detectors to detect smoldering deposits. In the case of electrostatic precipitators, there is a long list of protection measures including automatic sprinkler protection within the precipitator, interlocks to de-energize the precipitator and fans upon sprinkler actuation, inert gas purges prior to startup and shutdown for applications in which a hybrid flammable vapor-dust mixture may be present, and avoiding precipitators altogether in applications in which dust concentrations can exceed the MEC during normal operating conditions. Specific recommendations for fabric filter collectors include: 1) an automatic sprinkler system that upon actuation also stops the exhaust fan and rotary valve to prevent discharge of burning dust, 2) broken bag detectors that alarm and shut down the exhaust fan, and 3) proper grounding and bonding of bag cages, tubesheets, clamps, etc. to dissipate electrostatic charges, and 4) use of electrically conductive fabrics for dusts with a Minimum Ignition Energy of 3 mJ or less.

2.2 Blender Explosion Scenarios

The blending of particulates of two or more different compositions inherently involves the kind of inter-particulate friction and particulate-wall friction that causes electrostatic charge generation. If the particulate resistivity is sufficiently high, the electrostatic charge can continue to accumulate with correspondingly increasing voltage differences. Furthermore, if the blender wall is not well grounded, charge and associated high voltages can accumulate on the blender wall. If the particulate Minimum Ignition Energy (MIE) is sufficiently low, and if the eventual electrostatic discharge occurs in a location where combustible concentrations exist, the result is a dust explosion. This has occurred in a plastics manufacturing plant in a blender used to mix the primary polymer with various additives. The primary polymer had a resistivity of 2×10^{16} ohm-cm, a MIE of about 7 mJ, and a MEC of 20 g/m^3. The latter two values are lower than those of most organic powders. However, even with significantly larger MEC values, concentrations above the MEC should be anticipated toward the top of the blender during normal operation and throughout most of the blender volume during batch loading and unloading.

Eckhoff[4] has a detailed description of a 1973 aluminum dust explosion in a screw blender/mixer. A rubber lined steel tube surrounded the center screw of the mixer. Eckhoff states that the explosion was probably ignited by a propagating brush discharge due to electrostatic charge generation on the rubber lining within the grounded steel tube. The aluminum flakes had a specific surface area of 7.5 m^2/g, and Minimum Ignition Energy of only 1 mJ. Even though there was an attempt to nitrogen-inert the mixer, Eckhoff suggests that the volume within the steel tube was not inerted, and oxygen concentrations were also sufficiently high to support combustion. Thus, the explosion was initiated at the 3.3 m long enclosed screw, and then propagated as a flame jet into the 5.2 m^3 mixer.

Recommended blender explosion prevention measures described in the CCPS Guidelines book include blender construction and grounding considerations to minimize electrostatic charge generation, and blender design and operational considerations to reduce the hazards of frictional heating and tramp metal entry into the blender.

2.3 Dryer Explosion Scenarios

Overheating of particulate by a hot surface is by far the most likely ignition source in dryers. In some cases, the particulate accumulates on the hot surface and forms a smoldering nest, while in other cases the hot surface temperature is sufficiently high to directly ignite the suspended dust cloud. Although particulates near the dryer inlet may be too wet to be readily ignited, particulates exiting the dryer are both dry and often suspended in concentrations above the MEC.

Drogaris[7] cites an example of a fire and explosion in a batch rotating vacuum dryer used for drying a pharmaceutical powder. An operator had tested dryer samples on a number of occasions without any problems. After the last sampling, he closed the manhole cover, put the dryer under vacuum, and started rotation of the dryer. A few minutes later an explosion and flash fire occurred, which self-extinguished. No one was injured. Investigations revealed that after the last sampling, the dryer manhole cover had not been securely fastened. This allowed the vacuum within the dryer to draw air into the rotating dryer and create a flammable atmosphere. The ignition source was probably an electrostatic discharge (the Teflon coating on the internal lining of the dryer could have built up a charge). No nitrogen inerting had been used. After the incident, the following two precautions were instituted to prevent similar accidents from occurring in the future: 1) Nitrogen purging is carried out before charging or sampling of the dryer. 2) If the absolute pressure rises to about 4 psia, the rotation is stopped, an alarm sounds, and a nitrogen purge starts automatically.

Other dryer dust explosion hazard scenarios addressed in the CCPS *Guidelines*[1] include hybrid mixture formation due to vaporization of a flammable liquid, direct firing ignition sources, frictional sparks and overheating from failed bearings, misaligned fan blades, tramp metal, and electrical equipment not appropriate for atmospheres containing combustible dust clouds.

2.4 Grinder/Pulverizer Scenarios

Grinders, pulverizers, and other size reduction equipment inherently dissipate large energy inputs required to break up the particles. This energy dissipation inevitably causes heating of the particles and metal surfaces. Particles accumulating in the grinder can easily overheat, smolder, and ignite a dust explosion during grinder loading or unloading. Section 5.3.17 of the CCPS *Guidelines*[1] has two specific accounts of explosions ignited by friction-induced hot spots in grinder/pulverizers. Frictional heating of the housing of a grinder being used to produce 50 μm (300-mesh) silicon powder caused one explosion. Tramp metal was the apparent ignition source in another explosion that occurred in a hammermill being used for an intermediate stage powder with a low MIE and Autoignition Temperature (AIT).

Figure 1 shows evidence of the frictional or impact heating of hammers in a hammermill used to produce powdered sugar. The heating of sugar and metal in that vicinity of the mill ignited a sugar dust explosion that burned the hammermill operator who was responding to the sound of severe vibration due to tramp metal or a broken hammer in the mill.

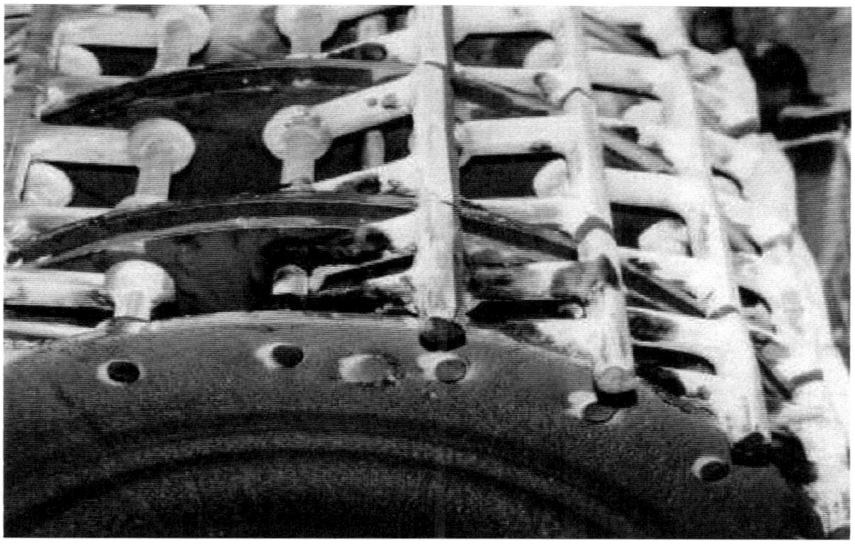

Figure 1: Scorched hammers in a hammermill.

Specific recommendations in the CCPS *Guidelines* book[1] for preventing grinder/pulverizer dust explosions include monitoring the mill motor current and having an interlock shutdown upon high current draw, use of magnetic separators to find and remove tramp metal before it enters the mill, and use of special enclosed mills to allow inerting of powders with extremely low MIE and AIT values.

2.5 Other Process Equipment

Besides dust collectors, blenders, dryers, and grinder/pulverizers, the CCPS *Guidelines* book[1] also describes fire and explosion incidents and specific explosion prevention measures for the following process equipment: bag openers, extruders, feeders and rotary valves, sampling systems, classifiers, silos and hoppers, size enlargement equipment, solids charging systems, and weighing systems. Appendix B of the book is an overview description of the equipment itself, including transport equipment and rail car and hopper truck loading and unloading stations as well as the previously mentioned process equipment. In addition to the explosion prevention suggestions, Chapter 6 of the CCPS Guidelines book discusses deflagration mitigation measures such as deflagration venting, deflagration suppression, and deflagration isolation systems.

3. Explosion Propagation to Connected Equipment

Many dust explosion incidents result in flame propagation through interconnected process equipment. The resulting explosion damage can extend far beyond the site of the originating explosion. The path for the explosion propagation is usually ducting used for pneumatic transport of particulates. In other incidents, the propagation path is enclosed or underground conveyor galleries/tunnels, sometimes leading to large, vulnerable silos.
Any approach to preventing explosion propagation needs to distinguish between the propagation of the ignition source and the propagation of the deflagration itself. If the ignition source is a smoldering nest or burning ember traveling through the ducting, properly designed/installed spark detection and extinguishing systems have been effective in preventing this scenario from escalating into a downstream deflagration. On the other hand, if the particulate loading in the ducting corresponds to a concentration above the MEC, a deflagration, rather than a mere smoldering ember/nest, can propagate through the duct. For example, a wood sander-initiated dust collector explosion incident involved a collector duct equipped with a spark detection and extinguishing system, and the deflagration overwhelmed the extinguishing system[8]. Flame from the dust collector was transported through the clean air return ducting back to the processing building, and caused scalp burn injuries to several employees.

Sometimes the dust explosion propagates back in the opposite direction of the air flow through the dust collection ducting. One such explosion was initiated by an electrostatic discharge in a drum situated under a dust collector. The drum was mounted on rubber casters such that it was electrically isolated from earth ground. Additionally, it was also isolated from the flexible ducting between the drum and collector discharge outlet. An explosion in the dust collector propagated back through the dust collection ductwork to the packaging room where it produced a secondary explosion with some structural damage but no injuries. The secondary explosion propagated into an adjoining room causing a third and the most violent of the three explosions. One of the correction measures taken at this facility is the use of drum grounding as illustrated in Figure 2. Other aspects of the sequence of explosions in this incident are described by Pickup[9].

Figure 2: Dust collection drum grounded to flexible duct.
(Courtesy of Gowan Milling, LLC)

Isolation of interconnected equipment to prevent full deflagration propagation requires some type of isolation system that can vent or quench both the flame and associated deflagration pressure. A test of one such system under development is shown in Figure 3. A later version of the device shown in Figure 3 is intended for use as a check valve to prevent reverse through dust collector ducting. Other isolation systems are described in Chapter 6 of the CCPS *Guidelines* book[1] and in NFPA 69[10]. Use of the systems also requires that the ducting be sufficiently strong to withstand the design flame speed and pressure associated with the isolation system certification.

Figure 3: Duct deflagration isolation valve being tested (Courtesy of Gowan Milling LLC).

4. Secondary Dust Explosions in Buildings

Perhaps the most devastating dust explosion scenario is the generation of a secondary dust explosion in the building surrounding the equipment in which some primary explosion occurs. The secondary explosion occurs when the blast wave emanating from the ruptured equipment/conveyor lifts the accumulated dust into suspension, and the flame from the primary explosion subsequently ignites the suspended dust cloud. The resulting devastation and casualties are associated both with the burning of building occupants and with the structural damage to the building.

One simple example of explosion propagation from process equipment into the surrounding building occurred in a powdered aluminum plant when somebody was trying to clear out caked aluminum deposits in a dust collector pickup line. The explosion propagated from the cleanout port, injuring two employees and destroying the steel framed process building[6].

A classic example of a devastating secondary dust explosion is the magnesium stearate explosion that occurred in 1976 in a plant manufacturing chewing gum. The magnesium stearate is applied as a lubricant on the freshly produced gum before it is cut into chewable pieces. The primary explosion occurred in the cutting machine several minutes after the machine started vibrating with sufficient intensity to generate a magnesium stearate combustible dust cloud. The rupture of the cutting machine generated a blast wave that displaced and lifted fugitive magnesium stearate powder from beams, ledges, and light fixtures, and the emerging fireball ignited the suspended cloud of powder. According to FM Data Sheet 7-76[8], the secondary dust explosion blew out windows on two sides of the building, demolished a cinder block wall about 3 m away from the cutting machine, and destroyed the equipment in the vicinity. The fireball and subsequent fires opened 166 sprinkler heads in the building. According to the New York City Fire Department account of this incident, six people were killed, and 50 other people suffered burn and blast wave injuries.

The extensive destruction and casualties that occurred in the January 29, 2003 explosion at the West Pharmaceuticals plant in Kinston, North Carolina, as indicated by the aerial view photograph shown here as Figure 4 are due in large part to a secondary dust explosion. The Chemical Safety Board investigators determined that polyethylene powder accumulations on the upper surface of the ceiling tiles on a suspended ceiling indicated in Figure 5 were dispersed either by either some undefined primary explosion or by some unidentified disturbance, and that the polyethylene dust cloud produced a deflagration in the space between the ceiling tiles and the concrete floor above[11].

Figure 4: Aerial view of the destruction and residual fire in the January 2003 dust explosion at West Pharmaceuticals.

The negative air pressure generated by the HVAC return duct inlets above the suspended ceiling apparently contributed to the collection and accumulation of polyethylene powder on the ceiling tiles. According to the CSB report[11], employees reported seeing accumulations that varied in thickness from 0.125 inch to 0.50 inch across most of the suspended ceiling. The milling and batchoff area under the suspended ceiling was reported to be relatively free of dust accumulations. However, the pressures produced by an incipient deflagration above the suspended ceiling probably blew polyethylene dust into the areas above and below allowing the deflagration to propagate in both the compounding area and milling/batchoff area of the plant.

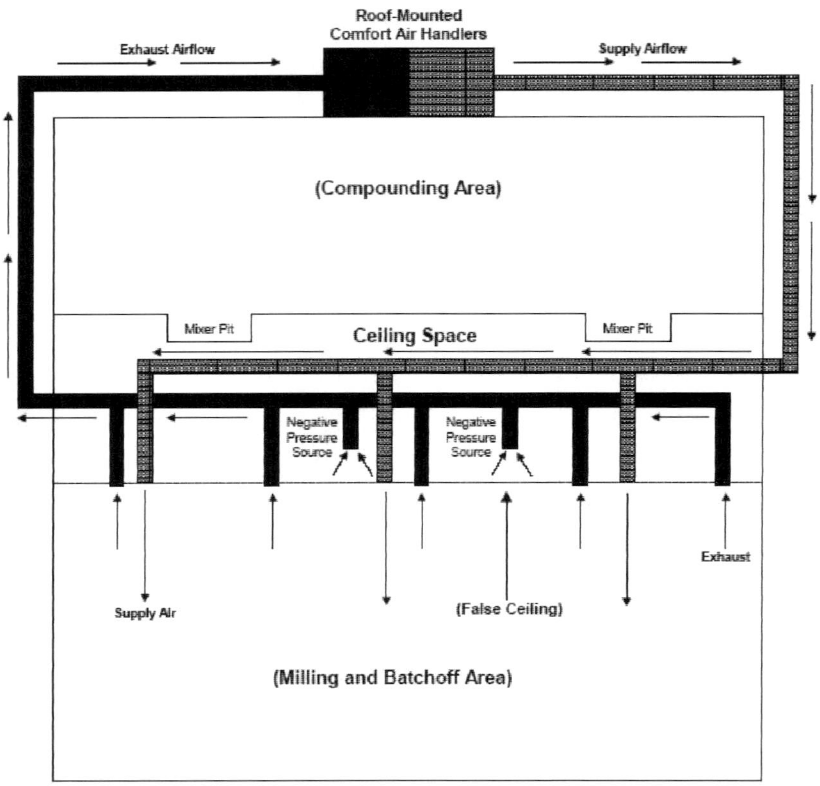

Figure 5: HVAC Ducting at West Pharmaceuticals Compounding and Milling Area (from CSB report)

The West Pharmaceuticals Kinston explosion produced 6 fatalities and 38 injuries, most from burns and blunt force trauma due to falling objects and collapsing walls. Other recent secondary dust explosion incidents with fatalities and multiple injuries include the Jahn Foundry phenolic resin explosion[12], the CTA Acoustics phenolic resin explosion (CSB July 8, 2003 press release), and the Hayes Lemmerz aluminum dust explosion (CSB November 5, 2003 press release).

One inherent issue in the prevention and mitigation of secondary dust explosions is the required level of housekeeping and other maintenance practices to minimize dust leakage and subsequent accumulations. Section 7.5 of the CCPS *Guidelines* book[1] discusses this issue and the pertinent recommendations in FM Data Sheet 7-76[8] and NFPA 654[13].

5. Dust Explosion Hazard Assessment and Protection Methods

Section 4.3.6 of the CCPS *Guidelines* book[1] provide descriptions of the laboratory test methods available for determining electrostatic charge and discharge hazards, while Section 4.3.7 describes the various dust cloud explosibility test methods. Table 2 summarizes the standardized test methods for dust cloud explosibility in terms of applications, advantages, and limitations. These test methods can be used in conjunction with flow charts and decision trees in the book to help establish appropriate explosion protection methods for specific materials and installations.

Numerous commercial testing laboratories are available for conducting the tests shown in Table 2 and other tests geared toward powder/dust explosibility and flammability. Appendix A of the Guidelines book lists 11 such laboratories in the U.S. and 11 more in other countries, and indicates which specific tests these laboratories perform.

6. Conclusions

Many reported dust explosions have originated in common powder and bulk solids processing equipment such as dust collectors, dryers, grinders/pulverizers, and blenders. Electrostatic discharges are frequently cited as the ignition source for dust collector and blender explosions, whereas particulate overheating is the most common ignition source in dryer explosions, and friction/impact heating associated with tramp metal or misaligned parts is probably the most frequent ignition source in grinder/pulverizer explosions.

Dust explosions are often exacerbated by propagation through ducting between process equipment, frequently via dust collector pickup and return ducting. More widespread use of effective deflagration isolation devices in such ducting would clearly be beneficial in mitigating the damage and injuries from these propagating dust explosions.

Secondary dust explosions in processing buildings probably cause the largest numbers of dust explosion fatalities and injuries. One crucial aspect of secondary dust explosion prevention and mitigation is greater awareness of good housekeeping and maintenance practices to prevent particulate leakage from equipment and subsequent accumulations of dust deposits in large areas of the buildings.

The CCPS *Guidelines for Safe Handling of Powders and Bulk Solids* is now available to provide guidance and elaboration on these issues. It also provides guidance on assessing dust explosion hazards for specific materials and facilities, and various flow charts and decision trees for utilizing the results of those assessments to determine the most appropriate explosion protection measures.

7. Acknowledgements

The authors wish to acknowledge the numerous contributions of the other CCPS subcommittee members responsible for producing Reference 1: John Bresland, Larry Briton, Warren Greenfield, Dave Herrmann, Dave Kirby, Melvin Nelson, Al Ness, Jeff Philiph, Gary Pilkington, Monica Stiglich, and also Walt Frank, Carla Hardy, Ronald Kersten, Roy Winkler, and Paul Wood.

Table 2: Dust Cloud Explosibility Test Methods

Test	Standard	Applications	Advantages	Limitations
Minimum Explosible Concentration (MEC)	ASTM E 1515 CEN prEN 14034-3	Prevention via dust concentration control, e.g. in pneumatic conveying	The measured MEC value can be compared to in-situ measurements of suspended dust concentrations in conveyors and other process equipment, and can be used to verify concentration is < MEC.	Measured MEC value is based on a pressure rise of one atmosphere above the pressure due to the igniter; explosions with smaller pressure rises can occur at concentrations below the reported MEC value.
Minimum Cloud Auto Ignition Temp: (MAIT)	ASTM E 1491 IEC 1241-2-1	Safe operating temperatures in heated process equipment.	MAIT is valuable data for both particulate material manufacturing and for post-production processing by other companies and facilities.	BAM (horizontal) oven yields lower MAIT values than Godbert-Greenwald (vertical) furnace. MAIT value depends on the residence time of the dust cloud in the heated equipment, and with the area of a heated surface.
Minimum Ignition Energy (MIE)	ASTM E 2019 IEC 1241-3	Electrostatic ignition hazard evaluations; FIBC material classes	MIE value determines precautions needed in silo/bag filling and other particulate handling operations.	Measured MIE value depends on dust cloud turbulence level as well as amount of inductance in spark generation circuit.
Maximum Explosion Pressure (P_{max}) and K_{st}	ASTM E1226 ISO 6184	Deflagration Containment, Deflagration Venting or Suppression	K_{st} is often considered the most important parameter to characterize dust material combustibility.	Measured K_{st} value depends on both ignition energy and dust cloud turbulence level as determined by time delay between dust injection and ignition.
Limiting Oxygen Concentration (LOC)	ASTM E 2079 CEN prEN 14034-4	Inerting per NFPA 69 and NFPA 654	Provides valuable explosion prevention data.	Measured LOC values vary with ignition energy and with particular inert gas used in test.

8. References

[1] Center for Chemical Process Safety, *Guidelines for Safe Handling of Powders and Bulk Solids*, AIChE, 2005.

[2] Febo, H., FM Global Personal communication to S.S. Grossel, 2001.

[3] Thornberg, W., "Personal communication from Mr. Thornberg of IRI to S.S. Grossel, 2001.

[4] Eckhoff, R., *Dust Explosions in the Process Industries*, Butterworth-Heinemann, 2nd Edition, 1997.

[5] Matsuda, T. and Yamaguma, M. "Tantalum Dust Deflagration in a Bag Filter," *J. of Hazardous Materials,* v A77, (2000) pp. 33-42.

[6] Tyldesley, A., Personal communication from Alan Tyldesley, HSE, Bootle, UK to S. S. Grossel (February 19, 2004)

[7] Drogaris, G. *Major Accident Reporting System:* Lessons *Learned from Accidents Notified.* Elsevier Science Publishers, 1993.

[8] FM 7-76, "Prevention and Mitigation of Combustible Dust Explosions and Fires," Loss Prevention Data Sheet 7-76. FM Global, 2001.

[9] Pickup, R. D. "Dust Explosion Case Study: Bad Things Can Still Happen to Good Companies," *Proc. Safety Prog.*, Vol. 20, pp. 169-172, September 2001.

[10] NFPA 69, "Standard on Explosion Prevention Systems," National Fire Protection Association, 2002.

[11] CSB, "Dust Explosion (6 Killed, 38 Injured)," Chemical Safety Board Investigation Report No. 2003-07-1-NC, September 2004.

[12] Zalosh, R., "A Tale of Two Explosions," Proceedings 34th Loss Prevention Symposium, AIChE, 2000.

[13] NFPA 654, "Standard for the Prevention of Fire and Dust Explosions from the Manufacturing, Processing, and Handling of Combustible Particulate Solids, National Fire Protection Association, 2000.

Paper TG004 - 3

Investigation Of A Combustible Dust Explosion At An Automotive Acoustic Insulation Manufacture A Case Study Of A U.S. Chemical Safety And Hazard Investigation Board

William Hoyle (*Speaker*), Giby Joseph,
Mark Kaszniak, Stephen J. Wallace
U.S. Chemical Safety Board
2175 K Street, NW, Suite 400
Washington, DC 20037-1809

Prepared for Presentation at
American Institute of Chemical Engineers
2005 Spring National Meeting
39th Annual Loss Prevention Symposium
Atlanta, Georgia
April 11-13, 2005

UNPUBLISHED

Investigation Of A Combustible Dust Explosion At An Automotive Acoustic Insulation Manufacture A Case Study Of A U.S. Chemical Safety And Hazard Investigation Board

William Hoyle (*Speaker*), Giby Joseph,
Mark Kaszniak, Stephen J. Wallace
U.S. Chemical Safety Board
2175 K Street, NW, Suite 400
Washington, DC 20037-1809

Abstract

The Chemical Safety Board investigated the explosion and fire that occurred at an acoustic insulation facility in Corbin, Kentucky in February 2003. Seven people died and several others were injured during the explosion. A smaller explosion led to cascading dust explosions that destroyed large portions of the facility. The proximity of multiple production lines to each other, design of the building, maintenance, and cleaning practices were all factors involved in the incident. A number of recommendations were made to various parties to minimize the chances of such explosions in the future.

Investigation Team: Francisco Altamirano, William Hoyle (lead investigator),Giby Joseph, Mark Kaszniak, Cheryl Mackenzie, Stephen J. Wallace, PE, CSP[1]

1. Introduction

The hazards of combustible dusts are not always recognized and addressed in industry. Although good practice guidelines exist involving design, housekeeping, and operation to prevent incidents; these guidelines are not always implemented in a way that decreases the risks of handling such materials. This paper includes the results of an investigation that the U.S. Chemical Safety and Hazard Investigation Board (CSB) conducted at a manufacturing facility for industrial and automotive parts.

On February 20, 2003, the CTA Acoustics manufacturing facility in Corbin, Kentucky was shaken by a series of explosions and a large fire that resulted in the death of seven workers and injuries to 37 others. The incident also caused extensive damage to the production area of the facility. As a precaution, nearby homes and an elementary school were evacuated, and portions of Interstate 75 was temporarily closed.[2]

---

[1] Mr. Altamirano and Mr. Kaszniak are Certified Fire and Explosion Investigators, a designation given by the National Association of Fire Investigators.

[2] Note: This paper presents a summary of the CSB investigation, and every effort has been made to represent the contents of the report approved by the Board. To read the full investigation report and find an expanded discussion of the issues presented here, refer to the CTA Acoustics Investigation report at www.CSB.gov.

2. Facility and Process Description

The Corbin facility manufactured industrial and automotive acoustical insulation products. The insulation was produced on five production lines that were located in the southern half of the plant. The production lines included material feeders, separating and mixing machines, curing ovens, and trimming equipment connected by conveyors.

2.1 General Description

Three raw materials—fiberglass, phenolic resin powder, and a nylon sheeting called "facing"—were used to manufacture the industrial and automotive acoustical insulation products. The phenolic resin powder was deposited onto fiberglass that was initially fed into the line. Air-suction dispersed the phenolic resin powder throughout the fiberglass to form a mat. Facing was applied on top and bottom of mat prior to it entering the oven. Residual resin powder that was not bound to the pad and small fiberglass fibers were removed by means of a vacuum system.

The mat was then cured in an oven by circulating hot air, causing the resin to evenly bind with the fiberglass and the facing. The cured parts were then trimmed and packaged before being shipped to customers.

Figure 1. Line 405 Process Flow

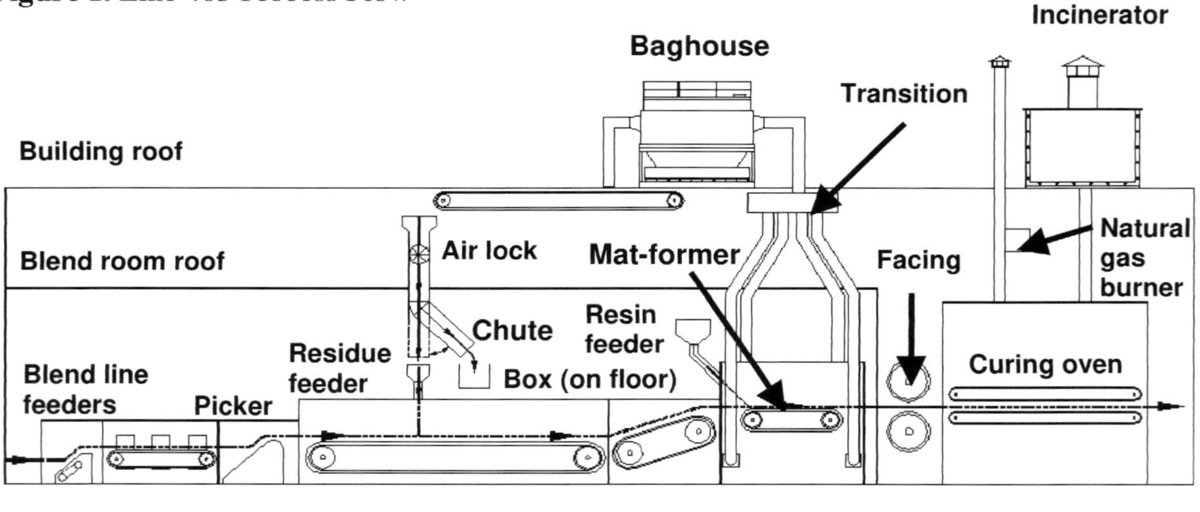

2.2 Dust Collection System

All the production lines were similar. This specific section discusses the line that the CSB determined was the likely origin point of the explosions (i.e., line 405). The suction air—along with excess resin powder and smaller fiberglass fibers—traveled from a machine called the "mat-former" via four vertical air ducts to a baghouse, which was located on the plant's roof. (See Figure 1.) The baghouse was a large rectangular metal box with a series of long cloth bags hung inside. Particulate-laden air entered on one side of the baghouse and passed through the bags, which acted like filters to remove fibers and resin dust. During this process, a dust cake formed on the exterior of the bags. The difference in pressure between the dirty and clean sides of the baghouse determined the amount of suction generated in the mat-former. Written procedures required that the line be shut down and the baghouse cleaned if the pressure gauge reading was above 6 inches, an amount recommended by the manufacturer.

Operators reported that when the bags clogged, an excessive amount of dust blew out of the base of the mat-former and into the plant. To keep the baghouse functioning efficiently, operations personnel would "shake the bags," by injecting high-pressure air into the bags while the process line was shut down, or manually strike the bags with a stick to remove the dust cake during daily shift line cleaning. Some residue that collected in the bottom of the baghouse was recycled to the process. The residue fell into a cardboard box placed under the chute on the blend room floor.

2.3 Curing Oven

Circulating fans and dampers directed heated air (260 to 290 degrees Fahrenheit (°F)) around the mat as it passed through the curing oven. As the fragile mat traveled through the curing oven, small pieces of the mat sometimes stuck to the flights and the drive chain or fell to the oven floor. This combustible material accumulated in the bottom of the oven. Four doors, measuring 4- by 4-foot and located at floor level, provided access to the oven for cleaning. Each door was hinged on its right side and was held shut by a latch and pin near the top of its left side.

A single natural-gas burner was mounted on the air inlet duct at line 405 several feet above the top of the curing oven. The burner incorporated combustion safeguards and interlocks to prevent fuel explosions. A thermocouple measured oven temperature. The thermocouple sent an electronic signal to a digital process controller, which had programmable set points and alarms for temperature settings.

3. Production Line Operation and Cleaning

The day shift started at 7:00 am and typically cleaned the lines at the beginning of each shift. Daily production line cleaning typically included tasks such as using compressed air to blow out areas where the fiberglass mats were formed, using a chimney sweep or compressed air to clean the vertical pipes leading to baghouses. As mentioned previously, waste from the baghouse dropped via the pipe chute into production area.

Material cleaned out of the processing line was typically dumped onto the plant floor and then swept up with brooms and deposited in a dumpster. Wall-mounted and portable pedestal fans were used to blow dust away from the operators as they cleaned around the lines. The fans increased the dispersion of combustible dust in the production area and resulted in the dust settling on overhead flat surfaces.

4. Incident Description

The facility's production lines had a history of small fires erupting in the ovens. Although no open flames were normally present in the ovens where the mats were fed, combustible materials accumulated within the ovens sometimes ignited. Operators routinely put out the fires with water by using garden hoses or with dry chemical fire extinguishers.

Four days prior to the explosion, the temperature controller on the line 405 oven was no longer regulating the temperature so it was switched to manual by the line operators. The oven temperature was regulated by opening and closing the access doors on the east and west sides of the oven.

The day shift began as normal on February 20 with operators performing a routine cleaning of the 405 production line. The previous night shift reported that the poor functioning of the baghouse and resulting lack of suction in the mat-forming equipment caused excessive release of dust from the production line. During cleaning that morning, an operator discovered that the transition leading to the baghouse was plugged. Some of the crew went to the roof to unplug it. The operators used poles and a compressed air lance to break up and blow the remaining material out of the transition. That material fell back through a chute into the production area. These activities and other line cleaning tasks - such as blowing out pipes with compressed air and dry sweeping floors - likely generated a cloud of combustible dust around line 405.

CSB concluded that the most probable initiating event was a combustible dust explosion at the line 405 oven. A fire likely developed inside the oven - probably due to a buildup and overheating of combustible material. Flames may have escaped through an open oven door; or a portion of the dust cloud may have entered the oven and ignited.

The pressure wave produced by the initial explosion stirred up dust that had accumulated on flat surfaces throughout the plant. This airborne dust provided fuel for fire and secondary explosions. The fire and explosions quickly spread over a wide area and destroyed the production area of the facility. In addition, firewalls separating the production area from other portions of the facility were partially collapsed or structurally damaged. Exterior plant walls and portions of the roof were also blown out as the pressure waves from the explosions vented out of the building

5. Hazards of Combustible Dusts Illustrated in this Incident

This incident illustrates the hazards of processing material that can become combustible dusts. Combustible dusts can be generated when solid combustible materials are handled or processed due to gradual reductions in particle sizes and/or segregation of particles. The standard from the National Fire Protection Association, NFPA 654, defines combustible dusts as any finely divided solid material 420 microns or less in diameter that presents a fire or explosion hazard when dispersed and ignited in air. The CSB conducted tests on material found in the processing area after the incident and determined the material characteristics as follows: Kst value of 165 bar-meters per second (Kst is a relative measure of the energy of dust explosion); Minimum Ignition Energy of 3 millijoules (MIE is a measure of how easy it is to ignite a material); and a Minimum Explosive Concentration of 50 grams per cubic meter (MEC is the lowest concentration at which a dust/air mixture will explode). Based on these results, the CSB concluded that dust particles of the phenolic resin would explode when suspended in air (in adequate concentrations) where there is confinement and an ignition source.

A dust fire occurs when the dust comes in contact with heat sources in the presence of oxygen. If the dust is suspended in a confined area, an explosion may occur. In this incident, the five parts of the dust pentagon came together at the same time to result in the explosion. Also, the particle size of the phenolic resin powder was very small, on the order of 10 to 50 microns, much lower than the threshold the NFPA considers a combustible dust. Also, the extensive damage that occurred in this incident occurred as a result of secondary and cascading dust explosions. This is a typical scenario in dust explosions, where an initiating event (such as a flash fire) lofts dust that is present on building surfaces into the air, which can then result in a large explosion.

6. Incident Analysis

The CSB investigation involved conducting numerous interviews, reviewing various documents, and testing physical evidence and material samples. Based on the findings and resulting conclusions, the investigation team determined the causes of the incident. The findings showed that the phenolic resin dust fueled the series of explosions. The design of the facility did not prevent the accumulation of dust or prevent the spread of fire and explosions throughout the facility.

Operating practices contributed to the likelihood of the incident. The lack of effective housecleaning policy allowed dust to accumulate in areas above the production line, and line cleaning procedures actually exacerbated the problem. Operating the oven with the doors open increased the likelihood that an unexpected fire inside the oven could contact material outside or that an external dust cloud could enter the oven.

The CSB also explored the effectiveness of organizations that had oversight for the facility and conducted inspections. Inspections conducted by regulatory agencies and insurance companies that had oversight for the facility either did not occur or failed to identify the hazard of combustible dusts. The team found that the Kentucky Occupational Safety and Health Program (OSH) conducted several comprehensive inspections of CTA but did not issue citations regarding the combustible dust hazard. Also, the State Fire Marshal's office has not inspected the CTA facility since its construction in 1972. Various insurance companies also inspected the CTA facility but did not identify phenolic resin dust as an explosion hazard.

6.1 Root Causes

Root causes are typically management system failures. If these systems had functioned properly, the incident would not have occurred. The team identified the following root causes.

Although the explosion hazard of dust was known by some facility managers, effective measures for preventing combustible dust explosions were not implemented. Information about the explosivity of the dust was not adequately communicated to the workforce and the principles of NFPA 654, as recommended by the phenolic resin supplier, were not utilized.

The procedures for cleaning and maintaining the facility production lines did not prevent the accumulation of unsafe levels of combustible dust on elevated flat surfaces. For example, the use compressed air and fans during line cleaning dispersed combustible dust and caused it to settle on elevated flat surfaces throughout the facility. The housekeeping program did not effectively remove these dust deposits.

The incident investigation program used by the facility was inadequate. For instance, the underlying causes of frequent fires in the oven were not determined and no corrective actions to prevent future fires were not implemented.

The design of the building did not address the hazards associated with combustible dusts. Neither the initial design nor later modifications adequately incorporated good design practices to minimize the hazards of combustible dusts. The use of flat surfaces where dust could accumulate were not minimized, firewalls and barriers to prevent the spread of fires and explosion damage were inadequate, and deflagration venting was not installed.

The supplier of the phenolic resin (Borden Chemical) did not explicitly convey the explosive hazards of the material to CTA Acoustics. The Material Safety Data Sheet (MSDS) noted that the resins used by CTA Acoustics were combustible dusts and referred the user to NFPA 654, but did not explicitly state that the material could be explosive. Also, the supplier did not communicated safety lessons from an earlier explosion at the Jahn Foundry (which was also using a Borden-supplied phenolic resin) to its customers.

6.2 Contributing Causes

Contributing causes are those causes that increased the likelihood or the consequences of the incident. If such a cause is removed the incident is less likely (though still possible) and the consequences should be less severe. The team identified the following contributing causes.

The design of the line 405 oven did not include adequate fire protection devices and automatic sprinklers.

The facility did not have effective procedures for evaluating the hazards associated with non-routine operating conditions on line 405. As an example, operating the oven with malfunctioning temperature control equipment and open doors was a nonroutine situation—the hazards of which were not recognized or controlled.

7. Recommendations

CSB developed the following recommendations from the findings and conclusions of this investigation. Recommendations are made to parties that can affect change to prevent future incidents. Recipients may include the facility where the incident occurred, the parent company, trade organizations responsible for developing good practice guidelines, or organizations that have the ability to broadly communicate lessons learned from the incident—such as trade associations and labor unions. Following is a summary of the recommendations. The full text of the recommendations can be found at www.csb.gov.

The CSB made a number of recommendations to CTA Acoustics. The CSB recommended that CTA develop a combustible dust safety program using good practice guidelines, such as NFPA 654. The provisions of this guideline include minimize surfaces where combustible dust could accumulate, ensure phenolic resin-handling facilities are designed to prevent the spread of fires or explosions, and preventing the unsafe accumulation and dispersion of combustible dust by frequently cleaning process areas. The CSB also recommended that CTA conduct hazard assessments of ovens to ensure that fire detection and suppression systems are adequate. Also, CTA should develop procedures to maintain safety during nonroutine operating conditions, such as the loss of oven temperature control. And finally, the facility should revise the incident investigation program to ensure that the underlying causes of incidents such as fires are identified and corrective actions implemented.

To CertainTeed, Inc., the company that designed, operated, and maintained the facility prior to its purchase by CTA, the CSB recommended that the organization evaluate their facilities that handle combustible dusts and ensure that good practice guidelines such as NFPA 654 are followed. And in a proactive measure to ensure that future facilities are designed safely, Certainteed should ensure that company design standards incorporate good engineering practices to prevent dust explosions.

The CSB recommended that the supplier of the phenolic resin, Borden Chemical, Inc., ensure that MSDSs for phenolic resins include, at a minimum, warnings that dust from these products can be explosive. Also, Borden should develop and distribute educational material, in addition to MSDSs, to inform customers of the explosion hazard of phenolic resin dust. And finally, Borden should communicate the findings and recommendations of this report to their customers that purchase phenolic resin.

The CSB recommended that the Kentucky Office of Occupational Safety and Health develop and distribute an educational bulletin on the prevention of combustible dust explosions, and enhance the training program for compliance officers regarding the recognition and prevention of combustible dust explosion hazards. To the insurers at FM Global, the CSB recommends that they incorporate the findings and recommendations of this report in their training of employees who conduct inspections at facilities that may handle combustible dusts.

To the Kentucky Office of Housing, Buildings, and Construction, the CSB recommended that they incorporate the findings and recommendations of this report in their continuing training of staff that interface with facilities that may handle combustible dusts. Further, the CSB recommended that the office identify sites that handle combustible dusts when facilities apply for new or modified construction permits, and use this information to help prioritize establishments that will be inspected by the fire marshal.

8. Conclusions

The incident that occurred at CTA Acoustics on February 20, 2003 was a tragedy that took the lives of seven people, injured several others, and costs millions of dollars in damage and lost revenue. The investigation by the U.S. Chemical Safety Board serves to highlight issues that allowed this incident to occur. The facility was not designed or maintained in a way to minimize the hazards of combustible dusts, and inspections by external organizations failed to adequately identify existing problems. The lessons from this investigation should be applied by any facility that is using similar material and faces the same type of hazards present at the CTA plant.

Paper TG004 - 4

Evaluation Of Dust And Hybrid Explosion Potential In Process Plants

Bernadette Nalda-Reyes
Ashok Ghose Dastidar
C. James Dahn
Safety Consulting Engineers, Inc.
2131 Hammond Drive
Schaumburg, IL, USA 60173
(847) 925-8100, sceinc@sceinc.com

Prepared for Presentation at
American Institute of Chemical Engineers
2005 Spring National Meeting
39th Annual Loss Prevention Symposium
Atlanta, Georgia
April 11-13, 2005

UNPUBLISHED

Evaluation Of Dust And Hybrid Explosion Potential In Process Plants

Bernadette Nalda-Reyes
Ashok Ghose Dastidar
C. James Dahn
Safety Consulting Engineers, Inc.
2131 Hammond Drive
Schaumburg, IL, USA 60173
(847) 925-8100, sceinc@sceinc.com

Abstract

Dust explosions are a serious hazard in the process industries. They have resulted in the destruction of process plants and equipment, injury to workers, and loss of process. The five factors required for a dust explosion are: 1) fuel, 2) oxidant, 3) confinement, 4) dispersion and 5) ignition source. These factors are often embodied in an "explosion pentagon". Disruption of any one of these factors can prevent a dust explosion from occurring. Before mitigation of a dust explosion can take place, one has to realize the full extent of the explosion potential of the dust/powder material in the intended process environment. This can be achieved by focusing on many ways these five components can come together in a specific process condition to form a pentagon.

The explosion potential of dusts and powders in a given process is an assessment of the hazard risk. This risk is comprised of two components – the probability of the explosion and the severity of the consequences. Both factors have to be fully realized to evaluate the dust explosion risk. Currently, standard test protocols are followed to establish a relative measure of these two factors. However, these standard test methods may not reflect the true process environment and interpretation of the results are required.

The purpose of the current work is to review the current test methodologies for assessing the magnitude of an explosion in a process environment and the ease at which the dust explosion can be initiated. It will also critique how they may be applied to "real world" process environments and potential mitigation steps to use to break up the "explosion pentagon".

1. Introduction

An explosion may occur when flammable gases, liquids or dusts are produced, processed or stored in process plants. The requirements for a dust explosion are often called the "explosion pentagon" [1]. The main elements, or components, of the "explosion pentagon" consist of the following: 1) fuel (like combustible dust, gas, or a hybrid mixture), 2) oxidizer (usually air), 3) degree of confinement, 4) dispersion/suspension and 5) adequate/sufficient ignition source. When the fuel and oxidizer are present in appropriate ratios, they comprised the flammable atmosphere. The confinement is usually the walls of the vessel, equipment or confined spaces where dust is dispersed. Dispersion may occur naturally in process like in pneumatic transport or may occur due to a process upset. Ignition sources occur due to causes such as electrostatic buildup, frictional or mechanical impact, and self-heating or hot surfaces. Mitigation of a dust explosion will require disruption of any one of these five factors.

An explosion can be prevented from occurring by either:

- Avoiding the development of explosible mixtures (combustible dusts, flammable gases) or
- Replacing the atmospheric oxygen with an inert gas, or
- Preventing the occurrence of effective ignition source.

Avoiding confinement or dispersion in a process environment is very difficult since these are inherent when operating industrial equipment.

One logical approach of determining the risk of a dust explosion is to evaluate the in-process powder (liquid and/or gas) for reactivity, ignitability and energy output. Once these aspects are established, an engineering hazard analysis can be conducted to determine the level of hazard and potential risk in typical processing applications.

2. Risk Of A Dust Explosion

In the chemical industry, there is great concern over loss prevention and mitigation of risk to process, equipment and labor. Explosions due to combustion of powders and dusts are a major source of loss in this sector. However, dust explosions are the least understood hazards facing the process industries today. There have been many explosions that have caused significant loss since the first reported dust explosion in 1785 in Turin, Italy [2]. The key to determine whether your process is at risk of a dust explosion is to have an understanding of the fire, explosion and thermal stability characteristics of your powders. It is preferable to determine the explosibility characteristics of an industrial dust by testing, because published data are for particular size distribution that may be different from the dust in question. The hazard risk comprising of the probability of the explosion and the severity of the consequences have to be realized to evaluate the dust explosion risk. ASTM tests are followed to establish a relative measure of these factors. The best method and guidelines for selection of tests to evaluate the hazards will be critical (see Figure 1).

2.1 Severity of Consequences

To determine the severity of consequences of a dust explosion the following ASTM Standard can be utilized:

- Explosion Severity Tests ASTM E 1226 (P_{max}, dP/dT_{max} and K_{St}). This test provides a relative measure of the explosion magnitude with respect to pressure output and rate of pressure rise as a function of powder concentration. It is important to remember that the overpressure may be slightly higher in actual industrial scale process (the test chamber is only 20 L and this limited size may not allow full flame development for some powders). In addition, the rate of pressure rise is heavily dependent on the dust turbulence level. The process environment under study may have lower, or greater turbulence, than that found in the standard test vessel. This turbulence difference should be considered when applying the test results to actual industrial dust explosion scenarios.

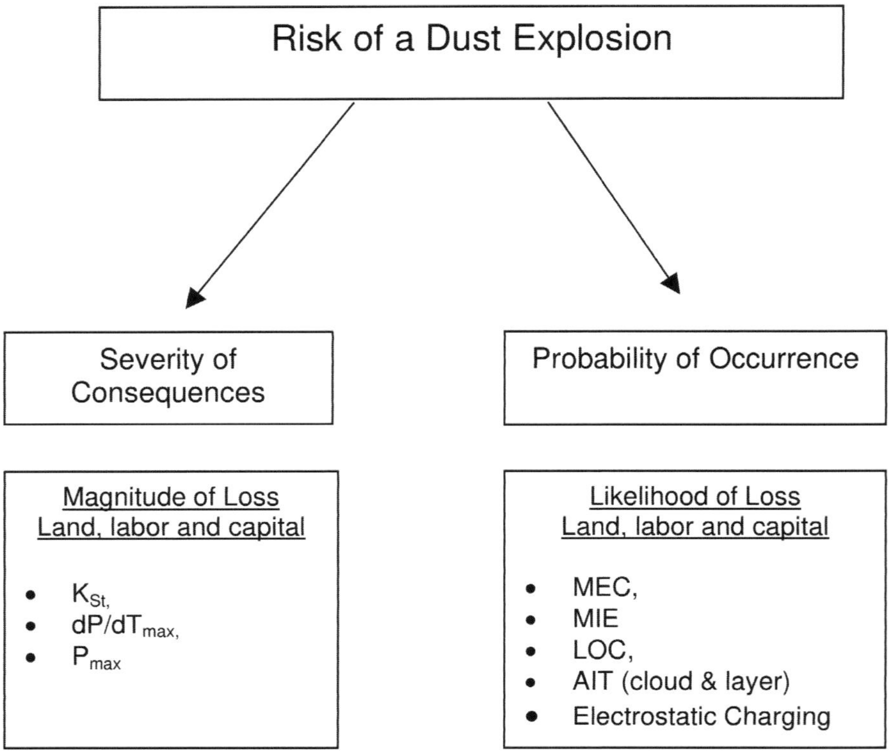

Figure 1. Risk of a dust explosion hazard is a product of the severity of the consequence and the probability of the loss occurring.

2.2 *Probability of Occurrence:*

If no information is available on the reactivity characteristics of the materials, standards (ASTM) for explosibility testing and analysis were written to determine the parameters such as:

- Minimum Explosible Concentration (MEC) ASTM E 1515. This test measures the amount of dust, when dispersed in air, required to propagate a dust explosion. This equates to the concept of lower flammable limit that is discussed with respect to gases. The lower the MEC the smaller that amount of dust required to create a dust explosion. Also, the lower the MEC the greater the likelihood that a sufficient quantity of dust will be lofted and made available for combustion during a process upset. It is important to note that the MEC is ignition energy dependent. Too energetic an ignition source may result in an unrealistically low MEC, or indicate that a dust is explosible when it is not.

- Minimum Ignition Energy (MIE) ASTM E 2019. This test determines, on a relative basis, the least amount of spark discharge energy required to initiate a dust explosion. The lower the MIE of your dust/powder, the easier it would be to ignite and, since small discharges occur more readily than big ones – the greater the likelihood that the process dust will be ignited by an electrostatic discharge. The MIE reported by this test method may vary depending on the test circuit used. The actual discharge geometry and circuit which may occur during a process upset or incident may not be similar to the test circuit in the standard. As a result your dust/powder may ignite at slightly lower (or higher energies) than determined by the test method.

- Limiting Oxygen Content (LOC). This test determines the least amount of oxygen required for explosion propagation through a cloud of dust. Again, as in the previous two test methods – the lower the LOC, the lower the amount of oxygen required for your dust to propagate combustion. Therefore, the lower the LOC level, the greater the likelihood that sufficient oxygen will be present to facilitate combustion making it harder to remove the "oxidant" component of the "explosion pentagon". An ASTM test is currently under development.

- Autoignition Temperature of Dust Cloud and Dust Layer (AIT) ASTM E 1491 and ASTM E 2021. These test methods measure the lowest temperature environment required to propagate combustion in a suspended cloud or in a layer. The lower these values the greater the likelihood that the minimum required propagation temperature conditions will be experienced, thus the greater the probability for combustion. These tests are also limited by the nature of the test apparatus used, like those above, and thus provide a good relative measure of your hazard risk, but actual industrial hazards may have slightly different risks.

- Electrostatic Charging (ESC) Tendency. This test determines the propensity of a powder/dust to accumulate a charge under flow conditions. The greater the charging tendency, and generated charge, the greater the risk for a dust explosion since it is more likely that discharges above the MIE will occur. This test is a relative measure and powder flow in industrial scenarios may differ slightly depending on the flow rates of the powder and the type of duct workings used for transport.

3. Hazard Analysis

Once the powder/dust/hybrid materials are characterized, a five-stage hazard analysis can be started in many situations. The materials reactivities may not be well defined prior to the hazard analysis, or safety audits. In these cases, process/facility evaluations can begin, but cannot be concluded until appropriate material reactivity is determined. A list of steps, which should be taken to evaluate fire/explosion hazards in process, is listed in Figure 2. The steps will facilitate an organized review of the dust explosion hazards present.

As an integral part of the investigation, two different formal analysis tools should be used for the task. Some of these tools are FMEA, FTA, and HAZOP. It is beyond the scope of this paper to discuss the details of these methodologies, however, each of them approach process safety uniquely and can complement each other. Additionally, they can act as a crosscheck for each other where one approach may even point out oversights made in the second.

Dust explosion hazards can be identified according to a specific process and equipment. A summary of explosion propagation, and initiation potential for various processes/equipment are shown in Table 1.

Table 1. Process and Equipment Dust Explosion Hazards

Process Equipment	Explosion Potential	Propagation Potential	Initiation Sources
I. Transport			
Air Conveying	High	High	ESD, Metal Sparks, Friction
Belt Conveyors	Low	Moderate	ESD, Metal Sparks, Friction, Heat
Elevators	High	Very High	ESD, Metal Sparks, Friction
II. Transfer			
Separators	High	Moderate	Heat, Friction, ESD
Cyclones	Very High	Very High	ESD, Metal Spark
Receivers			
Baghouse	High	Moderate	ESD, Heat Metal Spark, Hot Embers
Cartridge	High	Moderate	ESD, Heat Metal Spark, Hot Embers
Silos & Bins	Very High	Very High	ESD, Heat Metal Spark, Hot Embers
FIBCs*	Moderate	Moderate	ESD, Metal Sparks, Hot Embers
III. Grinding	Very High	Very High	Metal Sparks, Heat
IV. Blending	Very High	High	ESD, Metal Spark, Heat
V. Dryers			
Fluid Bed	Very High	High	Heat, ESD, Hot Embers
Spray	Very High	High	Heat, ESD, Hot Embers
Flash	Very High	High	Heat, ESD, Hot Embers
Belt	Moderate	Moderate	Heat, Friction, Hot Embers
VI. Screening	Moderate	Very High	ESD, Friction, Hot Embers, Metal Spark
VII. Granulating	High	Moderate	ESD, Friction, Heat, Hot Embers
VIII. Dust Collecting Receivers	Very High	Very High	ESD, Heat, Hot Embers, Metal Sparks
IX. Mixing Into Reactors	Very High	Very High	ESD, Metal Spark, Heat
X. Bagging & Filling FIBCs	High	High	ESD, Hot Embers, Metal Spark, Heat
XI.Emptying Bags/ FIBCs	High	High	ESD, Hot Embers, Metal Spark, Heat

**Flexible Intermediate Bulk Container*

A number of safety codes now address the dust/vapor explosion potential depending on the type of industry or operations. Codes of a general nature are listed as follows:

NFPA Codes:

- NFPA 65, 480, 481 for Combustible Metals and Metal Dusts
- NFPA 68, 69 for Explosion Protection Systems
- NFPA 91, 650, 654, 655 for Handling and Conveying of Dusts, Vapors and Gases
- NFPA 5000

Universal Building Codes:

- Requirements for Group H
- Occupancy, Section 307.1

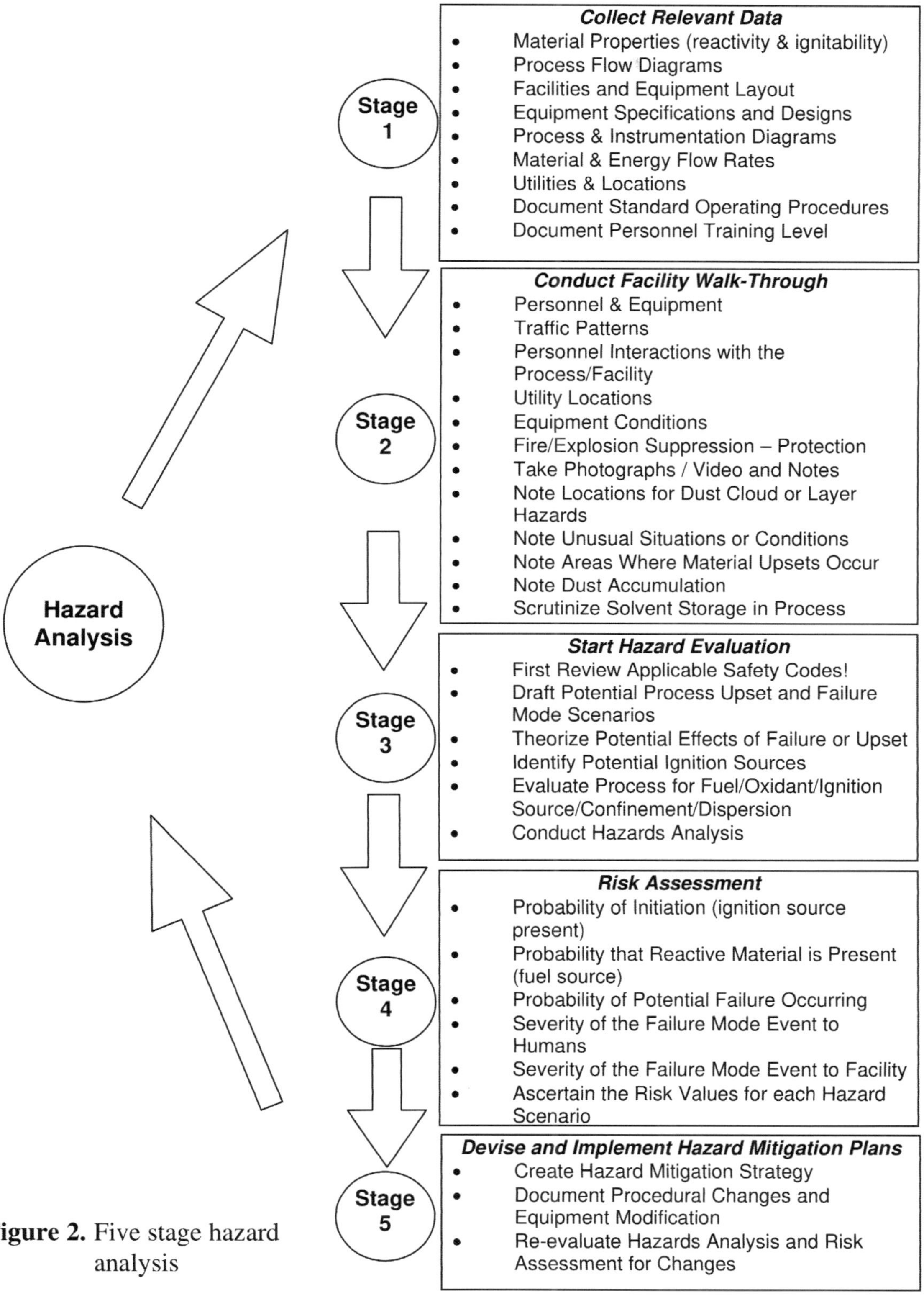

Figure 2. Five stage hazard analysis

4. Criteria For A Successful Hazards Analysis

A successful hazards analysis is when all the hazards have been identified. We know that we have identified all the hazards when the level of effort, knowledge and experience of the hazards analysts dictate the level of hazard identification. Another factor is related to the ease of getting process information and the level of openness or availability of plant personnel. This also relates to the conceived degree of hazard by management and plant personnel.

To be successful, the hazard analysis should include hazards associated with the following:

- Mechanical Items
- Each Equipment Items
- Electrical or Electro-Mechanical Items
- Electronic and Computer Control Items
- Hydraulics and Pneumatics
- Chemical Reactions, or lack of them
- Sequences of Operations
- Procedures, or lack of such procedures
- Utility Losses
- Facility Conditions
- Process Conditions
- Startup Sequence(s)
- Shutdown Sequence(s)
- Emergency Shutdown
- Maintenance Functions

The success of the hazard analysis can also be measured by whether the identified hazards are realistic and credible and the consequences are realistic.

Many times, plant personnel do not understand or accept the identified consequences. Recent plant incidents investigated by SCE indicated that a lack of thorough knowledge of the ignition and explosion behavior of dust-air mixtures permitted an explosion to occur. In one case, the ignition behavior and the explosion hazard of dust accumulation on ledges of trusses was not identified. The knowledge of a dust's minimum ignition energy, particle size distribution, electrostatic chargeability and resistivity would have allowed them to define the potential for a dust explosion. Once the potential was understood the necessary protective measures for their industrial operations could have been implemented.

5. Summary

Methods have been presented to evaluate the dust and hybrid explosion hazards in plants and process operations. Ways to evaluate hazards have been presented relative to material reactivity, ignitability and explosion output. Criteria for successful engineering hazards analysis have been identified. In powder operations, dust and hybrid fires and explosions can be minimized so that the potential is never near the operational conditions including startup, shutdown and emergency shutdown. Safety codes and testing have also been presented.

6. References

[1] Stephan C.R., Coal Dusts as a Fuel for Fires and Explosions, Mine Safety and Health Administration, Pittsburgh, PA, Report No. 01-066-90, 1990, 10pp.

[2] Eckhoff, R., *Dust Explosions in the Process Industries,* Second Edition, Butterworth-Heinemann, Cambridge, U.D. 1997

Dust Explosion Venting through Ducts

Erdem A. Ural
Loss Prevention Science and Technologies, Inc.
659 Pearl Street
Stoughton, MA 02072
Tel: +1-781-344-7656
erdem.ural@lpsti.com

Prepared for Presentation at
American Institute of Chemical Engineers
2005 Spring National Meeting
39th Annual Loss Prevention Symposium
Atlanta, Georgia
April 11-13, 2005

Dust Explosion Venting through Ducts

Erdem A. Ural
Loss Prevention Science and Technologies, Inc.
659 Pearl Street
Stoughton, MA 02072
Tel: +1-781-344-7656
erdem.ural@lpsti.com

Warning

This technical paper presents the results of research recently conducted by the author, and is issued for the purpose of peer review. Therefore, the equations provided herein should not be used for safety related design until the consensus review process has been completed.

Abstract

Explosion venting is often the lowest cost solution among available protection options. On the other hand, the burnt and unburnt material ejected from a vent opening can create a serious threat to personnel or to vulnerable equipment located near the vent. In situations where such a threat exists, installing a vent duct to redirect the discharge to a safe location is a common solution.

Through various factors discussed in the paper, the presence of the vent duct increases the resistance to vent discharge. Several public guidelines such as NFPA 68, prEN 14491, and VDI 3673 provide differing methodologies to account for this effect. Besides, currently available public methodologies are limited to idealized straight vent ducts only.

This paper reviews available test data and our contemporary understanding of the phenomena involved in deflagration venting through ducts. Then, an improved calculation method is proposed. The new method eliminates the problems identified with the published methods. It can also handle real life vent duct installations which may have obstructions such as bends, elbows, bird screens or rain covers.

Vent duct effects are accounted for using simple algebraic correlations. Improved correction factors for vent cover deployment pressure (Pstat), and enclosure aspect ratio (L/D) are also developed and presented in this paper. These new correlations are also applicable to simple vents without ducts.

The practical use of the correlations are illustrated by solving two sample benchmark problems.

Depending on the details of the venting application, the new correlations may require smaller or larger vent areas as substantiated by available test data. Most importantly, the new correlations expand substantially the number of practical applications that can be solved with explosion venting.

1. Introduction

Various explosion protection options are discussed in NFPA 69 and NFPA 68. Users often select the venting solution because of its low initial and recurring costs[1]. Explosion venting is also attractive because design equations are in the public domain. Thus, anyone can check to verify in the field whether a vent is properly sized. Most importantly, any scientist or practitioner can review the design equations, identify problems, or propose fixes and improvements as the author is doing in this paper.

Owing to its great practical importance, vent ducts have been the subject of a number of experimental and analytical studies. Most noteworthy is the experimental work conducted at the Health and Safety Executive in UK, which was reported by Lunn et al (1988a), and later by Hey (1991).

Analytical studies spanned a broad range in complexity, all the way from partial differential equations in computational fluid dynamics software to simple algebraic data correlations. A greatly simplified model (Ural, 1993), utilized ordinary differential equations to capture all the salient features of the data but is too complicated for use in the standards. Lessons learned from that model are discussed in Section IV of this paper and are used to develop the improved algebraic vent duct correlation presented in this paper. Earlier, Tamanini (1995) had used the conceptual framework of the Ural (1993) model to develop a vent duct correlation that is used by the Factory Mutual DustCalc model.

Presently, there are four widely used methodologies for vent duct design: VDI 3673, CEN (prEN 14491), NFPA 68, and FM Global DustCalc. The first three are publicly available, while the fourth is proprietary.

Since the NFPA 68, "Guide for Venting of Deflagration" is being modified for conversion to a standard as soon as 2007, a closer scrutiny of the vent sizing equations was warranted. This paper focuses on the sizing of explosion vents for enclosures with or without vent ducts when:

- the enclosure is initially at nearly atmospheric pressure,
- explosible dust cloud can fill the entire enclosure, and
- the vent cover has negligible mass.

---

[1] *For example, the design, installation, maintenance, and decommissioning of an explosion suppression system may cost 10 times or more when compared to the cost of an explosion vent for the same application.*

Explosion venting correlations are simple algebraic expressions, given in the most general form as:

$$\mathbf{Av = fcn(V, Pred, Pstat, Kst, Pmax, (L/D)_{enclosure}, (L/D)_{Vent\ Duct}, etc.)} \qquad \mathbf{(1)}$$

where:

Av: vent area (m^2)
V: enclosure volume (m^3)
Pred: maximum pressure of the vented explosion (barg)
Pstat: vent cover deployment pressure (barg)
Kst: volume scaled maximum rate of pressure rise in a standard unvented enclosure (bar-m/s)
Pmax: maximum pressure rise in a unvented enclosure (barg)
$(L/D)_{enclosure}$: Aspect ratio of the enclosure
$(L/D)_{Vent\ Duct}$: Vent duct length to diameter ratio.

The choice of units shown above are the default units in NFPA 68, and are used throughout this paper.

The parameters Kst and Pmax are unique properties of the particular dust posing the explosion hazard, and are determined using a standard test method (ASTM E 1226). Theoretically, Pmax indicates how much energy dust releases during an explosion, while Kst relates to how fast the explosion propagates, or to the energy release rate per unit flame surface area during the explosion propagation.

In common practice, equation (1) is used three different ways:

(i) to calculate necessary vent area for a specified enclosure strength (Pred) and dust Kst,
(ii) to calculate necessary enclosure strength (Pred) when the vent area and dust Kst are known, and
(iii) to calculate the maximum dust Kst that can safely be handled in a particular installation.

Modern venting standards such as NFPA, VDI and CEN take a modular approach to the development and expression of equation 1. More specifically, they each contain the following components:

A. Basic Equation applicable to simple compact enclosures,
B. Correction factor for Pstat when it is different than 0 or 0.1 barg,
C. Correction factor for enclosure L/D when it is different than 1 or 2,
D. Correction factor for the presence of vent ducts.

Finding a solution that will satisfy all these correlations at the same time, often requires the use of a trial and error method. Practical benchmark problems described in Figure 1 will be used to illustrate how the existing correlations, as well as the new correlations presented in this paper, are used.

2. NFPA 68-2002 Equations: Background And Related Issues

2.1 Basic Equation for simple compact enclosures

The 2002 edition of NFPA 68 provides a single basic equation to be used for both low strength (Pred < 0.1) and higher strength enclosures. The derivation and discussion of the form of the current equation

$$\mathbf{Av} = \alpha \cdot \mathbf{Kst} \cdot \mathbf{V}^{3/4} \cdot \sqrt{\mathbf{Pmax}/\mathbf{Pred} - 1} \qquad (2)$$

can be found in Ural (2001). The value of the adjustable coefficient α was selected to be 10^{-4} in order to achieve the best fit[2] to the VDI data tables provided by Radant[3]. These data tables do not provide the actual data, instead they were created by Dr. Radant's team from an analysis of the actual test results. All the tests were conducted with plastic vent covers nominally rated at Pstat = 0.1 barg. Ural (2001) and Ural (2003) provided extensive comparisons of equation (2) predictions to the results from the previous NFPA 68 edition.

2.2 Correction factor for Pstat when it is different than 0 or 0.1 barg

It is generally considered that higher vent cover deployment pressures tend to create additional turbulence inside the enclosure which effectively enhances the Kst. The Pstat correction factor in the 2002 edition was adopted from Factory Mutual's DustCalc model, and is given in the form:

$$\mathbf{Kst_{effective}} = \mathbf{Kst} \cdot (1 + 1.75\mathbf{Pstat}) \qquad (3)$$

Thus, the basic equation with the Pstat correction was obtained:

$$\mathbf{Av} = \alpha \cdot \mathbf{Kst} \cdot (1 + 1.75\mathbf{Pstat}) \cdot \mathbf{V}^{3/4} \cdot \sqrt{\mathbf{Pmax}/\mathbf{Pred} - 1} \qquad (2a)$$

Since this Pstat correction starts at Pstat = 0, α was reduced to 0.85 10^{-4} to match the Radant's table at Pstat = 0.1 barg. This Pstat correction factor is substantially different from that used in the European standards and in the older editions of NFPA 68. It is also is also known to be responsible for large differences observed in low Kst – high Pstat applications (Ural, 2003).

A comparison of the Equation (2a) predictions with data is shown in Figure (3a). The variable ranges in the test matrices can be found in Table I. Equation (2a) is seen to underestimate the data when the Pstat is less than 0.1 barg, which might be a problem for building vent designs.

2.3 Correction factor for L/D of elongated enclosures

In situations where the vent opening is provided at the end of an elongated enclosure, internal flow created by the vent discharge distorts and enlarges the flame front beyond that which would

---

[2] *The equations in the 2002 edition of NFPA 68 were designed to provide the best fit to the data, unlike those in the previous editions which attempted to envelope most data.*

[3] *See Table I for a summary of the data sources (and the parameter ranges) used in this study*

be experienced in a compact enclosure. To keep the reduced pressure the same, an additional vent area, $\mathbf{\Delta Av}$, beyond that predicted by Equation (2a) is required for elongated enclosures. The correction correlation in the 2002 edition of NFPA 68 is

$$\frac{\mathbf{\Delta Av}}{\mathbf{Avo}} = 1.56\left(\frac{1}{\mathbf{Pred}} - \frac{1}{\mathbf{Pmax}}\right)^{0.65} \mathbf{Log}\left(\frac{\mathbf{L}}{\mathbf{D}} - 1\right) \qquad (4)$$

where Avo is the vent area calculated for the same volume compact enclosure calculated using Equation (2a). This formula is used for enclosure L/D ranging from 2 to 6. Additional vent area requirement increases with enclosure L/D, and decreases with increasing Pred.

As will be discussed in the section III, this equation is found to be responsible for unnecessarily large vent area requirements for low Pred applications, and for unrealistic trends[4] in vent duct calculations.

Previous editions of NFPA 68 used the same L/D correction factor used in VDI and CEN standards, which is given by:

$$\frac{\mathbf{\Delta Av}}{\mathbf{Avo}} = \left(0.758 - 4.305 \cdot \mathbf{Log(Pred)}\right)\ \mathbf{Log}\left(\frac{\mathbf{L}}{\mathbf{D}}\right) \qquad (4a)$$

This equation forces the enclosure L/D correction to zero for applications where Pred ~ 1.5 barg, and is not used for higher Pred applications. New edition of VDI 3673 allows this equation to be used for L/D values of up to 20 so long as the calculated vent area does not exceed the enclosure cross sectional area.

2.4 Correction factor for the presence of vent ducts.

Presence of a vent duct tends to impede the vent discharge thus increasing the reduced pressure. If the enclosure can not tolerate the Pred increase due to vent ducts, a larger vent area is required. The 2002 edition of NFPA 68 uses a formula developed by Siwek (1998):

$$\frac{\mathbf{P'red}}{\mathbf{Pred}} = 1 + 17.3\left(\frac{\mathbf{Av}}{\mathbf{V}^{0.753}}\right)^{1.6}\left(\frac{\mathbf{L_{duct}}}{\mathbf{D_{vent}}}\right) \qquad (5)$$

where P'red, and Pred are the reduced pressures when the vent duct is present, and absent, respectively.

The latest CEN and VDI standards also uses Equation (5), except for the vent diameter D_{vent} in the denominator of Equation (5) is omitted in the VDI 3673.

[4] *Such as P'red calculated by Equation (5) increasing with increasing Av. Same problem has also been reported for VDI 3673 calculations.*

Siwek correlation also claimed that duct lengths beyond

$$\mathbf{Ls} = 3.764\,\mathbf{Pred}^{-0.3724} \tag{6}$$

impose no additional pressure increase. Therefore, if Lduct > Ls, Lduct is replaced with Ls value calculated from Equation (6). In the latest CEN and VDI standards Equation (6) has been increased by 20% to

$$\mathbf{Ls} = 4.564\,\mathbf{Pred}^{-0.37} \tag{6a}$$

VDI 3673 claims that "the influence of the vent duct upon the pressure increase is most pronounced when the flame propagation from the secondary explosion in the vent duct reaches the velocity of sound." This statement contradicts the science of gas dynamics. The magnitude of this "maximum contribution" duct length can also not be explained with the limited quantity of dust/air mixture available for explosion. Therefore, this concept has been eliminated with the new correlation.

2.5 Comparisons with Data:

Predictions of the NFPA 68-2002 are compared to the HSE vent duct data in Figure (2). The current edition is seen to underpredict substantially the majority of the data. The concept of "maximum contribution" duct length and Equations (6) or (6a) are also not supported by the data shown in Figure (2).

2.6 Sample calculations using NFPA 68-2002:

BENCHMARK PROBLEM A

Necessary details of this straight duct application for elongated enclosure are shown in Figure 1. The reduced pressure Pred to be calculated must satisfy, simultaneously, Equations (2a), (4), (5) and (6).

First, calculate the reduced pressure Predo if the vent duct were not present. The requirement is that Av + ΔAv (calculated from equations (2a) and (4) respectively) must be equal to the vent area of 1.77 m^2 specified in Figure 1. A trial and error procedure converged to Predo = 0.655 bar-g.

Second, calculate the maximum contribution duct length from Equation (6) and compare it to the actual duct length. Using Predo = 0.655 barg, Equation (6a) predicts Ls = 4.41 m, which is smaller than the 12 m actual duct length.

Finally, equation (5) is used to calculate reduced pressure P'red when the duct is present. Inserting Lduct = 4.41 m, Pred = 0.655, and other design parameters from Figure 1, the reduced pressure is calculated to be 1.3 barg.

BENCHMARK PROBLEM B

Since the benchmark problem B involves elbows and a rain cover, it cannot be solved using the NFPA 68 equations. For the same reason, the benchmark problem B cannot be solved using the equations in VDI and CEN standards either.

3. Improving Pstat And Enclosure L/D Correction Factors

As can be seen from Section II, and the example problem above, vent duct models build upon the basic equation and the enclosure L/D correction equation. In order to improve the reliability of the new correlations this work has started by fixing the identified problems associated with the Pstat correction factor, and the enclosure L/D correction factor.

3.1 New Pstat correction factor:

As seen in Table I, effect of Pstat on reduced pressure can be deduced from the HSE data without ducts documenting test results for Pstat of 0, 0.05, 0.1, 0.2 and 0.5 barg. This data set, as well as the Radant set (for Pstat = 0.1 barg only), are plotted in Figure (3a) in a coordinate system selected to isolate the Pstat effect from Equation (2a). The ordinate parameter α was calculated using Equation (2). This plot suggests that Pstat has little or no effect on the Pred (or Av) when the Pstat is less than or equal to 0.1 barg. Figure (3a) also indicates that the proposed curve $\{\alpha = 10^{-4}\ (1+1.54\ Pstat^{4/3})\}$ represents the data trend better than the $\{\alpha = 0.85*10^{-4}\ (1+1.75\ Pstat)\}$ form used in the 2002 edition of NFPA 68.

Therefore, the basic equation for compact enclosure with Pstat correction should be replaced with the following equation:

$$\mathbf{Av} = 10^{-4} \cdot \mathbf{Kst} \cdot (1 + 1.54 \cdot \mathbf{Pstat}^{4/3}) \cdot \mathbf{V}^{3/4} \cdot \sqrt{\mathbf{Pmax}/\mathbf{Pred} - 1} \qquad (7)$$

The numerical coefficients, 10^{-4} and 1.54, are not non-dimensional, and are consistent only with the default units shown after equation (1). Equation (7) requires an approximately 15% larger vent area for buildings when compared to that required by the 2002 edition of NFPA 68.

A comparison of the data with the predictions of Equation 7 is provided in Figure (3b). The plus or minus 40% data scatter seen in both Figures (3a) and (3b) is not uncommon in dust explosion tests.

A statistical analysis revealed that the following formula

$$\mathbf{Av} = 1.1 \cdot 10^{-4} \cdot (\mathbf{Kst} + 315 \cdot (\mathbf{Pstat} - 0.1)) \cdot \mathbf{V}^{3/4} \cdot \sqrt{\mathbf{Pmax}/\mathbf{Pred} - 1} \qquad (7a)$$

provides a better fit to the majority of the data. In applications where Pstat is less than 0.1 barg, Pstat = 0.1 barg should be used with Equation (7a). Equation (7a) was not used in the vent duct analysis because it underestimates the Pstat effect for high Kst dusts (more specifically, Lunn's aluminum flake data point (Kst = 630 bar-m/s).

The only problem seen in Figure (3b) involves Lunn's low Kst data set, which consists of test results for 40 to 60 weight % mixtures of coal dust (Kst = 144 bar-m/s) and stone dust (inert). Mixture Kst values measured in a 20 liter sphere ranged from 15 to 56 bar-m/s. A portion of this data clearly does not group with other test data obtained for single fuels. Some scientists have attributed this behavior to the non-uniformity of the bulk mixture, or to the non-uniformity of the suspended dust cloud at the moment of ignition in the deviant tests. In fact, the deviant data groups well with the other tests if the Kst of pure coal dust is used in place of the mixture Kst values reported in Lunn et al (1988b).

3.2 New Enclosure L/D correction factor:

The cause of a number of difficulties reported by the NFPA 68 users has been traced to Equations (4) and (4a). Therefore, the large L/D enclosure venting data have been re-analyzed to develop an improved correlation. The parameter ranges of the data used in this analysis are listed under the column heading "L/D data" in Table I.

The highlights of the analysis shown in Figure (4) indicate that the correlation:

$$\frac{\mathbf{\Delta Av}}{\mathbf{Avo}} = 0.6 \cdot \left(\frac{\mathbf{L}}{\mathbf{D}} - 2 \right)^{0.75} \cdot \mathbf{exp(-0.95 \cdot Pred^2)} \qquad (8)$$

provides a much better representation of the data. Since the data went up to L/D = 6.25, this equation might be extrapolated for enclosure L/D values of up to 8. When the enclosure L/D<2, ΔAv/Avo is not defined and should be taken as zero.

Figure (4b) shows that the enclosure L/D correction given in the 2002 edition of NFPA 68 is overly conservative when compared to the data for L/D = 3. Also, it seriously over-predicts the L/D effect when the Pred is smaller than 0.4 barg, and larger than 1 barg. In fact, the L/D correction tends to infinity as the Pred approaches zero. This behavior does not seem to be supported by available data.

Figure (4c) shows that the enclosure L/D correction given in VDI 3673, and in earlier editions of NFPA 68 is overly conservative when the Pred is smaller than 0.3 barg, and also tends to infinity as the Pred approaches zero. Figure (4c) also shows that VDI correction factor is not conservative when Pred is greater than 1.2 barg, but the error is small in this range.

4. Physics Of Deflagration Venting Through Ducts

A description of the phenomena is provided in Ural (1993). As seen in Figure 5 reproduced from that paper, deflagration initiates in the enclosure and proceeds as an unvented explosion until the internal pressure reaches the vent deployment pressure. After that moment, the vent cover opens, and the unburnt fuel-air mixture starts spilling into the duct. Later in the process, the fireball grows to a size sufficiently large to ignite the fuel air mixture spilled into the duct. The ignition timing of this secondary explosion could have a significant effect on the maximum overpressure developed in the protected vessel and in the duct. Inside long ducts, flame can run out of the dust/air mixture and get extinguished. Factors expected to affect the maximum explosion pressure are:

- inertia of the fuel/air mixture occupying the duct,
- flame speed and flame acceleration in the enclosure as well as in the duct,
- the fact that fuel/air mixture in the duct is limited to that originally in the enclosure,
- flow resistance,

- wave propagation and interactions in the duct, and
- choking of the duct due to flow resistance or combustion heat release.

The last mechanism is not expected to be important for most practical duct applications since there is a substantial pressure drop at the inlet of the duct, and the quantity of the fuel/air mixture in the duct is limited to that originally existed in the enclosure[5].

As seen from the gas explosion data comparisons in Figure 6, Ural (1993) model captures all salient features of pressure development. This model suggests that the first three factors are dominant, while the flow resistance is if secondary importance for most straight unobstructed ducts.

Forward and backward traveling compression and expansion waves are known to be important for flame propagation in pipes and ducts closed at the ignition end. An analysis has shown that this mechanism would not play as significant a role in most practical vent duct geometries for the following three reasons:

1) Duct inlet is open to the enclosure so that any backward traveling compression wave is reflected from the inlet as an expansion wave. Compression waves generated by reflections from the internal surface of the enclosure are likely to have lost their coherence (i.e. get longer and flatter).
2) Large convective velocity in the duct induced by the vent discharge flow severely impedes the backward wave travel .
3) Large convective velocity in the duct is also responsible for rapid sweeping of the flame zone through the duct. Thus, compared to the closed end scenario, flame has much less time to accelerate in the vent duct.

In fact, Hey (1991) reports that in the HSE 18.5 m^3 enclosure tests, highest pressures were generally measured in the enclosure, and not in the duct.

The analysis further showed that there are three important time scales that need to be considered when the data are correlated:

1) Time scale of explosion in the enclosure, Tc
2) Enclosure discharge time scale, Tdisch
3) Duct inertia time scale, Th

As was done the work described in Ural (2001), these time scales have been calculated using a simple model for the low Pred asymptote[6], and have been expressed explicitly in terms of the practical vent design parameters. Because both the data and theory are ambiguous about the Pmax effect, Pmax values are not shown in the time scale expressions below.

---

[5] *Vent covers are placed at the inlet of the duct. Good housekeeping practices should be designed to avoid accumulating large quantities of dust in the duct.*
[6] *The strategy used in both papers is to focus on the more practical low Pred regime while forcing the equations to go to the correct asymptote at high Pred values.*

The simple analysis reveals that the combustion time scale is:

$$\mathbf{Tc} \sim \frac{\mathbf{V}^{1/4}}{\mathbf{Kst} \cdot (1 + 1.54 \cdot \mathbf{Pstat}^{4/3})} \cdot \sqrt{\mathbf{Pred}}$$

where the symbol ~ is used to designate proportionality. This expression accounts for the $V^{1/12}$ enhancement[7] of the Kst (or the burning velocity) with the integral length scale of turbulence as discussed in Ural (2001). It also accounts for the enhancement of the Kst due to vent deployment pressure discussed above.

Similarly, the enclosure discharge time scale is calculated as

$$\mathbf{Tdisch} \sim \frac{\mathbf{V}}{\mathbf{a_o} \cdot \mathbf{Av}} \cdot \sqrt{\frac{\mathbf{Po}}{\mathbf{Pred}}}$$

where, a_o is the sound speed in the unburnt mixture, and Po is the atmospheric pressure.

Duct inertia time scale is taken as the period of the Helmholtz oscillations observed in Ural (1993):

$$\mathbf{Th} \sim \frac{1}{\mathbf{a_o}} \cdot \sqrt{\frac{\mathbf{V} \cdot \mathbf{L_{duct}}}{\mathbf{Av}}}$$

where L_{duct} is the duct length. This equation can also be derived by considering the enclosure and duct as a simple spring-mass system where the mass is that of the mixture occupying the duct, and the spring coefficient is the resistance force created by the enclosure content (due to compression or expansion) per unit travel of the duct content.

This analysis suggests that the data correlation should be expressed as:

$$\frac{\mathbf{Pred}}{\mathbf{Predo}} = \mathbf{g}\left[\frac{\mathbf{Th}}{\mathbf{Tdisch}}, \frac{\mathbf{Tc}}{\mathbf{Tdisch}}\right] = \mathbf{g'}\left[\mathbf{E_1}, \mathbf{E_2}\right]$$

or

$$\frac{\mathbf{Av}}{\mathbf{Avo}} = \mathbf{f}\left[\mathbf{E_1}, \mathbf{E_2}\right] \qquad (9)$$

where Av and Avo are the vent area requirements for a given Pred in the presence and in the absence of the vent duct, respectively. Correlating parameters are defined as:

$$\mathbf{E_1} = \frac{\mathbf{Av} \cdot \mathbf{L_{duct}}}{\mathbf{V}} \qquad (10)$$

[7] i.e., $V^{1/4} = V^{1/3} / V^{1/12}$

and

$$\mathbf{E}_2 = \frac{10^4 \cdot \mathbf{Av}}{\mathbf{Kst} \cdot (1 + 1.54 \cdot \mathbf{Pstat}^{4/3}) \cdot \mathbf{V}^{3/4}} \qquad (11)$$

Note that both E_1 and E_2 are non-dimensional, while the coefficient 10^4 has a dimension in units consistent with the standard units listed after Equation (1) above. Where it is more convenient, and only for compact enclosures, the parameter E_2 can also be calculated from Pmax and Pred values (see Equation (7).

Flow resistance caused by skin friction and fittings can also be incorporated into equation (9) simply by using the fluid mechanics concept of equivalent discharge coefficient C_d[8], which turns it into:

$$\frac{\mathbf{Av}}{\mathbf{Avo}} = \mathbf{f}\left[\mathbf{E}_1, \mathbf{E}_2\right] \cdot \mathbf{K}^{1/2} \qquad (9a)$$

where the total flow resistance coefficient K is defined on the static pressure drop ΔP from the enclosure to the duct exit at a given the duct flow velocity U, i.e.

$$\mathbf{K} \equiv \frac{\Delta \mathbf{P}}{\frac{1}{2} \rho \mathbf{U}^2} \qquad (12)$$

Additional information on the flow resistance due to frictional and local losses is provided in Section VI.

5. New Vent Duct Correlation

Equation (9a) has been used to correlate the Lunn et al (1988) data and Hey (1991) data for central ignition. The result is the following correlation for vent duct design:

$$\frac{\mathbf{Av}}{\mathbf{Avo}} = \left(1 + 1.18 \cdot \mathbf{E}_1^{0.8} \cdot \mathbf{E}_2^{0.4}\right) \sqrt{\mathbf{K} / \mathbf{Ko}} \qquad (13)$$

where Ko is the nominal flow resistance coefficient corresponding to the venting tests without ducts. All test enclosures are equipped with some form of an attachment to hold the vent cover, which effectively presents a chamfering or a rounding effect to the incoming flow. The flow resistance coefficient also depends on the ratio of vent area to enclosure cross-section area. For the sake of simplicity, a fixed Ko = 1.5 value (corresponding to Cd = 0.8) has been used in the data analysis and hence this value is built into the new correlation given by Equation (13). Other parameters have been defined in Section 4.

---

[8] *In an ad hoc fashion, $1/C_d = K^{1/2}$ accounts for how much vent area increase is required to maintain the same discharge flow rate at the same Pred.*

It is important to point out that the new correlation has been developed as a correction on the vent area rather than a Pred correction as in Siwek's correlation. This deliberate choice eliminates the need for the "maximum contribution" duct length concept built into the current standards either by Equation (6) or Equation (6a) which is clearly not supported by the data shown in Figure (2). As will be seen in the data comparisons below, Pred versus duct length curves predicted by Equation (13) automatically level off as the duct length increases.

6. Calculation Of K, The Total Flow Resistance Coefficient

The total flow resistance coefficient expressed by the factor K in Equation 13 includes both the internal skin friction with the duct as well as local resistance caused by fittings such as elbows, or rain covers mounted on the duct.

6.1 Duct Skin Resistance:

Duct friction component of the flow resistance coefficient is given by:

$$K_{df} = f \cdot \frac{L}{D_h} \tag{14}$$

where L is the total duct length, Dh is the hydraulic diameter, and f is the duct friction factor. For practically all vent ducts, the Reynolds number is so large that fully rough flow regime will be applicable. In this regime, the friction factor is only a function of the ratio of the internal duct surface effective roughness (ε) to duct diameter. Duct friction factor can thus be calculated using the equation:

$$f = \frac{1}{\left[1.14 + 2 \cdot \mathrm{Log}\left(\frac{D_h}{\varepsilon}\right)\right]^2} \tag{15}$$

The effective roughness for smooth pipes and clean steel pipes are typically 0.0015, and 0.046 mm, respectively. Recognizing that the pipes used repeatedly in combustion tests will be corroded, author assumed $\varepsilon = 0.26$ mm for Lunn's pipes, a value which corresponds to a typical surface roughness of cast iron. In his analysis, Lunn used f = 0.025, which corresponds to an effective surface roughness range of 1.3 to 2.8 mm according to Equation 15, which appear to be too large to be credible.

6.2 Local resistance coefficients:

Extensive design data are available in references such as Idelchik (1996) or the latest ASHRAE Handbook.

When using flow resistance data from reference books, it is important to understand which velocity scale, and whether static or total pressure was used in the definition of the coefficient. If different than the definition given by Equation (12), then the reference book value must be properly adjusted. Otherwise, serious errors may arise in section where cross-sectional area changes. An example of rain cover data is shown in Figure 7. The coefficient shown in this figure is defined on the total pressure. In order to convert these values to the definition given in Equation (12), one must be subtracted from all the values. From Figure 7, it is easy to see that a standard rain cover must be placed more than 0.8 duct diameters away from the end of the duct to avoid additional flow resistance and associated Pred penalty.

For elbows employed in the tests, Lunn assumed:

$$\mathbf{K_{45}} = 0.35 \text{ for } 45^{\circ} \text{ elbows,} \tag{16a}$$

and

$$\mathbf{K_{90}} = 1.12 \text{ for } 90^{\circ} \text{ elbows.} \tag{16b}$$

Reference books (e.g. Idelchik, 1996) also show that:

$$\mathbf{K_{inlet}} = 1.5 \tag{16c}$$
$$\mathbf{K_{exit}} = 0 \tag{16d}$$

are appropriate for simple duct inlet and straight duct exit, respectively.

6.3 Total flow resistance:

The total flow resistance is calculated by summing the individual contributions discussed in this section. For example, the flow resistance for the Lunn et al (1988a) test that employed 0.5 m duct diameter, two 45° elbows, and duct L/D = 30, the contributors and the total resistance are calculated as follows:

$\mathbf{K_{inlet}} = 1.5$
$\mathbf{K_{df}} = 0.017*30 = 0.5$
$\mathbf{2*K_{45}} = 2*0.35 = 0.7$
$\mathbf{K_{exit}} = 0$

and

$\mathbf{K = 2.7}$

It is seen that, even for this case involving the smallest vent area and the longest duct in the test matrix, duct inlet loss dominates the flow resistance. This is the reason why the exponent of the "friction loss parameter" is artificially small in the correlation built into the FM DustCalc correlation, which ignored the duct inlet flow resistance.

7. Comparisons With Data

7.1 Data repeatibility and reproducibility:

As discussed in Section III, a plus or minus 40% scatter (or a factor of 2 band) for data is not uncommon in dust explosion tests. This is due various factors which will not be discussed here. Ideally, one would have to perform a number of repeat tests at each condition to obtain the statistics. Due to limited research budgets, repeats are performed only in the small scale tests.

Fortunately, there is some duplicate test data are available for the HSE 18.5 m^3 enclosure with ducts. Lunn et al (1988a), and Hey (1991) reported nominally identical coal dust tests with straight vent ducts without any vent cover. In order to quantify the data scatter issue for the HSE tests, Hey (1991) results[9] are plotted against Lunn et al (1988) results for identical test conditions. The majority of the duplicate test data are seen to agree with each other within a factor of 2.3. Unfortunately, the effects being studied here is well within the factor of 2.3 apparent data scatter observed for coal dust (Kst = 144 bar-m/s). Generally, the data scatter decreases as the Kst increases.

7.2 Comparisons with Lunn et al 18.5 m3 enclosure data:

Figures (9) through (11) show plots of the measured and calculated explosion pressures for coal dust, aspirin dust, toner, and aluminum flakes against the duct length. Explosion pressure for zero duct length is calculated using Equation (7). Calculations shown in these figures were made using the test parameters for each data point. Calculated points were then connected with straight lines[10].

While reviewing the data comparisons, it should be kept in mind that Equation (13) establishes the slope of the calculated points in Figures (9) through (11), whereas Equation (7) is responsible for the offset from the L axis. Within the large data scatter discussed in Section 7.1, both Equations (7) and (13) are seen to capture the data and the salient trends remarkably well.

Figure (12) compares the calculated results to all 18.5 m^3 enclosure data reported by Lunn in three separate plots for straight ducts, for ducts with one or two 45° elbows, and for ducts with a 90° elbow, which. Once again, within the scatter inherent in the data, the agreement is quite good.

7.3 Comparisons with 1 m3 enclosure data provided in Radant's tables:

Radant's tables listed data for 1 m^3 enclosure equipped with straight duct lengths of 0, 2, 4, and 6 meters. Data for three different Kst values (100, 200, and 250 bar-m/s) and two different vent/duct areas (0.0707, and 0.196 m^2) were provided. Since, Pmax and Pstat values were not listed, 9 barg and 0.1 barg were assumed, respectively, in the calculation of explosion pressures. Calculated explosion pressures are compared to measured explosion pressures in Figure (13). Data shown with double-circles designate tests with zero duct length. Equation (7) and (13) are seen to underpredict a portion of the data, especially for no-duct tests. This observation tempted some scientists to increase the Kst values listed by Radant to get the no-duct calculations to match the experimental results. While scientifically not a good practice, this approach was seen to reduce the data scatter, and significantly improved the degree of conservatism of the calculated pressures. Unfortunately, Radant's tables were not accompanied by information on experimental, procedural, and analysis details that are necessary to perform such an independent assessment of the data integrity.

---

[9] *The data points were digitized from the plots published by Hey (1991).*

[10] *The sharp Pred change seen at small duct lengths for ducts with elbow is a reflection of the fact that L=0 corresponds to no elbow.*

7.4 Comparisons with Lunn et al 20 liter enclosure data:

Lunn et al (1988a) also reported a large data set for a 20 liter sphere equipped with varying straight duct lengths. Coal dust, polyethylene dust, aspirin dust, toner, and aluminum dust were used in these tests. Analysis of this data is not provided here since explosion venting standards are applicable for enclosure volumes that are larger than 100 liters.

8. Additional Considerations

8.1 Validation and Extrapolation:

New equations developed in this paper have been validated using data that span the ranges listed in Table I. On the other hand, the explosion venting standards permit vent designs over a much wider range of parameters. Until such time the data to cover all parameter ranges become available, the new equations can not be extrapolated without going through a consensus approval process as in the development of NFPA standards. On the other hand, judicious use of physical reasoning employed in the process of correlating parameter selection provides some level of comfort for extrapolation.

8.2 When The Duct Area Is Different Than The Vent Area:

Equation (13) is valid for cases where the duct diameter is equal to the vent diameter. Small differences between the two may be justified using engineering analysis. If the entire duct length is of uniform but smaller diameter than the vent diameter, the duct area should be used as the vent area in Equation (13). For larger ducts, article by Hey (1991) may be consulted. However it should be kept in mind that Hey did not test large dust concentrations where the secondary explosion effects in the enlarged duct will be more pronounced.

8.3 Effect of Ignition Location:

HSE data sets used to select the adjustable coefficients of Equation (13) employed central ignition with 30 grams of black powder. On the other hand, venting theories suggest larger and larger Pred values as the ignition location is moved away from the vent. Hey (1991) reported exploratory tests with coal dust using an igniter placed in the front, center, and the rear of the enclosure. Test results generally confirm the theoretical expectations.

8.4 Limiting The Duct Length To Avoid DDT:

Flame accelerations and Deflagration to Detonation Transition (DDT) are known to occur in pipes and ducts when the mixture is ignited at the closed end. Chapter 8 of NFPA 68 provides some guidance to avoid such problems. On the other hand the discussion in Section IV suggests that DDT may be more difficult to achieve in a vent duct arrangement. Nevertheless, it will be prudent to limit both the duct length and the duct L/D to those used in the experiments with perhaps a small allowance for extrapolation.

First of all, it is important to recognize that the mass of dust air mixture in the duct is limited by what originally was in the vessel. So, there is a limit to the duct length that can be occupied with the dust-air mixture. A simple analysis shows that the dusty duct length can be estimated with the following equation:

$$\mathbf{L_{dusty} = (Pmax - Pred) \cdot \frac{V}{Av}} \qquad (17)$$

This expression is overly conservative because it was derived by ignoring the pressurization inside the duct. It can be reduced substantially by performing a case specific engineering analysis. Then the effective duct length for DDT evaluation will be the smaller of the Ldusty and the actual duct length, L, i.e.,

$$\mathbf{Leff = min(L, Ldusty)} \qquad (17a)$$

Permitting a modest 10% extrapolation allowance on the Lunn et al (1988a) parameter range, and assuming, rather arbitrarily, a linear sensitivity on the Kst, the following limitations might be considered:

$$\mathbf{L_{eff}} \le \frac{11{,}000}{\mathbf{Kst}} \qquad (18)$$

and

$$\frac{\mathbf{L_{eff}}}{\mathbf{D}} \le \frac{10{,}000}{\mathbf{Kst}} \qquad (19)$$

8.5 How To Apply The New Duct Correlation For Elongated Enclosures

Some practical applications require placement of the vent and the duct at the end of an elongated enclosure. In such cases enclosure L/D and vent duct correction equations may be applied sequentially so that:

$$\mathbf{Av = Av_{Eq(7)} \cdot \left(\frac{Av}{Avo}\right)_{Eq(13)} \cdot \left[1 + \left(\frac{\Delta Av}{Avo}\right)_{Eq(8)}\right]} \qquad (20)$$

where $Av_{Eq(7)}$ is the vent area calculated from Equation (7), and $(\Delta Av/Avo)_{Eq(8)}$ and $(Av/Avo)_{Eq(13)}$ are the elongated enclosure and vent duct correction factors given by Equations (8) and (13), respectively.

It should be emphasized that author is not aware of any data which either supports or disproves the validity of the combination formula given in Equation (20). The adjustable coefficients of Equation (13) were chosen to fit the Lunn et al (1988a) data from a single enclosure which had an L/D ratio of 1.6. On the other hand, VDI 3673 claims “experimental studies have proven that the influence of vent ducts decreases markedly with increased length diameter ratio of the vessel." Siwek (1998) refers to an unpublished work where Bartknecht observed no vent duct effect for an enclosure L/D=6, and enclosure strength of 0.25 barg. Without access to the actual test parameters for this Bartknecht data point, it is impossible to assess whether this claim is valid or whether it is an artifact of the unnecessarily large enclosure L/D correction employed by the VDI 3673 when the Pred is less than 0.3 barg (see Figure 4c).

8.6. Implication For Building Vents Without Duct:

Form of the flow resistance factor in Equation (13) implies that the required vent area is inversely proportional to the discharge coefficient (or proportional to the square root of the flow resistance coefficient). This dependence has been validated with data comparisons shown in Figures (9) through (12).

Test data used to validate Equation (7) had been obtained in enclosures equipped with some sort of fitting to hold vent covers in place. It is likely that the discharge coefficients for the test enclosure vents ranged from 0.8 to 0.9 (i.e. K = 1.5 to 1.2). On the other hand for most building vents the building wall thickness is a tiny fraction of the effective vent diameter. Thus a discharge coefficient of 0.6 (K=2.78) corresponding to sharp edge orifices will be more appropriate.

It is therefore recommended that building vents areas calculated using Equation (7) should be increased by one third to adjust for the effect of the reduced discharge coefficient (i.e. 0.8/0.6).

9. Solution Of The Sample Problems With The New Equations

Problem statements are provided in Figure 1. The reduced explosion pressure Pred to be calculated must satisfy Equation (20) which, simultaneously, makes use of Equations (7), (8), and (13).

9.1 Solution of the Benchmark Problem A

1) Compute the Friction factor for the problem
   - From Eq. (15) f = 0.013
   - From Eq. (14) Kdf = 0.107
   - Kinlet = 1.5
   - Kexit = 0
   - K = 1.607
2) Assume a Pred value = 1 barg
3) From Eq. (7) Av = 0.735 m^2
4) From Eq. (8) DAv/Avo = 0.39
5) From Eq. (10) E_1 = 0.85
6) From Eq. (11) E_2 = 6.37
7) From Eq. (13) Av/Avo = 3.28
8) From Eq. (20) Av = 3.355 m^2

9) Go back to step 2 above until the Av calculated in step 8 is equal to the specified vent area of 1.77 m^2.

A trial and error process (or the goal seek button in Excel) satisfy the requirement in step 9 when Pred = 1.8 barg.

As expected, this value is larger than the Pred = 1.3 barg calculated in section 2.6 using the 2002 edition of NFPA 68.

10) Equations (17) through (19) show that there is no DDT propensity for this particular application.

9.2 Solution of the Benchmark Problem B

1) Compute the Friction factor for the problem

    From Eq. (15) f = 0.013
    From Eq. (14) Kdf = 0.107
    Kinlet = 1.5
    Kelbows = 2.4
    Kexit = 0.75
    K = 4.757

2) Assume a Pred value = 1 barg
3) From Eq. (7) Av = 0.735 m^2
4) From Eq. (8) DAv/Avo = 0.39
5) From Eq. (10) $E_1 = 0.85$
6) From Eq. (11) $E_2 = 6.37$
7) From Eq. (13) Av/Avo = 5.65
8) From Eq. (20) Av = 5.77 m^2
9) Go back to step 2 above until the Av calculated in step 8 is equal to the specified vent area of 1.77 m^2.

A trial and error process (or the goal seek button in Excel) satisfy the requirement in step 9 when Pred = 3.52 barg.

10) Equations (17) through (19) show that there is no DDT propensity for this particular application.

10. Summary And Conclusions

The new method presented in this paper is comprised of a set of simple correlations and procedures, and brings forward significant improvements over those available in the existing standards. The new method is applicable to simple vents without ducts, vents with ducts, compact enclosures, and elongated enclosures.

The new method eliminated a number of issues associated with the existing methods. It also identified applications, such as building vents, where additional vent area is necessary.

Unlike those published the existing standards, the new method can be applied to real vents that may be equipped with objects such as elbows and rain covers that impose additional flow resistance. The new method also expands the enclosure L/D limitation in NFPA 68. Thus, the new method significantly increases the number of explosion protection applications that can be treated using the lower cost venting option.

11. Acknowledgements

Partial funding for this work has been provided by Hosokawa-Micron Powder Systems, and by Procter & Gamble Company.

12. References

CEN Standard, "European Standard prEN 14491 - Dust explosion venting protective systems," European Committee For Standardization, Brussels, Belgium, June 2002.

Hey (1991), M., "Pressure Relief of Dust Explosions through Large Diameter Ducts and Effect of Changing the Position of the Ignition Source," *Journal of Loss Prevention in the Process Industries*, v. 4, pp. 217-222, 1991.

Idelchik (1996), I.E., "Handbook of Hydraulic Resistance," 3rd Edition.

Lunn (1988a), G., Crowhurst, D., and Hey, M. "The Effect of Vent Ducts on the Reduced Explosion Pressures of Vented Dust Explosions," *Journal of Loss Prevention in the Process Industries,* v. 1, pp. 182-196, 1988.

Lunn (1988b), G. A., Brookes D. E. and Nicol A., "Using the Kst nomographs to estimate the venting requirements in weak dust-handling equipment," *Journal of Loss Prevention in the Process Industries*, 1988, v. 1, pp. 123-133, 1988.

NFPA 68 (2002), "Guide for Venting of Deflagrations," National Fire Protection Association, Quincy, MA.

NFPA 69 (2002), "Standard on Explosion Prevention Systems," National Fire Protection Association, Quincy, MA.

Radant Silo Data, confidential data tables submitted for NFPA 68 committee use (Date unknown).

Siwek (1998), R., "New Findings on Explosion Venting," 9th International Symposium on Loss Prevention and Safety Promotion in the Process Industries, 1998.

Tamanini (1995), F., "An Improved Correlation of Experimental Data on the Effect of Ducts in Vented Dust Explosions," 8th International Symposium on Loss Prevention and Safety Promotion in the Process Industries, 1995.

Ural (1993), E.A. "A Simplified Method for Predicting the Effect of Ducts Connected to Explosion Vents," *Journal of Loss Prevention in the Process Industries*, v. 6, pp. 3-10, 1993.

Ural (2001), E.A. "A Simplified Development Of A Unified Dust Explosion Vent Sizing Formula," *Process Safety Progress*, v. 20, no.2, pp.136-144, 2001.

Ural (2003), E.A. "NFPA 68: How the revised explosion venting guidelines affect your plant," *Powder and Bulk Engineering*, June 2003.

VDI 3673, "Pressure venting of dust explosions," Verein Deutscher Ingenieure, Düsseldorf, Germany, November 2002

VDI (Radant) Data, confidential data tables submitted for NFPA 68 committee use (1993).

Table I
The data sets and the test variable ranges used in this study.

	Basic Equation			Enclosure L/D	Vent Ducts	
	VDI (Radant)	HSE no duct	HSE no duct low Kst	Misc.	HSE with duct	VDI (Radant) Duct
V (m^3)	2.4 to 250	18.5	18.5	10 to 500	18.5	1
Enclosure L/D	1 to 1.55	1.6	1.6	3, 6 and 6.25	1.6	?
Av (m^2)	0.086 to 8.3	0.196 to 0.95	0.196 to 0.95	0.25 to 12	0.196 to 0.95	0.0707 to 0.196
Pstat (bar-g)	0.1	0 to 0.5	0 to 0.5	0.1 or less	0 to 0.5	0.1 (?)
Kst (bar-m/s)	206 to 322	144 to 630	15 to 56	100 to 280	144 to 630	100 to 250
Pmax (barg)	8.7 to 9.2	7.3 to 10	6.5 to 7	8 to 9	7.3 to 10	9 (?)
Pred (bar-g)	0.25 to 2.4	0.12 to 5	0.12 to 1.25	0.1 to 2.0	0.12 to 7.1	0.1 to 4.4
Duct Length (m)					0 to 20	0 to 6
Duct L/D					0 to 43.6	0 to 20
45° elbows					0 to 2	0
90° elbows					0 to 1	0
REFERENCES See Previous Page	VDI (Radant) Data	Lunn et al (1988a)	Lunn et al (1988b)	Confidential	Lunn et al (1988a) and Hey (1991)	VDI (Radant) Data

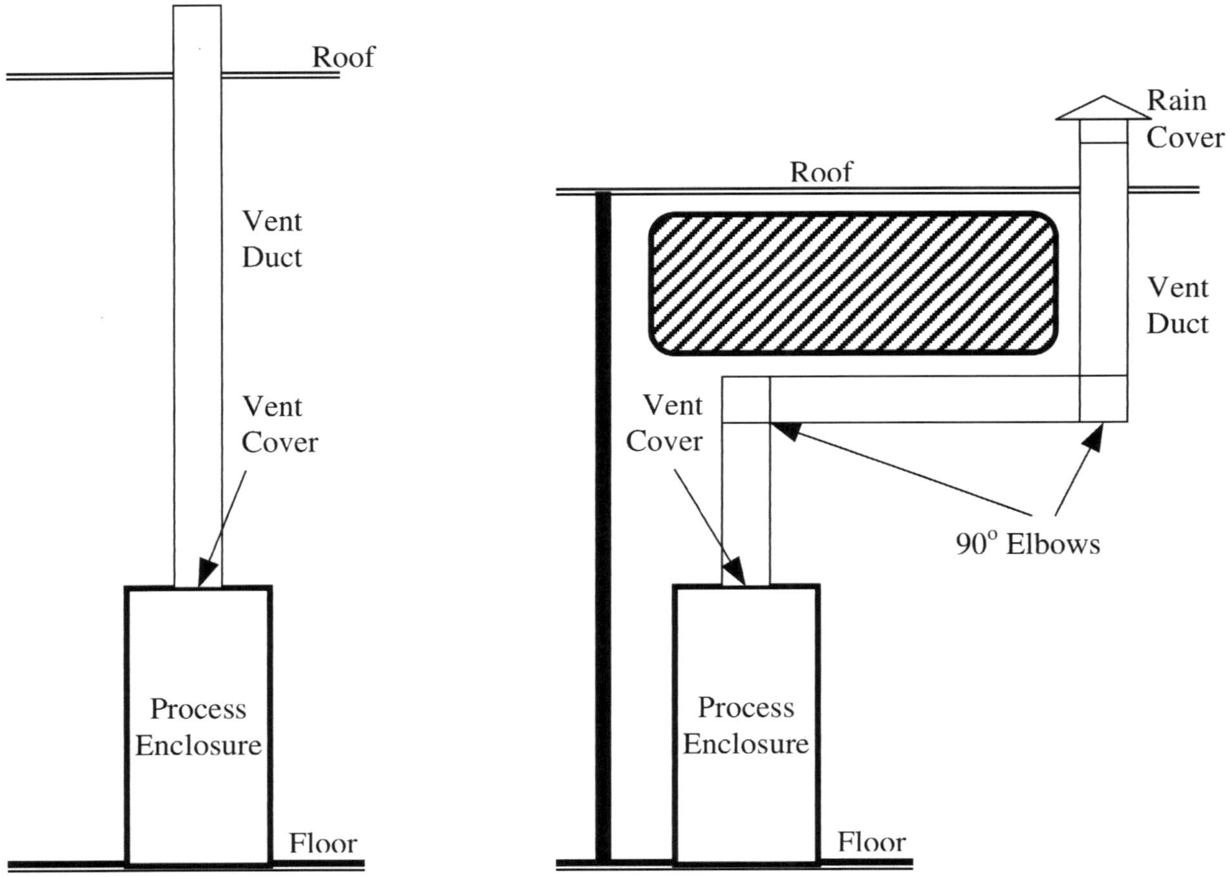

Benchmark Problem	**A**	**B**
GIVEN:	above left	above right
Enclosure Volume, V (m^3)	25	25
Enclosure L/D	4	4
Vent Diameter, Dv (m)	1.5	1.5
Duct Diameter, Dv (m)	1.5	1.5
Av (m^2)	1.77	1.77
Pstat (barg)	0.25	0.25
Kst (bar-m/s)	200	200
Pmax (bar)	8	8
Duct Length (m)	12	12
Duct effective roughness, ε (mm)	0.26	0.26
Elbows	none	2 X 90°
Elbow Flow Resistance	N/A	2 * 1.2 = 2.4
Rain Cover Flow Resistance	N/A	0.75
CALCULATE:	**Pred**	**Pred**

Figure 1. The two benchmark problems considered in this study.

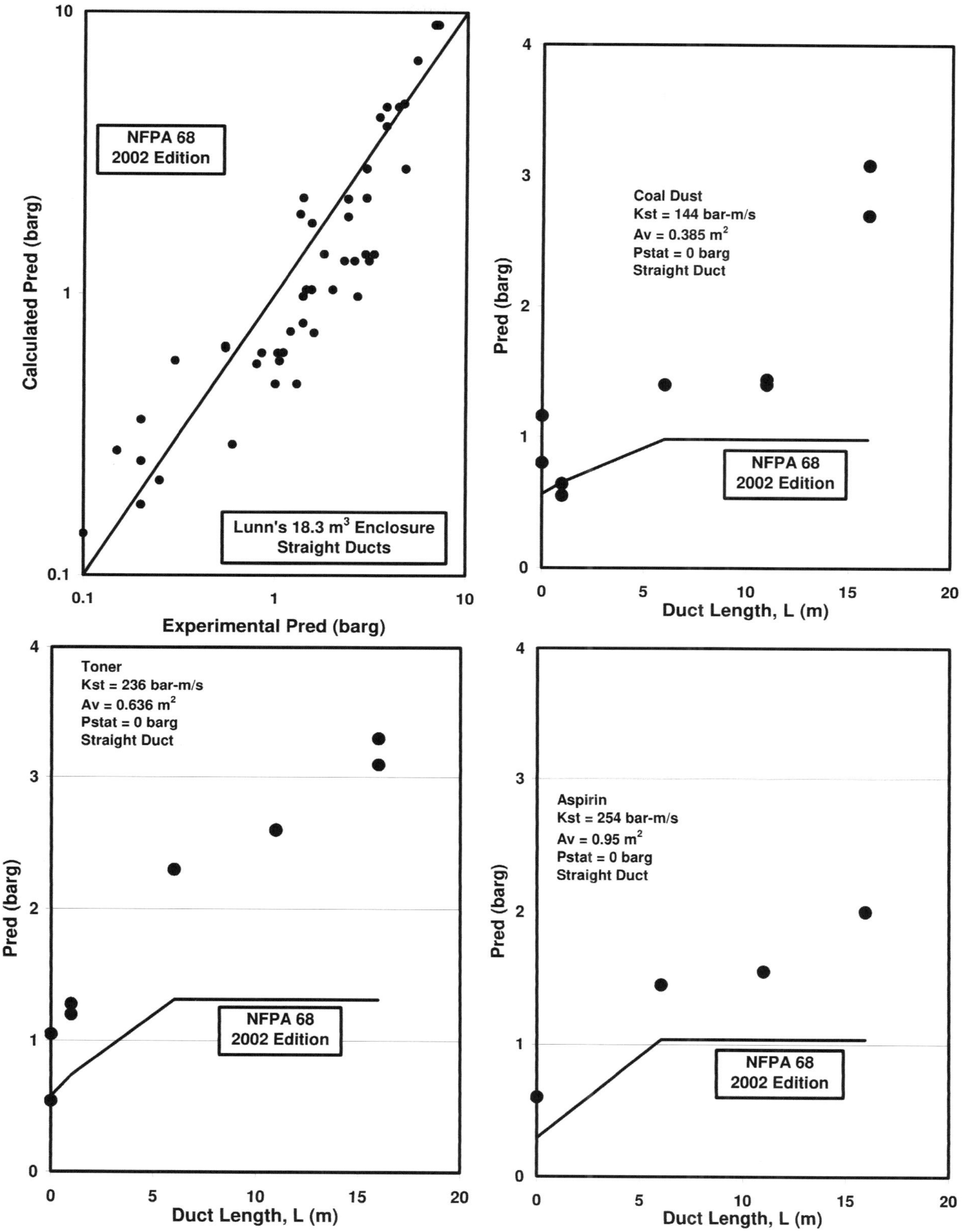

Figure 2. Comparisons of the NFPA 68-2002 predictions with the HSE vent duct data.

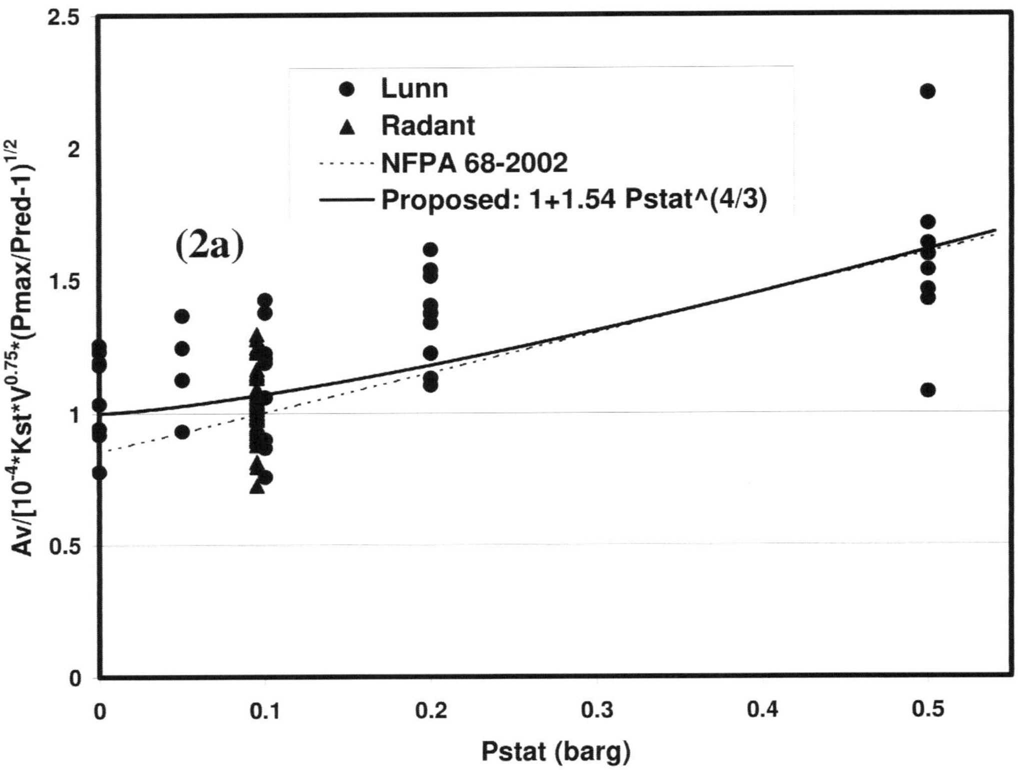

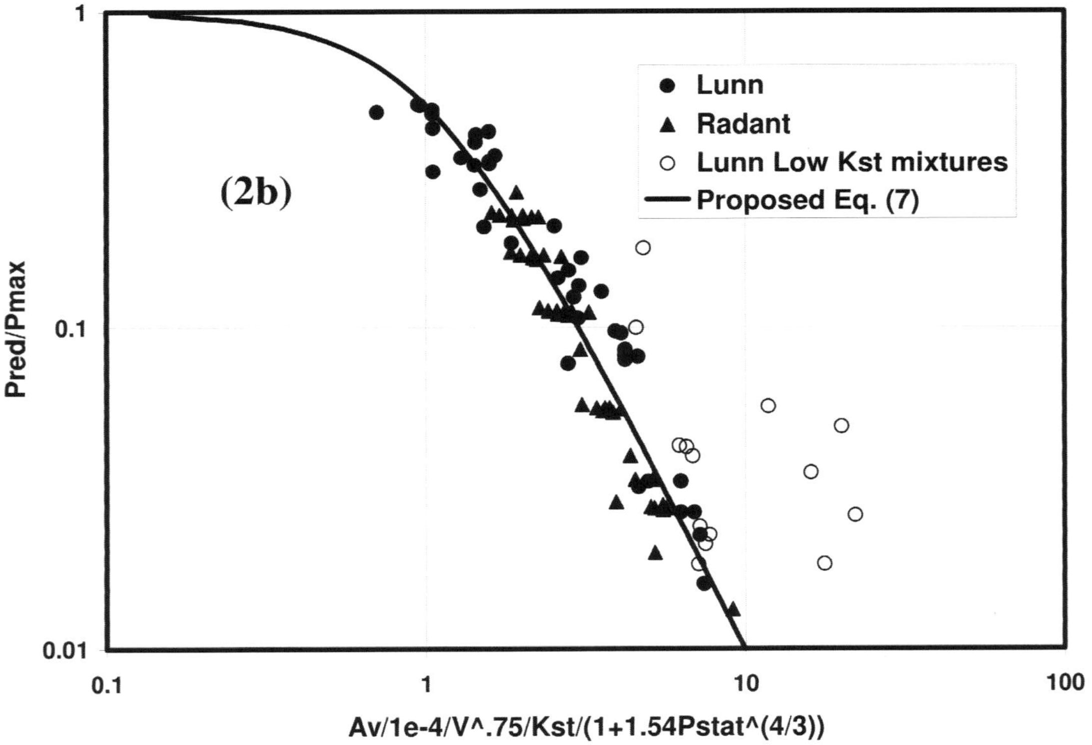

Figure 3. Comparison of the data with the predictions of the proposed compact enclosure venting equation.

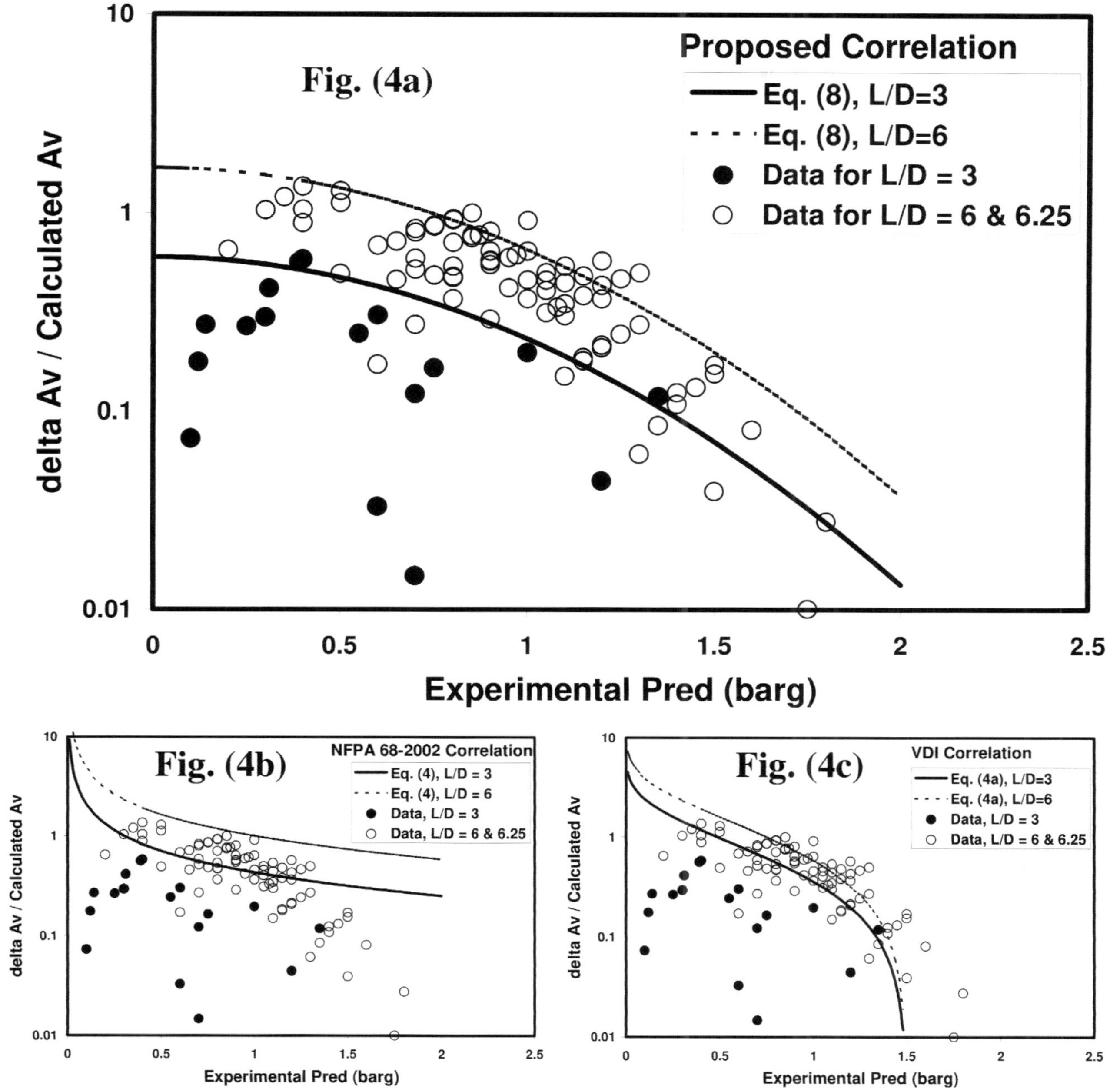

Figure 4. Comparison of the data with: (a) the predictions of the proposed enclosure L/D correction factor (Eq.(8)), (b) the predictions of the NFPA 68-2002 enclosure L/D correction factor (Eq.(4)), and (c) the predictions of the VDI 3673 enclosure L/D correction factor (Eq.(4a)).

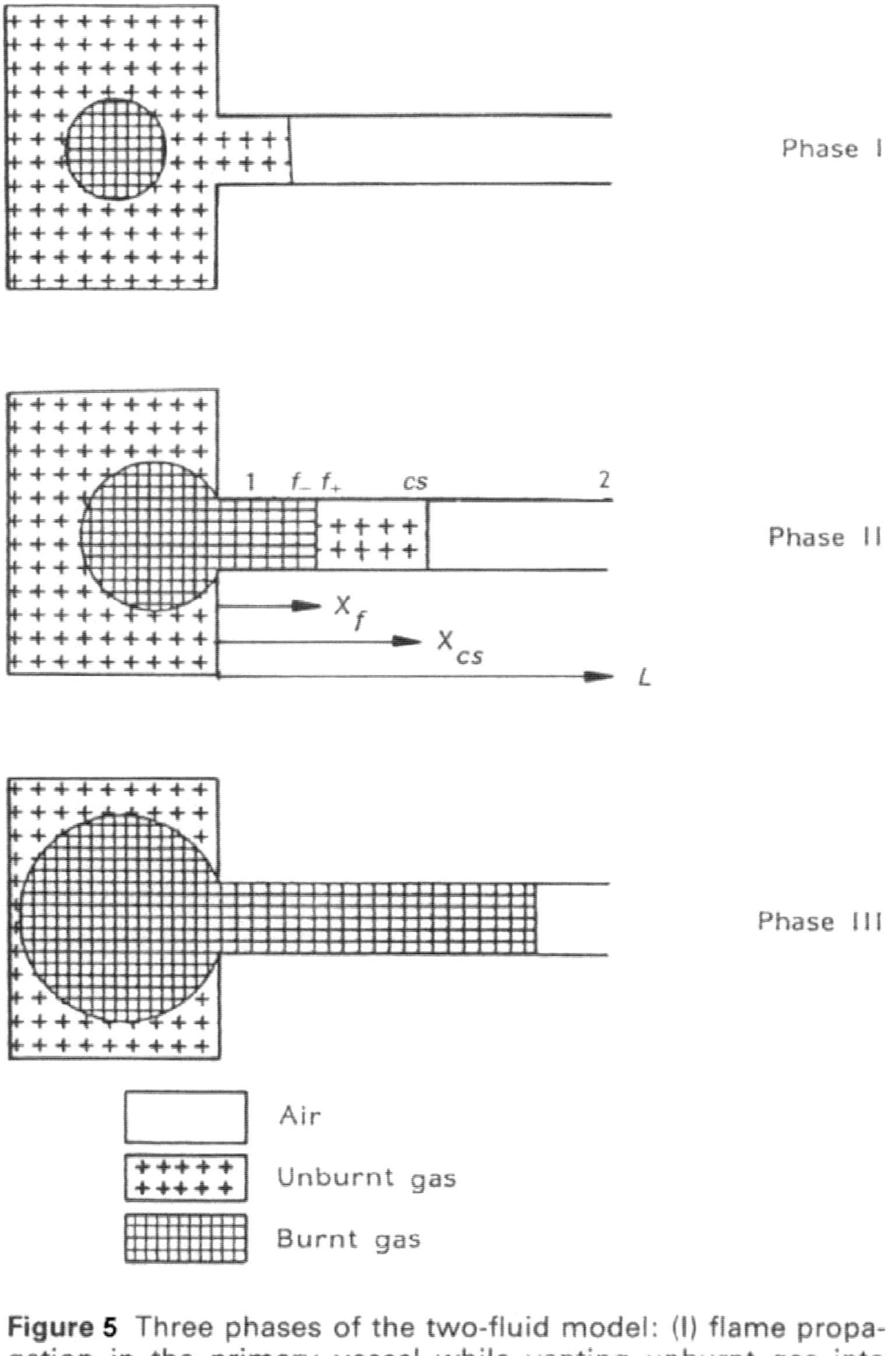

Figure 5 Three phases of the two-fluid model: (I) flame propagation in the primary vessel while venting unburnt gas into duct; (II) flame propagation in the vessel and in the duct while venting burnt gas; and (III) flame propagation in the vessel while venting burnt gas.

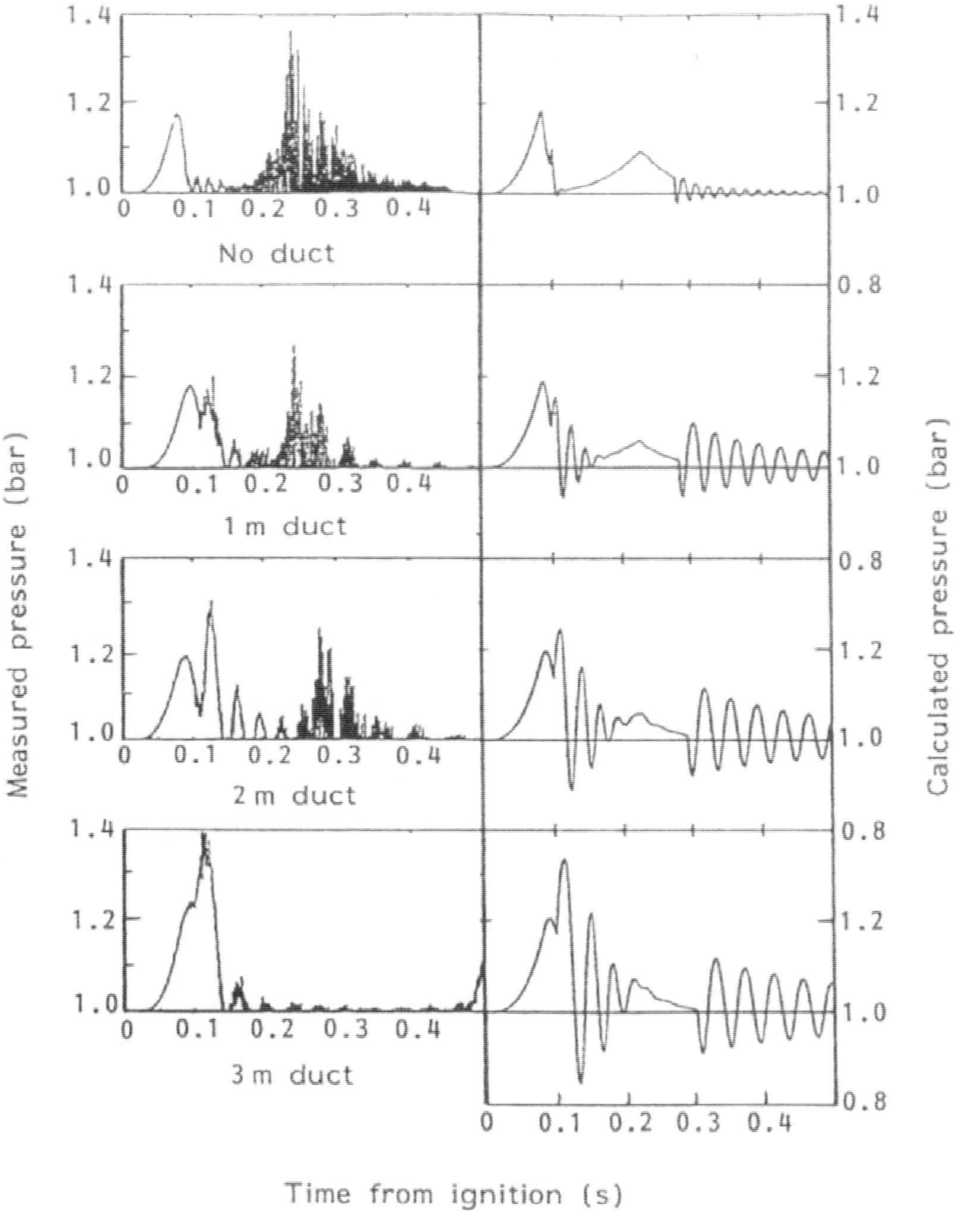

Figure 6 Comparison of the experimental and calculated pressure–time traces for DeGood and Chatrathi[13] tests with different duct lengths. Sub-atmospheric pressures were not recorded in the experiments. High frequency acoustic oscillations are not modelled. The experimental traces were provided courtesy of DeGood and Chatrathi.

Straight and circular exhaust vents[12] $Re = w_0 D_0/\nu > 10^4$	Diagram 11-20

No. of curve	Vent type	Scheme
1	With plane baffle	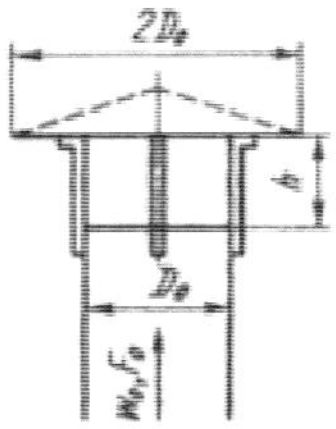
2	With split canopy	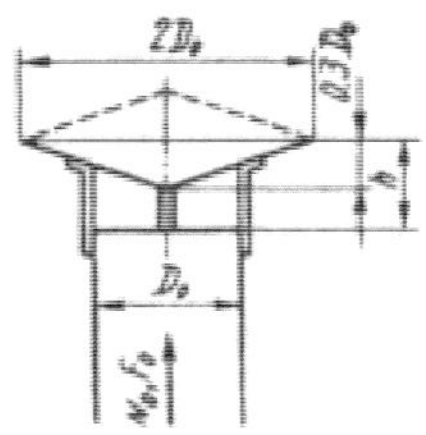
3	With hood	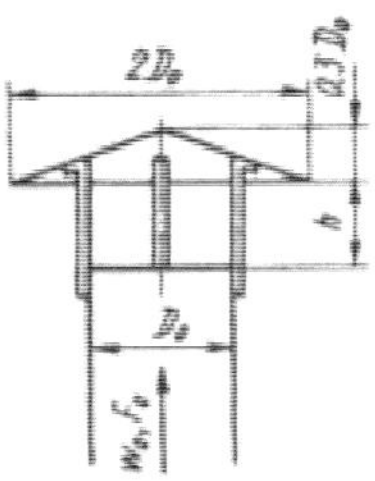
4	With hood and split canopy	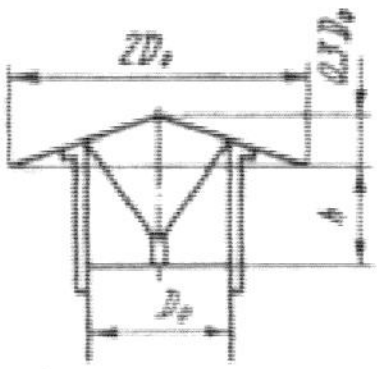
5	With diffuser and hood	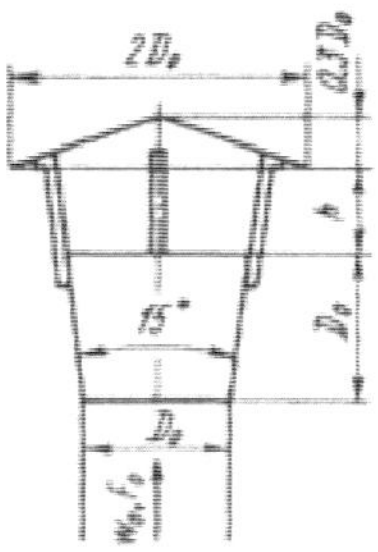

Resistance coefficient $\zeta = \frac{\Delta p}{\rho w_0^2/2}$

Values of ζ

	h/D_0				
Vent	0.10	0.20	0.25	0.30	0.35
1	–	–	3.40	2.60	2.10
2	–	–	–	–	–
3	4.00	2.30	1.90	1.60	1.40
4	–	2.90	2.30	1.90	1.70
5	2.60	1.20	1.00	0.80	0.70

	h/D_0				
Vent	0.40	0.50	0.60	0.80	1.0
1	1.70	1.40	1.20	1.10	1.00
2	3.50	2.00	1.50	1.20	1.10
3	1.30	1.15	1.10	1.00	1.00
4	1.50	1.30	1.20	1.10	1.00
5	0.65	0.60	0.60	0.60	0.60

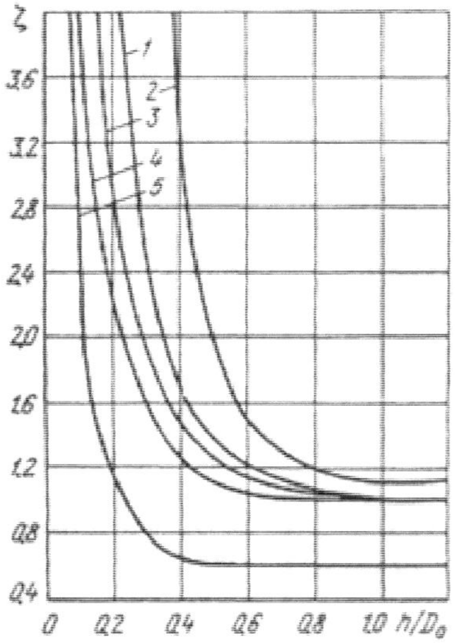

Figure 7. Flow resistance data for rain covers (reproduced from Idelchik, 1996). The resistance coefficient on this figure is defined on the total pressure, not on the static pressure.

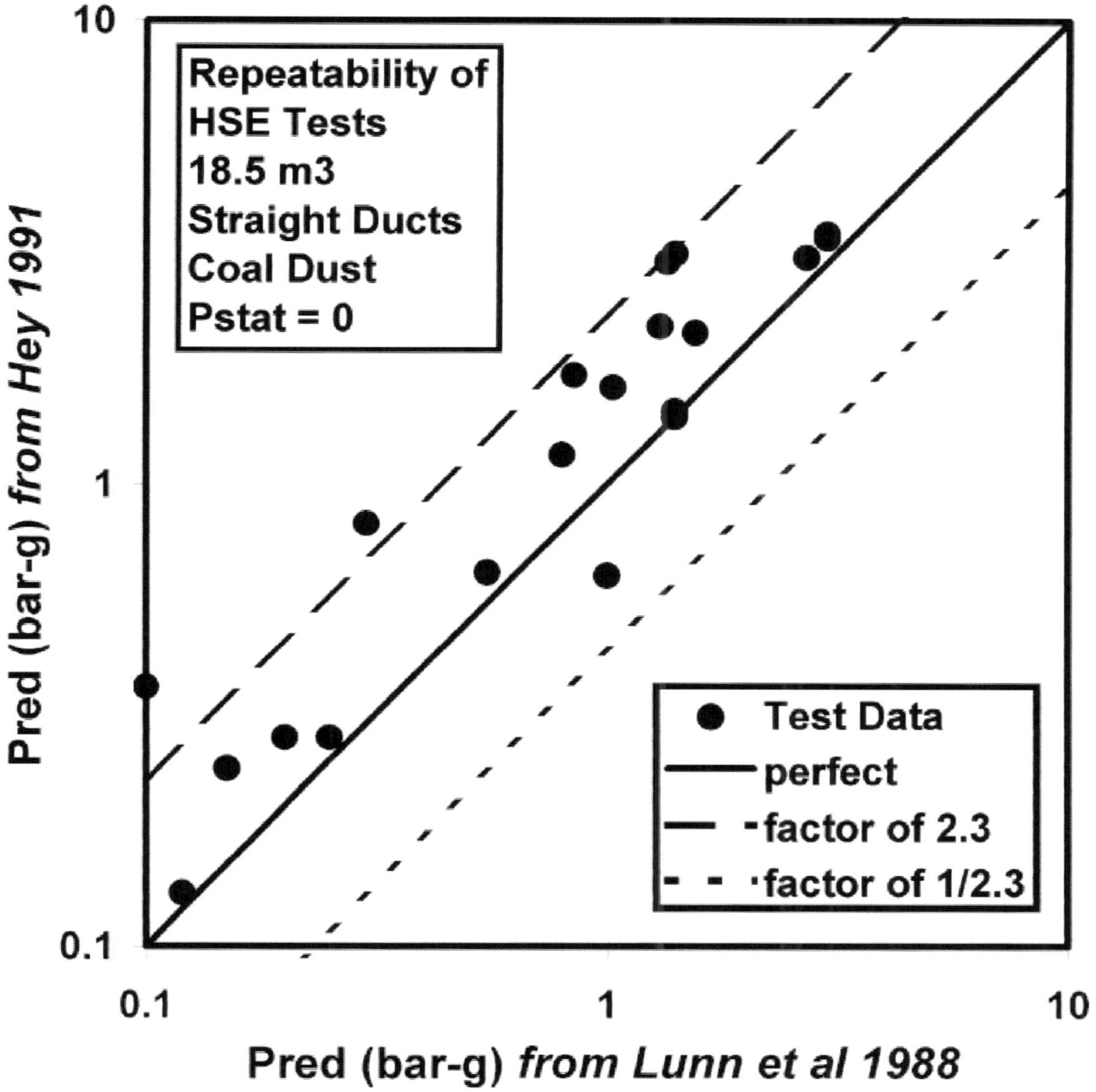

Figure 8. Repeatibility of HSE 18.5 m^3 vented enclosure explosion data.

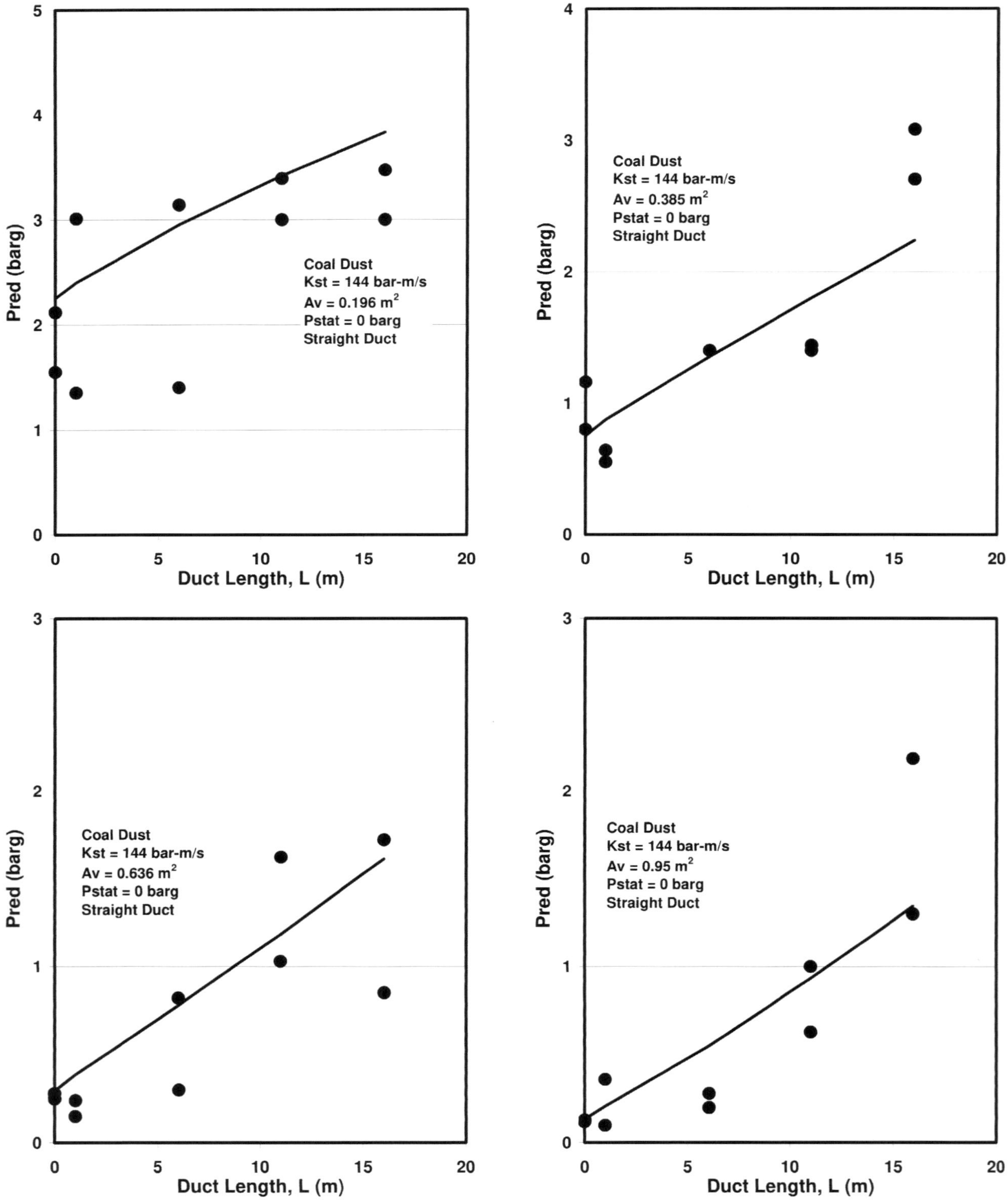

Figure 9. Detailed comparisons of the coal dust explosion pressures measured in Lunn's 18.5 m^3 enclosure (symbols) to those calculated using Equations (7) and (13) (connected with solid lines).

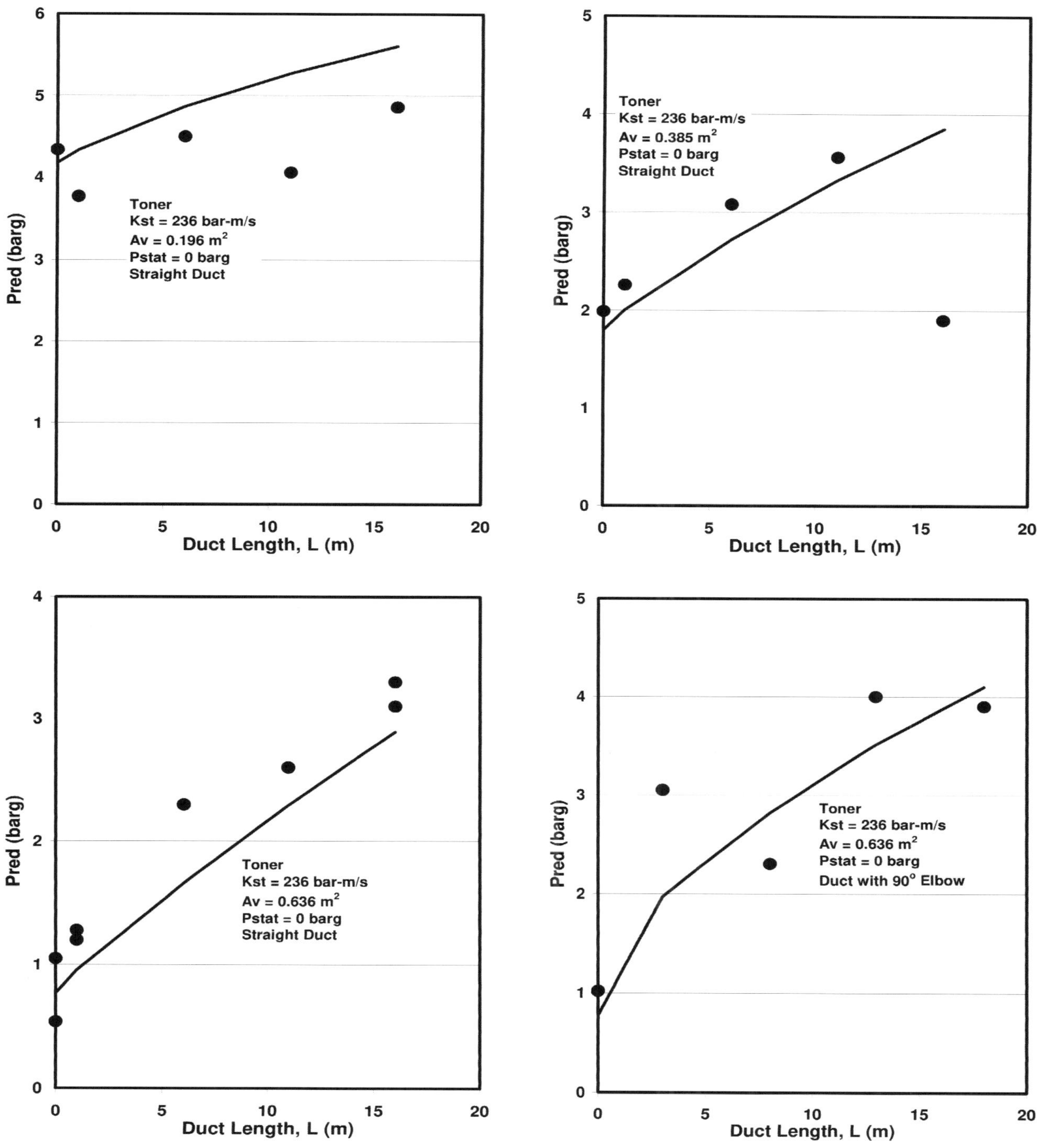

Figure 10. Detailed comparisons of the toner explosion pressures measured in Lunn's 18.5 m^3 enclosure (symbols) to those calculated from Equations (7) and (13) (connected with solid lines). Data for straight ducts as well as ducts with 90° elbow are shown.

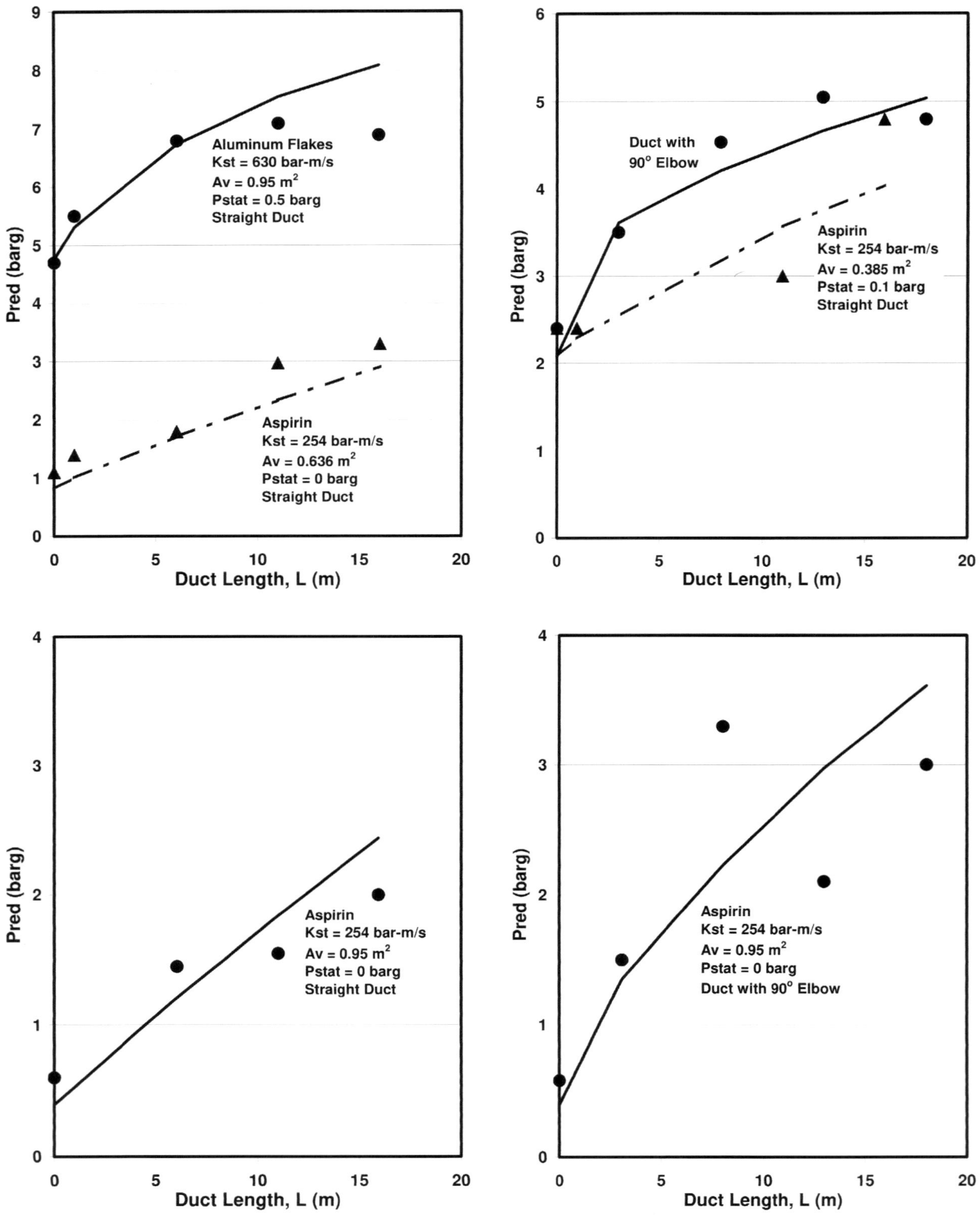

Figure 11. Detailed comparisons of the aluminum flakes and aspirin dust explosion pressures measured in Lunn's 18.5 m^3 enclosure (symbols) to those calculated from Equations (7) and (13) (connected with solid lines). Data for straight ducts as well as ducts with 90° elbow are shown.

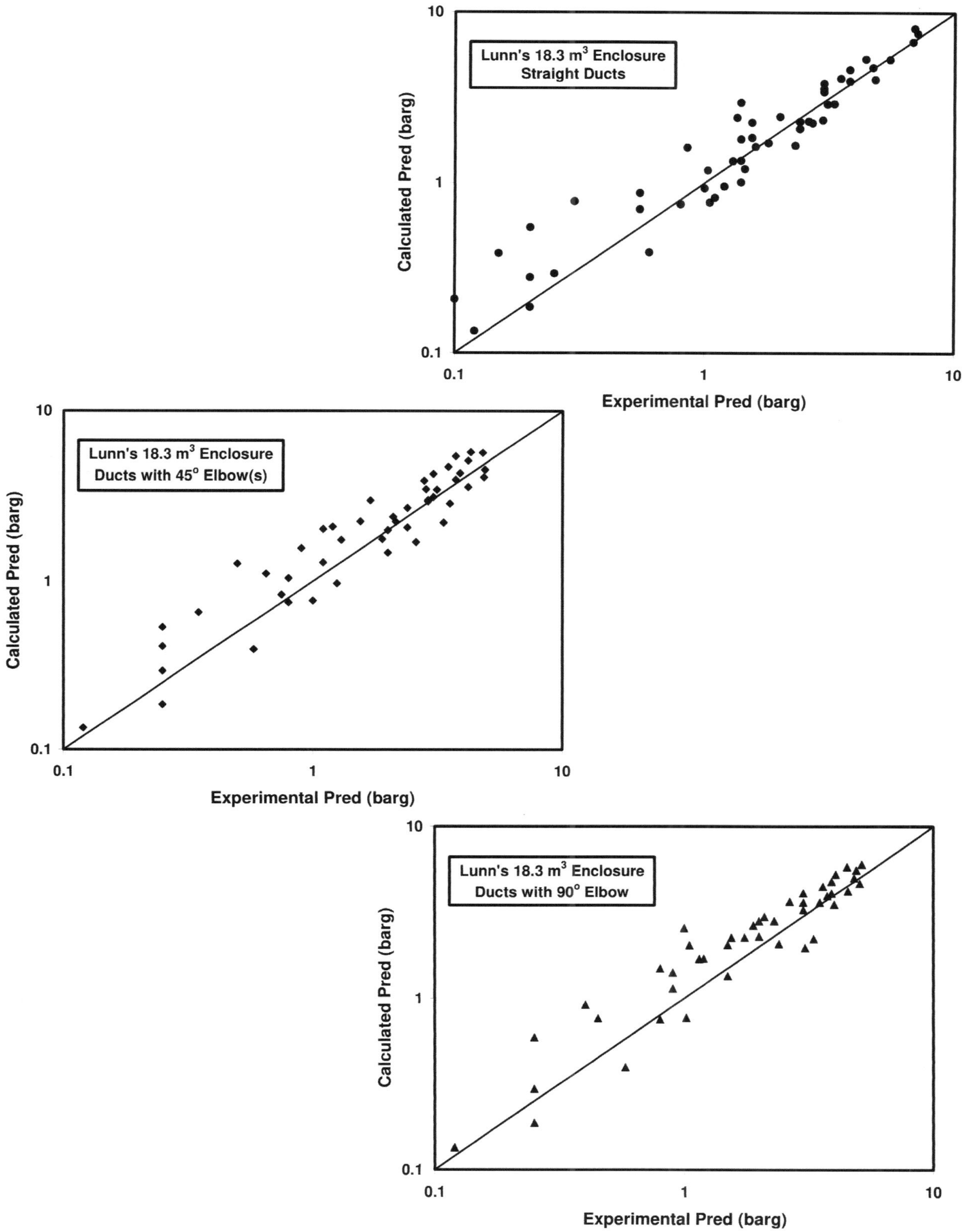

Figure 12. Comparison of the explosion pressures measured in Lunn's 18.5 m^3 enclosure to those calculated from Equations (7) and (13). Top: Straight ducts, Middle: Ducts with one or two 45° elbows, Bottom: Ducts with a single 90° elbow.

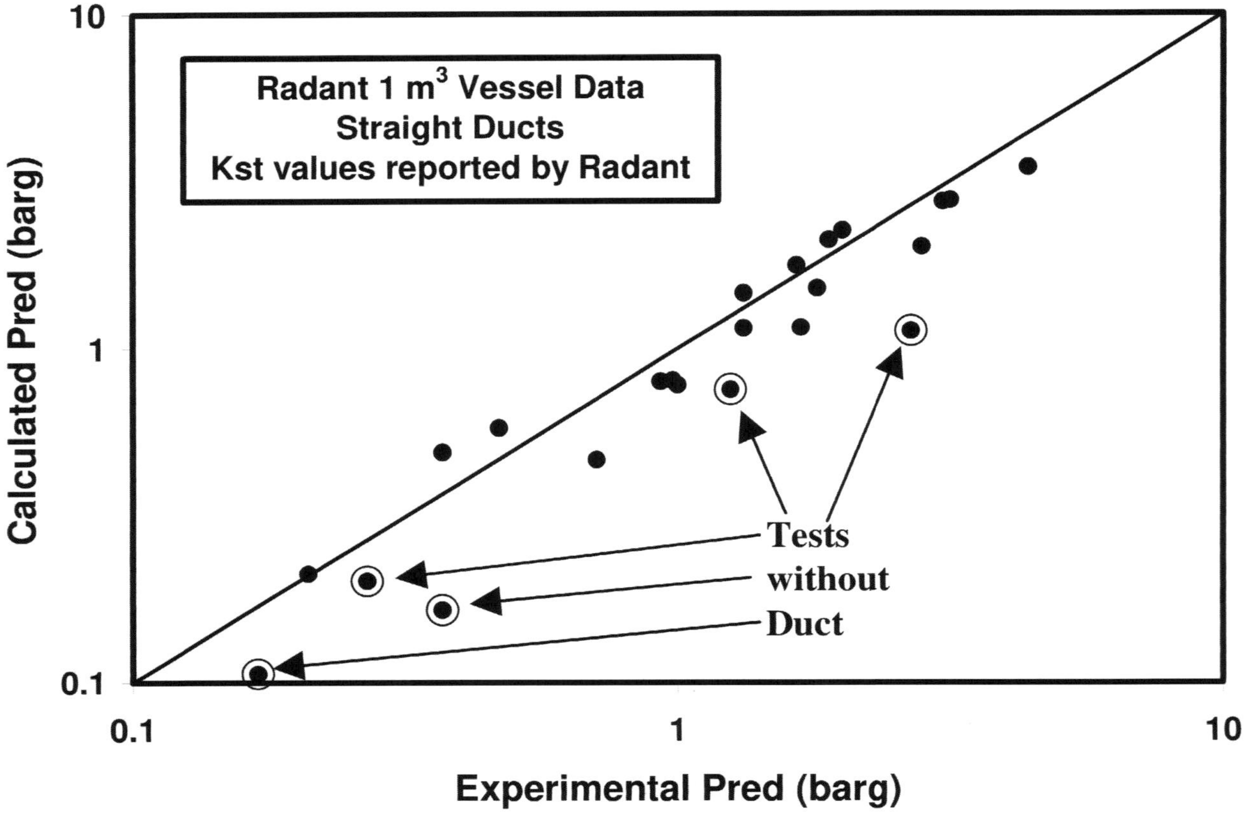

Figure 13. Comparison of the measured explosion pressures to those calculated from Equations (7) and (13). Designated with double circles are the data for tests without a duct.

Paper TG004 - 6

Combustible Dust Hazard Study

Angela Blair ***(speaker)***
Bill Hoyle
U.S. Chemical Safety Board
2175 K Street, NW, Suite 400
Washington, DC 20037-1809

Prepared for Presentation at
American Institute of Chemical Engineers
2005 Spring National Meeting
39th Annual Loss Prevention Symposium
Atlanta, Georgia
April 11-13, 2005

UNPUBLISHED

Combustible Dust Hazard Study

Angela Blair ***(speaker)***
Bill Hoyle
U.S. Chemical Safety Board
2175 K Street, NW, Suite 400
Washington, DC 20037-1809

Abstract

The U.S. Chemical Safety and Hazard Investigation Board (CSB) investigated three fatal dust explosions that occurred in industrial facilities in the United States in 2003 that killed 14 people and injured dozens more. Some of the underlying regulatory, engineering and safety issues raised during these investigations prompted the Board to initiate a study to assess combustible dust hazards in industry.

CSB objectives for the combustible dust hazard study include:

- Analyzing combustible dust incidents that have occurred in the United States over the past 20 years (1984-2004)
- Determining if there is a need for OSHA to interpret, amend, or issue guidelines to more effectively address combustible dust hazards in its Hazard Communication Standard
- Comparing how state governments currently address combustible dust hazards in fire and building codes and determining if improvements may be warranted
- Determining if a need exists for a new OSHA safety standard on combustible dusts in general industry
- Assessing training received by fire and building code enforcement officials and OSHA compliance officers on combustible dust hazards.

Specific topics resulting from prior CSB public hearings on the combustible dust hazard study will be presented.

A group discussion will follow where CSB will actively seek further input on these efforts from interested parties and stakeholders in the audience.

91g Discussion on Dust Hazard Study

Angela Blair, Bill Hoyle

Input from the audience on the dust hazard study.

Insidious Explosion Hazards in the Mining Industry

Derrick M Tjernlund PE, CFEI
Steven J Luzik, PE, CFEI, CMSP
MSHA, Approval and Certification Center
Box 251, Industrial Park Rd. Triadelphia, WV 26059

Prepared for Presentation at
American Institute of Chemical Engineers
2005 Spring National Meeting
39th Annual Loss Prevention Symposium
Atlanta, Georgia
April 11-13, 2005

UNPUBLISHED

Insidious Explosion Hazards in the Mining Industry

Derrick M Tjernlund PE, CFEI
Steven J Luzik, PE, CFEI, CMSP
MSHA, Approval and Certification Center
Box 251, Industrial Park Rd. Triadelphia, WV 26059

Abstract

Explosions have occurred in industry throughout modern history. Most everyone, from those that either work in or manage industrial sites, to the general public, associate the word "explosion" with the large-scale catastrophic events that cause multiple fatalities and widespread property damage. These are the events that grab the front page headlines of newspapers or make the top stories of the evening television newscasts. Under the radar of these large-scale explosions are a number of small-scale events that occur on a frequent basis. These events often result in injury or loss of life without serious property damage, or represent near misses that few people removed from the accident ever hear or read about.

This paper discusses the Mine Safety and Health Administration's (MSHA) experience with these types of explosions that include incidents involving vehicle tire maintenance, industrial pumps, refining processes, and acetylene tanks. Identification of root causes and contributing factors are included, as well as a reemphasis of safety training, task evaluation, and outreach programs that the industry may consider to prevent future occurrences.

1. Introduction

On a cold February morning in 1999, just after midnight, a railroad tank car filled with liquid hazardous waste-derived fuel suddenly exploded at a cement plant in north central Indiana. The tank shell, weighing approximately 19 tons, was launched up and over two rotary cement kilns at a departure angle of approximately 60 degrees or greater. The shell finally came to rest about 800 feet from where it started. The explosion touched off a large fire in the waste fuel handling area of the plant that lasted for more than four hours and required the effort of numerous local fire departments and the crash-fire-rescue equipment from a nearby military base before it was brought under control. The emergency required evacuation of surrounding communities and resulted in the response from four federal agencies: the Mine Safety and Health Administration (MSHA), the Environmental Protection Agency (EPA), the Federal Railroad Administration (FRA), and the National Transportation Safety Board (NTSB). Needless-to-say, the story grabbed front page headlines for several days in regional newspapers. Yet, despite the news worthiness of the incident, there were miraculously no injuries or fatalities, due in large part, to the time of day of the event when few employees were on site.

In the early morning hours of July of the same year, another devastating explosion occurred at a bauxite ore processing plant in Louisiana when a number of very large pressure vessels containing superheated caustic slurry ruptured, destroying the central part of the facility. In this event, over 20 workers were injured, some seriously, with two receiving extensive, nearly fatal burns. The domed top of one pressure vessel, weighing more than 7,000 pounds, was thrown approximately 3,000 feet from where it started by the force of this BLEVE[1] event. Nevertheless, as with the railroad tank car, there were no fatalities in this event. A team of investigators, engineers, and mine emergency response specialists from MSHA, along with environmental specialists from the EPA, responded. The news worthiness of this event was also well documented.

In between these two events, on a warm sunny morning in Mississippi, a mine worker died when the impeller hub of the slurry pump he was working on exploded. He was struck by a metal fragment weighing less than one pound and died within minutes of receiving his injuries. Because this site too was a Mine Safety and Health Act regulated property, an MSHA investigation team responded to the accident. However, because of the limited nature of the event, only MSHA responded. Unfortunately, small, relatively un-newsworthy fire/explosion events such as the last one described occur far too frequently.

This paper presents a number of such events experienced by the mining industry in an effort to raise awareness about them. These fire/explosion hazards can be considered "insidious" because the hazards leading to these events often go unrecognized by the general work force (and not just the workforce within the mining industry). Industry itself may sometimes take too little notice of such events; hence, the lessons that could be learned never get delivered to those who could be exposed to similar hazards---hazards hidden in misunderstanding, inadequate knowledge, careless attitudes, or unsound practices.

2. Background

Fires and explosions have long been a deadly menace to those who work in industry. Nowhere was this truer than in the mining industry, where decades of history conveyed the deadly nature of unsafe coal mining practices which claimed thousands of miners' lives. During the Twentieth Century, miners, mine operators, and various government agencies worked diligently to halt this tragic loss of life. The U.S. Bureau of Mines, established in 1910 as a branch of the U.S. Department of the Interior, spent the better part of the century building a massive body of knowledge through research and investigation---a body of knowledge describing both the causes of and preventative measures needed to remove mining from the top of the list as the nation's most dangerous livelihood.

In 1969, Congress passed the Federal Coal Mine Health and Safety Act, which established significant binding safety and health standards for the coal mining industry. In 1977, Congress reorganized these efforts by passing the Mine Safety and Health Act which created the Mine Safety and Health Administration within the U.S. Department of Labor. MSHA's role is defined similar to that of OSHA[2], but with the authoritative mandate to specifically regulate and promote the health and safety of the nation's miners not only in coal mines, but also in metal and nonmetal mines.

---

[1] Boiling Liquid Expanding Vapor Explosion
[2] Occupational Safety and Health Administration

Over the years, the combined efforts of MSHA (and its predecessor agencies), along with efforts by NIOSH[3] (and its predecessor, the U.S. Bureau of Mines), have turned the nation's mines into some of the safest workplaces in the country. Yet despite these impressive accomplishments, MSHA has continued to observe a number of incidents that have resulted in single fatalities involving explosion events. In examining and investigating these accidents, one common theme has come to light; in all of these events, the hazards seemed hidden from the victims. Those involved (both victims and witnesses) often appeared to misunderstand what an explosion was, how it could happen, and how to prevent it.

Perhaps the first problem is that there are many different ideas as to what constitutes an explosion. The large scale, "headline grabbers" are easy to accept. A pump impeller blowing apart is less obvious.

NFPA[4] 921 defines an explosion as[1]:

> *The sudden conversion of potential energy (chemical or mechanical) into kinetic energy with the production and release of gases under pressure, or release of gas under pressure. These high pressure gases then do mechanical work such as moving, changing, or shattering nearby materials.*

While this definition is certainly a sound technical one, and a definition spoken in the language understood by those of us in the field of science and engineering, it is not necessarily a good one for the general work force.

The authors propose a more generic version for safety training and planning purposes that may better alert workers to examine their assigned work tasks <u>before</u> proceeding with the task. Our definition for an explosion is:

> *A violent release of pressure energy; often from a bursting container, vessel, or enclosure; causing, or capable of causing, death, injury, or damage to equipment or property.*

This definition is similar to a simple definition put forth by the U.S. Bureau of Mines as[2]:

> *A sudden release of pressure or energy.*

Our proposed definition utilizes the technical simplicity of the Bureau definition, while retaining emphasis on the potentially catastrophic nature of the event to cause death, injury, or damage, as implied by the NFPA definition.

MSHA classifies explosion-type accidents[5] into one of two categories[3]:

1. exploding vessels under pressure
2. ignition or explosion of gas or dust

---

[3] National Institute of Occupational Safety and Health

[4] National Fire Protection Association

[5] These classifications do not include accidents involving blasting activities and explosives.

What follows are examples of 11 mining accidents that fall into these two categories. These case histories resulted in fatalities and/or injuries. One important aspect in workplace understanding of explosion hazards is the ability to recognize the conditions under which pressure energy can be created. This energy can be generated by combustion; by other chemical reactions; by external application of energy such as flame or heat, by electrical energy, by mechanical energy generated by a rotating shaft; or by the structural failure of a vessel which is already normally under pressure, or thought to be under pressure.

3. Discussion of Accidents

Case History No. 1: Coal preparation plant; fatal exploding-vessel-under-pressure accident, Buchanan County, Virginia

Overview: A 54-year-old miner was fatally injured when a clean coal filter drain pump used for pumping a slurry of water and coal fines exploded at a coal preparation plant. The victim was standing approximately eight feet away at the pump's on/off switch while the pump was running, when the volute cover plate blew off and struck him. An examination of the pump after the accident revealed that the discharge pipe was plugged full of coal fines. This was not the first occurrence of such a pump failure at this facility as a similar failure had occurred 10 years earlier.

Findings and Conclusions: The pump became overheated when the water content in the slurry apparently decreased, causing a consequent rise in the specific gravity of the slurry and resultant decrease in pump performance. This permitted the suction and discharge lines to and from the pump to become clogged with coal fines, causing a total blockage of the pump. This caused the pump to become a closed pressure vessel hydraulically full of water/coal slurry. The slurry trapped within the pump continued to absorb energy from the pump as it continued to run for up to 35 minutes after the discharge became plugged. This transfer of energy into the entrapped slurry caused a significant rise in temperature and pressure of the contents, eventually resulting in explosive failure of the pump volute cover plate. The presence of welds at the access cover attachment points materially contributed to the mode and potentially the timing of the pump failure. The accident occurred because the discharge line to the pump became clogged while the pump continued to be operated. The confined coal slurry within the pump volute was consequently overheated, resulting in a pressure buildup to the point the access cover plate blew off.

Causal Factors:

- Company's failure to properly assess the hazards associated with overheated pumps.
- Company's failure to install pump casing temperature monitoring on all pumps of similar duty when the first pump failed ten years previous.
- Company's failure to establish a procedure for starting the pump from a safe location.
- A secondary causal factor was the presence of welds at the volute cover attachment points that materially contributed to the mode and potentially the timing of the pump failure.

Case History No. 2: Metal/Nonmetal sand and gravel; fatal exploding-vessel-under-pressure accident, Comal County, Texas

Overview: A 28-year-old maintenance worker was killed and two coworkers injured when a large off road tire exploded while the three workers were trying to remove the brake drum from the rim in an effort to replace the drum. The tire/rim assembly had been removed from the rear axle of a Caterpillar model 631G pan scraper and laid down on the ground with the drum side of the rim facing up. The tire had not been removed from the rim (nor depressurized) prior to commencing the repair work. The brake drum was bolted to the inside of the rim, but workers had trouble getting the drum to break loose even after removing all of the bolts and were using a sledge hammer to try and break it free. Approximately an hour before the accident, workers had used an oxyacetylene torch to cut the brake drum into two pieces in an effort to loosen it. When this failed, the workers took a break for lunch.

Within minutes after returning to the tire to continue the work, the tire exploded. One worker had just turned away from the tire to get a tool from his utility truck parked next to the tire when the explosion occurred. This worker did not witness the explosion but was knocked to the ground. The second injured worker was believed to have had a pry bar in his hands and was standing next to the tire. He was knocked unconscious and did not regain memory of what happened prior to the explosion. The worker killed was believed to have been standing directly next to, or on top of the tire. He had been holding a sledge hammer and may have begun hammering on the brake drum in an attempt to loosen it. He was killed instantly and both of his arms were severed at the shoulders by the force of the explosion. Significant shredding of the upper side wall of the tire had occurred as the steel belt cables were ripped from around the bead cable. The tire and rim had been propelled high enough into the air to allow the assembly to land up-side-down from its original orientation. The tire had also been turned inside out and pieces of its' sidewall had been propelled up to 165 feet from the original location.

Findings and Conclusions: Physical examination and Gas Chromatography/Mass Spectroscopy (GCMS) testing of the gas in the other rear tire from the scraper indicated it was pressurized with air to approximately 50 psig. Both tires were of the same make and model and had similar purchase and maintenance histories. An examination of the rim assembly indicated that the torch flame had directly impinged on the rim in the vicinity of the tire bead sufficiently to "nick" the rim during the attempt to cut the brake drum in half. It is believed that the use of the torch transferred sufficient heat through the rim steel to ignite the inner liner material in contact with the rim at this location. This resulted in a smoldering fire of the inner liner material. Examination of the tire's interior surfaces confirmed this.

It is known that smoldering tire rubber generates copious quantities of incomplete products of combustion that are, in and of themselves, combustible. Over the course of the delay of work during the lunch break, the tire likely filled with a sufficient quantity of combustibles to create an ignitable mixture within the tire. The recommencement of work after lunch may have dislodged a hot smoldering ember from the lining material that fell into the mixture, igniting it. The extensive shredding of the tire sidewall and the severe injuries to the worker killed were both indicative of a rapid rate of pressure rise and high peak pressure typically associated with a combustion explosion. The accident occurred due to the unsafe application of a cutting torch flame to the tire/rim assembly. Application of heat to a rim can only be done safely if the tire is first completely removed from the rim prior to conducting this work.

Figure 1. Case History 2–tire explosion

Figure 2. Case History 2–charred inner liner of tire

Causal Factors:

- Company's failure to establish maintenance procedures and training that mandated the need to always completely remove tires from rims before performing any cutting or welding to a rim assembly.
- Use of air instead of an inert gas such as nitrogen for pressurizing tires.
- Failure of the company to develop an emergency plan for immediately isolating any tire/rim assembly suspected of being heated. It is recommended that this isolation establish a no entry zone of at least 300 feet in all directions around the tire for at least 24 hours.

Case History No. 3: Metal/Nonmetal surface gypsum operation; fatal ignition or explosion of gas or dust, Sevier County, Utah

Overview: A 63-year-old quarry maintenance supervisor, with six years mining experience, was fatally injured while he and a coworker were replacing the seals on the rear suspension cylinder (shock absorber) of a haulage truck. The suspension cylinder assembly was mounted on the rear axle of a 35-ton, Model 769C Caterpillar haulage truck. While dismantling the suspension cylinder, a small threaded valve broke off. After unsuccessful attempts to remove the threaded portion of the broken valve, penetrating oil was sprayed into the valve port and heat was applied with a torch. An explosion occurred in the cylinder and the rod was forcefully ejected, striking the victim.

Findings and Conclusions: The cylinder rod measured approximately 25 inches long, seven inches in diameter, and weighed approximately 114 pounds. The upper mounting pin for the cylinder had been removed from the cylinder, allowing it to be rotated to a horizontal position. The nitrogen and hydraulic oil normally in the cylinder had been removed as part of the maintenance effort. As the nitrogen charging valve was removed, the lower threaded section broke off in the cylinder. The port within the broken portion of the valve tip was drilled out to a larger diameter to facilitate the use of an "easy-out" and compressed air from a 140 psig supply was used to blow the valve remnant free of material. An aerosol penetrating oil was applied to loosen the valve threads. An oxyacetylene torch with a welding/brazing tip was then used to heat the valve remnant.

Closed cup flash point testing conducted on samples of the hydraulic oil collected from the cylinder determined the flash point to be 374 degrees F. Similar tests conducted on a sample of the penetrating oil determined the flash point to be 175 degrees F. Tests were also conducted to determine if the use of compressed air would create enough internal pressure within the suspension cylinder to discharge the rod. No significant internal pressure was detected when 140 psi of air was supplied through the nitrogen charging valve. The compressed air escaped from several cylinder assembly ports, preventing any measurable increase of air pressure above atmospheric. These ports provided adequate venting for the applied compressed air, but not for the rapid rate of pressure rise due to ignition of the mixture of air and penetrating oil vapors.

Tests were conducted on the suspension cylinder assembly to recreate the accident by spraying the penetrating oil through the nitrogen valve, followed by the injection of compressed air, and then application of an oxyacetylene torch flame in a manner similar to what was done by the victim. During the tests, an ignition resulted within the suspension cylinder, resulting in the rod being forcefully ejected from the cylinder.

This accident occurred when an ignition source was applied to a combustible vapor contained with a vessel.

Figure 3 Case History 3–suspension cylinder explosion

Causal Factors:

- Company's failure to establish a job safety analysis involving tasks associated with components potentially containing combustible materials.
- Company's failure to provide adequate training in the service tasks the victim was assigned to perform.

Case History No. 4: Metal/Nonmetal sand and gravel; fatal exploding-vessel-under-pressure accident, Itawamba County, Mississippi

Overview: A 42 year old crusher operator with 2 years mining experience was fatally injured at a sand and gravel operation when he was hit by a fragment from an exploding slurry pump impeller. The victim and a coworker were attempting to remove the worn impeller from a pump in an effort to replace it. The impeller attached to the pump shaft by directly threading onto it. The end of the impeller was cast such as to form a cone-shaped hub that enclosed the end of the impeller shaft to protect the shaft from the abrasive nature of the slurry. The impeller metal was of a type that had a very high hardness to resist abrasion, but was also very brittle.

Efforts were made to unscrew the impeller from the shaft but were not successful. This may have been due to an incorrect assumption as to the direction of the threads. The two workers then hammered the sides of the impeller, breaking most of it off. However, the central threaded cone section of the impeller remained in place. The victim then proceeded to heat the impeller hub with the flame from an oxyacetylene torch. His coworker stated that this heating was applied for several minutes and that the surface of the hub turned a dull red. At that time, the victim turned off the torch and set it down while still leaning down and standing directly in front of the impeller hub. The hub suddenly exploded, blowing the cone tip off the hub and into the chest of the victim, causing fatal internal injuries.

Findings and Conclusions: An examination of the replacement impeller revealed that the cone hub section formed an enclosed cavity of approximately 3.7 cubic inches (60 milliliters) in volume. Examination of the damaged impeller revealed traces of a grease-like material on the threads. The pump manufacturer's literature indicated the recommended use of anti-seize compound on these threads. GCMS testing of small samples of grease recovered from the threads confirmed the presence of hydrocarbons.

It is believed that upon heating the impeller hub, some of the grease was vaporized and mixed with air trapped in the hub cavity. The auto-ignition temperature of the vapor/air mixture was likely reached by continued heating to the point of turning the impeller metal red. Normally, small cavities are not prone to development of significant pressures when internal combustible fuel/air mixtures are ignited because of the high surface area to volume ratios that exist. A high surface area to volume ratio encourages rapid heat transfer between the developing flame front and the sides of the container, resulting in a reduced flame temperature, and thus reduced pressure. However, in this case, the walls of the cavity were heated to a significant temperature, eliminating most of any heat sinking capability. The pump manufacturer's maintenance literature warned that the application of heat to the impeller created an explosion hazard. An examination of the pump frame also revealed an attached label warning against the application of heat to the impeller. The warning label had become covered with hardened rock dust over a period of time, completely obscuring it from sight. The accident occurred when a worker applied heat from a torch to an enclosed cavity containing an ignitable material.

Figure 4: Case History 4–slurry pump impeller explosion

Figure 5: Case History 4– replacement impeller, shown for reference.

Figure 6: Case History 4–cross sectional sketch of exploded impeller

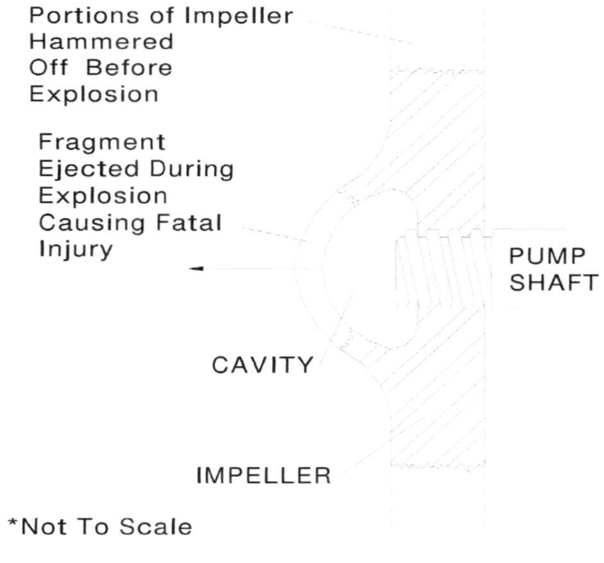

Figure 7: Case History 4–impeller pieces broken off using hammer before explosion.

Causal Factors:

- Company's failure to review the pump manufacturer's maintenance literature, which clearly warned against the application of heat to the impeller because it could create an explosion hazard.
- Company's failure to utilize the maintenance literature, which clearly identified the correct direction of rotation for removing the impeller, and identified use of a special tool needed to make removal of the impeller easier.

Case History No. 5: Metal/Nonmetal sand and gravel; fatal ignition or explosion of gas or dust accident, Fond du Lac County, Wisconsin

Overview: A 38-year-old foreman with 11 years experience was fatally injured at a limestone operation while trying to thaw out a frozen, in-ground truck scale with enclosed sump. The truck scale was scheduled to be returned to service after sitting during the winter. The March weather was still cold enough to prevent natural thawing of a 2-inch thick layer of ice in the bottom of the sump. The ice impeded the movement of the scale lever arms.

The foreman began his morning shift by putting tarps over the open edges of the scale in an apparent attempt to hold heat in the sump. He then placed a 20 pound propane cylinder down into the sump through an access hatch. The cylinder was equipped with a "weed burner" style torch head equipped only with a manual shutoff valve. The foreman placed the lit torch and cylinder down into the sump and left for most of the remainder of the shift. In the late afternoon, he returned to the scale with a 100 pound propane cylinder equipped with a second weed burner torch. One worker driving out of the quarry passed the victim in his truck as he headed toward the scale; however, no one was near the scale when the explosion occurred. Witnesses down in the quarry heard the explosion and saw several plates of the truck scale fly high enough into the air to be seen above the quarry spoil piles.

Findings and Conclusions: It is believed that the smaller propane torch self extinguished sometime during the day due to consumption of oxygen within the sump, before the cylinder was empty. The unlit torch continued to feed unburned propane gas into the sump space, and, along with the migration of fresh air into the sump after the flame extinguished, created an ignitable mixture. Upon returning with the second torch, the foreman apparently lit it and attempted to place it into the sump area. This ignited the gas/air mixture that had accumulated in the sump space, causing the explosion. The scale's six steel cover plates, weighing approximately 480 pounds each, were propelled up or away from the scale, with one plate landing approximately 100 feet from it. The victim was believed to be standing on the plate equipped with the access hatch and was killed instantly by the upward force of the plate. The investigation also revealed that the victim had used this same procedure on previous occasions when assigned the task of preparing truck scales after winter weather. The accident occurred when a worker applied a torch flame to an enclosed cavity containing an ignitable gas/air mixture.

Figure 8: Case History 5–truck scale explosion

Causal Factors:

- Company's failure to conduct a job safety analysis involving a procedure using a flammable gas.

Case History No. 6: Metal/Nonmetal cement plant; fatal exploding-vessel-under-pressure accident, Little River County, Arkansas

Overview: A 27-year-old waste fuel technician suffered fatal burns when a liquid waste fuel pumping system suffered a catastrophic overpressure within the system. The overpressure ruptured a casing on a waste fuel grinder, releasing substantial quantity of flammable fuel and covering the victim in the flammable liquid which was then ignited by an electrical arc from a power cable that was severed by the initial explosion. The accident occurred at a cement plant which burned liquid hazardous waste derived fuel in the plant's two rotary cement kilns.

Liquid fuel was received either by railcar or truck trailer and stored in vertical steel cone roof tanks within a large steel framed building. Two pumping trains took suction from either of two 100,000 gallon storage tanks and passed the fuel to the kiln burner floor for injection into either of the two kilns. Each pumping train consisted of nominal six inch steel piping, two variable-speed centrifugal pumps connected in series, followed by a pass-through vertical solid waste grinder, and then onto the kilns. The grinder was driven by a vertically mounted, direct coupled, 3 horsepower, 480-volt AC electric motor. The suction side of each pumping train had a number of pipe bends and low/high points capable of impeding flow and acting as air traps. When the pumps were started, a process computer took control of the pumps and normally ramped them up to full speed in several timed steps. The waste fuel technician was in close proximity to the pumps and grinder during the start up procedure. When it became apparent that the pumps were not moving any liquid, the technician began the process of opening bleed vents and may have opened a possibly closed discharge valve downstream of the grinder. During this effort, the explosion and fire occurred.

Findings and Conclusions: During a typical start up operation following a shutdown period, the pumps had occasionally failed to pick up a liquid prime from the storage tank because of entrapped vapors in the suction piping, resulting in zero pressure output. Small bleeder valves were installed on the downstream piping to allow for venting of trapped vapors within the piping during these occasions.

On this particular occasion, the pumps had already ramped up to full speed when the waste fuel technicians began the process of bleeding vapors from the system. A possibly closed supply valve downstream of the grinder may also have been opened during this process. At approximately that time, the first pump suddenly picked up the liquid supply from the tank and discharged it into the second pump. The second pump, in turn, suddenly boosted the pressure further, resulting in a large surge of liquid into the grinder which acted as a flow restriction. This caused a "water hammer like" pressure surge in the grinder. The severe overpressure spike catastrophically ruptured the grinder casing, forcing it upward and discharging flammable liquid throughout the area, including onto the waste fuel technician. The upward thrust of the ruptured grinder pulled the energized motor leads out of their conduit, causing a large electrical arc and igniting the liquid fuel. The fuel technician was engulfed in the ensuing fireball. A heat activated, deluge foam-water sprinkler system controlled the fire but could not respond fast enough to protect the technician from receiving second and third degree burns over 90 percent of his body.

Causal Factors:

- Company's failure to conduct an adequate chemical process safety management risk assessment of this operation.
- Company's failure to establish a safe location for workers during start up or during other transitional periods of operation.

Case History No.7: Coal surface operation; fatal ignition or explosion of gas or dust accident, Kanawha County, West Virginia

Overview: A 42-year-old contract mechanic, with 19 years mining experience, received fatal burns to the upper part of his body. He was preparing to use starting fluid to clean coal dust and oil from around the top of two motor mount bolts on a front-end loader. During this process, the aerosol can containing the starting fluid contacted an energized battery relay terminal and the frame of the machine, causing an electrical arc that burned a hole through the side of the can. The starting fluid ignited and sprayed through the hole and onto the victim, causing his clothes to become engulfed in flame. He ran around the front of the loader to a nearby service truck where another mechanic laid him on the ground and extinguished the flames. The victim was transported to a burn center where he later died as a result of his injuries.

Findings and Conclusions: Examination revealed that the cap (rubber insulating boot) provided to cover the energized battery relay terminal located inside the engine compartment was deteriorated (swollen from age and exposure to heat and oils) and was not covering the terminal. The accident occurred because a flammable fluid was being used near potential ignition sources. The ignition source, an energized battery relay terminal, was exposed because its insulating cap had deteriorated over time. When the side of the pressurized metal can containing the flammable liquid contacted both the energized terminal and the grounded mounting bracket, a high amperage electrical arc occurred that melted a hole through the side of the can. The starting fluid ignited and sprayed through the hole onto the victim, causing fatal burn injuries to his upper body.

Causal factors:

- Company's failure to maintain electrical equipment in adequate operating condition.
- The victim was using a highly flammable fluid near potential ignition sources.

Case History No. 8: Coal surface operation; fatal exploding-vessel-under-pressure accident, Buchanan County, Virginia

Overview: A 28-year-old utility man with 10 years of mining experience was fatally injured when unburned acetylene gas being discharged from a torch into a drum ignited. The fatally injured miner was holding an unlit oxygen/acetylene torch, through which he was delivering acetylene into an empty 55 gallon antifreeze drum when the drum exploded. Earlier in the day, the victim demonstrated the explosive nature of acetylene to a new mine employee by using a trail of starting fluid to ignite a bag filled with acetylene. At the time of the accident, he was intending to demonstrate a larger explosion. The drum was placed inside a large unattached Caterpillar, Model 992-G, front-end loader bucket, which concealed its view and shielded the blast from nearby buildings.

A mechanic who was standing next to the victim, and two other mine employees who were approaching the bucket at the time of the explosion, were also injured. The fatally injured victim received massive fatal head injuries when he was struck by debris from the explosion.

Findings and Conclusions: The accident occurred due to the explosion of an acetylene-oxygen-air mixture inside the 55 gallon drum. Forces and heat from the explosion injured four miners, one of whom died from his injuries. Ignition energy was most likely an electrostatic charge that built up on the torch tip from the rapid release of non-burning acetylene through the tip orifices. At the time of the accident, the fatally injured miner was using the unlit torch to intentionally fill an empty antifreeze drum with acetylene, and possibly oxygen, to demonstrate an explosion to a new employee. During this process, the torch tip neared the surface of the drum, which was grounded by contact with the front-end loader bucket, permitting an electrostatic arc that ignited the gas mixture inside the drum. Contributing to the accident was mine management's failure to establish, maintain, and enforce an effective safety program addressing horseplay and other hazards associated with mining activities, such as those related to the use of acetylene.

Figure 9: Case History 8 – empty 55-gallon drum explosion

Causal Factors:

- A safety program, adequate to deter horseplay and misuse of equipment, was not in place.
- The victim introducing acetylene/oxygen into the 55-gallon drum apparently without fear of reprisal on the company's part.

Case History No. 9: Surface of an Underground Coal Mine-; fatal exploding-vessel-under-pressure accident, Claiborne County, TN.

Overview: The victim was part of a two man crew performing repair work on a piece of mining equipment on the surface of an underground coal mine. The torch operator could not get the flame hot enough to perform the repair. The victim turned the acetylene regulator valve completely clockwise in an attempt to provide more gas to the torch. When this failed to solve the problem it was decided that the acetylene cylinder was depleted. The two man crew went to the supply building and retrieved a new cylinder of gas. The victim hooked up the new tank and his helper took his place onboard the mine scoop along with the torch. They were separated by about seven feet at that time. The torch operator tried to light the torch but was unsuccessful. The victim then stated that he may have backed the acetylene regulator out too far and was going to check it. It is unclear exactly what transpired after that but the torch operator stated that the victim went back over to the tank and performed an adjustment of some kind and told the torch operator to try to light the torch again. The torch operator could not say for certain if he struck his lighter but thought that he did. A violent explosion occurred at that time which dismembered the victim who was straddled over the Size B [40 cu. Ft.] acetylene tank. Examination of the remains of the tank suggested that a detonation had taken place.

Findings and Conclusions: It is believed that oxygen gas back-flowed into the acetylene supply hose and through the acetylene regulator as the acetylene cylinder became depleted. Absence of pressure gauges prevented the victim from recognizing the fact that the cylinder was nearing depletion. It also prevented him from realizing that he was dispensing the fuel gas at unsafe pressures above 30 psig. Absence of a check valve in the fuel line allowed oxygen to migrate back into the fuel line and regulator. Ignition is believed to have occurred from either the striking of the flint or adiabatic compression of trapped gases as the acetylene regulator was turned in clockwise by the victim. In either case, this primary ignition of the acetylene/oxygen mix in the fuel line is believed to have propagated into the acetylene cylinder and caused a much more violent secondary detonation within the cylinder. Testing and evaluation of the acetylene regulator confirmed that the flame path from the regulator into the tank would have exceeded the minimum quenching distances reported for acetylene gas in oxygen.

Causal Factors:

- Both the acetylene and oxygen regulators were being used without pressure gauges.
- Both the acetylene and oxygen pressure regulator diaphragm handles were turned almost completely clockwise allowing delivery of these gases at unsafe pressures.
- Check valves were not provided on either the fuel or oxygen lines between the torches and the hoses.
- Flashback arresters were not provided between the regulators and the delivery hoses.
- The acetylene and oxygen lines were not properly purged after changing the acetylene cylinder

Case History No. 10: Event: Nonfatal Fatal ignition or explosion of gas or dust accident, Gonzales County, TX.

Overview: An unplanned ignition of a fine organoclay dust (organically treated clay) occurred at a manufacturing plant while workers were in the process of troubleshooting a closed fluid bed batch dryer. The dryer was provided with a filter bag dust collection system which collected the fines. Periodically during the drying cycle the bags were shaken to allow collected dust to fall back into the drying bed. Hot air for drying was provided by a gas-piloted burner and air fan. The dryer was unexpectedly shutting down a few minutes into the drying cycle. The dryer was opened for troubleshooting purposes but the filter bags above the drying bed were not shaken prior to opening. Because of this, they were partially filled with collected dust. In order to facilitate troubleshooting, the bags were mechanically raised above the fluid bed. At that time an unexpected release of dust occurred from the bags. A flash fire ensued as a result of this ignition and five men suffered burns of varying degrees.

Findings and Conclusions: A leaking solenoid valve which allowed the gas pilot flame to remain on after shutdown of the dryer was found during the investigation. This provided a probable ignition source for the accumulations of dust within the combustion chamber, causing this material to smolder. Either this smoldering material, or the uncontrolled pilot flame itself, may have provided an ignition source for the dust cloud created when the workers failed to perform the shake cycles on the bags prior to opening of the dryer.

The unplanned ignition of organic clay dust occurred when a large quantity of the dust, used as a thickening agent in the manufacture of inks, dyes and paints, was released from the filter bags after the dryer was opened for troubleshooting purposes. It is highly probable that this cloud was ignited by a gas flame at the pilot burner that failed to extinguish after shut down of the burner due to a leaky gas pilot solenoid valve. The presence of both the fine dust cloud above the lower flammable limit and an ignition source of sufficient energy resulted in the ignition of the cloud and a flash fire. The construction of the dryer building and the geometry of the dryer allowed for adequate venting, preventing the build up of significant pressure.

Causal Factors:

- Company's failure to adequately maintain the combustion safeguards and fuel train appurtenances for the gas fired burner.
- Company's failure to establish a mandatory shakedown of the filter bags before any attempt to open the dryer under any conditions.

Case History No. 11: Metal/Nonmetal limestone hydration; nonfatal exploding-vessel-under-pressure accident, Clark County, Nevada.

Overview: A 45-year-old hydrator operator was seriously injured at a quicklime hydration plant when a piston-type slurry pump exploded while the operator was attempting to divert pump flow from a reaction tank to an exterior discharge bin. The valve used to divert flow was directly adjacent to the pump. The operator was struck by the pump cover plate when it blew off while he was attempting to operate the transfer valve.

Findings and Conclusions: The hydration reaction between the calcium oxide and water is substantially exothermic. In this operation, calcium oxide (quick lime) was mixed with water in an open chamber (mixing box) feeding a two cylinder piston pump. The pump discharged the reacting mixture through piping into a reactor vessel, where a vapor space could be used to control the pressure of the reaction. The suction side of each piston cylinder had a ball check valve that normally (but not always) prevented backflow out of the pump into the mixing box.

In this accident, the piping between the pump discharge and the transfer valve had plugged up, preventing flow out of the pump. If plugging of the pump discharge piping occurred, the operator received a pressure alarm and/or high amperage alarm on the pump motor, or heard the characteristic sound of the pump malfunctioning as flow backed up. Under those conditions, he had to manually change the position of a three-way valve on the pump discharge piping to redirect the back flow from the reactor vessel to the open bin. This was necessary to prevent the pressure and slurry in the reactor vessel from backing up into the pump, and out of the mixing box if the ball check valves failed. This event was not unusual since the mixture had a tendency to cause scaling and occlusion of the piping from time to time.

In this event, the reacting mixture was trapped in the pump. Heat from the reaction continued to be absorbed within the entrapped mixture and created increased pressure. During the accident, the pump was believed to have been hydraulically full. With no way to control the reacting mixture within the pump, dangerous internal pressure developed causing the pump lid to blow off, resulting in impact injury to the worker's arm and steam and thermo-chemical burns to his skin. A thermo-chemical analysis assuming isochoric and adiabatic conditions in the pump predicted that pressures in excess of 100,000 psig were possible under these conditions. This accident occurred because the pumping system was not capable of allowing for control of the chemical reaction rate within the entrapped mixture. Systems intended to handle chemically reacting materials must be designed using recognized chemical process safety management principles and technologies.

Causal Factors:

- Company's failure to conduct an adequate chemical process safety management risk assessment of this operation.
- Company's failure to establish a safe location for workers during start up or during other transitional periods of operation, such as plugged piping.

4. Methodology to Prevent Future Occurrences

The common thread in all of the above accidents is the failure of those doing the work to recognize the conditions created to permit an explosion to occur. The lack of recognition arises from a misunderstanding as to how pressure energy can be created or how it can be intensified beyond the structural design specifications of equipment. This misunderstanding, in turn, is often the result of poor procedures and inadequate training.

A dedicated effort in three distinct areas is needed to reduce the likelihood of repeating tragedies like those presented in this paper. These three areas are: 1) safety training, 2) job task evaluations, and 3) outreach programs. These three areas of effort can be effectively utilized as part of a safety culture approach taken by management.

4.1 Safety Training

Training is a fundamentally important component in any program to prevent deaths and injuries in the work place. Because job tasks and the hazards associated with them can be so diverse, it is difficult (if not impossible) to propose a simple, all-inclusive program for all those that are employed in any given operation.

One area that seems to need more work is educating personnel in the recognition of potential energy, and the role it can play in causing accidents. Certainly, such training will benefit the traditionally defined maintenance person. However, those charged with developing such programs within the work place must be cognizant of all those job positions which may expose the worker to potentially dangerous sources of potential energy while performing their assigned duties. In smaller operations in particular, care must be taken to identify jobs that may require a "jack-of-all-trades" role of the worker.

4.2 *Job Task Evaluation*

Formal safety analysis of all job tasks is slowly become more widely practiced within industry. An effective job safety analysis (JSA) coupled with training, can be a powerful tool in reducing tragic accidents often associated with traditionally unrecognized hazards. If not already completed, workplace safety officials should conduct reviews of all job assignments and tasks to prioritize them in terms of conducting JSA's. Any tasks that involve potential energy such as the application of high temperature heat; electricity; use of chemicals; the removal, repair, or installation of equipment, or is associated with pressurized fluids are all candidates for a red flag---a red flag that implies the need to immediately conduct a JSA.

4.3 *Outreach Programs*

Experience can often be an unforgiving teacher. Hence, industry must set a priority to communicate within itself to alert the workforce to unforeseen hazards. MSHA, for example, uses both its web site (www.MSHA.gov) and direct mailings to stakeholders in an effort to alert miners, mine operators, manufacturers, and others involved with the mining industry to accidents occurring within the industry. However, the workforce and those charged with safety management and training within industry must encourage the use of, and provide input to, outreach programs if the message is to truly get to those who need to hear it.

5. Conclusion

An aggressive, effective safety program utilizing a variety of methodologies is needed to eliminate the unseen insidious explosion hazards that can endanger those in the workplace. Training, job safety analysis, and communications are some of these important methodologies. Job safety analysis of all industrial work activities is particularly necessary in order to identify hidden potential explosion hazards. Outreach programs must be delivered to the grassroots personnel, and must be designed to affect the attitude of personnel toward working safer.

Emphasis should be placed on each worker, supervisor, manager, and trainer to develop and exercise a "questioning attitude" in approaching all work tasks. A questioning attitude is a mindset that must replace the all-too-often mentality of "I didn't think, I didn't see, I didn't know." It is a skill only sharpened with use. This will serve to enhance the awareness of workers who may otherwise face the insidious explosion hazards that still claim the lives of the nation's most precious industrial resource, its workers.

6. References

[1] NFPA 921 – 2001, Guide for Fire and Explosion Investigations, National Fire Protection Association, One Batterymarch Park, Quincy, MA, 02169

[2] Kuchta, Joseph M, "Investigation of Fire and Explosion Accidents in the Chemical, Mining, and Fuel-Related Industries – A Manual", Bulletin 680, U.S. Bureau of Mines, 1985

[3] MSHA Handbook Series, Handbook PH00-I-5, Accident/Illness Investigations Procedures, (http://www.msha.gov/readroom/handbook/handbook.htm)

Two Large Losses: Refinery Fire And Vapor Cloud Explosion At A Natural Gas Processing Plant

Jeffrey Yuill, P.Eng.
Starr Technical Risks Agency, Inc.
145 Wellington Street West,
Toronto, Ontario, CANADA M5J 1H8
Jeff.Yuill@starr-tech.com

Prepared for Presentation at
American Institute of Chemical Engineers
2005 Spring National Meeting
39th Annual Loss Prevention Symposium
Atlanta, Georgia
April 11-13, 2005

December 1, 2004

UNPUBLISHED

Two Large Losses: Refinery Fire And Vapor Cloud Explosion At A Natural Gas Processing Plant

Jeffrey Yuill, P.Eng.
Starr Technical Risks Agency, Inc.
145 Wellington Street West,
Toronto, Ontario, CANADA M5J 1H8
Jeff.Yuill@starr-tech.com

Abstract

This paper discusses two large losses.

The first loss involves a fire at a lube oil complex which forms part of a larger refinery site. The fire was the result of a failure of a portion of the flare header. The fire resulted in one death and $10 million dollars worth of physical damage.

The second loss involves a vapor cloud explosion at a natural gas processing facility. A release of a large amount of ethane and heavier hydrocarbons occurred for eight to 10 minutes before the cloud was ignited. The explosion resulted in eight deaths and $136 million in physical damages.

Details for much of this paper is based on reports provided by internal reports from adjusters and other specialists working on our behalf, references are not being submitted to ensure the anonymity of the losses.

1. Fire At A Refinery

South America, December 15, 1995

1.1 Background

The oil refinery where the incident took place is owned by the principal oil company operating in that country. Through its subsidiaries the company is involved in all phases of the oil industry including exploration, transmission, refining and retail sales. The company has three wholly-owned refineries, the largest being the subject of this case study and rated at 180,000 bbls/day.

The refinery was originally constructed in 1925 with a crude processing capacity of 11,340 bbls/day. Over the years, the capacity was increased with the addition of a topping unit, a FCC and delayed coker, a petrochemical complex and a lube oil complex.

The lube oil complex where the incident took place consists of the following units:

- Lube vacuum distillation - 26,566 bbls/day. Column bottoms are directed to the propane de-asphalting unit. Side draws from the column are directed to the furfural unit and then the MEK dewaxing unit for additional processing.
- Propane de-asphalting (PDA) - 10,064 bbls/day. LP gasses have the ability to precipitate asphaltic and resinous materials from crude residues while the lubricating oil components remain in solution. All liquified hydrocarbon gasses have this property to a marked extent, but propane is used due to its relatively low cost and its ease of separation from lubricating oils.
- Furfural unit- 10,600 bbls/day
- Solvent/MEK dewaxing.

The units were constructed in 1969 by a European engineering firm. All units are located in a common process block measuring 203 feet by 362 feet to the battery limits.

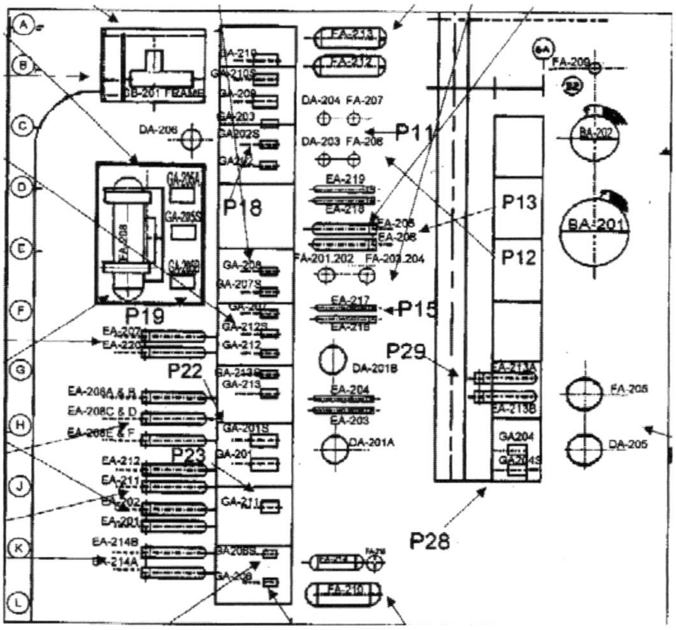

Figure 1: Plan of the PDA unit

1.2 Events

A chronological sequence of events is as follows. All times are local.

December 15, 1995:

- 4:40 PM- A loud hissing noise is heard by outside operations staff followed quickly by a white-yellowish cloud in the vicinity of flash drums FA-201/FA-202. A fire ensues almost immediately followed by an explosion. The fire is immediately reported to the control room who initiate a shutdown of the entire refinery site.

- ~4:43 PM- The refinery's own fire department arrives, responding with two fire engines. Soon afterwards a total of 13 pumper trucks arrive from the surrounding towns to fight the fire. The incident consisted of a main fire and 3 localized secondary fires. The fires not only involved the PDA unit but spread to the furfural and lube vacuum distillation units as well, due to inability of the sewer system to remove the volume of water and hydrocarbons. The fires are fought with a combination of water and foam.
- 5:25 PM- Fire declared under control.
- 6:10 PM- All fires extinguished except for propane being released from towers DA-201 A/B (1st and 2nd stage extractors) which is allowed to intentionally burn to avoid any vapor cloud explosions.

December 17, 1995

- 9:15 AM- Propane which was being purposely burned is exhausted.

1.3 Outcome

Two employees were seriously injured, one subsequently died.

Figure 2: View Of Damaged Piperack Between Pipe Columns G And H

Equipment damage is as follows:

- **Heaters** - The asphalt heater (BA-201) and gas oil heater (BA-202) reveal little damage. Melting of non-metallic components such as sight glasses, plastic tubing and wiring is not apparent.

- **Vessels** - Damage to vessels and heat exchangers is limited to loss of insulation and cladding as well as the loss of instrumentation cables.
- **Piping** - Piping between column lines C and K receive the high degree of damage. Some spalling of the fireproofing over structural steel supports occurs. The heat caused several other hydrocarbon lines (between 3 and 6 inches in diameter) to rupture increasing the combustible load present.
- **Pumps** - Pumps, associated electrics, and instrumentation are all severely damaged between column lines E & H. Limited damage occurs to the concrete pump bases.
- **Compressor** - The compressor (GB-201) located on the south-west corner of the process block is undamaged.

Figure 3: Overall View Of Fire Damaged Area With The Arrow Showing The Location Of The Ruptured Flare Line

1.4 Cause

The fuel that fed the fire came from a ruptured section of flare line inside the PDA unit. The rupture started at a small imperfection adjacent to a weld in the reducing section where the flare line increased from 14 inches in diameter to 20 inches in diameter. The rupture progressed into both the 20 inch and 14 inch portions of the line.

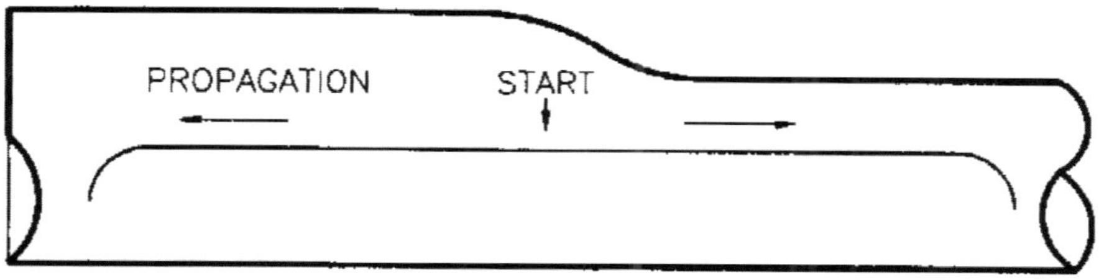

Figure 4: Flare Line

The reducer was not a forged fitting, rather was fabricated from a 20 inch diameter pipe and then slotted, trimmed and compressed to 14 inches and joined together by several longitudinal welds.

The metal composition is analyzed after the incident and determines:

Material	% Carbon	%Manganese	%Sulfur
Main pipe	0.19	0.33	0.07
Reducer	0.20	0.63	0.05

The flare line was refrigerated by mixed phase propane that was vented from pump GA-205B into the flare line.

At the same time, an apparent failure of the distributive control system allowed the high flow of heating oil to pass through several heat exchangers (EA-217, EA-218 and EA-219). The increased flow resulted in a higher temperature on the process side of the exchangers with an increase in the system pressure.

The pressure resulting from this temperature excursion caused various relief valves to open, venting into the flare header that was also receiving the mixed phase propane vented from pump GA-205B.

When the hot hydrocarbons from the relief valves hit the cold propane, they caused a very rapid increase in vapor load in the flare header.

Figure 5: Photograph Of The Flare Line After It Was Removed From The Field

The flare line contained about 15 m^3 of liquid which was lying at the bottom of the pipe, allowing the material being vented from the pump to pass over the liquid with little resistance. However, when the relief valves vented hot material into the line, the velocity lifted the liquid resulting in plug flow and because of the inertia of this liquid resulted in a high pressure spike.

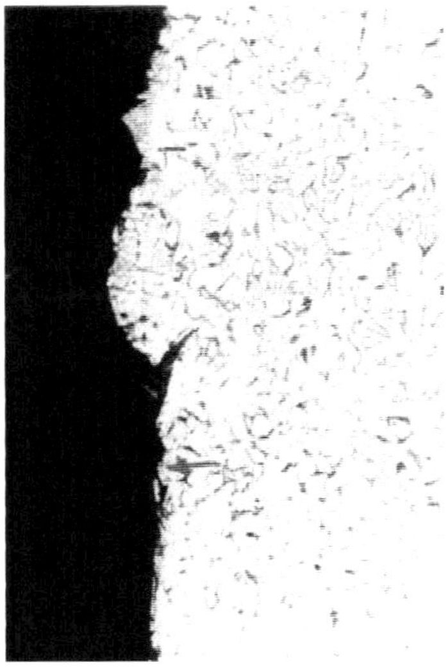

Figure 6: Micrograph Of Interior Of Pipe At Weld (X 200)

With the reducer cooled below the brittle zone, the pressure spike caused the metal in the reducer to fail by rapid fracture. After progressing a short distance, the crack entered a more ductile zone and finished the failure by ductile failure.

With both propane and bright stock emanating from the failed pipe, this mixed phase caused a large static charge which probably ignited the mixture.

In addition the storm water sewer system was not large enough to remove the fire protection water and spilled hydrocarbons which resulted in the carry over of burning hydrocarbons into the vacuum unit.

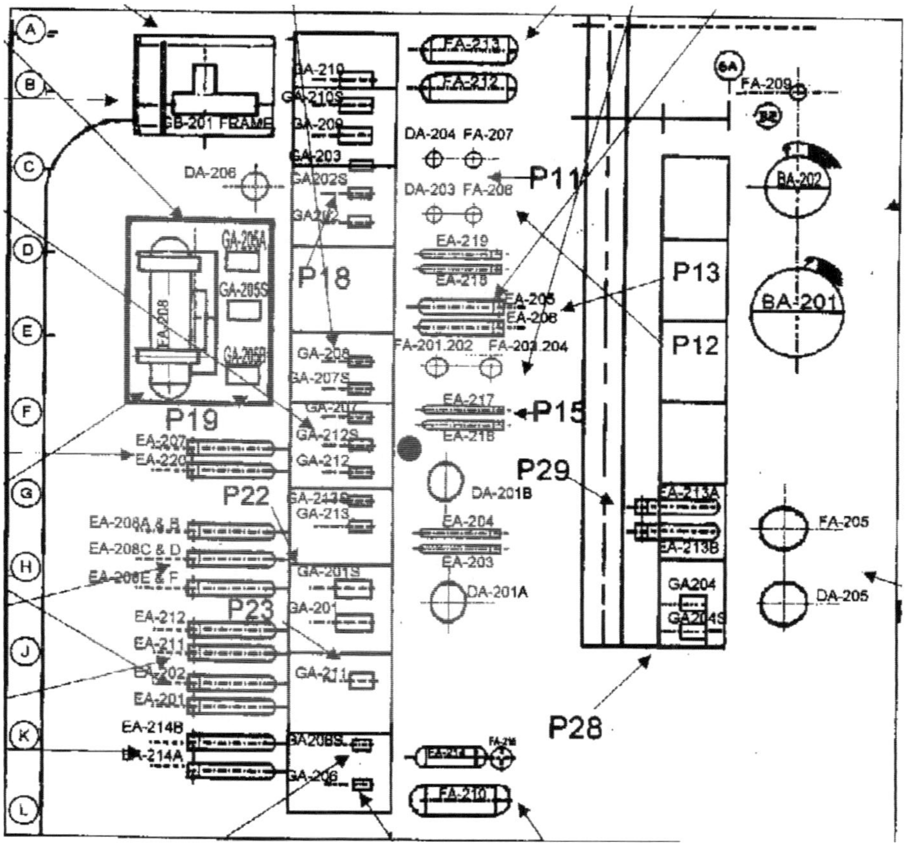

Figure 7: View Of PDA Unit Shaded Area - Piping & Steel Replaced Red Dot Indicates Approximate Position Of Flare Line Rupture

1.5 Damages

Physical damages were in the amount of US$10 million in 1995 dollars. A breakdown of some of the damages include:

Civil/mechanical repairs	$5.7 million
Instrument repairs	$1.6 million

Electrical repairs $512,000

The loss of profits associated with the downtime was not insured (it was assumed by the company directly). Total downtime is unknown but repairs did not start until April 1996.

1.6 Lessons Learned

The following major lessons can be drawn from the loss

- **Drainage -** The need to safely remove fire water and hydrocarbons to limit damages. This can be accomplished via drainage and/or an arrangement of diversions/swales and sloping to safely remove the material.
- **Engineering Design -** The original design of the flare header should have considered the auto-refrigeration effect of the propane and the metallurgy modified accordingly. Additionally, a proper transition piece should have been utilized instead of a shop assembled unit.

2. Vapor Cloud Explosion at a Natural Gas Processing Plant

Central America- July 1996.

2.1 Background

The facility is owned by the national oil and gas company. Through its subsidiaries the company is involved in all phases of the oil industry including exploration, oil refining, gas processing and petrochemicals. The natural gas production unit of the company had eight gas processing operations.

The site where the incident took place is occupied for the processing of sour natural gas and sour condensates. The gas plant is located on 700 acres and was originally constructed in 1974. The plant is rated as follows:

- Sour gas treating- 1350 million scf/d
- Sour condensate treating- 48,000 barrels/day
- Cryogenic separation- 1450 million scf/d
- Liquid fractionation- 105,000 barrels/day

Products from the site are sweet, dry natural gas, ethane, propane, butane, light natural gasoline (naphtha), sweet stabilized condensate and sulfur.

Processing is conducted using 10 desulfurization units to remove hydrogen sulfide and 5 cryogenic units which lowered the gas temperature to –130°F. Cryogenic plants #3, #4, #5 and #6 were constructed in 1978 and were each rated at 150 million scf/d. Cryogenic unit #3 was relocated off-site several years before the loss. Cryogenic plants #1 and #2 were constructed in 1981 and are each rated at 500 million scf/d.

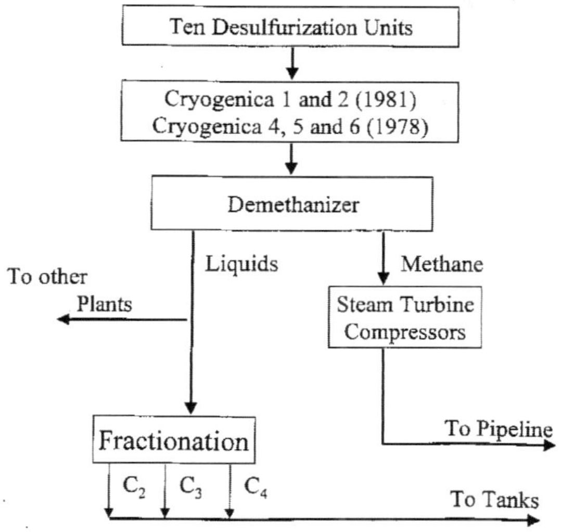

Figure 8: Process Block Flow Diagram

All of cryogenic plant #1 sat up to 15 feet below the perimeter of the road surrounding the unit.

The pump station where the incident occurred consists of a concrete slab sitting about 16 inches above grade. A concrete roof structure supported by 6 concrete columns covered the pumps. There are no walls. The pump station is located in the south-east corner of the Cryogenic #1 process unit.

The pump station consists of two, 10 stage centrifugal pumps (BA-110A and BA-110B) each driven by a 500 HP electric motor mounted vertically above. Each pump is used alternatively to increase the line pressure from 400 psi to 700 psi. Both the suction and discharge piping was 16 inches in diameter.

The pumps receive their input from Vessel FA-108 located about 200 feet from the pumps. The vessel holds a hydrocarbon mix of 42% ethane and 58% light hydrocarbons (propane to gasoline) at 90°F and 400 psi.

A number of means of isolation was provided both upstream and downstream of the pumps. Upstream, a manually operated valve was provided at the discharge of vessel FA-108. A motorized plug type valve was provided just prior to each of the booster pumps (BA-110A & B), this valve could be operated from the control room and from the valve itself. These motorized actuators were installed in 1995 as part of a process hazard analysis. Each of the actuators featured a selection dial with:

- Off (allows for local manual operation only)

- Remote
- Local

The selection dial could be locked-in-place at the actuator.

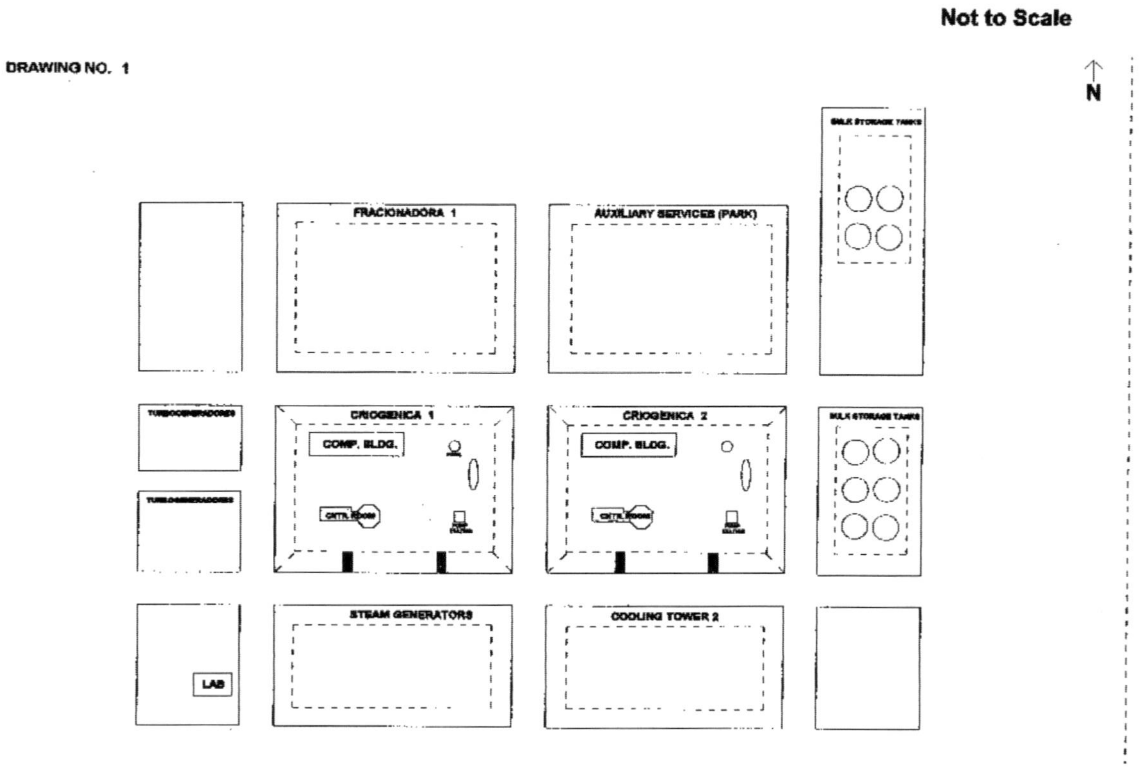

Figure 9: Simplified site plan

Downstream of the pumps, an in-line check valve followed by a manually operated valve is provided. The isolation valves that operated from the control room were not coupled to the DCS system. Only four staff on shift had authorization to operate any of the valves (even during emergencies).

There were two different types of safe work permits in use at the site:

- General
- Potentially dangerous- This type of work order required the review of several safety oriented prerequisites and ultimately the signature of an Operations Engineer

Figure 10: Valve Actuator Similar To That Involved In The Loss (Finger Points At The Locking Mechanism)

Some of the fixed fire protection and detection systems consist of:

- Four electric motor driven and two diesel engine drive fire pumps capable of delivering 14,000 gpm. Manual start.
- A remote, manually actuated deluge system over some of the process pumps (one head each for pumps BA-110A & B).
- Monitor nozzles spaced around the perimeter of each process unit.
- Combustible gas detection throughout the process units. Five gas detectors are provided within cryogenic plant #1, gas detection was specifically provided over pumps BA-110.

Fire protection consists of 14 dedicated firefighters per shift. Mobile firefighting equipment consists of three pumper trucks and a site inventory of 7200 gallons of protein foam.

2.2 Events

A chronological sequence of events is as follows. All times are local.

- **May 1-** Pump BA-110B in cryogenic plant #1 started as part of the normal practice of rotating pumps every four months.

- **July 25-** Pump BA-110B trips electrically. Further investigation finds a mechanical seal located between the vertically oriented motor and the pump beneath leaking. A decision is made to fix the seal and then determine why the motor tripped. Pump BA-110A is started until repairs are completed.

- **July 26**

  9:00 AM- 'Potentially dangerous' type of work order is issued. Separate work orders are issued for the seal replacement, blind flange installation and electrical repair.

  11:00 AM- After safety department attendance, mechanical seal replacement initiated by disconnecting the suction & discharge lines and then inserting blind flanges. Between 8 and 13 people are in attendance during this procedure.

  1:00 PM- Reattached suction & discharge lines. Shaft coupling disconnected.

  6:00 PM- Mechanical seal replaced. Electrician(s) investigate cause of motor failure and replace an electrical switch in one of the substations.

  7:00 PM- Suction & discharge lines disconnected. Blind flanges removed and replaced with 'flex-steel' seals. The shaft coupling had not as yet been reconnected.

  7:20 PM- All of the discharge line bolts have been replaced. The 20 suction-line bolts were in the process of being replaced. Flanges did not line up correctly. All but 3 bolts had been replaced (albeit only hand-tightened) when the **SUCTION VALVE OPENS**. It should be noted that there were no operation staff on hand at the repair site at this time.

  ~7:28 PM to 7:30 PM- Easterly wind drifts vapor cloud toward cryogenic plant #2

  Cloud covers eastern third of cryogenic plant #1 and western two-thirds of cryogenic plant #2. Cloud was most likely ignited by a process furnace in cryogenic plant #1 (located in the north-east quadrant of the unit). Prior to the explosion, personnel try to disperse vapor with monitor nozzles, however, the flow is weak. It should also be noted that plant personnel are told to stand-by to shutdown entire plant site only. **VAPOR CLOUD IGNITES**.

  ~7:32 PM- Vessels EA-1109 & EA-1110 in cryogenic plant #2 BLEVE.

  10:30 PM- Ensuing fire under control. In total 2100 to 2600 gallons of foam is used

2.3 Outcome

The explosion results in the death of eight employees and injury to 40 others. Damages were extensive and are summarized as follows:

- **Cryogenic plant #1-** Much of the plant suffers heavy blast damage. The control room is destroyed. The east end of the compressor building is damaged by fire. Many product pumps are damaged or destroyed.

- **Cryogenic plant #2-** This unit is the center of the blast and receives the brunt of the damage. The entire process unit is essentially destroyed. The entire compressor building and its contents are destroyed when the roof collapses onto the compressor deck and the floor collapses onto the condensers. The control room is totally destroyed.

- **Utilities-** Heavy blast and fire damage occurs to several electrical substations. Even underground high-voltage electrical cable are damaged to the point of requiring replacement. There is blast damage to the duct work associated with four of the process boilers. Two cooling tower receive blast damage but are repairable. A portion of the water treatment plant also receives damage.

Figure 11: Aerial photograph
Arrow 1- Pump BA-110B
Arrow 2 - Cryogenic unit #2

Figure 12: Aerial photograph
Arrow #1- Cryogenic unit #1
Arrow #2- Cryogenic unit #2

2.4 Cause

Following examination of the scene, coupled with statements provided by witnesses, the explosion and ensuing fire were the result of the opening of the BA-110-B suction line valve while the downstream 16 inch pipeline was open for repairs.

Due to the death of several key witnesses (control room board operator, shift engineer and chief engineer) as well as the degree of damage, the reason why the valve opened could not be determined with certainty. Three possibilities exist:

- **Human error-** Someone inadvertently, accidentally or through miscommunication, either locally or remotely, opened the suction valve.
- **Electrical malfunction-** Due to an electrical malfunction, the valve opened without specific instructions from anyone.

- **Mechanical stress-** Due to ongoing repairs, the handling of the components in the repair area resulted in the activation of the motorized valve's relay resulting in its opening.

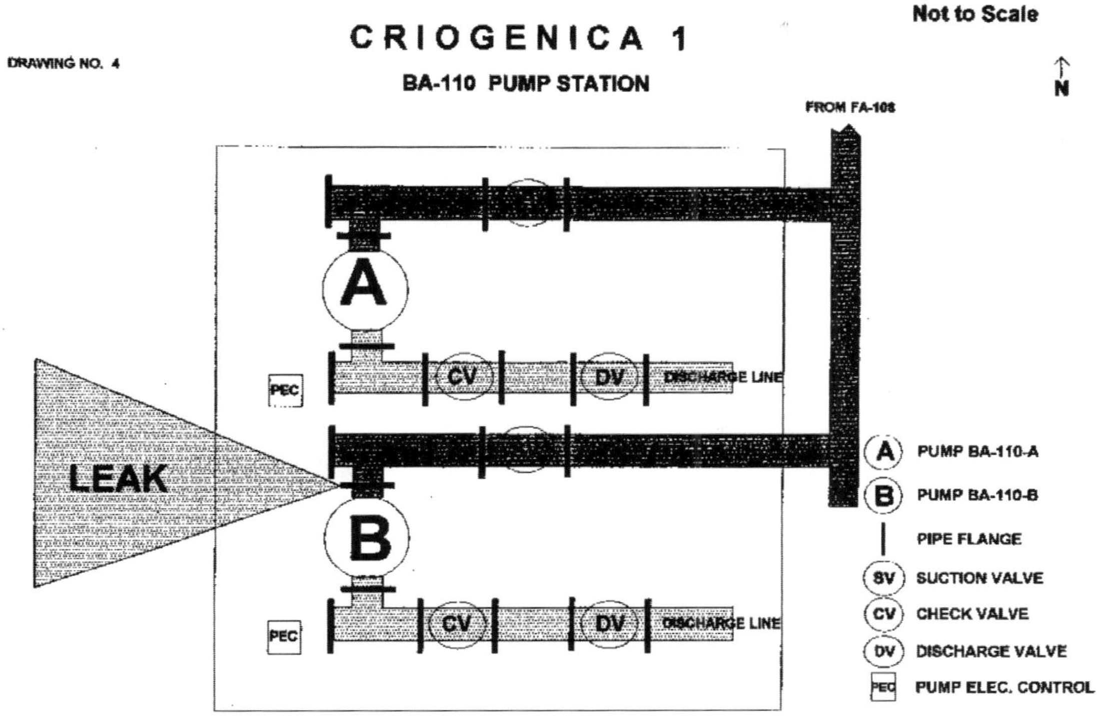

Figure 13: BA-110 Pump station

2.5 Damages

Physical damages were in the amount of US$137.6 million in 1996 dollars. A breakdown of some of the damages include:

Cryogenic unit #1	$27.5 million
Cryogenic unit #2	$75.7 million
Auxiliary services	$23.1 million
Contracts, debris removal etc	$11.4 million

The numbers stated above are gross amounts and do not include deductibles or salvage.

The loss of profits associated with the downtime was not insured (it was assumed by the company directly). It did however, require 19 months to rebuild cryogenic plant #2 which was built at another site.

Figure 14: View of Suction side of pump BA-110B

Figure 15: Suction valve for pump BA-110B

2.6 *Lessons Learned*

The following major lessons can be drawn from the loss

- **Maintenance/Safe Work Procedures -** Maintenance practices should be augmented to include to locking out of critical control valves when hydrocarbon or high energy steam lines are opened. Many maintenance staff were unaware of the status of the motorized or manual valves during their activities.
- **Training of Maintenance Staff -** All maintenance staff need to understand the process and specifically the location of alternative isolation valves. Only senior staff and operators were aware of the location of the manual isolation valve near FA-108.
- **Procedures -** All staff need to be able to operate valves in the event of an emergency. This item ties back to training as well.
- **Emergency Procedures -** Emergency procedures should be revised. Plant staff had no recollection of the plant alarm being sounded during the 8 to 10 minute period prior to the explosion. Additionally, operators were told to only 'standby' to shut down the operation during the same period.
- **Control Rooms -** The control rooms for each unit should be relocated to an area not subject to damaging overpressures or reinforced to sustain the impulse of an overpressure. The majority of deaths occurred to staff occupying the control rooms.
- **Fire pumps -** The fire pumps should be automated to start on pressure drop. Several of the operators and maintenance staff attempted to operate monitors prior to the explosion but only a 'trickle' of water was available.

U.S. Chemical Safety Board Investigation:
Positive Pressure Control Room Failure

Lisa Long – Speaker
Johnnie Banks
Mike Morris
John Murphy, P.E.
Stephen Wallace, P.E., CSP
U.S. Chemical Safety Board
2175 K St, NW
Washington, DC 20037

Prepared for Presentation at
American Institute of Chemical Engineers
2005 Spring National Meeting
39th Annual Loss Prevention Symposium
Atlanta, Georgia
April 11-13, 2005

UNPUBLISHED

U.S. Chemical Safety Board Investigation: Positive Pressure Control Room Failure

Lisa Long – Speaker
Johnnie Banks
Mike Morris
John Murphy, P.E.
Stephen Wallace, P.E., CSP
U.S. Chemical Safety Board
2175 K St, NW
Washington, DC 20037

Abstract

On July 20, 2003, a release of chlorine gas from the Honeywell chemical plant in Baton Rouge, Louisiana, resulted in the hospitalization of eight plant workers and required residents within a half-mile radius to shelter in their homes. The chlorine was released when a cooler failed allowing chlorine to enter a coolant system not designed for chlorine. Plant operators noticed the release and attempted to stop it. Response efforts were halted when the chlorine entered the unit's positive pressure control room. Operators in the control room were overcome by the chlorine, and had to evacuate the plant prior to determining and stopping the source of the leak or shutting the process down. The U.S. Chemical Safety Board investigated the incident. This paper presents details of the positive pressure control room failure and its consequences, discusses available guidance and standards for positive pressure control rooms, and details the inadequacies in this particular positive pressure system.

1. Chlorine Release

At 3:10 am on July 20, 2003, operators inside the control room noticed a chlorine odor. One operator went outside to investigate and observed a leak near the coolant system pumps. Plant personnel outside the control room quickly realized the need for additional help and protection to stop the leak. However, before they could take any action, the chlorine concentration both inside and outside of the control room became overwhelming. Operators experienced difficulty breathing and were forced to evacuate the area before they could determine why chlorine would be leaking from the coolant pumps. The process was still running when they were forced to leave the area.

2. Incident Response

By 3:25 am, plant personnel had evacuated to the main plant gate. A level II incident (i.e., effects inside the plant, with potential for offsite effects) was reported to local authorities through the East Baton Rouge I-notification system. By 3:29 am, the incident was raised to level III (i.e., effects outside the plant).

Several operators from the unit were exposed to chlorine during the release; they were given oxygen in the first-aid building. The site water deluge towers[1] were turned on to suppress chlorine vapor, and additional off-duty plant personnel were called into the plant to help stop the release.

Plant emergency response personnel activated the incident command system and began attempts to stop the leak. However, the chlorine that entered the control room had corroded the process control system and rendered it inoperable. Response personnel had to manually shut down the process. Once it was shut down, they realized that the chlorine was still being released, and the chlorine railcar (the source of the chlorine feed) would have to be isolated to stop the leak.

The chlorine leak was finally stopped at approximately 6:46 am, when the emergency response team manually closed the railcar valve. As a result of this incident, eight employees were transported to the hospital for medical evaluation and treatment.

The exact amount of chlorine released could not be determined. Honeywell reported a maximum estimated release of 14,400 pounds to the Louisiana Department of Public Safety and Corrections, Office of the State Police. This estimate was calculated based on pre- and post-incident estimates of chlorine railcar weight. The release may have been less if some of the chlorine dissolved in the coolant or remained in the system piping. Additionally, the amount of chlorine that drifted offsite may have been less than the amount released because of mitigation by the deluge tower water spray.

3. Community Impact

At 4:00 am, the local fire department issued a shelter-in-place advisory for all residents within a 0.5-mile radius of the plant. Community sirens were activated, and the auto dialer was used to notify these residents. The advisory remained in effect for 3 hours. At 7:09 am, the fire department downgraded the incident to level I (i.e., effects inside the plant only).

At the time of the release, the predominate wind speed was 3 to 5 miles per hour (mph) from the southwest. Chlorine concentrations in air were not measured until 5:30 am—at which time, Honeywell recorded a concentration of 1.5 parts per million (ppm) at the main gate. At 7:13 am, LADEQ arrived and sampled air near the facility; these measurements showed less than 0.35-ppm chlorine outside the plant fenceline.[2]

CSB interviewed emergency responders, who reported noticing a strong chlorine odor over 1.5 miles away, in the vicinity of Plank and Prescott Roads. As they proceeded west on Chippewa Street toward the site, they stated that the odor was much less pronounced. A few citizens also noticed a strong chlorine odor on Interstate 110, north of the Honeywell facility. Ten members of the general public sought treatment at the local emergency room—three initially and seven

[1] Water deluge towers spray water from a nozzle. Studies indicate that the water spray can suppress chlorine vapors, reducing the concentration of chlorine outside the area of release (The Chlorine Institute, Inc., 1990).

[2] These measurements are indicators of concentrations at a specific location and time and do not necessarily represent concentrations at other locations and time.

throughout the day. They complained of headaches and sore throats. They were either treated or observed and then released from the emergency room.

4. CSB Investigation

CSB conducted a full investigation of this incident. A detailed report with root causes and recommendations will be available on CSB's website (www.csb.gov) once the investigation has been voted on and approved. This paper presents details of the positive pressure control room failure and its consequences, discusses available guidance and standards for positive pressure control rooms, and details the inadequacies in this particular positive pressure system.

4.1 Design and Maintenance of Positive Pressure Control Rooms

The control room had been converted into a positive pressure control room in 1998. Plant personnel had complained of odors getting into the control room; and an engineering study had identified the potential infiltration of hydrogen fluoride, a highly hazardous substance used in the plant.

A divisional engineering group managed the positive pressure control room conversion project. The control room was designed to provide short-term protection for personnel during a chemical release so that they could safely shut down the process prior to evacuating. Escape respirators, located in the control room, were to be used for operators to evacuate after shutting the process down.

The design of the positive pressure control room included the following features:

- Air intake piped to pull air from the highest point in the plant.
- Sealing of the room to allow the heating, ventilating, and air conditioning (HVAC) system to maintain a positive pressure.
- Manual shutoff of the HVAC system intake.
- Audible alarms to indicate low pressure.

During the July 20 release, chlorine entered the control room. CSB investigators found that it was drawn into the building through holes and gaps in the HVAC intake ducts located on the roof. Some holes appear to have been drilled in the duct and not plugged, while others resulted from gaps in joints. Duct tape had been used to seal some of the gaps, but it became dry and brittle over time. Figures 1 and 2 show some of this deterioration. Figure 3 shows the location of the coolant system chlorine leak.

Figure 1: Condition Of Duct Tape Used To Seal Gaps.

Figure 2. Hole In Ductwork.

Figure 3. HVAC intake system on control roof (foreground). Arrow points to location of coolant system chlorine leak.

The positive pressure control room system did not protect personnel or equipment during the July 20 chlorine release. When interviewed by CSB, engineers stated that they were unaware of any standards for design and maintenance of control rooms to withstand toxic gas releases.

The following deficiencies in the positive pressure control room system contributed to its failure:

- There was no maintenance program for the control room, including HVAC ductwork.[3]
- Duct tape was used to seal some joints in HVAC ductwork. This tape eventually became brittle and exposed gaps in the joints.
- There were numerous entrances and exits from the control room—none of which had a double door air lock system.
- There were no toxic gas alarms on the HVAC intake system. Alarms could have warned the operators to use the escape respirators and exit the control room.
- Although the positive pressure control room system was listed as a safeguard in the hazard analysis, it was not designated as critical equipment or maintained as such.[4]

---

[3] When asked, Honeywell indicated that there was no preventive maintenance on the positive pressure control room system, including ductwork.

CSB researched available industry standards and guidance for control rooms, and found that existing standards aimed at the chemical manufacturing industry do not adequately address design and maintenance for protection against toxic chemical releases. Current guidance is summarized below:

- ***American Petroleum Institute (API) Recommended Practices***: API develops standards for petroleum refineries. However, API Recommended Practices 550, 551, and 752 provides direction in the area of positive pressure control rooms that can also be helpful to the chemical industry (API, 1977; 1995; 1999).
- ***National Fire Protection Association (NFPA)***: NFPA 496, Standard for Purges and Pressurized Enclosures for Electrical Equipment (1998), contains only a few recommendations for pressure and air velocities. It was not intended to address the protection of personnel.
- ***Instrument, Systems, and Automation Society (ISA)***: Several ISA standards include design recommendations to protect process measurement and control systems; however, none of the standards address protection of personnel.
- ***U.S. Nuclear Regulatory Commission (NRC)***: Although NRC regulatory guides are not intended for the chemical industry, they may provide relevant information in the area of positive pressure control room design and maintenance (USNRC, 2003a; 2003b; 2001). NRC staff use the guides to implement specific regulations, to evaluate specific problems or postulated accidents, and to review applications for permits and licenses.

The United Kingdom Chemical Industries Association provides some guidance on control room design and maintenance, including recommendations for inspecting and maintaining HVAC systems and for entrance and exit design. In *Loss Prevention in the Process Industries*, Lees (1996) briefly covers topics such as the need for an airtight design, positive pressure requirements, and possible use of toxic gas monitors.

As outlined above, U.S. guidance and standards for designing controls rooms in chemical manufacturing facilities to protect against toxic releases are limited. If the positive pressure control room system involved in this incident had been better designed, documented, and maintained, it is likely that the operators would have been better protected and able to more expeditiously stop the July 20 chlorine leak.

At the time of this writing, the Chemical Safety Board has not yet voted this on investigation report. More information regarding CSB's final report and recommendations will be available during this paper's presentation at the 2005 AIChE Loss Prevention Symposium, and can be found on CSB's website (www.csb.gov).

This paper has not been independently approved by the Board and is published for general informational purposes only. Any material in the paper that did not originate in a Board-approved report is solely the responsibility of the authors and does not represent an official finding, conclusion, or position of the Board.

---

[4] In its citations, OSHA considered the positive pressure control room to be covered by the PSM Standard.

The 27 March 2003 Billy-Berclau Accident: A Technical And Organisational Investigation

Jean-christophe Lecoze, Nicolas Dechy, Samantha Lim, Emmanuel Leprette, Ruddy Branka
INERIS (Institut National de l'Environnement Industriel et des Risques)
Parc Alata, 60550 Verneuil-en-Halatte, FRANCE
Tel: 33 3 44 55 66 77, ineris@ineris.fr

Abstract

Following a dynamite explosion in a pyrotechnic plant in Billy Berclau, France, which caused the death of 4 persons and injured 13 persons, the Environmental Ministry appointed INERIS (National Institute For Environmental Safety) to carry out an investigation. The purpose of the investigation was to consider technical and organisational issues related to the origins of the accident as well as the damages of the accident. This paper will introduce the results of the investigation with a focus on its organisational side.

The organisational requirement within that investigation along with the technical dimension is a first in France in the chemical industry and was enforced partly by the SEVESO II European Union directive context. The context of the Toulouse disaster played its role also. Indeed this directive requires the implementation of a safety management system, dedicated to major hazard prevention for certain plants, according to their risk level (based partly on the quantity of hazardous materials stored). The plant where the accident occurred was under that regulation and had to implement such a management system. The need for an understanding of the management system's underlying root causes in that accident was therefore high.

INERIS has been developing research programs on the organisational sides of major hazard prevention for few years and could therefore use its expertise in that field for the investigation.

In a first part of the paper, the accident will be presented, through a chronology of the events, damages and through the technical causes that have been identified and retained as most probable during the investigation. This first part is a preliminary step for understanding the case itself and to get into the second part of the paper.

The second part will discuss the organisational investigation, from a methodological point of view and then from the findings. The methodological aspect will question the ability to get into the complexity of the organisational life. Organisations are complex systems that need to be looked at with appropriate lenses. These lenses are influenced by the investigator experience and background and it is part of the methodology to address this issue. The level of opening of the organisational investigation is another strong methodological issue. As organisations are complex systems, they interact at all levels, and their understanding implies opening the system to many individuals involved inside and outside the boundaries of the organisation. That is another important issue of the investigation. The outcomes of the investigations are presented according to different levels of the system and are discussed.

1. Introduction And Context Of The Investigation

Following a dynamite explosion in a pyrotechnic plant in Billy Berclau, France, which caused the death of 4 persons and injured 13 persons, the Environmental Ministry appointed INERIS (National Institute For Environmental Safety) to carry out an investigation. The purpose of the investigation was to consider technical and organisational issues related to the origins of the accident as well as the damages of the accident. This paper proposes to describe some related issues around the investigation, with a focus on its organisational side.

1.1 Context Of The Investigation

A dynamite explosion, in a pyrotechnic plant in Billy Berclau, France, occurred on 27 march 2003 at 6h16, and caused the death of 4 persons and injured 13 persons. Fires resulting from the explosion were quickly extinguished by emergency services.

After the event, several investigations were launched by Authorities:

- the litigation inquiry carried out by the Police and the Justice,
- the Labor Ministry carried out its investigation,
- the investigation of the company Nitrochimie S.A. that merged with the Health and Safety committee: their focus was to find out the causes of the explosion,
- the administrative inquiry of IGE-CGM-IGAS (Inspection Générale de l'Environnement, Conseil Général des Mines and Inspection Générale des Affaires Sociales) set up by four Ministries: Environment, Defence, Labor and Industry; the investigation had to focus on circumstances of the accident, causes (technical, human and organisational), safety measures, recommendations that could lead to regulation changes, and should use expertise of INERIS and IPE (Inspection des Poudres et Explosifs); the report would be made publicly available (on the web in April 2004),
- the investigation of INERIS, launched by the Environment Ministry, focusing on technical and organisational dimensions of the accident.

The general objective of administrative inquiries carried out by IGE, IGAS and CGM, that are organisations within their Ministries, is to focus on the circumstances (in case of accidents, pollution,...) and on the local application of the law by companies and local control authorities. The investigation of INERIS brought (final report was issued in September 2003) the parallel inquiry of the IGE-CGM-IGAS, a technical and expertise point of view, independent of the other investigations (litigation, company...). It was also used as a support for the local control authority (DRIRE) to require implementation of safety measures from the company to get authorization to re-start.

This investigation occurred in a broader context following the Toulouse disaster that occurred the 21st September 2001. As a reminder, several investigations were launched by stakeholders (a few more than in this case), and provided numerous findings, lessons and recommendations for changes (Dechy et al, 2004). Disasters are usually seen as catalyst for changes and the Environment Ministry was designing a new law on industrial risk prevention (Law of 2003-699 of 30th of July 2003) and discussing it with stakeholders at the time of this event.

After Toulouse disaster, other decisions of the Environment Ministry were taken to upgrade the existing funding for carrying research at our Institute on the learning from experience, accident investigation methodologies and the integration of human and organisational factors into the major hazard prevention. This investigation was the first one to use most of these developments. As a reminder, INERIS has activities for certifying explosives and hazardous materials (for transportation) in France and EU.

1.2 Objectives, Contents And Management Of The Investigation

The objectives of the investigation were set up by the risk bureau on chemical risk (SEI/BRTICP) and the BARPI (that runs the ARIA database on industrial accidents) within the Environment Ministry and were accepted by the scientific director of INERIS.

The objectives were to:

- estimate the effects of the explosion (inside the site and in the environment), its strength, its eventual domino effects, the differences with the safety studies scenarios,
- find the origin, the circumstances and the possible technical and organisational root causes that should be investigated; for the technical causes, the nature, quantities, reactions of material involved, interactions with the process have to be investigated; organisational factors should be investigated also with regards to the Seveso II regulation and its safety management system requirement,
- formulate technical recommendations on process, workshop design and organisational recommendations such as cleaning of units, risk analysis, learning from experience, etc.

To fulfil these requirements, INERIS management appointed a multidisciplinary team of 5 main investigators (the Institute has 150 people working in accidental risk prevention) that were competent in the field of pyrotechnic process, the pyrotechnic and Seveso II regulations and its safety management requirements, the inspection process, technical investigation, materials sensitivity testing, explosion and pressure waves effects, human and organisational factors. The reports were reviewed by two top level managers (technical director of the Accidental Risk Division and the technical director of the Certifying Division) and approved by the scientific director.

The first visit consisted in collecting facts, photos of the damages, material samples, and documents and were followed, during 3 months, by several interviews of process operators, frontline managers, maintenance technicians, safety managers, top management of plant, top management of the group and control authorities.

2. The Context Of The Accident

2.1 The Company

The company, Nitrochimie S.A., is owned by a French group (SAEPC, Société Anonyme d'Explosifs et de Produits Chimiques) quoted to the stock exchange market of Paris and by the Davey Bickford Smith English group. The turnover was 120 M€ in 2001. The company Nitrochimie S.A. owns two sites in France: the Billy-Berclau site in north of France that produces explosives such as dynamites, blasting gelatine, powder dynamites, and ammonium nitrate fuel oil (ANFO) with 80 to 90 people, and another site (with the Research and Development laboratories) that manufactures blasting emulsions, expansive cement, and monomethylamine nitrate.

The plant is on a site of 750 000 m², on Billy-Berclau and two other cities and its first neighbours are at 700 m. The plant was created a hundred years ago in 1904 to produce nitroglycerine based explosives for the coal mines nearby. It was destroyed during WWI and rebuilt in 1920 with two areas for different manufacturing purposes: one for ANFO and one for dynamites. In 1991, the nitrate activity was stopped and the number of employees was reduced from 170 to 117. Ten years later, in 2003, they were 84 permanent workers and 4 to 6 interim workers.

The explosives manufacturing rates per year were roughly 6 500 tons of dynamites, 8 500 tons of ANFO and requires the consumption of 4 000 tons of sulfonitric mix, 240 tons of DinitroToluène, 10 700 tons of AN and 550 tons of light fuel. These manufacturing explosives to be sold on the EU market needed to be approved according to EU Directives specifications. INERIS does certify explosives in the EU and since 1997 certified those of Nitrochime S.A. and the last time was in May 2001 for 15 different dynamites and 2 ANFO.

2.2 The Safety Related Regulations

These manufacturing activities were under two main type of safety-related regulations:

- The labor regulation and its pyrotechnic regulation: a specific regulation from the labor Ministry was set up in 1979 (Decree n°79-846) and implemented with positive results measured by reduced rates of accident. It required specific prevention measures to be implemented. An emphasis of the investigation was on a comparison between observed effects and estimated effects from the safety cases,
- The environment regulations and its Seveso II regulation (96/82/EC) transposed in 2000 in France: indeed this Directive requires the implementation of a safety management system, dedicated to major hazard prevention for certain plants, according to their risk level (based on the quantity of hazardous materials stored). The plant where the accident occurred was under that regulation for 3 years and had to implement such a management system. The need for an understanding of the management system underlying root causes in that accident was therefore high.

2.3 *The Process*

There exist several kinds of dynamites that differ in rate of nitroglycerine or nitroglycol. To manufacture dynamite gelatine, four main steps are briefly described:

- Nitration (in a workshop, continuous process) of a mix of glycerine/glycol by a mix of nitric and sulfuric acid to get nitroglyceroglycol (NCG)
- Paste or gelatine manufacturing: it can be done in two different ways: an automated mixing bath holder called Tellex operated at distance in a safe place - or by batch in a tank called Guedu where NCG and adsorbents are poured by operators and the blast gelatine is removed after it has been automatically mixed. The Tellex produces 3 tanks in a row (that are made of polyethylene) for a total amount of 135 kg of dynamite gelatine. These tanks are equipped with caps and 4 wheels. The Guedu produces 3 tanks for a total amount of 120 kg. Then the tanks are driven on a trolley to an intermediate workshop storage n°49 that has a storage capacity of 1 300 kg. From this storage, tanks are automatically driven following a cable, to the processing workshop n°50 to manufacture cartridges of dynamites. These workshops are all separated by barricades of 5 to 6 meters high.
- The automated processing of gelatine to dynamite cartridges at workshop n°50: The inlet of the tanks into the feeder (see figure 1) is automated when the process is started. After the extrusion of the gelatine into the die (photo 1 and 2) and its cut at the roller, the gelatine is rolled in a wax paper before introducing it in the cartridge. The cartridge, processed in the Rollex machine (rate of 900 kg/hour), are transferred on a conveyor to be packed in boxes in a neighbouring workshop n°52. This neighbouring workshop has the video surveillance reporting screen (and recorder) for operators to monitor the process once it is started.

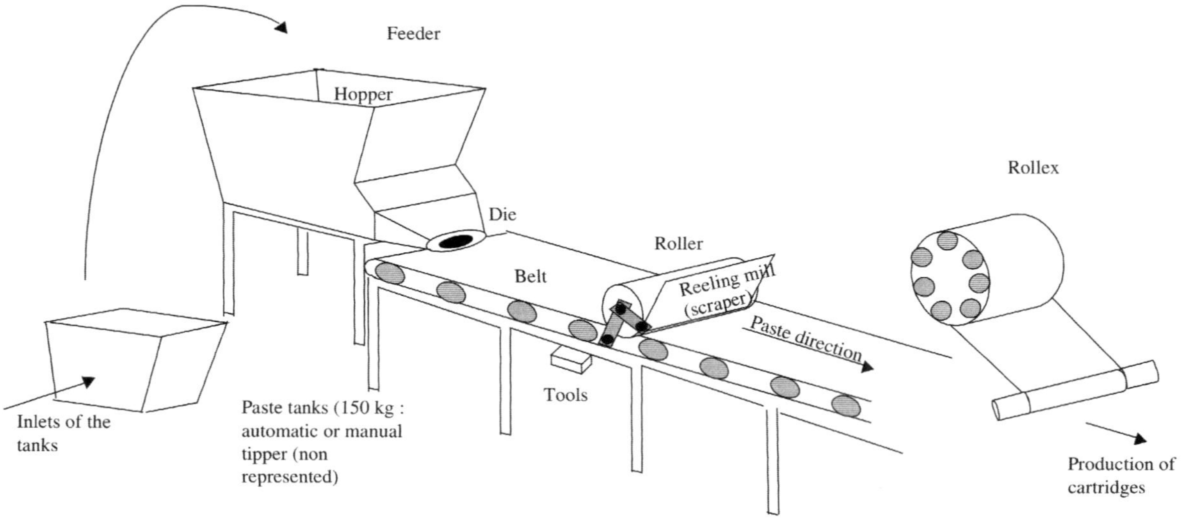

Figure 1: The Rollex Machine And The Feeder

Photo 1: View Of The Feeder

Photo 2: Inside The Feeder (Post-Accident Experiments By The Company)

2.4 Organisation To Operate The Process Involved In The Explosion

Several people are involved on site in the operation and management of the high risk process of dynamites cartridges manufacturing. The key actors were:

- the operator of the Rollex machine: its role is required on the Rollex at the starting of the process; to prepare the automated processing state, its actions are to pour the tanks into the feeder, realising the checks and mechanical adjustments, verifying the size of the cartridge, the level of the conveyor belt, the level of the roller before moving to workshop n°52 to monitoring the process on the video screen; the starting of the automated process is known as a critical operation for the operator's life due to the vicinity with the paste processing into the mechanical machine;
- the assistant of the operator of the Rollex, that helps but does not make the adjustments
- the cleaner of the pyrotechnics waste for the all site,
- a mechanical technician of the maintenance team,
- 3 frontline managers and 1 production manager

The 3 frontline managers and the production manager are responsible for scheduling the production on a monthly basis and then on a weekly basis. The shifts are from 5h-12h and to 12h-19h or from 7h-15h.

2.5 A Summary Of The Chronology

In the days before the accident, a decision was taken to recycle a specific blast gelatine (27 tons of Supedopex) that was stored in UK after the Irish subsidiary decided to cancel the production of these explosives. The gelatine would be recycled in the process for the manufacturing of a specific type of dynamite called Dynaroc 6 after a testing phase. These tests started at the beginning of March 2003.

Because it is not a current product in Billy-Berclau, a new recipe had to be created. The results then have to be approved by the internal laboratory to confirm the quality and safety specifications. The production started three days before the accident.

In the following chronology, there are two frontline managers (A and B): A is responsible for the nitration and mixing process steps and B is responsible for the processing of gelatine to cartridge manufacturing.

The day before the accident, some problems of homogeneity of the mix of Superdopex gelatine and other reactants had delayed the production. The frontline manager B decided to start the shift work at 5 a.m. instead of 7 a.m. to process stocks of paste. In parallel, some testings for a new dynamite called Frangex, have resulted in 8 boxes of 20 kg to be recycled in the production. The operator of the Rollex and its frontline manager B, are unaware of these testings.

So, the day of the accident, the rollex operator arrived at 5h15 in the workshop n°50 (notice that some steps of the chronology have been identified with the video recorder). He looked for gelatine to be processed into the pyrotechnic area and met some operators. From 5h20 to 5h30, he started the rollex machine, made the adjustments, processed some gelatine and packed two boxes (50 kg of cartridges were found after on the workshop 52). There were two other operators scheduled but they were not there that day: one did not wake up, and another switched with a colleague and was not aware of the schedule change. During a half hour some discussions and actions involving the rollex operator and the frontline managers are performed to find two other operators.

After being informed that he will be helped, the rollex operator came back at 6 a.m. to start the process. He faced problems with processing the gelatine through the die. It required him to do a lot of operations (mechanical, cleaning) on the feeder, the plasticating screw, the die, the belt, the roller, etc.

A few minutes before the accident, the assistant joined him. A few seconds before it, the cleaner came into the workshop and took the pyrotechnic paper garbage. He was supposed to start at 7 a.m. but showed up early that day, and with the agreement of frontline manager B, was authorised to start his work of unit cleaning. Also the mechanical technician for maintenance, for some operating reasons (he had forgotten a tool) was walking and passing in front of the tunnel linking workshop n°49 and n°50 when the explosion occurred at 6h16'34.

3. The Effects Of The Accident

3.1 Damages To Structures

The explosion produced extensive damages in the plant and in the nearby village of Billy-Berclau. The methodology of data collection and damage assessment was the same as followed for the investigation of the Toulouse accident by INERIS for IGE (Mouilleau and Dechy, 2002). Some principles of theories on shock wave propagation (Wright J.K., 1961, and Baker W.E. et al, 1983) were used for damages assessment. As a reminder, this damages investigations involves uncertainties and methodological biases that lead, for example, to several TNT equivalent estimates for the explosion of Toulouse (Dechy and Mouilleau, 2004).

Workshop n°50, where the explosion took place, was totally destroyed (photo 3).

Photo 3: Workshop 50, where the explosion occurred ΔP > 100 kPa

Photo 4: Crater in Workshop 50 ΔP > 100 kPa

Photo 5: Damages in workshop 49 (30 m from Workshop 50) - ΔP = 10 –15 kPa

Photo 6: Damage to a building, 370 m from Workshop 50. (dislodged tiles and broken panes) - ΔP = 1-2 kPa

The heavy Rollex parcelling machine was thrown against a barricade, barricades were partially collapsed, and two small craters (about 1 m diameter) were dug in the concrete at two locations were dynamite charges were stored on the ground (photo 4). Nearby workshops within about 100 m from the explosion suffered extensive damages (photo 5), but mainly to light structures (roofs, sidings, wood beams). Barricades isolating each workshop from the others proved once again to be an efficient protection safety barrier against the explosion. In the far field, minor damages to light structures were observed (photo 6).

Most buildings in the plant suffered from dislodged tiles and asbestos-cement panes, broken windows and panes. Similar effects were noticed in Billy-Berclau village, about 1000 m distant from the explosion epicentre.

It was not mentioned any domino effect. The whole dynamite charge present in workshop n°50 exploded, but the explosion did not propagate to nearby workshops where high quantities of explosives were also stored.

3.2 Analysis Of The Damages

Overpressure estimates were made on 34 damage points related to typical overpressure level or ranges available in the literature (Mouilleau and Dechy, 2002; Cozzani and Salzano, 2004). The ranges of estimated overpressures recorded are plotted versus distance from the explosion epicentre (figure 2).

The screen effect provided by barricades between workshops was taken into account in the assessment methodology, according to the following rules recommended by the French Pyrotechnic Regulations (figure 3):

- within a distance twice the height of the barricade, the overpressure level at ground is divided by 4 to 6, compared with the overpressure that would be measured without barricade.
- At a distance between twice and four times the height of the barricade, the overpressure level at ground is divided by 2 to 4, compared with the overpressure that would be measured without barricade.

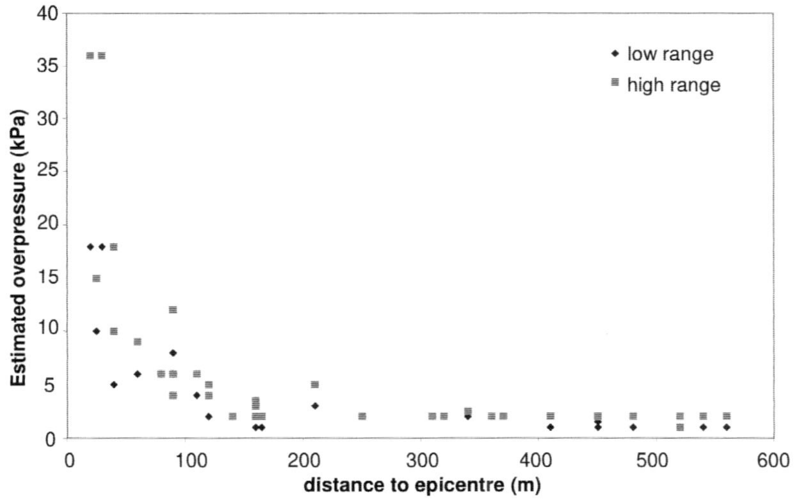

Figure 2: Overpressure Estimations Vs Distance To The Epicentre

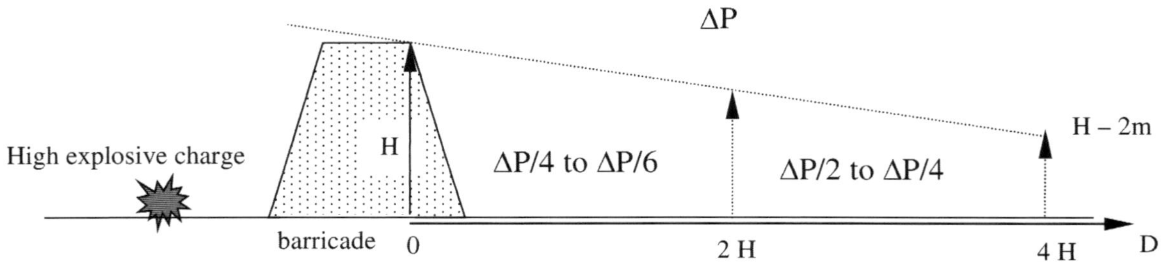

Figure 3: Screen Effect Of A Barricade (Adapted From The French Pyrotechnic Regulations)

As one would expect, the estimates show a decrease in the overpressures the greater the distance measured from the epicentre. In the pyrotechnic area, the shock wave propagation was clearly disturbed by reflection and diffraction phenomena in crossing the barricades, that complicates the damage interpretation (see Wenzel A.B. and Bessey R.L., 1969). After correction by the screen effect, a mean overpressure level from 5 to 15 kPa seems plausible. Outside the pyrotechnic area (beyond 250 m from workshop n°50), the overpressure level seemed about 1 to 2 kPa.

From the very first analysis of the damage points, it has been deduced that the TNT equivalent of this explosion turned out to be between 70 to 130 kg. These values correspond to the medians of the weight values calculated from the overpressure estimates respectively on the low side and on the top side. Considering an energetic equivalent of 1.1 kg TNT for 1 kg dynamite, that would provide a dynamite charge from 60 to 120 kg. However, a discrepancy was noticed in TNT equivalents from inside and outside the pyrotechnic area. The analysis of the inside damage points would suggest a dynamite charge from 50 to 100 kg, whereas the outside damage points would give 100 to 300 kg.

This disparity should be explained by considering:

- in the pyrotechnic area, the overpressure wave disturbance caused by the barricades;
- in the far field, the difficulty to quote accurately damage to light structures, and the possible disturbance by external phenomena, such as the wind speed and the wave reflection by the lower atmospheric layers (Grant et al, 1967, and Reed, J.W., 1973).

Those estimates can be compared with the real explosive charge present in workshop n°50 when the explosion occurred (figure 4). This has been assessed to 580 kg, distributed into:

- two dynamite tanks on hold at the entrance of Workshop 50: 260 kg,
- dynamite in the feeder: 150 kg,
- one tank discharge near the belt: 90 kg,
- 4 boxes of Frangex dynamite: 80 kg

The first explosion has probably occurred in the feeder containing 150 kg of dynamite. Applying the French Pyrotechnic Regulations rules of calculation, the detonation in the feeder would have been transmitted by pressure wave to other close dynamite charges within a distance of no more than:

$$D = 0.5.Q^{1/3}$$

where Q is the explosive mass.

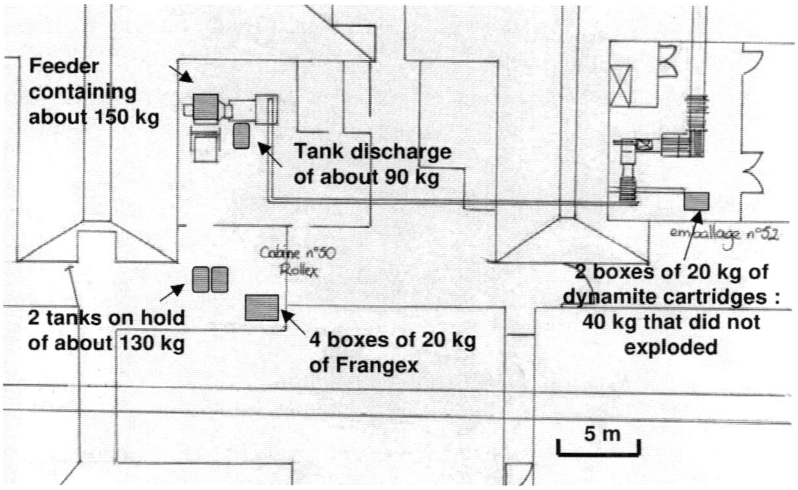

Figure 4: Gelatine Quantities At The Moment Of The Explosion

Thus, it could have been transmitted by pressure wave to the tank discharge closed to the feeder. Regarding the other charges distant from about 10 m from the feeder, the detonation would have rather been propagated by missile projection. Barricades must have prevented the explosion transmission by missile projection from workshop n°50 to nearby workshops.

However, the real total dynamite amount in workshop n°50 turns out to be quite larger than the one estimated by the damage analysis. This gap could be explained considering that several successive detonations generated independent overpressure peaks which did not cumulate. Many witnesses mentioned a "long, dull noise" which could emphasise this assumption.

It could be admitted that two main charges were located in Workshop 50:

- 240 kg near the feeder (feeder + tank discharge)
- 340 kg at the entrance (2 tanks + 4 Frangex dynamites boxes).

If the explosion was transmitted by a missile impact from the first charge to the second one, the speed of transmission would have been about 1000 m/s, meaning that the second charge would have detonated about 10 ms after the first one. As the overpressure pulse of a detonation has a typical period of 1 ms, the global pressure wave signal would have been a succession of independent overpressure peaks that would not have cumulated, at least in the near field. Therefore, the overpressure damages would have been caused by the detonation of the biggest charge (340 kg) rather than the total charge (580 kg).

One can also consider that not two but four independent dynamite charges, as identified on figure 3, generated four successive detonations and four independent overpressure peaks. Thus, the overpressure damages in the near field would have been caused by the detonation of 150 kg of dynamite in the feeder. This hypothesis would agree with the TNT equivalent estimate of 50 – 100 kg from the near field damages interpretations.

Beyond the first barricades, very complex waves diffraction and reflection phenomena should have occurred, and pressure signal should have partially recombined. That could explain the discrepancy between TNT equivalent estimates from the damage analyses in the pyrotechnic area and in the far field.

As a conclusion, it seems that the overpressure effects were not generated by the massive explosion of the whole dynamite charge in workshop n°50, but were most probably caused by several successive and independent detonations whose overpressure signals might have partially recombined in the far field, according to shock waves propagation theories.

3.3 Effects To People

Fourteen accident at work notifications were registered by the company during the days following the explosion. Casualties are located on figure 5. Among them are:

- 4 deaths,
- 2 minor injuries by blasting onto the ground,
- 1 minor injury to ears,
- 1 minor injury by ceiling fall,
- 5 psychological shocks,
- 1 disease,

Notice that five people located within 100 m from the explosion epicentre were not injured.
The four deceased people were located in workshop n°50 and for one of them, in the communicating tunnel between workshops n°49 and n°50. Other people located in the nearby workshops suffered minor injuries, highlighting the protection efficiency against pressure waves offered by barricades between workshops.

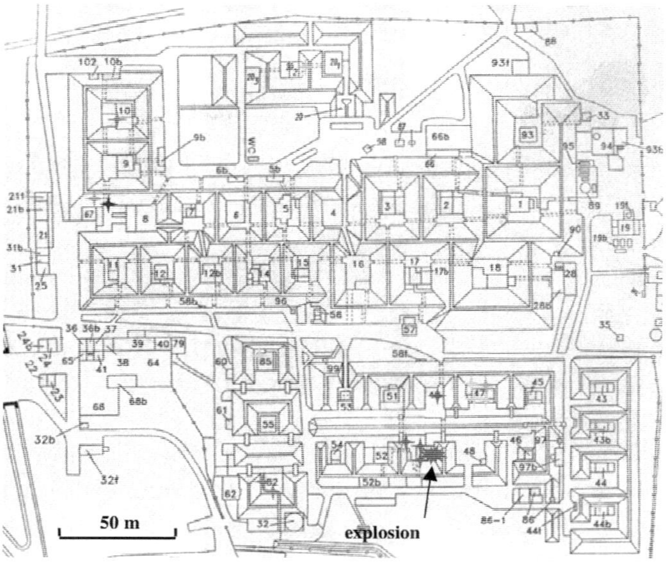

Figure 5: Victims location
+ death, + injury by blasting onto the ground, + injury to ears + injury by ceiling fall,
+ psychological shock, + disease, + no injury

4. The Technical And Direct Causes Investigation

The technical analysis of the investigation was performed by using an analytical tool (fault tree) to identify in a structured way the probable causes of the explosion of the workshop n°50. The analysis with a fault tree is an inductive-type method used to determine the technical causes and some human actions leading to an accident.

The building of this fault tree was based on:

- the information gathered during the interviews,
- the tests on the samples of the explosive that was packed and stored in the workshop n°52 at the time of the accident,
- on the analysis of the damage caused by the explosion,
- on the expertise of INERIS in its knowledge about the phenomena of explosives explosion.

The possible location of the explosion had to be determined as a starting point in order to be relevant in our assumption in the remaining analysis. Thus, based on data from the damage plotting, the chronology and the video recorded file, the feeder was identified as the most probable location of the explosion.

Then, the investigation focused on the nature of the blasting material in the feeder. The material in hand here is obviously dynamite but many questions arose (see chronology):

- What kind of dynamite was it ?

- Did the dynamite meet the required technical specifications ?
- Or was it more sensitive than the required technical specifications ?

About the type of dynamite, there were two possibilities: the Dynaroc 6 made on a regular mode at the Tellex or the Dynaroc 6 made with the Superdopex gelatine on a manual mode.

Two charges were put into the hopper: the first one was some Dynaroc 6 made of Superdopex gelatine produced the day before the accident. But, the origin of the second container is still unknown. Because of the sequence of events, it certainly came from one of the storage buildings, but the inventory made after the explosion did not help much to draw clear conclusions.

Concerning the suitability of the dynamite to the technical specifications, the tests carried out at INERIS on the samples collected from the company, if one considers them as representative of what was in the feeder at the time of the explosion, helped to conclude that the explosive was adequate and not more sensitive to shock or to friction than the dynamite usually produced by Nitrochimie S.A..

Yet, even if the explosive was adequate, its sensitivity can be questioned as, for example, the gelatine had to remain a long time (one night) in the feeder and the die. The gelatine was hard to process (see chronology) that had an effect on the exposure of operator to risk. Many factors may have increased the sensitivity of dynamite:

- the formulation of the gelatine can be inadequate,
- the viscosity of the gelatine can be off-specification,
- the absorbents can have a high acidity rate,
- exudation can occur,
- the dynamite can be unstable,
- the gelatine can have an inadequate homogeneity.

Consequently, except the issue of acidity, all the factors were kept for different reasons. For example, an exudation could have occurred in the die due to the long stay of the paste in it. Besides, even if it is difficult to set a relation between an inadequate homogeneity of the gelatine and its level of explosibility, this factor cannot be turned down.

There is a minimum amount of energy required to trigger the explosion and thus knowing the paste sensitivity appears to be of high interest. Some of the ignition sources could have been present but not powerful enough. Several ignition sources were identified:

- heat action by an external fire,
- electrical, electrostatic arc,
- external pressure wave,
- missile impacts,

- mechanical action from the process or something else.

The approach consisted in making an inventory for each type of ignition source of the situations or circumstances where they could have appeared given the equipment, the labor organization, the task performance itself in the workshop, etc.

According to the available data, only the mechanical action was kept as probable. Among the possible causes of a mechanical ignition, there were possibilities of friction due to different metallic parts between the container, the feeder, the hopper or the tipper or the friction of a foreign matter on the feeder. This last option was kept as the most probable.

Because not enough evidence was collected, many options are still conceivable. The foreign matter could have been introduced:

- in the hopper by the operator, notably when he threw a ball of gelatine (that fell into or close to a tool box placed along the Rollex belt) to the feeder a few seconds before the explosion,
- in the hopper by the fall of an object of the tank when unloaded
- in the hopper by the fall of an object from above the feeder
- in the tank during the manufacturing of the dynamite, the transportation of tank on the trolley to the workshop n°50 or in the workshop n°50
- in the hopper before the paste was loaded in.

As a preliminary conclusion of that step of the investigation, there are still some uncertainties about the exact technical causes of the explosion. It would have been interesting to continue the tests on the gelatine used at the moment of the explosion to clarify the link between its properties (viscosity, exudation, …) and its sensitivity to explosion. Besides, there is still this uncertainty about the origin of the second tank.

The most probable scenario is the presence of a foreign matter in the feeder either brought inadvertently by an operator or due to the fall of objects from the container or the hopper. But the timing and the circumstances where it could have been introduced are still unspecified. The internal company investigation concluded also that the most probable scenario is a foreign matter in the feeder.

However, this was not the only issue as a matter of a fact that four people died in the explosion, introducing the risk factor of the exposure of operators. Limiting exposure is a key thing in the pyrotechnic industry. In that accident, the circulation issue appeared therefore as a key element of the understanding of the severity of the human losses. The cleaner, the maintenance operator as well as the assistant of the main operator were not supposed to be where they were in these particular circumstances, at the time of the explosion.

With this most probable technical scenario assumption and the circulation problem, we could get into the issue of why the organization was not able to prevent this scenario. Safety barriers and controls to prevent this scenario and its severity, which was a known risk by the company, were analysed in their organizational context.

5. The Organisational Investigation

Three main points are discussed in the following part: the methodological aspects of the organisational investigation and the background models, the outcomes and a discussion around them.

5.1 Methodological Issues

5.1.1 Rationale For Investigating Organisations

Organisations are complex systems. They imply a high number of interactions, at all levels, from the shop floor to the board committee but also the outside of the organisation, with individuals located in its environment (regulators, economic partners). These interactions are rich; they involve the nature of the social world. This social world is made of situation of power between people, of cultural dimensions, of social construction of reality, of norms, of values, of rule. Organisations also can be seen through several lenses like the financial one, the human resources one, the economical one, the institutional one, the historical one, the societal one... Organisations are therefore a subject of research for various disciplines like economy, politics, sociology, anthropology, psychology, psychosociology, all of them applying different filters. This makes the organisational life a quite dense subject of study (J.Rasmussen and I.Svedung (1997, 2000) suggested a representation of the socio-technical system where many of these features are introduced and linked).

Qualitative studies investigating on large organisations but even on small organisations deal with the problem of collecting and analysing data for interpretation, and also deal with the ability to consider various possible angles and dimensions, across levels.

However, with investigations, one helping thing is that the focus following an accident is given. Indeed, the accident sequence reveals some inadequacies that are the starting point of the investigation. The aim of the investigator is therefore to give a context to the events. This implies to go back into the history of the organisation in order to better describe in which circumstances the decisions and actions of the individuals were taken. The idea of going back in history is to find out more about the conditions that "*shaped the landscape in which the accident unfold itself*" (J.Rasmussen I. Svedung, 2000) further back from the close events, like human errors and technical failure leading to the accident.

The purpose of an organisational investigation could then be said to describe the events with a broad scope, to bring some explanation (defining some causal or rather conditional links between the events and a context) and sometimes to build a predictive model from it. The latter has not been performed so far many times. It is the work of researchers, social scientists, who conceptualised further their interpretation of the organisational dimensions, applying social science techniques of theorising from grounded data, and single cases.

What is actually performed is represented in this figure n°6.

As said, it is nevertheless not such a common thing to theorise from accidents. Thus, when we go through the existing literature on the organisational side of accidents, it shows that there are actually several types of possible approaches for investigating.

We classified them in three types (JC Lecoze 2005).

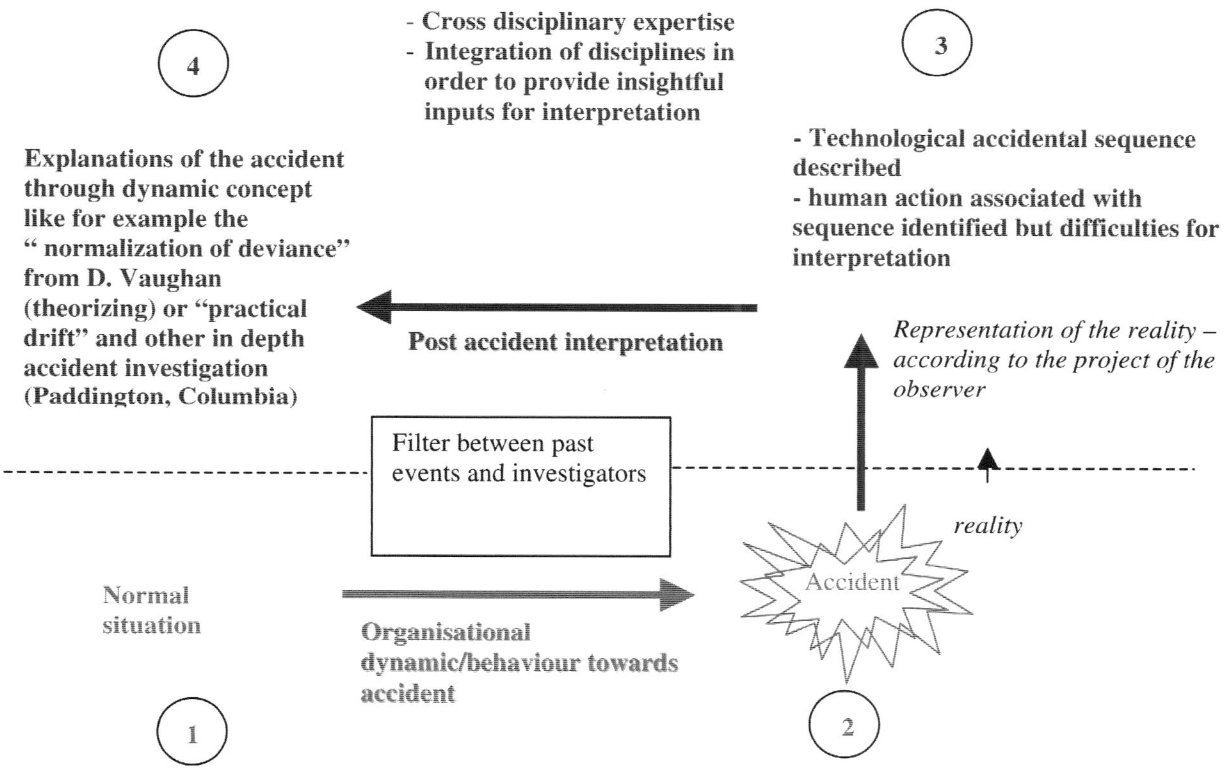

Figure 6: Purpose And General Process Of Investigating (Lecoze Et Al 2002)

5.1.2 Three different approaches of investigations

First, these works already mentioned with research and theorizing perspectives (i.e D.Vaughan, 1996, S.Snook, 2001), secondly the commissions set up after disasters (i.e: Columbia, 2003, Ladbroke grove, 2000) and then the structured root causes analysis (i.e MORT, STAMP).

All of these three types must face the bias of forcing the data.

The "forcing the data" concept comes from one of the difficulty of the social science theorizing process. In social science, theories can be grounded, generated and inferred from the data collected through fieldwork. These are not to be generated purely by a logico-deductive approach. However, in front of the richness of the organisational life, theories can be validated through strategy consisting in forcing the data to fit the theory. This is a symptom that also could be called the "data that should fit the model" rather than the "model that should fit the data" approach (from Glaser and Strauss, 1967).

The three types of investigating techniques deal with it differently.

Pre defined models, in structured root causes analysis, apply a "data that should fit the model" approach but are made for systematic recommendation purposes and are useful in that sense. Commission and research work are more descriptive and explanatory, and do not specifically rely on pre defined model. However, they also introduce underlying models and theories to bring some light on the data extracted from the investigation.

These three types involve different relationship with the underlying models that they use for treating the data, but also different time constraints, techniques, resources (number of people, competences, expertise etc), scope (micro-meso-macro levels) and objectives (describing, explaining, making recommendation).

5.2 Methodology Used For The Case

5.2.1 The Underlying Models

In the investigation performed for the Billy Berclau accident, the approach was based on an investigating process that was half way between a structured root cause analysis (i.e MORT, implying focused questions and pre-defined safety management system) and a more descriptive approach based on open interviews and openness of the system with micro-meso-macro perspective as well as a historical approach.

The investigation report made clear in a introductory part that several organisational dimensions would be investigated, with references to works in the field, like the High Reliability Organisation (i.e Roberts 1990, Rochlin 1996, 1999, Weick, 1987), Normal Accident theory (C.Perrow, 1999), Incubation period (B.Turner, 1978), Normalisation of deviance (D.Vaughan, 1996), Latent and active failure (Reason, 1990, 1997) as well a the idea of organisational behaviour towards accidents (Rasmussen and Svedung, 2000).

Also, it has been stated that some dimensions were investigated and articulated with the help of a type of matrix as suggested by Glandon and Waring (1998). This matrix contains the following dimensions:

Risk contexts
Organisational environments
Economies and market
Public policy, legislation and regulation
Social and political climate

	Technology
	History, operating territories and conditions
Human Factors	
	Culture
	Power relations, political processes and decision making
	Perception, cognition and meanings of success
Formal coping arrangements	
	Risk management
	Risk assessment
	Management systems
	Approaches to change

Table 1: Thematic table linking theoretical aspects of risk with empirical studies (A.Waring and A.I. Glendon, 1998)

All of the dimensions can be investigated with related theories in the background, models or works in the safety field (like in perception, power in companies, safety management systems etc).

The idea was not to validate any of the frameworks suggested by any theories and concepts but rather to explicitly mention that they were in the mind of the investigator when collecting and analysing the data. In her article dealing with the heuristic of case analysis (1992), D. Vaughan has debated this point of the pre-existence of models in mind when investigating and theorizing.

The importance of the references and underlying models in organisational investigation is high. It has been quite rightly stated that an investigator could find what he/she wanted to find in accidents.

5.2.2 The link between technical and organisational factors

The technical anchoring of the organisational investigation has been performed through the barriers and controls revealed through the technical sequence. Twenty three barriers and controls were identified based on the most probable scenario retained. The barriers from B1 to B12 were aimed at decreasing the likelihood of foreign object presence in the feeder, B13-B16 were designed for preventing the explosion of the feeder in case of a foreign object friction and the B17-B23 decreased the severity of the human loss in case of accident. Each of the barriers were investigated in detailed through interview and then replaced in a broader context.

To illustrate it in a simple way, the investigation has been performed like the next figure n°7 is suggesting it (adapted from a representation used for MORT, Johnson, 1980 and mixed with Waring and Glendon, see table 1).

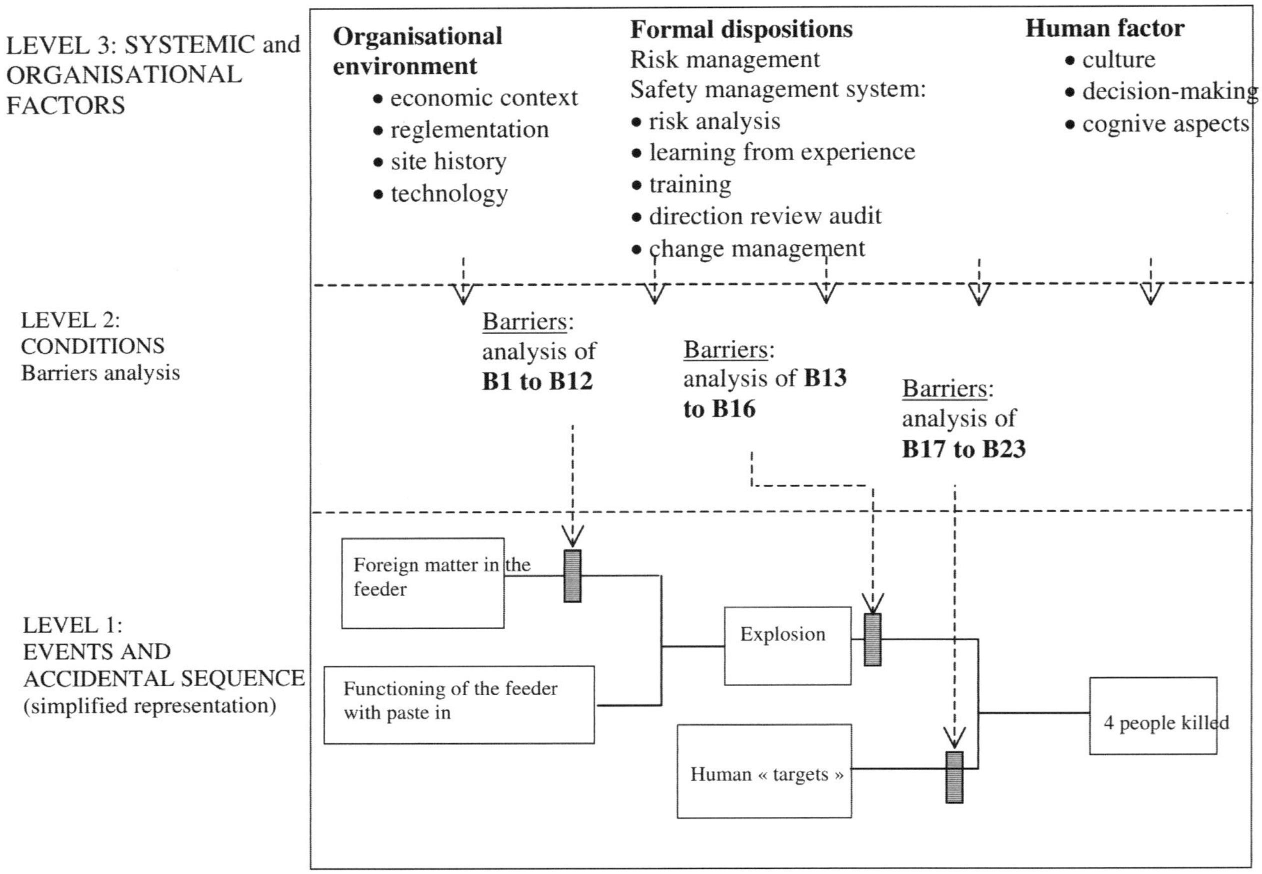

Figure 7: Link Between Technical And Organizational Factors In The Billy Berclau Case

The first level is the accident sequence with the inadequacies of the barriers and controls, the second deals with the conditions for these barriers and control not to fulfil their functions and the third level, which brings some insights on the overall context influencing the conditions.

In fact, this investigating process shares some similarities with what has been described for example in Fortune and Peters (1995:110):

- Pre-analysis
- Identification of significant failures (similar to level 1)
- System selection (similar to level 2)
- System modelling (similar to level 2)
- Comparisons (similar to level 2 and level 3)
- Further analysis (similar to level 3)
- Synthesis

The investigation consisted in a series of 30 interviews (70% for the levels 1 and 2, 30% for the level 3) with people from the shop floor to the board committee at the corporate level. An extensive investigation focusing on the documents provided by the company describing quality and safety management systems as well as safety cases, and the records of the activity of learning from experience, safety meeting, risk analysis, etc was also performed. The use of a book about the culture and history of the company celebrating a century of existence was used and very helpful for getting to know better the company and its history.

5.3 Outcomes Of The Investigation, Lessons To Learn

A certain numbers of findings made are quite consistent with some already mentioned influencing organisational factors, in other investigations like in A. Hopkins, (2001) but also general lessons learnt throughout accident (Llory et al., 2004) and were similar to accident models mentioned earlier in this paper. Ninety recommendations were made for the company to improve the level of safety. They were made at several levels of the system, from the technical level to the management and regulatory level.

Here is a selection and discussion of relevant aspects that were identified.

Some lessons at the level of the authorities and regulations (risk contexts Public policy, legislation and regulation, in regard with table 1):

The regulations are increasingly dense for industries, as much from their contents as from their overlapping. The pyrotechnical site was liable to many regulations concerning safety under the pyrotechnics regulation depending on the health and safety regulations but also under the environmental regulation (the classified installations in France and Seveso II Directive in EU) dealing with major hazards. The site was classified high threshold concerning the transposition of the SEVESO II directive. In this case, there is an overlapping of two regulations which however both deal with the prevention of major hazard.

These regulations have nevertheless basic differences. One of these difference relates to prescriptive nature for one (the law on pyrotechnics safety) and the other by objectives (in particular the SMS within the framework of the application of the SEVESO II directive).

Consequently, a time for integrating the new regulation was required. This adaptive time is all the more important than the culture of the industry has been shaped by the regulations applied for more than two decades. In the case of pyrotechny, the decree of 1979, very detailed and seen as an effective regulation, which allowed notable improvements of safety on sites at the time of its implementation, had somehow played this shaping role.

Another one of these differences relates to the risk assessment practices and calculations of safety distances, which is not without causing administrative issues. Indeed, the two regulations have different thresholds, different classifications. This brings confusion and a difficulty of treatment of the safety reports and a difficulty in meeting regulations requirements. Concretely, at the level of the plant, this also resulted in an important volume of administrative work which has been regularly growing since the nineties with the increase of the environmental regulation and requirements. This risk context took part in diverting the safety department from its operational activities.

The diversity of the regulations implies also the diversity of controls. This can be a good thing, but can also be a bad thing when the links between the parts have consequences for the consistency of the whole. In particular, it is necessary to take into account that this company faced variations in terms of number, objectives and quality of inspection. An inspectorate like the IPE (Inspection des Poudres et Explosives - Explosive and Powder Inspectorate) has the means and experience for understanding well the operations of the installations and can bring a useful and constructive help for risk management, while keeping a control and regulative function. This is therefore, not a surprise to find clearly stated in the regulation the possibility for the assistance for in site inspection and safety reports by the IPE for the DRIRE and the labour inspectorate (in charge of work safety regulation as well labour regulations). Consequently, in case of numerous updating of safety reports, the company has different strategies available for dealing with the submission of the reports, to one of the control authorities. In that case, this precise situation occurred and revealed consequently several non-conformities.

This general context of regulation overlapping with sometimes different objectives and contents, with multiple control authorities, creates a complex regulatory situation for which the company must organise its resources in order to cope specifically with it. These remarks do not, of course, justify non-conformity, but remind about the sometimes dense and complex regulatory contexts that organizations are facing, and shed some light on the difficulties that they meet through their attempt to comply fully with them. The pyrotechnical site faced a situation from this point of view for which a great part of its safety operational activity was diverted to treat these administrative tasks involved.

Authorities in France work actively on this issue of overlapping regulations.

At the level of the organization of the company and site (human factors and formal coping arrangement, in regards to table 1)

- ✓ *Changes in organizations*

Changes in organizations must be done with a careful attention given to the potential impact of these on the safety practices at the operational level. Usually, the impact of changes is approached for technical issues but much less in terms of organizational issues. It is clear that the impact can't be measured the same way. The impact of organizational changes is clearly not easy to assess, but some measures should be taken before and after. In this accident, important changes were done at the level of the site top management and at the level of the technical corporate department. These two changes impacted on the operational safety of the site. Indeed, following the changes, the safety department was not able anymore to play its safety role against the production pressure, as it had been working for years. Thus, it is believed that any changes that imply an evolution in the way the system sustains its level of safety through a balance between safety against the natural production pressures, increases the risk of operating outside the boundaries of safe practices. The evolution that was identified through the investigation in this case concerned the way information was available at the level of the board committee. This information was important for high levels in order to hear about the level of safety at the operational level. This situation was not without reminding the "trickle down effect" described by D.Vaughan about the Challenger accident (1997) when she describes the effect of political, economical and organisational arrangements made at the top of the system that slowly and gradually spread across the lower levels.

✓ *The cultural dimension*

Culture is seen as an important concept today for explaining how companies operate. However it is still a scientific concept that is strongly debated. For the purpose of the investigation, it appeared however relevant for some aspects. We retained about culture that, according to E.Schein, (1992), it is:

" a pattern of shared basic assumptions that the group learned as it solved problems of external adaptation and internal integration, that has worked well enough to be considered valid and, therefore, to be taught to new members as the correct way to perceive, think, and feel in relations to this problems"

The history of the company is hundred years old and the site is one of the oldest. A long tradition of operators has been working on the site, a dynasty as they would call them (a term that was found in the document celebrating the century of the company). This long history, through the events and the technology used, shaped the practices and maintained a tradition in the way the blasting gelatine was handled. Despite the new technology, requiring less manual handling of the paste, some remaining practices described as "pratique à l'ancienne" equivalent for "old time practices" were still in use. They consisted in regularly manipulating the paste with bare hands. This kind of old practices interacting with more recent practices required by the new technology can interact inappropriately with safety principles introduced with the technology. In the accident, the video revealed a behavior of the operator taking a piece of gelatine that fell on a tool box and putting it directly back in the feeder. This is not without reminding an old way of dealing with the paste.

It was also found interesting to talk about safety culture in our case as it is today a widely shared concept (or rather assumption) in the research and practice though also still very debated. There is again no consensual definition among scientists. Here, for practical purposes, we can retain the HSE (UK Health and Safety Executive) definition about safety culture as being:

"The product of individual and group values, attitudes, competencies, and patterns of behaviour that determine the commitment to and the style and proficiency of, an organisation's Health and Safety programmes. Organisations with a positive safety culture are characterised by communications founded on mutual trust, by shared perceptions of the importance of safety, and by confidence in the efficacy of preventive measures"

As a guidance through the concept of culture the HSE gives these items related to the dimensions of safety cultures for which some are relevant in this accident case:

- **Organizational commitment and communication**
- **Line management commitment**
- Risk taking behaviour and some contributory influences

- Personnel role
- Workmates' influence
- Competence
- Supervisors' role
- Permit to work systems
- **Reporting of accidents and near misses**

In bold the dimensions that appeared illustrative, as items representative of aspects of safety in the site. The line management commitment and organisational commitment appeared through the interview and documents analysis to be less than adequate to maintain the activities against natural and expected drifts of safety practices (this point is highlighted below). However, the concept of culture is not used further here as it tends to give a deterministic vision of the organisation. A vision which implies that culture creates the conditions for people to behave the way they behave although individuals have also a strategic and adaptive dimension, with their own intentions, that are never fully determined by outside constraints (it is one of the difficulties for predicting organisational behaviour as a whole).

✓ *Safety management (formal coping arrangements in regards with table 1)*

From a more structural point of view of the organisation, it was clear that the SMS required by SEVESO II directive was currently being under implementation, but had some shortcomings in the way it was thought and designed. A context for this difficulty to fully apply the safety management principles has already been stressed with the existence of a prescriptive regulation.

Some are the shortcomings pointed out during the investigation:

- the principle of defences in depth (also called the barrier and control approach) extracted from the risk analysis was not fully applied and only limited to a small number of defences. This strategy was implemented in order to focus only on the last defence, called the ultimate defence principle. The idea for the company was not to identify too many defences which is very heavy afterwards in terms of management, maintenance, inspecting and auditing. Moreover only technical defences were included and not behavioural ones (like the circulation issue). It is a problem not to make barriers and controls more explicit for a risk control perspective. This was a serious flaw in the way the system could have been thought. Also, it appeared that the risk analysis sessions were not opened to operators. This can be surprising as it is a widely known practice within industries but was eventually less surprising in a prescriptive regulatory context where everything was already codified and written, ready to be applied.

- The learning from experience is a key activity that must help operators and managers to think and work on the conditions leading to the incidents. These conditions can be technical as well as organisational. In this case, the organisational dimensions of the incident were not so much analysed, as the safety department was no longer involved into the learning sessions following the recent changes in the management of the plantAS a consequence, there was more focus more on technical rather than organisational issues. A long list of warnings concerning foreign objects in the last 3 years triggered some concerns about the cleaning within the installations. The management had therefore planned to improve cleaning practices.

- The training was also a concern for the company who had to manage in a flexible manner the production for coping with the low visibility of the market. This implied recruitment of temporary workers, which created the problem of the following up on training. Training must also be used as a way of reminding people about safety principles, to prevent the work routine from becoming unsafe, according to the law. The company had not been very cautious on the follow-up for all operators on the refresher sessions.

- This aspect of routine introduces the idea of dynamic drifting (as for example described in S.Snook, 2000). This consists in shifting at the level of the operators from a rule based approach to a more task based approach, where the individual can locally cross the safe boundaries of the system (J.Rasmussen, 1997), and therefore can create risk, but also being exposed to risks. Micro deviations can install in the organisation with no impact on the production as a whole but create a kind of deviance within the system in terms of safety (D.Vaughan, 1996, 1999). As dynamic and evolving, it is of no surprise to discover that operational practices defined by rules, confronted with multiple dimensional and sometimes conflicting aspects, are moving towards adaptive tasks. It appeared, with retrospect, that what was observed in this accident, had obvious common points with what J. Rasmussen (1997) described:

- "In any well designed work system, numerous precautions are taken to protect the actors against occupational risk and the system against major accidents, using a 'defence-in-depth' design strategy. One basic problem is that in such a system having functionally redundant protective defences, a local violation of one of the defenses has no immediate, visible effect and then may not be observed in action. In this situation, the boundary of safe behaviour of one particular actor depends on the possible violation of defenses by other actors. Therefore, in systems designed *according to the defence-in-depth strategy, the defenses are likely to degenerate systematically through time, when pressure toward cost-effectiveness is dominating."*

Although the pressure towards cost effectiveness didn't show to be so important through the analysis of the economical constraints of the company, an internal pressure for production was however clearly stated through the interviews. In the accident the various drifts around the rules involved the shift schedule that was modified without front line managers knowing about it but also the early start of the installations without the required number of people, the issue of circulation in the pyrotechnic area and the cleaning of installations. They all appeared, with hindsight all to be outside the rules defining the way things should have been performed. However, they couldn't be seen as individual and isolated mistakes of operators in a traditional sense, but had to be seen as a general behaviour moving outside expected practices and safe boundaries, and involving all levels within the organisation, that allowed this situation to occur.

K.Weick (1987) gave an insightful comment about the problem of drifting and crossing the safe boundaries. He suggests an interesting explanation for the reasons why safe boundaries are crossed. It also had some quite strong echoes with our case, particularly with the issue of circulation, with the installations cleaning but also the specific morning situation:

"Reliability is invisible in the sense that reliable outcomes are constant, which means there is nothing to pay attention to. Operators see nothing and seeing nothing presume that nothing is happening. If nothing is happening and they continue to act the way they have been, nothing will continue to happen. This diagnosis is deceptive and misleading because dynamic inputs create stable outcomes"

Operators regularly do things to cope with situations in order to be able to make their way out of conflicting goals but also unexpected situations, and therefore can cross, without totally noticing it, the boundaries of safe operations. In this accident for example, without any adaptation and with a strict respect of the rules, the operations wouldn't have started that day as two operators were missing at the beginning of the shift.

Deviance or drifts can sometimes be sustained in the long run, as the normal(ized) practices, in some specific organisational, economical and historical contexts. Companies should acknowledge this situation and work on these micro adaptations through the design of organisational means where people can talk about them and find way to overcome the risky ones. These kind of dynamic aspects have clearly to be addressed specifically with an appropriate organisational support. This would need to be experienced and developed in the future, for safer operations.

- The safety indicators related to occupational health and safety (LTI for Lost Time Injury) were very good. The company had a past records of no accidents for four years. This led to the conclusion, following the accident, that these types of indicators do not make clear how good a company manages the major hazard prevention. This is also very consistent with what has been identified by A. Hopkins in his investigation of the Longford plant gas accident (2001). Indicators of these types can be misleading and give a wrong picture of the safety state of a plant, in terms of major hazard prevention. Efforts should therefore be made to indicate the level of safety through the concept of barriers for example, where specific links with the major hazards are possible. A quality type of approach for indicators can be used for that purpose, which could be seen maybe as a bit expensive but that is believed to be worthwhile implementing. The limits of such systems should however also be stated, and complemented for example with softer approach where people can deal with deviation from expected practices which have no direct and measurable impact on safety, and are sometimes necessary in light of conflicting or constraining goals.

- The last point discussed here from this case is the usefulness of cross sites auditing. In the company, this practice had been gradually decreased throughout the years. It seemed that this decrease has revealed to be detrimental to the level of safety. It is believed, that quite consistently with some ideas derived from the normalisation of deviance concept, that external auditing from competent individual is a great help for getting out and pointing out established practices at odds with what is being done outside or known outside as being safe. Control authorities can of course fulfil this mission but internal cross auditing between plants make the process of exchanging maybe sometimes easier, and more open to discussion.

6. Conclusion

The multidisciplinary approach developed in this accident investigation covered a wide range of issues from technical to organisational ones. Damages analysis, technical cause analysis, material tests, chronology, barriers analysis and organisational investigation all gave complementary insights, in order to better understand the accident. The damages analysis, chronology and material test supported the technical cause analysis and the barriers approach helped to create a link with the organisation. There are of course many aspects that could be developed further. But the time spent on the case helped to get deep enough to generate insights.

The methodological approach developed for the organisational investigation has been described as half way between a type of structured root cause analysis and a more descriptive and contextual approach looking for conditional links between facts, using the existing literature on the subject as the background models for guiding the interpretation. The methodology has the potential for being applied again in other accident investigations with organisational dimension, but will be also practicable for smaller investigations.

The findings about the organisational side of this accident have some similarities with in depth investigations dealing with this level of analysis. Some of them have been described in the outcomes discussed in this paper.

7. References

Baker W.E., Cox P.A., Westin P.S., Kulesz J.J., Strehlow R.A., 1983: Explosion hazards and Evaluation, Elsevier, New York.

Columbia Accident Investigation Board, 2003, Vols. II-VI of Final Report.

Cozzani and Salzano, 2004, The quatitative assessment of domino effects caused by overpressure Part I, Probit models, Journal of Hazardous Materials A107 (2004) pp 67-80.

Cullen, Lord. 2000 The Ladbroke Grove Rail Inquiry Part 1 and 2 reports, HSE books, her majesty stationery Office, Norwich, U.K.

Dechy N. and Mouilleau, Y. 2004, « Damages of the Toulouse disaster, 21st september 2001 », Proceedings of the 11th International Symposium Loss Prevention 2004, Praha, 31 May - 3 June 2004.

Dechy N., Bourdeaux T., Ayrault N., Kordek M.-A., Le Coze J.-C., 2004, First lessons of the Toulouse ammonium nitrate disaster, 21st september 2001, AZF Plant, France, Journal of Hazardous Materials 111 - July 2004, (special issue on JRC-ESReDA seminar on Safety accident investigation, Petten, the Netherlands, 12-13 May 2003)

Fortune J. Peters G. (1995) Learning from failure, the system approach. John Wiley and Sons.

Glaser B. G. Strauss A. L. (1967) The discovery of grounded theory: strategies for qualitative research. Aldine de Gruyter.

Grant, R.L., Murphy J.N., Bowser M.L., 1967: Effect of weather on sound transmission from explosive shots, Bureau of Mines Report of Investigation 6921

Hopkins A. 2000. Lessons from Longford: the Esso Gas Plant Explosion. Sydney. CCH Australia.

Lecoze J-C., Lim S., Intégration des aspects organisationnels dans le retour d'expérience: l'accident majeur, un phénomène complexe à étudier, pour le Ministère de l'Ecologie et du Développement Durable, 2002, downloadable on http://www.ineris.fr

Lecoze J-C, 2005, Organisation and disaster: from lessons learnt to theorising, forthcoming

Llory.M, Dien Y, Montmayeul R. (2004) Organisational investigation methodology and lessons learned. Journal of Hazardous Materials 111 – July 2004

Mouilleau Y. and Dechy N., 2002. « Initial analysis of the damage observed in Toulouse after the accident that occured on 21st of september on the AZF site of the Grande Paroisse company », *International ESMG Symposium, Nürnberg, Germany 8-10 October 2002, on Process safety and industrial explosion protection*

Perrow C. (1999) Normal accident theory, living with high risk technology, second edition

Rasmussen, J. (1997) Risk management in a dynamic society: a modelling problem, in safety science vol 27 n°2/3, p 183-213.

Rasmussen J. Svedung I. (2000). Proactive risk management in a dynamic society. Swedish Rescue Service Agency. Karlstad.

Reason.J. L'erreur humaine. Presse Universitaire de France. Collection Le Travail Humain. 1993.

Reason J. 1997. Managing the risk of organisation accident. Ashgate.

Reed, J.W., 1973: Distant Blast Predictions for Explosions, Minutes of the Fifteenth Explosives Safety Seminar, Dpt of Defense Explosives Safety Board, Washington D.C., vol. II, pp. 1403-1424

Roberts K.H. (1990) Some characteristics of one type of high reliability in organisation. Organisation Science. 1(2). 160-176.

Rochlin G.I. (1996) Reliable organisations: Present research and future directions, journal of contingencies and crisis management, volume 4, june.

Rochlin G.I. (1999) Safe operation as a social construct. Ergonomics. Volume 42. N° 11. 1549-1560.

Schein.E, Organisational culture and leadership, second edition, Jossey-bass, 1992.

Snook, S.A. (2000) Friendly fire, the accidental shootdown of US black hawks over northern Iraq, Princeton university press.

Turner B. Pidgeon N. (1997) Made-man disaster: the failure of foresight. Second edition, Butterworth-Heinmann.

Vaughan D. (1992) Theory Elaboration: The Heuristics of Case Analysis, in Charles Ragin and Howard S. Becker (eds.), What is a Case?: Exploring the Foundations of Social Inquiry. Cambridge University Press.

Vaughan D. (1996) The Challenger launch decision: risky technology, culture and deviance at NASA, University of Chicago Press, Chicago.

Vaughan, D. (1997) The trickle down effect: policy decisions, risky work and the challenger tragedy, California Management Review. vol 39. n°2.

Weick K. (1987) "Culture as a source of high reliability" , California Review of Management.

Waring A, Glendon A.I. (1998) Managing risk, critical issues for survival and success into the 21st century. Thompson Learning.

Wenzel A.B., Bessey R.L., 1969: Barricaded and Unbarricaded Blast Measurements, Contract n° DAHC04-69-C-0028, Subcontract 1-OU-431, Southwest Research Institute, San Antonio, Texas

Wright J.K., 1961: Shock tubes, John Wiley & Sons Inc., New York

Reports of INERIS on the Billy-Berclau accident (to download on http://www.ineris.fr)

R. Branka, N. Dechy, J.-C. Lecoze, E. Leprette, S.Lim, Intervention de l'INERIS après le sinistre survenu le 27 mars 2003 sur le site de Billy-Berclau de la Société Nitrochimie, Rapport final, Septembre 2004, 30 pages. (*global summary report*)

S. Lim, J.-C. Lecoze, Chronologie de l'accident du 27 mars 2003 survenu à Billy-Berclau sur le site de Nitrochimie, Rapport Annexe 1, MEDD/DPPR, août 2003, 15 pages. (*chronology*)

R. Branka, Examen de l'explosif en cause dans l'accident du 27 mars 2003 survenu à Billy-Berclau sur le site de Nitrochimie, Rapport Annexe 2, MEDD/DPPR, Juin 2003, 8 pages. (*material tests*)

N. Dechy, E. Leprette, Analyse des dommages observés à Billy-Berclau (62) après le sinistre du 27 mars 2003 survenu sur le site de la société Nitrochimie, Rapport Annexe 3, MEDD/DPPR, Août 2003, 63 pages. (*Damages investigation*)

S. Lim, J.-C. Lecoze, Détermination de l'arbre des causes de l'accident du 27 mars 2003 survenu à Billy-Berclau sur le site de Nitrochimie, Rapport Annexe 4, MEDD/DPPR, Août 2003, 24 pages. (*technical analysis*)

J.-C. Lecoze, S. Lim, Analyse organisationnelle de l'accident du 27 mars 2003 survenu à Billy-Berclau sur le site de Nitrochimie, Rapport Annexe 5, MEDD/DPPR, Août 2003, 133 pages. (*organisational investigation*)

Reports of IGE-IGAS-CGM on the Billy-Berclau accident (to download on http://www.ladocumentationfrancaise.fr/brp/notices/044000226.shtml)

Y. Lecointe, J.-P. Chassine, H. Legrand, P. Monsel, Accident survenu à l'usine Nitrochimie de Billy-Berclau (Pas-de-Calais), Inspection Générale de l'Environnement, Conseil général des Mines, Inspection Générale des Affaires Sociales, Mars 2004.

Handling Chemicals in Small Containers

Albert Ness
Rohm and Haas
3100 State Road
Croydon, PA, 19021
aness@rohmhaas.com

Richard Gibson
Rohm and Haas
4800 State Route 12
Elma, WA. 98541
rgibson@rohmhaas.com

Prepared for Presentation at
American Institute of Chemical Engineers
2005 Spring National Meeting
39th Annual Loss Prevention Symposium
Atlanta, Georgia
April 11-13, 2005

January, 2005

UNPUBLISHED

Handling Chemicals in Small Containers

Albert Ness
Rohm and Haas
3100 State Road
Croydon, PA, 19021
aness@rohmhaas.com

Richard Gibson
Rohm and Haas
4800 State Route 12
Elma, WA. 98541
rgibson@rohmhaas.com

Abstract

This paper describes three incidents involving buckets or other small, open containers that have occurred at Rohm and Haas facilities. In one, a reaction between Sodium and water propelled a bucket with considerable force, resulting in an injury. In another, a fire occurred when the lid a plastic container containing a liquid peroxide became contaminated during filling and was blown off. In the last, a solid peroxide and several different initiators and catalysts were staged on a pallet for use in batch reactors. A fire was caused by mixing of small spills from the containers on or under the pallet used to hold them. The fire caused severe damage and temporary shutdown of a reactor train.

1. Introduction

Trevor Kletz once wrote that, in a chemical plant, buckets are more dangerous than a book of matches![1] This paper describes three incidents involving buckets or other small, open containers that have occurred at Rohm and Haas facilities.

In the first, an operator was injured when a bucket previously containing sodium had not been handled and cleaned by the normal procedures. The bucket was then exposed to water from a heavy rain, and the resulting reaction propelled the bucket into the operator's leg with considerable force.

In the second, a fire occurred when the lid of a plastic container containing a Cumene Hydroperoxide (CHP) was blown off. A contaminated container was used, resulting in a reaction of the contaminant with CHP.

In the third, a fire started on or under a pallet used to stage small containers of several different initiators and catalysts. Over time small spills of a solid peroxide and other materials from the containers accumulated on a cable tray located under the pallet and the resulting mixture finally caught fire. The fire caused severe damage to electrical and instrumentation wiring, resulting in a temporary shutdown of a reactor train.

---

[1] Kletz, Trevor, "Improving Chemical Engineering Practices; a new look at old myths of the chemical industry" second edition, Hemisphere Publishing Corporation, New York, 1990.

2. Bucket Propelled by Sodium - Water Reaction

2.1 Incident Description

At a plant that uses sodium, small pieces of waste metallic sodium are generated during line removals and clean-outs, and each time an empty sodium rail car is disconnected. The waste sodium pieces were kept in the sodium handling building in a stainless steal bucket filled with mineral oil until enough pieces were generated for reclamation. The reclamation process involved dissolving the pieces of metallic sodium in clean, dry methanol, with a water analysis of less than one percent. This operation was done in a stainless steel bucket. When the sodium had fully reacted in the methanol, as determined by the absence of sodium "chunks", the mix was transferred from the bucket to a sample recovery pot.

Earlier in the week of the incident, an operator successfully dissolved sodium in methanol in two other buckets, and noticed nothing unusual. On the day of the incident, the operator came across a bucket that contained 2-3" of a sludge material. He asked co-workers about the material and what to do with it. The co-workers were not sure where the sludge material in the bucket came from so they collectively recommended adding methanol to ensure that there was no sodium remaining in the sludge. The operator added methanol to the bucket, placed a lid on it, and placed it on a steam-out pad. About 6-7 hours later the operator returned to the bucket and assumed that, if there was any sodium in the sludge, it would have reacted and transferred the contents of the bucket into the recovery pot. After completing the transfer, the operator noticed there was still residual material left in the bucket. Thinking that all the sodium had reacted, he proceeded to an effluent building to dump the sludge into an effluent pit and wash the bucket out. The operator did not see any sign of reaction when the sludge from the bucket was dumped into the pit. The operator placed the bucket down on the walkway and went to get a hot water hose to rinse the bucket out. At this time an extremely heavy rain started, igniting the contents of the bucket. While attempting to cover the bucket, he knocked it over and its contents spilled onto the wet pad. The resulting reaction propelled the bucket into the operator's leg, resulting in an injury.

2.2 Findings

The investigation found that the bucket still contained sodium even after a 7 hour hold with methanol to recover the sodium, because the sodium was under a layer of dolomite and did not contact the methanol.

The sodium - dolomite mixture was left over from an earlier maintenance activity, and the bucket did not get properly labeled.

Although the operators discussed what to do with the "unknown" material, an adequate EHS management of change review was not conducted for this unusual situation.

2.3 Actions Taken

The sodium reclamation procedure was halted immediately.

The plant contacted the sodium supplier for advice on reclamation procedures, and discovered that the supplier sent their waste to a hazardous waste treatment facility. After that, the plant also started to do this and stopped doing the reclamation process.

For the possibility that some sodium reclamation may have to be done, the procedure was revised to include the following learning's: ensure all containers are properly labeled, do not perform the procedure outdoors when it is raining, ensure all the sodium chunks are dissolved by stirring until no more chunks are left, and, if any unusual situation arises, ensure that an EHS review is conducted.

Several training sessions were held on how to properly label the buckets, how to prepare the sodium waste for disposal and on the importance of management of change and EHS reviews.

3. Container Fire due to Contamination

3.1 Incident Description

This incident involved two operating areas in a plant which we will call Area 1 and Area 2 for convenience. Area 1 supplies an intermediate product to Area 2. Cumene Hydroperoxide (CHP) is a raw material for the intermediate. Storing and handling of CHP, as well as some other small charges, are done by Area 2. The procedure is to have the operators from Area 2 weigh up the small charge of CHP into a labeled plastic container with a hand pump and transfer the containers to Area 1 when needed.

At about 8:00 PM, the Area 1 operator located the plastic containers labeled for CHP use and contacted the Area 2 operator to fill the containers for a batch to be run early the next morning. The hand pump normally used was broken, so another one was located and washed out with water from a flush drum. The containers were covered and stored at Area 2. At approximately 11:15 PM, one of the CHP containers was found smoking. Pressure had blown the top off the container. The fire brigade was called and directed water onto all of the containers. The containers and the drum from which they were pumped were all moved to a safe storage area outside and the top bungs loosened to safely vent any possible reaction. No further reactions occurred.

3.2 Findings

The Area 2 operator noted that he had observed dark debris on the outside of one of the containers, and that it looked like Ferrous Sulfate. After the containers were filled, the operator also noted that the contents of the container looked darker than the other containers, but believed it was an effect from the debris on the outside of the container.

The MSDS notes that the CHP is incompatible with acids, oxidizing agents, reducing agents, oxidation catalysts, copper, lead and iron. Additionally, when CHP and Ferrous Sulfate are entered into the NOAA *Chemical Reactivity Worksheet*[2], it returns the entries; "Explosive when mixed with oxidizing substances", "Heat generated from chemical reaction may initiate explosion", "May cause fire" and "Heat generation by chemical reaction, may cause pressurization" for the combination.

The investigation also found that when the handling of CHP containers by the two areas began, clear responsibility for maintaining the containers was not established.

The investigators concluded that the most probable cause was contamination with Ferrous Sulfate from the outside of the container. A possible cause was contamination from residual chemicals in the flush drum, which is used to flush lances for the drum pump used for monomers, but the investigators considered this less likely.

3.3 Actions Taken

Clean replacement containers were obtained.

A dedicated storage area for the empty and full CHP containers that protects them from contamination from other chemicals was established.

A new pump dedicated to CHP service was obtained.

The operator training was updated and repeated to ensure all operators have an understanding of the hazards of CHP and to emphasize that containers for storing chemicals should be free of contamination inside and out.

The process operating procedures were updated to establish clear ownership of the procedural steps in making this product, including the maintaining and cleaning of the containers.

The process operating procedures were updated to limit the time frame for staging containers of CHP.

4. Fire from Oxidizer Spills

4.1 Incident Description

Small charges for several batch processes were weighed into small containers and staged near the reactors on a pallet. Among the materials handled in these containers were solid organic peroxide, a persulfate and other small charges. Figure 1 shows a re-creation of the staged pallets.

A fire occurred in the process building on the pallet of oxidizers containing the peroxide and persulfate. The sprinkler system activated, but its effectiveness was limited because a cable tray was between the sprinklers and the pallet. The fire caused extensive damage to electrical power,

---

[2] *Chemical Reactivity Worksheet*, U. S. National Oceanic and Atmospheric Administration, http://response.restoration.noaa.gov/chemaids/react.html

control and instrumentation wiring, resulting in considerable downtime in a reactor train. Figure 2 shows the actual pallet after the fire.

Figure 1: Re-Creation of Pallet with Pre-Weighed Materials in Process Area

4.2 Findings

The investigation revealed that, in the past, there had been spills of the different materials because buckets tipped over and cracks had been reported in the buckets. The floor was an open grate and a cable tray was located underneath the staging area, creating a location for the spills to collect because it was difficult to clean up.

The investigators concluded that the most likely cause of the fire was a spill of the peroxide from a bucket onto a cable tray below the grating which contained other, incompatible materials, particularly some activators that had spilled over time. The peroxide reacted with other materials starting a fire.

The location of the pallet near cable trays and junction boxes led to the extensive damage.

Contributing to the incident was a lack of detailed knowledge of how the specific materials would react if mixed. There was a general understanding that the materials should not be mixed, but not in-depth knowledge of what would happen. This led to a practice of staging incompatible materials needed for several batches on the same pallet.

Figure 2: Remains of pallet after fire

4.3 Actions Taken

All chemical residues on pipes and equipment below the grating on the floor of the process building were cleaned up.

All existing containers for pre weighed materials were inspected and replaced as needed. Procedures were established to inspect all containers within one year and adjust the frequency of the inspection or implement a replacement policy based on the results.

Separate storage areas for of catalysts and activators, away from critical equipment and potential contamination sources were set up.

Separate, dedicated, color coded pallets for storing incompatible materials were obtained and used.

Spill containment was provided for all pre-weighed, packaged materials staged on the process floor.

Limits on the amount of materials that could be pre weighed and staged on the process floor were established.

The peroxide containers were replaced with new, color-coded buckets with lids with gaskets.

5. Conclusions and Recommendations

These incidents have several things in common:

- Open buckets or other small containers were used for handling reactive chemicals.
- Contact occurred between the material in the containers and other, incompatible material(s) through spills or contamination.
- The corrective actions rely heavily on administrative procedures.

The incidents show that whenever small containers or buckets are used to handle small charges, the potential for contamination or inadvertent mixing exists. To prevent these kinds of incidents, look for ways to avoid use of small containers. If this is not possible:

- Determine the potential hazards of contamination. Constructing a reactivity matrix is a good way to do this (see the reference to the Chemical Reactivity Worksheet for a good source for beginning to construct a reactivity matrix).
- Provide training to personnel on the potential hazards of contamination and ignition of the materials being handled.
- Provide separate areas for weighing and staging incompatible materials such as activators and initiators.
- Use dedicated containers and handling equipment, such as pumps and pallets.
- Clearly label the containers, and consider color coding them.
- Provide lids for containers.
- Establish procedures for cleaning and inspecting of containers.

Paper TN014 - 3

When Risk Becomes Reality:
Formosa Plastics' Response to a Plant Explosion

Robert Thibault
Formosa Plastics
9 Peach Tree Hill Road
Livingston, NJ 07039
rthibault@fpcusa.com

Prepared for Presentation at
American Institute of Chemical Engineers
2005 Spring National Meeting
39th Annual Loss Prevention Symposium
Atlanta, Georgia
April 11-13, 2005

UNPUBLISHED

When Risk Becomes Reality: Formosa Plastics' Response to a Plant Explosion

Robert Thibault
Formosa Plastics
9 Peach Tree Hill Road
Livingston, NJ 07039
rthibault@fpcusa.com

Abstract

At 10:45 p.m. Central Time on Friday, April 23, 2004, an explosion ripped through Formosa Plastics' polyvinyl chloride plant in Illiopolis, IL. The explosion effectively destroyed the plant, leaving five people dead, several others injured, caused the evacuation of thousands of people for two square miles and generated headlines in media around the world.

Formosa's response to the incident has been called a "textbook example of how to handle a crisis." This presentation will outline the steps Formosa Plastics took before, during and after the crisis to help reduce the impact on victims and their families, the community, the company and its employees and customers.

AUTHOR INDEX

AUTHOR INDEX